THE BENZODIAZEPINES

MONOGRAPHS OF THE MARIO NEGRI INSTITUTE
FOR PHARMACOLOGICAL RESEARCH, MILAN

THE BENZODIAZEPINES

EDITORS:

S. Garattini, M.D.
Director, Mario Negri Institute
  for Pharmacological Research
Milan, Italy

E. Mussini, M.D.
Head, Department of Biochemical
  Pharmacology
Mario Negri Institute
  for Pharmacological Research
Milan, Italy

L. O. Randall, Ph.D.
Director, Department of Pharmacology
Research Division
Hoffmann-La Roche, Inc.
Nutley, New Jersey, U.S.A.

Raven Press • New York

International Standard Book Number 0-911216-25-1
Library of Congress Catalog Card Number 78-181304

PREFACE

Since the discovery of chlordiazepoxide hydrochloride in 1957, the benzodiazepines have had a tremendous impact on the treatment of disorders of nervous origin. These drugs have become among the most widely prescribed in medicine. More significant is the widespread scientific investigation that has been generated by this unique class of drugs.

Sufficient material on the benzodiazepines had accumulated in the literature to justify the organization of this symposium. It was the first opportunity to bring together an international group of investigators to attempt the correlation of the chemical, pharmacological, and clinical data. Various authors discussed the structure-activity, general pharmacology, neuropharmacology, psychopharmacology, metabolism, biochemical effects, and clinical activity of the benzodiazepines. In view of the volume of work done in this area, we were compelled to limit the number of topics and the speakers in the program.

The discussions during the symposium revealed many deficiencies in our understanding of the mechanism of action of these centrally active drugs. Their effects in various biological systems should be investigated thoroughly in order to determine with greater precision the particular indications and limitations for each drug in this group. Although there are many benzodiazepine derivatives which appear to exert similar action, they may show differences that could be of fundamental significance if evaluated properly. The discovery of related compounds with different spectra of activity can be predicted.

We hope that the interesting and stimulating character of this symposium has been transmitted to the pages of this volume.

The editors wish to express their appreciation to all the session chairmen, participants, and staff of the "Mario Negri" Institute who contributed to the success of this symposium. We express our gratitude to Zula Melup of the Mario Negri Institute and Virginia B. Martin of Raven Press for their skillful editorial assistance.

THE EDITORS

CONTENTS

Neurobiochemistry

CHARACTERIZATION OF ACTIVITY: EXPERIMENTAL AND CLINICAL LEVEL

Antianxiety Effects

INTERACTION OF BENZODIAZEPINES WITH OTHER DRUGS

LIST OF CONTRIBUTORS

BALZAR ALEXANDERSON
Department of Pharmacology, Karolinska Institutet, Stockholm 60, Sweden.

ELLSWORTH L. AMIDON
   See D. S. Robinson

*ALLEN BARNETT
Department of Pharmacology, Biological Research, Schering Corporation, Bloomfield, New Jersey 07003.

*G. BARTHOLINI
Research Department, F. Hoffman-La Roche & Co. Ltd., CH-4002 Basel, Switzerland.

DANIEL E. BATES
   See L. A. Gottschalk

TRUDY C. M. BEELEN
   See E. van der Kleijn

BARRY D. BERGER
   See L. Stein

HERMAN BIRCH
   See L. A. Gottschalk

*MARIALUISA BOZZA-MARRUBINI
Servizio di Anestesia e Rianimazione "G. Bozza," Reparto di Terapia Intensiva e Rianimazione, Ospedale Maggiore di Milano, Ospedale CA' Granda, 20100 Milan, Italy.

*UTE BRAUSCH
Physiologisches Institut der Universität, Humboldalle 7, 34 Göttingen, Germany.

---

* The names preceded by an asterisk are those of the first authors.

*A. Breckenridge
Department of Medicine, Royal Postgraduate Medical School, Hammersmith Hospital, London W. 12, England.

Claire G. Cable
    See L. A. Gottschalk

Mary Carskadon
    See W. C. Dement

*G. B. Cassano
Istituto di Clinica Psichiatrica, Università di Pisa, 56100 Pisa, Italy.

Paolo Castrogiovanni
    See G. B. Cassano

Silvana Consolo
    See H. Ladinsky

L. Conti
    See G. B. Cassano

*Leonard Cook
Department of Pharmacology, Hoffmann-La Roche Inc., Nutley, New Jersey 07110.

Hans Corrodi
Department of Pharmacology, University of Göteborg, Göteborg, Sweden.

Arnold B. Davidson
Pharmacology Department, Smith Kline and French Laboratories, 15th and Sp. Garden, Philadelphia, Pennsylvania.

*José M. R. Delgado
Department of Psychiatry, Yale University School of Medicine, 333 Cedar Street, New Haven, Connecticut 06510.

*William C. Dement
Department of Psychiatry, Stanford University School of Medicine, Stanford, California 94305.

*Alberto Di Mascio
Director of Psychopharmacology, Department of Mental Health, Commonwealth of Massachusetts, Boston State Hospital, Boston, Massachusetts 02124.

CONCEZIO DI ROCCO
   See G. F. Rossi

*WALTER B. ESSMAN
Department of Psychology, Queens College of the City University of New York, Flushing, New York 11367.

JOHN W. FIORE
   See A. Barnett

H. FIRTH
   See I. Oswald

EUGENE W. FLEMING
   See L. A. Gottschalk

W. FRIIS
   See A. D. Rudzik

KJELL FUXE
Department of Histology, Karolinska Institutet, Solnavägen 1, S-104 01 Stockholm 60, Sweden.

D. M. GALLANT
   See R. Guerrero-Figueroa

JOYCE GALLANT
   See R. Guerrero-Figueroa

*SILVIO GARATTINI
Director, Istituto di Ricerche Farmacologiche "Mario Negri," Via Eritrea 62, 20157 Milan, Italy.

*K. FRIEDERICH GEY
F. Hoffmann-La Roche & Co. Ltd., Grenzacherstrasse 124, 4002 Basel, Switzerland.

*LOUIS A. GOTTSCHALK
Department of Psychiatry and Human Behavior, California College of Medicine, University of California, Irvine, California 92664.

AMALIA GUAITANI
Laboratory of Isolated Organ Perfusion.
   See S. Garattini

PIETER J. M. GUELEN
    See E. van der Kleijn

C. GUERRERO-FIGUEROA
    See R. Guerrero-Figueroa

*ROBERTO GUERRERO-FIGUEROA
Department of Psychiatry and Neurology, Tulane University School of Medicine, New Orleans, Louisiana.

W. HAEFELY
    See M.-A. Monachon

I. HAIDER
    See I. Oswald

H.-D. HENATSCH
    See U. Brausch

J. B. HESTER
    See A. D. Rudzik

ERIC HODDES
    See W. C. Dement

*LEO E. HOLLISTER
Stanford University School of Medicine, Stanford, California 94305.

M. JALFRE
    See M.-A. Monachon

*SAMUEL C. KAIM
Alcohol and Drug Dependence Service, Veterans Administration Central Office, Washington, D. C. 20420.

*ANTHONY KALES
Department of Psychiatry, College of Medicine, the Pennsylvania State University, the Milton S. Hershey Medical Center, Hershey, Pennsylvania 17033.

H. KELLER
    See G. Bartholini

E. K. KILLAM
    See K. F. Killam

*KEITH F. KILLAM
Department of Pharmacology, Division of the Sciences Basic to Medicine, School of Medicine, University of California at Davis, Davis, California 95616.

*HERBERT LADINSKY
Laboratory of the Cholinergic System, Istituto di Ricerche Farmacologiche "Mario Negri," Via Eritrea 62, 20157 Milan, Italy.

R. LAVERTY
Department of Pharmacology and Experimental Therapeutics, Johns Hopkins University School of Medicine, 725 North Wolfe St., Baltimore, Maryland 21205.

S. A. LEWIS
    See I. Oswald

*PETER LIDBRINK
Department of Histology, Karolinska Institutet, Solnavägen 1, S-104 01 Stockholm 60, Sweden.

ODD LINGJAERDE, JR.
Psykiatrisk Institutet, Universitetet I Oslo, Vinderen, Oslo 3, Norway.

*A. LONGONI
Istituto Carlo Erba per Ricerche Terapeutiche, Via C. Imbonati 24, 20159 Milan, Italy

GIULIO MAIRA
    See G. F. Rossi

V. MANDELLI
    See A. Longoni

FRANCA MARCUCCI
Laboratory of Biochemical Pharmacology.
    See S. Garattini

*ADRIANO MARINO
Istituto di Farmacologia, Università di Bari, 70100 Bari, Italy.

M. MATSUZAKI
    See K. F. Killam

MARIO MEGLIO
    See G. F. Rossi

M.-A. MONACHON
Department of Experimental Medicine, F. Hoffmann-La Roche & Co. Ltd., CH-4002 Basel, Switzerland.

*PAOLO LUCIO MORSELLI
Laboratory of Clinical Pharmacology, Istituto di Ricerche Farmacologiche "Mario Negri," Via Eritrea 62, 20157 Milan, Italy.

G. B. MUSCETTOLA
    See P. L. Morselli

EMILIO MUSSINI
Department of Biochemical Pharmacology.
    See S. Garattini

ERNEST P. NOBLE
Neurochemistry Laboratory, Department of Psychiatry and Human Behavior, California College of Medicine, University of California at Irvine, Irvine, California 92664.

LARS OLSON
    See P. Lidbrink

M. ORME
    See A. Breckenridge

*IAN OSWALD
Senior Lecturer, University Department of Psychiatry, Royal Edinburgh Hospital, Morningside Park, Edinburgh EH 10 5HF, Scotland.

GIUSEPPE PERI
    See H. Ladinsky

I. PESSOTTI
Dpto. Genetica, Facult. Medicina, Ribeirao Preto, Brasil.

L. PIERI
    See G. Bartholini

*PAOLO PINELLI
Clinica Neurologica, Via Palestro 3, Universitá di Pavia, Pavia 27100, Italy.

G. F. PLACIDI
  See P. L. Morselli

A. PLETSCHER
  Research Department, F. Hoffmann-La Roche & Co. Ltd., CH-4002 Basel, Switzerland.

*LOWELL O. RANDALL
  Department of Pharmacology, Research Division, Hoffmann-La Roche Inc., Nutley, New Jersey 07110.

*G. REGGIANI
  Clinical Research, F. Hoffmann-La Roche & Co. Ltd., CH-4002 Basel, Switzerland.

*KARL RICKELS
  Psychopharmacology Research Unit, Department of Psychiatry, University of Pennsylvania, Philadelphia, Pennsylvania 19104.

*JÜRG RIEDER
  Department of Experimental Medicine, F. Hoffmann-La Roche & Co. Ltd., CH-4002 Basel, Switzerland.

NICO V. M. RIJNTJES
  See E. van der Kleijn

MARCELLA RIZZO
  See P. L. Morselli

DONALD S. ROBINSON
  Department of Pharmacology, University of Vermont College of Medicine, Given Medical Building, Burlington, Vermont 05401.

*GIAN FRANCO ROSSI
  Istituto di Neurochirurgia, Università Cattolica del Sacro Cuore, Via Pineta Sacchetti 526, 00168 Rome, Italy.

*ALLAN D. RUDZIK
  CNS Research, The Upjohn Company, Kalamazoo, Michigan 49001.

*RUTH RUSHTON
  Department of Pharmacology, University College London, Gower St., London WC1E 6BT, England.

MARTIN B. SCHARF
  See A. Kales

*Morton A. Schwartz
Research Division, Hoffmann-La Roche Inc., Nutley, New Jersey 07110.

Adriana Selenati
    See M.-L. Bozza-Marrubini

Folke Sjöqvist
Department of Clinical Pharmacology, the Medical School, 581 85 Linköping, Sweden.

Harvey Smythe
    See W. C. Dement

G. A. Sneed
    See R. Guerrero-Figueroa

*Larry Stein
Wyeth Laboratories, P.O. Box 8299, Philadelphia, Pennsylvania 19101.

Hannah Steinberg
Department of Pharmacology, University College London, Gower St., London WC1E 6BT, England.

*Leo H. Sternbach
Director of Medicinal Chemistry, Research Division, Hoffmann-La Roche Inc., Nutley, New Jersey 07110.

Gordon E. Stolzoff
    See L. A. Gottschalk

R. N. Straw
    See A. D. Rudzik

Bengt Strindberg
    See J. Vessman

C. Student
    See U. Brausch

Anders Sundwall
    See J. Vessman

J. Tagney
    See I. Oswald

KOHSI TAKANO
   See U. Brausch

A. H. TANG
   See A. D. Rudzik

*KENNETH M. TAYLOR
Department of Pharmacology, University of Otago Medical School, Dunedin, New Zealand.

M. TOMKIEWICZ
   See R. Rushton

REGINA L. ULIANA
   See L. A. Gottschalk

*LUIGI VALZELLI
Department of Neuropsychopharmacology, Istituto di Ricerche Farmacologiche "Mario Negri," Via Eritrea 62, 20157 Milan, Italy.

*EPPO VAN DER KLEIJN
Department of Clinical Pharmacy, Sint Radboud Ziekenhuis, Katholieke Universiteit, Heyendael, Nijmegen, The Netherlands.

*JÖRGEN VESSMAN
Department of Clinical Pharmacology, the Medical School, 581 85 Linköping, Sweden.

GERNOT WENDT
   See J. Rieder

*H. WESSELING
Instituut voor Klinische Farmacologie der Rijksuniversiteit, Bloemsingel 1, Groningen, The Netherlands.

C. DAVID WISE
   See L. Stein

VINCENT P. ZARCONE
   See W. C. Dement

TON L. B. ZUIDGEEST
   See E. van der Kleijn

CHEMISTRY OF 1,4-BENZODIAZEPINES AND SOME ASPECTS OF THE STRUCTURE-ACTIVITY RELATIONSHIP

Leo H. Sternbach

Chemical Research Department, Hoffmann-La Roche Inc., Nutley, New Jersey

I. INTRODUCTION

In this chapter, I will present a short historical review of our work which ultimately led to the first synthesis of the pharmacologically active 1,4-benzodiazepines. I shall also describe the most important routes of synthesis that resulted in compounds of this type and discuss some problems of the structure-activity relationship in this series.

Our interest in tranquilizers started in the mid-fifties, shortly after the first representatives of this group of drugs proved to be of remarkable clinical value. Since we were chemists at heart, we planned to attack this problem completely empirically, considering mainly the chemically attractive features of such an approach. We decided to select a relatively unexplored class of compounds and to prepare novel members belonging to this group, in the hope that some of these products might exhibit the desired pharmacological properties.

In the search for such a group of compounds, we considered a class of heterocycles which had been the subject of the author's studies in the thirties, at the University of Cracow in Poland (Dziewoński and Sternbach, 1933*a*, 1935*b*). These were the compounds of the type shown in Fig. 1.

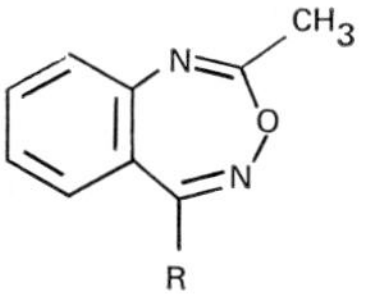

FIGURE 1

They were known since 1891 and were called in the German literature

4,5-benzo-hept-1,2,6-oxdiazines (Auwers and von Meyenburg, 1891). These compounds were relatively readily accessible, crystallized very well, and were expected to lend themselves to a multitude of variations and transformations.

A literature search revealed that, in the 20 years since my investigations in Cracow, nothing had been published about the chemistry or the biological properties of compounds of this group.

In 1955, we started our new experimental work and decided to synthesize compounds of this type, which, by treatment with amines, could be readily converted into products possessing basic side chains. Figure 2 shows a

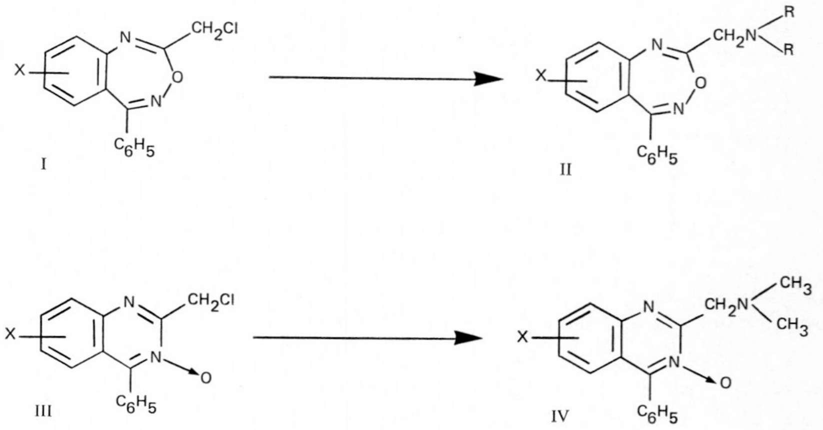

FIGURE 2

product of this type (I) and its transformation into type II, which was our original aim. The reaction products, we hoped, might have interesting properties, since it is a fact known to medicinal chemists that basic groups frequently impart biological activity.

Shortly after our studies had started, we began to have serious doubts about the structure of these heptoxdiazines of types I and II. Particularly, the results of hydrogenation experiments were very unexpected. The oxygen was removed with great ease and the products formed in good yield were quinazolines. Additional chemical studies showed beyond any doubt that the so-called heptoxdiazines did not possess the postulated structure but were in fact quinazoline 3-oxides, as shown in III and IV (Fig. 2) (Sternbach et al., 1960).

The interesting novel structure of these compounds encouraged us to continue our work on synthesis on an even broader basis. We synthesized a number of quinazoline 3-oxides, represented by the general formula I (Fig. 3), and transformed them by treatment with various amines. Secondary amines gave the expected substitution products of type II (Fig. 3). The

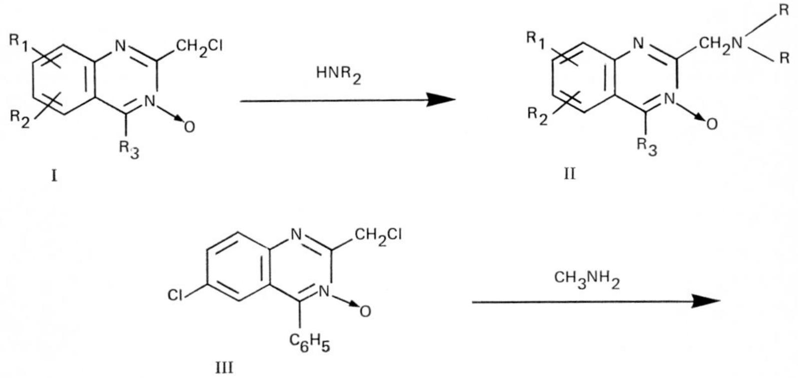

FIGURE 3

reaction occurred with ease and yielded products that crystallized well; however, to our disappointment, the pharmacological properties were rather uninteresting.

II. SYNTHESIS OF BENZODIAZEPINES BY RING ENLARGEMENT (CHLORDIAZEPOXIDE)

The results were quite different when primary amines or ammonia were used in the reaction with compounds of type I (Fig. 3). The first product of this type was obtained on treatment of the quinazoline derivative III with methylamine.

The reaction product possessed, as found by Dr. Randall, very interesting properties in the tests which were generally used for the preliminary screening of tranquilizers and sedatives. These tests, which will be discussed in more detail elsewhere in this volume, indicate muscle-relaxant, taming, sedative, and anticonvulsant properties.

Table I shows the first screening results and the comparison of the properties of this new compound with those of meprobamate, chlorpromazine, and phenobarbital (Sternbach et al., 1964).

The figures show the dosage (mg/kg) of the drug which was required

TABLE I

Compound	Incl. screen	Foot-shock	Cat	Pentylene-tetrazol	Electroshock	
					Max.	Min.
New compound	100	40	2	18	92	150
Meprobamate	250	250	100	150	200	167
Chlorpromazine	17	20	2.5	42	150	600
Phenobarbital	120	80	10	75	18	90

to achieve the desired effect. High figures therefore indicate low activity, and low figures high activity.

In all tests the new compound was considerably more potent than meprobamate; in most of the tests it surpassed phenobarbital; and in some tests it was even more active than chlorpromazine. In addition, the compound showed a pronounced taming effect in monkeys and a very low toxicity (620 mg/kg) in mice.

These attractive pharmacological properties also caused an intensification of our chemical investigations. The physical and physicochemical properties indicated that it could not be the substituted quinazoline 3-oxide,

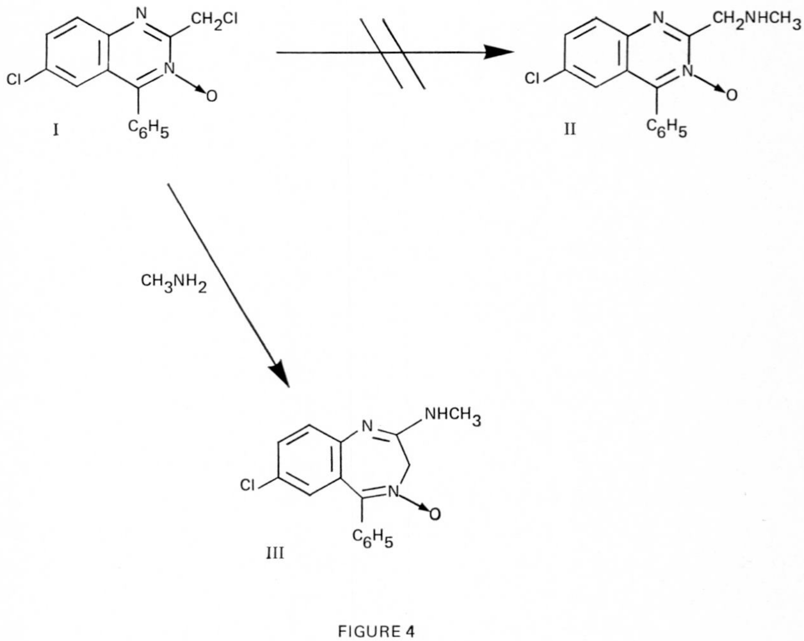

FIGURE 4

formula II shown in Fig. 4. The ensuing chemical exploration showed, indeed, a completely different structure (Sternbach and Reeder, 1961*a*). The heterocyclic 6-membered ring present in I had undergone a transformation, and the reaction product had the structural formula III, containing a 7-membered novel hetero-ring.

The formation of this compound can be best explained by the sequence of reactions shown in Fig. 5 (Sternbach, 1971). The methylamine does not

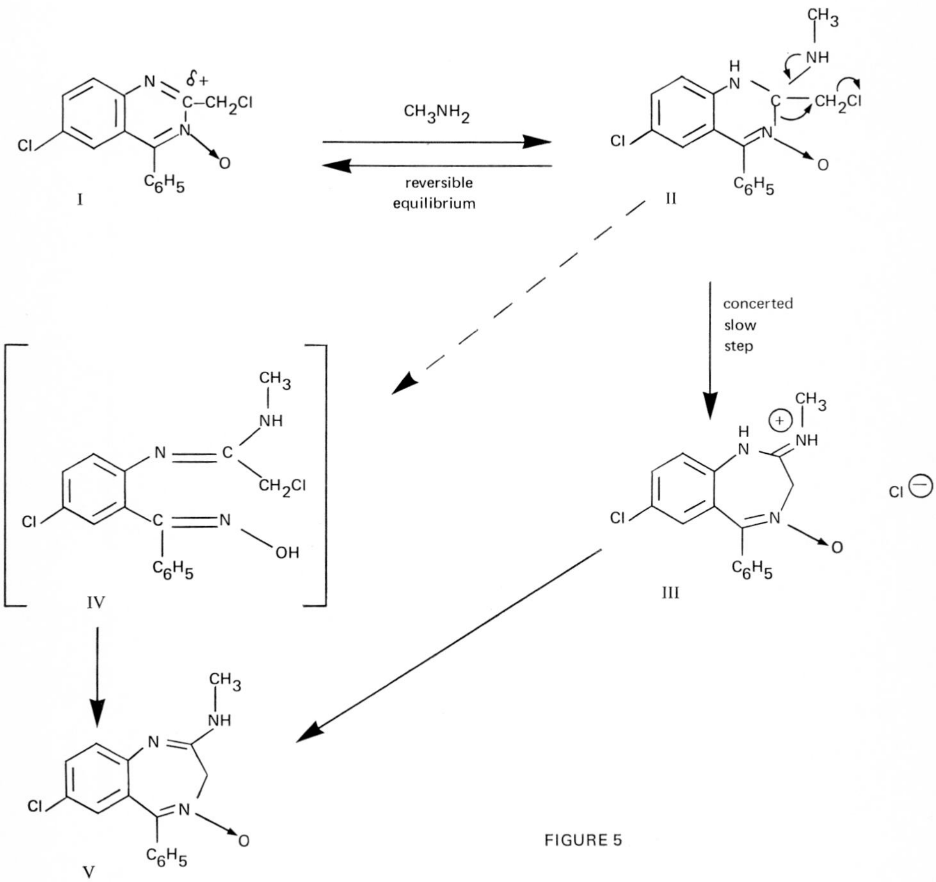

FIGURE 5

react with compound I in the normal manner. It does not replace the reactive chlorine atom as might be expected, but it does attack the quinazoline N-oxide molecule in the 2-position, which possesses a positive charge.

This causes the formation of an intermediate of type II which is ultimately transformed via III or IV into the 1,4-benzodiazepine derivative V.

The interesting and promising pharmacological properties of this compound, which I had mentioned before, resulted in the synthesis of a number of closely related compounds obtained by the reaction of the quinazoline 3-oxide I with ammonia and various primary amines, as shown in Fig. 6.

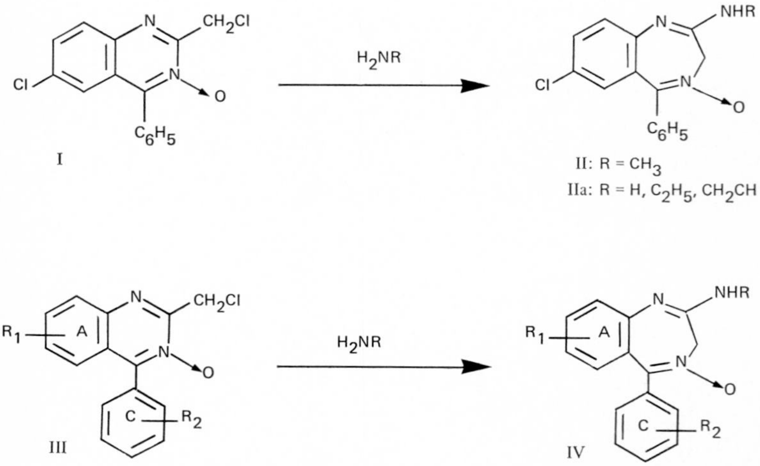

I

II: R = CH$_3$
IIa: R = H, C$_2$H$_5$, CH$_2$CH = CH$_2$, etc.

III   IV

R$_1$ and R$_2$ = halogens, CF$_3$, nitro group, etc., in different positions of the rings A and C.

FIGURE 6

However, none of the products thus obtained proved to be superior to the methylamino derivative II.

In our experiments, we also used other quinazoline 3-oxides (III) possessing various substituents in rings A and C (see e.g., Sternbach et al., 1961; Saucy and Sternbach, 1962), but also in this case we did not obtain products which were substantially more active or differed significantly from compound II.

We, therefore, concentrated our efforts on this product. Its pharmacology was explored thoroughly, and, after completion of the toxicological studies, the clinical investigation was started under the direction of Dr. Hines in 1957. These efforts culminated in its introduction in 1960 under the trademark LIBRIUM. The generic name which was later generally accepted was chlordiazepoxide.

III. TRANSFORMATIONS OF CHLORDIAZEPOXIDE

Since the water-soluble hydrochloride used in our clinical studies was extremely bitter, we investigated other forms which, we hoped, would lend themselves to the preparation of a pharmaceutically acceptable elixir. A suspension of the quite insoluble, finely pulverized base itself was still unsuitable. The pharmacologically highly active acetyl derivative II shown in Fig. 7, which was prepared in connection with these studies, also proved

FIGURE 7

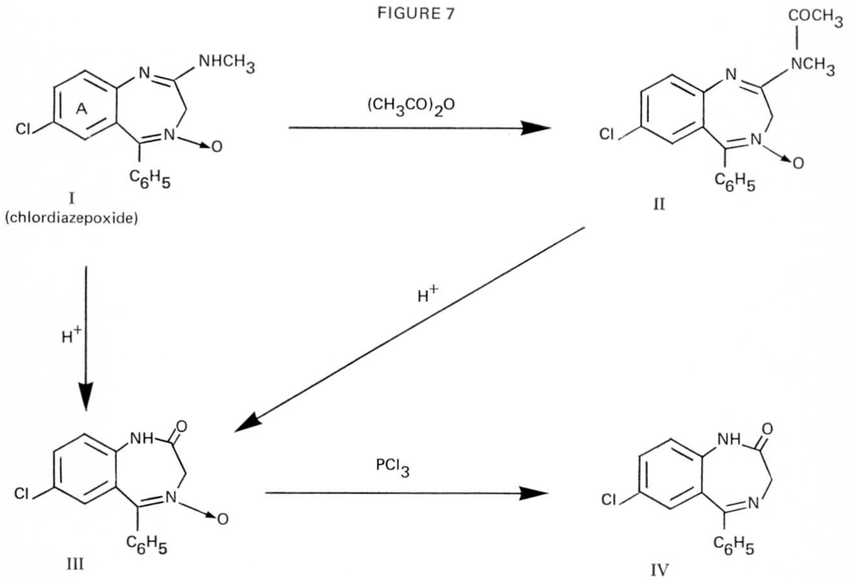

Compound	Incl. screen	Foot-shock	Cat	Pentylenetetrazol	Max.	Min.
I	100	40	2	18	92	150
II	100	20	2	15	150	150
III	75	40	1	6	52	400
IV	75	20	1	6	25	61

to be too bitter despite its low solubility.

Stability studies of I and II (Fig. 7) led, however, to the very important finding that the substituent in the 2-position was readily removed by acid hydrolysis. The decomposition of I occurred slowly at room temperature in an aqueous solution of the hydrochloride. After a day or two, the solution became turbid and the product formed was the lactam III. It is very interesting and probably meaningful that this compound is also one of the major metabolities of chlordiazepoxide. When the acetyl derivative II was hydrolyzed with acid, the transformation occurred with a quantitative yield within a few minutes (Sternbach and Reeder, 1961*b*).

Rather unexpectedly, this lactam III showed the same pharmacological activity as I. A further transformation of this product consisted in the removal of the N-oxide oxygen, which also had no detrimental effect on the pharmacological properties; quite to the contrary, the activity seemed to be enhanced, as can be seen in Fig. 7. Thus, it turned out that some of the features apparently so characteristic of chlordiazepoxide were not needed at all for its pharmacological activity. The N-oxide function and particularly the basic substituent, which had been the cornerstone of our initial working hypothesis, proved to be only unnecessary adornments.

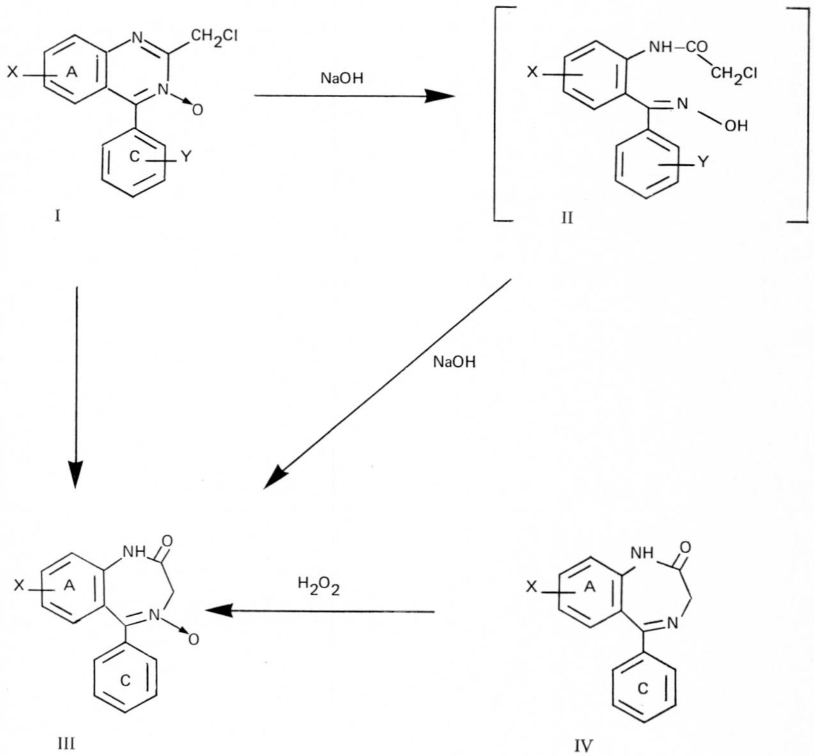

FIGURE 8

IV. METHODS FOR THE SYNTHESIS OF
5-PHENYL-1,4-BENZODIAZEPINONES

Our subsequent studies were concentrated first on the investigation of synthetic methods leading to compounds of type III and were then broadened into the intensive study of compounds of type IV.

N-Oxides of type III were most readily accessible by treatment with alkali of 2chloromethylquinazoline N-oxides of type I, as shown in Fig. 8.

This reaction proceeds very probably via an intermediate of type II, as indicated by further additional chemical studies (Stempel et al., 1965). It is quite general, occurs with great ease, and gives excellent yields. It was used for the synthesis of a number of lactam N-oxides, with various substituents in rings A and C. Another method leading to such N-oxides is the oxidation of lactams of type IV. However, these N-oxides, in general, did not prove themselves to be exceptionally attractive.

But they undergo, however, an interesting rearrangement on treatment

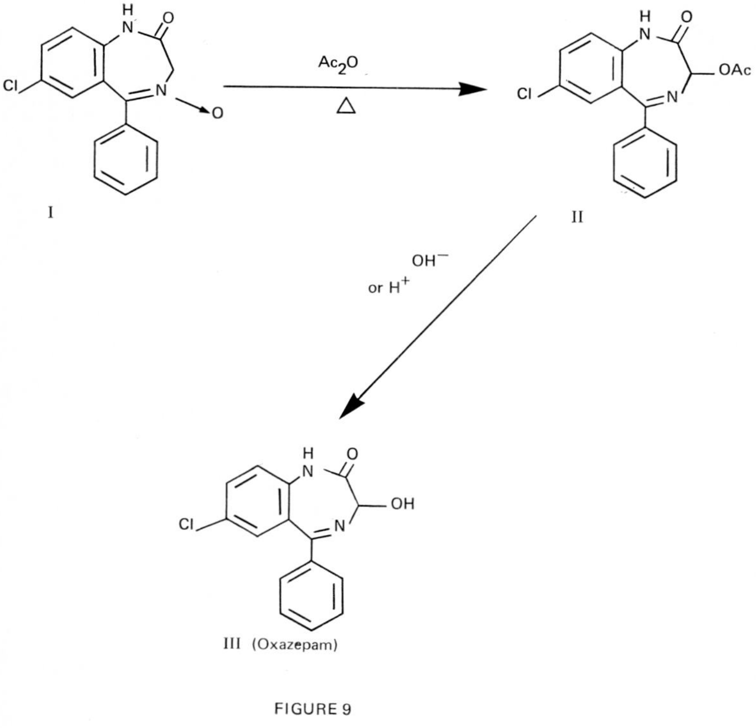

FIGURE 9

with acid chlorides or acid anhydrides, as shown in Fig. 9 (Bell and Childress, 1962). This transformation, the so-called Polonovsky rearrangement, results (Fig. 10) in the formation of the 3-acetoxy derivative II which on mild hydrolysis yields the biologically active 3-hydroxy derivative III. This rearrangement of benzodiazepine N-oxides was studied by two research

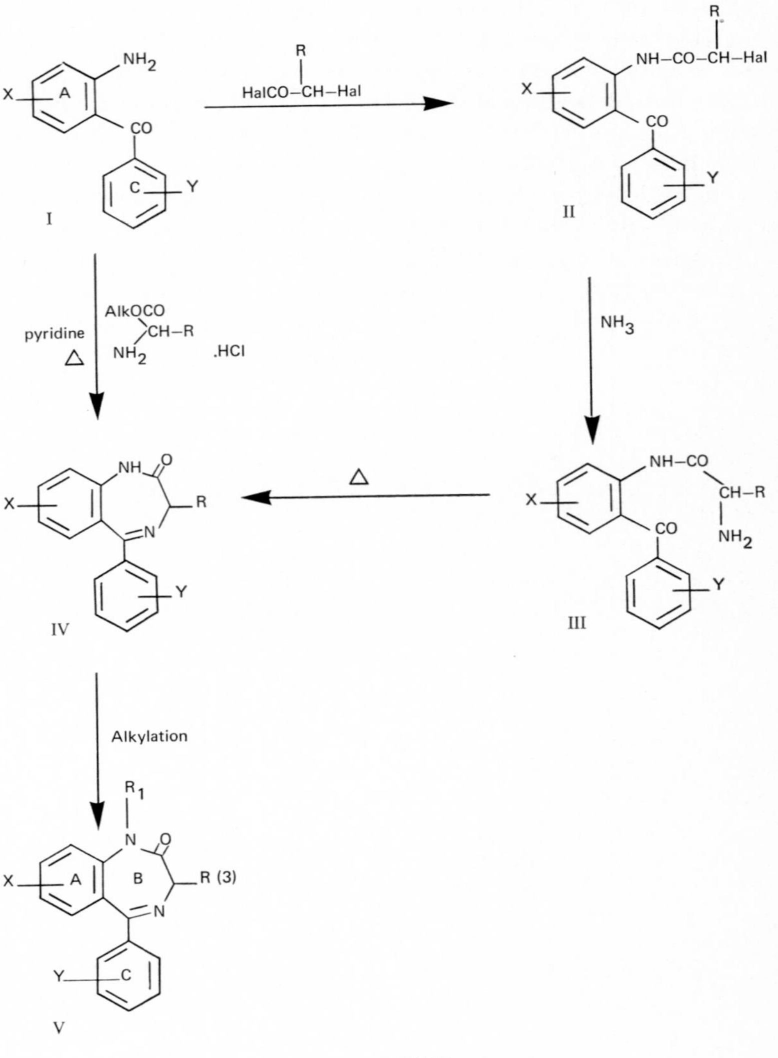

FIGURE 10

teams, the Wyeth group (Bell and Childress, 1962) and investigators at Hoffmann-La Roche. The hydroxy derivative III, which resembles chlordi-azepoxide in its activity, was introduced by Wyeth in 1965. It bears the generic name oxazepam, and is marketed in the United States under the trade name SERAX, and in Germany as ADUMBRAN and PRAXITEN.

The simple benzodiazepinones without the N-oxide oxygen, however, were, as mentioned before, pharmacologically more interesting than the N-oxides I or their rearrangement products (III), and they were, therefore, most intensively studied. The search for alternative syntheses of these rela-tively simple compounds resulted in various routes leading with good yields to the desired products. The most frequently used two methods are shown in Fig. 10 (Sternbach et al., 1962).

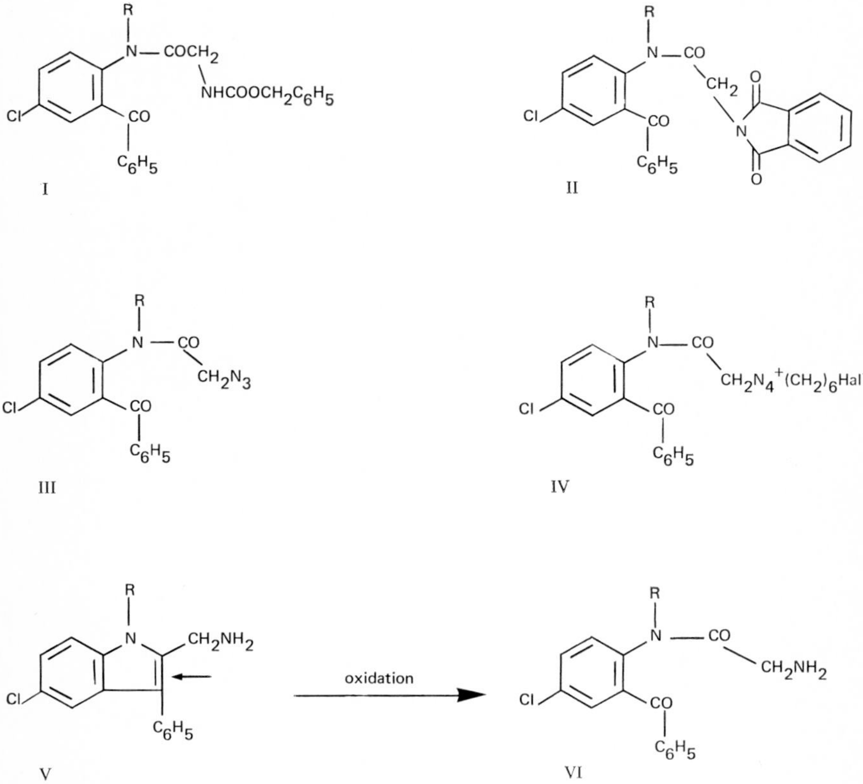

FIGURE 11

As can be seen, in both cases, *o*-aminoketones are used as starting materials. Treatment of the appropriately substituted aminobenzophenone with a haloacetyl halide yielded a compound of type II which, on treatment with ammonia, gave the benzodiazepinone IV via an amino derivative III. Another extensively used method was the treatment of an aminobenzophenone with an amino acid ester hydrochloride in pyridine, leading directly from I to IV. The first method generally gave better overall yields of up to 70 and 80%, although it involved more steps. The second method facilitated the synthesis of benzodiazepinones bearing substituents at position 3, since many *α*-amino acids bearing various additional substituents on the *α*-carbon are commercially available.

Other routes for the construction of the 7-membered ring, which were developed subsequently, consisted in the use of the intermediates possessing a protected or potential glycine moiety (Stempel and Landgraf, 1962; Podesva, 1966; Yamamoto et al., 1968; Petersen and Lakowitz, 1969). Figure 11 shows some of these intermediates.

Particularly ingenious was the approach used by Japanese chemists at the Sumitomo Company, who synthesized indole derivatives of type V, which, on oxidation of the double bond (indicated by an arrow in Fig. 12), formed the aminoacetamido compound that cyclized under the reaction conditions.

One of the first benzodiazepinones which showed a very high biological activity and an interesting pharmacological profile is the 1-methyl derivative illustrated in Fig. 12. It was synthesized in 1959 and, after the appropriate

FIGURE 12

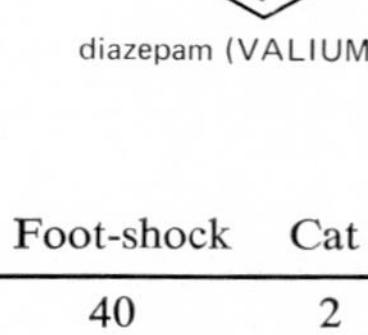

diazepam (VALIUM)

Compound	Incl. screen	Foot-shock	Cat	Pen.	Max.	Min.
Chlordiazepoxide	100	40	2	18	92	150
Diazepam	30	10	0.2	1.4	6.4	64

toxicological and clinical studies, was given the generic name diazepam; it was introduced near the end of 1963 under the trademark VALIUM.

V. STRUCTURE-ACTIVITY RELATIONSHIP

The interesting properties of this new compound led to a broad program concerned with the synthesis and study of the biological properties of 1,4-benzodiazepinones (Sternbach et al., 1968). We have prepared more than 2,000 1,4-benzodiazepines of various types, which include more than 1,000 benzodiazepinones; the preparation involved the synthesis of almost 4,000 intermediates and by-products.

A. Substitution in position 7

One of the first facts which became apparent when the structure-activity relationship was investigated was the paramount importance of substituents in the 7-position. Only compounds bearing a substituent in this position were highly active. In addition, the character of the substituent was also of great importance. Electron-withdrawing substituents generally imparted high activity, whereas electron-releasing groups decreased the activity considerably. This is illustrated in Table II.

TABLE II

Substituents in ring A

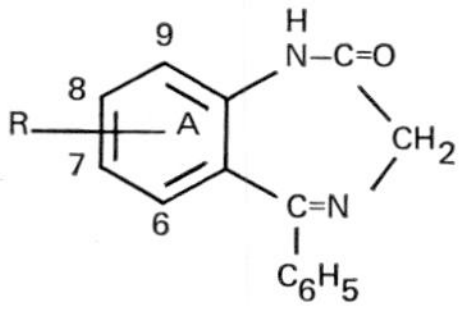

Compound	R	Incl. sc.	F.S.	Cat	Pentylene-tetrazol	Max.	Min.
1		150	100	>30	>800	30	75
2	7-F	150	100	4	>800	23	100
3	7-Cl	75	20	1	6	25	61
4	7-Br	25	20	0.5	1.7	3.7	200
5	7-NO$_2$	15	5	0.1	0.5	8.4	132
6	7-CF$_3$	10	10	0.1	1	5	21
7	8-Cl	>500	>100		334	150	800
8	9-NO$_2$	500	>100		>800	400	>800
9	7-CH$_3$	>500	>100		175	167	>800

From this, it may be seen that the 7-unsubstituted product is very weak;

the 7-fluoro derivative also is not very potent, but the potency increases considerably with the electron-withdrawing properties of heavier halogens and, particularly, with the nitro and trifluoromethyl group. As mentioned before, the electron-releasing methyl group, present in the last compound, causes a significant activity decrease.

The data for muscle relaxation and sedation in the cat are not available in every case since the less active compounds were, in many instances, not submitted to this test.

It is interesting to note that the potency in the unanesthetized cat is generally paralleled by the activity against the pentylenetetrazol (METRAZOL) shock. This suggests that the anti-pentylenetetrazol activity, in this series of compounds, indicates mainly muscle-relaxant or sedative properties.

B. Substitution in ring C

The substitution in ring C also played an important role, as can be seen in Table III. It was studied most extensively in the 7-chloro and 7-nitro

TABLE III

Compound	R	Incl. sc.	F.S.	Cat	Pentylene-tetrazol	Max.	Min.
1	H	75	20	1	5.9	25	61
2	2'-F	40	5	0.05	0.08	1.5	50
3	2'-Cl	100	40	0.1	0.4	13	115
4	2'-CH$_3$	150	100	0.5	7.7	21	183
5	2'-OCH$_3$	300	100	1	7.5	30	133
6	3'-F	200	40	2	4.2	20	667
7	3'-OCH$_3$	>500	>100		1.8	102	>800
8	4'-F	>500	>100		>800	50	>800
9	4'-Cl	>500	>100		>800	300	>800

derivatives. Table III also shows some of our results with the 7-chloro compounds.

The first compounds to be prepared were those bearing a substituent at the *para*-position of ring C. This change resulted in an almost complete disappearance of activity, in contrast to the high activity of compound 1. Despite this rather disappointing development, we decided to synthesize an electronically equivalent compound, bearing an *ortho*-substituent in ring C. This compound had chlorine at the *ortho*-position and was quite unexpectedly very active, being even more potent than the unsubstituted product. 7-Nitro-, bromo-, cyano-, and trifluoromethyl derivatives showed quite analogous results. This led to an extensive study of the effect of *ortho*-substitution in ring C. It was found that a fluorine or chlorine atom at that position had a very strong positive effect, whereas other substituents caused no significant change or even decreased the activity. It is interesting to note that electron-releasing substituents (compounds 4 and 5) did not completely abolish the activity.

Meta- and *para*-substituents had, generally, an undesirable effect. As mentioned above, this was most pronounced in compounds bearing a *para*-substituent in ring C.

Two halogens in the C ring, in both *ortho*-positions, again caused high activity, as was observed in the 2',6'-difluoro- and dichloro-derivatives. Di-substitution in *ortho*- and *para*-positions decreased the activity almost to zero.

The general conclusions concerning the effects on the activity of substituents in rings A and C are shown in Fig. 13.

FIGURE 13. Effect on activity of substituents in rings A and C.

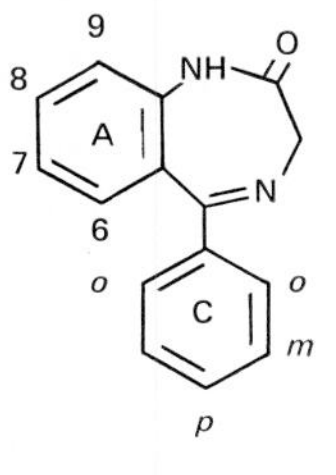

Ring A

Positive: Electron-withdrawing substituents at position 7. (Halogen, CF₃, NO₂, CN)

Negative: Electron-withdrawing substituents at positions 8 and 9. Electron-releasing substituents in position 7.

Ring C

Positive: Fluorine and chlorine at *ortho*-position, also di-substitution in both *ortho*-positions.

Negative: Any substituent at *meta*- or *para*-position.

TABLE IV.

Effect of methyl group at position 1

Compound	1	7	o	Incl. sc.	F.S.	Cat	Pentylenetetrazol	Max.	Min.
1		Cl		75	20	1	5.9	25	61
1a	CH₃	Cl		30	10	0.2	1.4	6.4	64
2		NO₂		15	5	0.2	0.5	8.4	132
2a	CH₃	NO₂		10	2.5	0.1	0.6	1.5	190
3		N(CH₃)₂		>500	>100	10	42	367	667
3a	CH₃	N(CH₃)₂		150	40	1	5.9	92	333
4		CN		75	40	0.5	1.3	34	675
4a	CH₃	CN		20	5	0.5	1.2	14	200
5		NO₂	F	4	2	0.05	0.96	20	200
5a	CH₃	NO₂	F	1	0.8	0.02	0.12	12	345

TABLE V.

Compound	R_1	R_2	Incl. sc.	F.S.	Cat	Pentylenetetrazol	Max.	Min.
1	CH_3	H	30	10	0.2	1.4	6.4	64
2	C_2H_5	H	75	20	1	4	8.3	600
3	$CH_2CH(CH_3)_2$	H	200	50	20	8.5	50	>800
4	$C(CH_3)_3$	H	>500	>100		>800	600	>800
5	$CH_2CONHCH_3$	H	450	80	20	0.5	15	85
6	CH_3	Cl	40	10	0.1	0.42	15	83
7	$(CH_2)_2N(C_2H_5)_2$	Cl	475	100	10	22	150	440

C. Substitution in position 1

Table IV illustrates the effect of the methylation of benzodiazepinones in position 1.

The effect of a substituent at this position was studied very extensively. It was found that, quite generally, an increase in activity occurred when a methyl group was introduced, as can be seen in Table IV.

The last compound shown in Table IV combines all the features known to enhance activity, namely a nitro group at position 7, the *o*-fluoro substituent in ring C and a methyl group at position 1. Indeed this combination resulted in an extremely potent compound, possibly the most active benzodiazepinone derivative known at present. It is interesting to note that compound 3, bearing an electron-releasing substituent at position 7 and showing a very low activity, became quite active after methylation. It might be appropriate to mention at this point that the effects observed in the benzodiazepine series were generally additive. Products possessing more than one of the features known to enhance the activity were almost invariably more active than the compounds having only one of those characteristics. On the other hand, it was also found that changes which had a detrimental effect on activity exerted the same influence on very potent and on less potent compounds.

The action of substituents other than methyl at position 1 was also studied, and it was found that heavier substituents generally had a detrimental effect. This is shown in Table V, where the activities of the 7-chlorobenzodiazepinones 1, 2, and 3 are compared, or 6 with 7. A very significant difference exists between isomers 3 and 4. Whereas compound 3 behaves as expected, isomer 4 is completely inactive at the doses tested. We are inclined to ascribe this to differences in the metabolism of these two compounds. Dr. Schwartz's studies which are still underway might produce an answer to this question.

Another compound shown in Table V has a rather interesting pharmacological spectrum. It is the "methylamino-acetyl" derivative (5) which is equipotent to the methyl derivative diazepam (1) in the anticonvulsant tests, whereas its sedative and muscle-relaxant properties are 10 to 100 times weaker. This led us to consider it as a particularly valuable anticonvulsant. However, preliminary clinical results did not bear out our hopes.

VI. BENZODIAZEPINES OF VARIOUS TYPES

While our studies on synthesis and structure-activity of the benzodi-

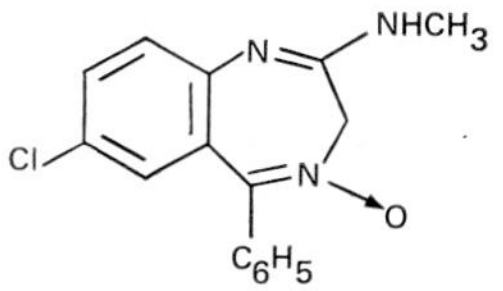

I. chlordiazepoxide

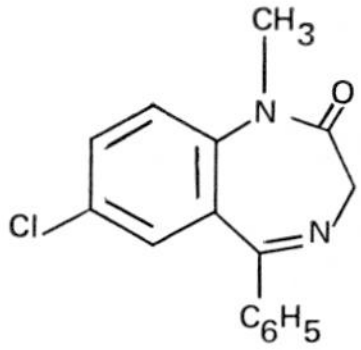

II. diazepam

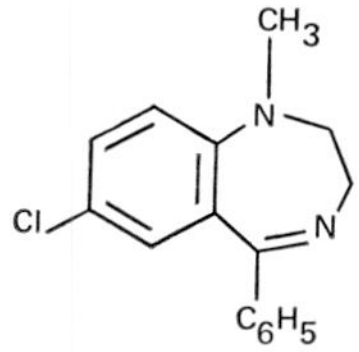

III. medazepam

FIGURE 14

azepinone series were underway, we devoted part of our time also to the synthesis of other types of 1,4-benzodiazepines. Of particular interest were compounds which had, instead of the amino group of chlordiazepoxide or the carbonyl function present in diazepam, two hydrogens in the 2-position. A representative of this type is compound III shown in Fig. 14.

These products were quite readily accessible by various routes as, for example, by the direct or indirect reduction of benzodiazepinones (Metlesics et al., 1963; Sternbach et al., 1963; Archer and Sternbach, 1964). Compounds of type III were less potent than the corresponding benzodiazepinones, but seemed, nevertheless, of considerable interest because of the structural difference. One of these products (III), medazepam, was introduced under the trademark NOBRIUM and, despite, its metabolic similarities to diazepam, seems to show clinical differences.

The above discussion has been limited to benzodiazepinones with a phenyl group in the 5-position because such compounds were generally biologically most active and were, in addition, most readily accessible. However, as shown by the example in Fig. 15, a large number of benzodiazepinones bearing other substituents in the 5-position were also synthesized and studied pharmacologically by us and other teams of scientists. Compounds without a substituent in the 5-position were also investigated.

The greatest obstacle was in every case, as mentioned before, the

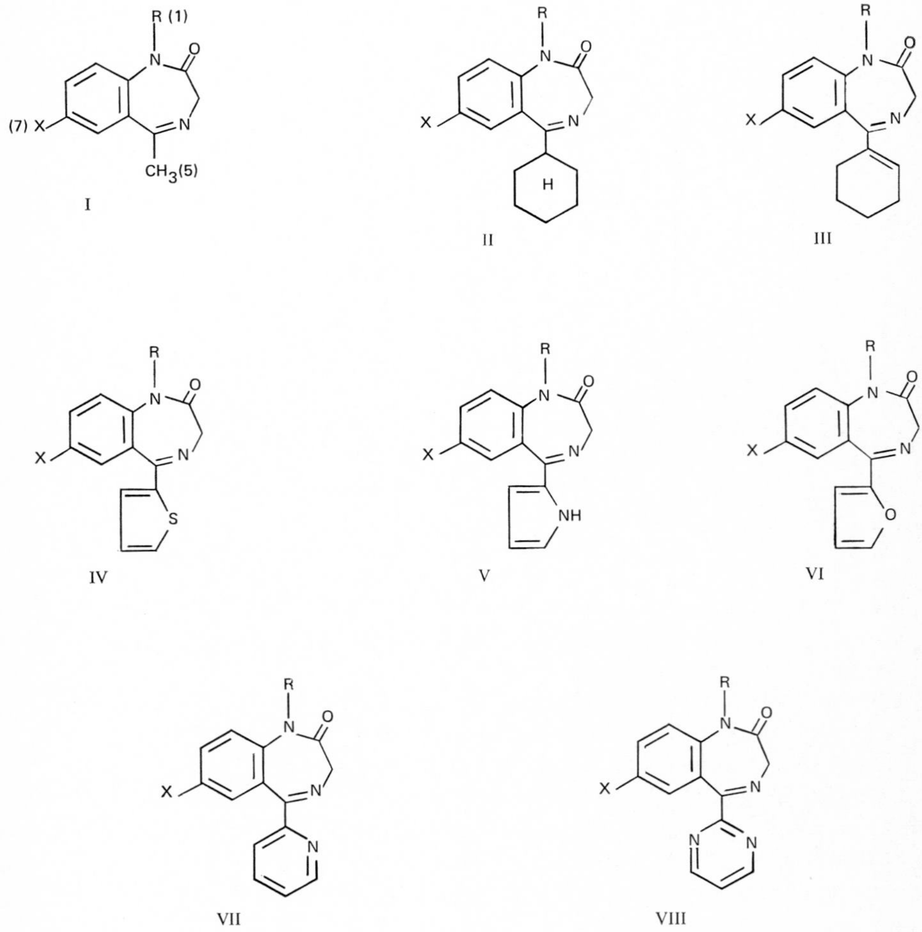

FIGURE 15

preparation of the appropriate aminoketone needed as starting material. Based on our experience, most of the compounds which were studied pharmacologically by us and also by other investigators bore a pharmacophoric substituent in the 7-position, in most cases a chlorine atom, and preferably a methyl group in position 1. Very few were of interest and some of them were rather difficult to prepare. Exceptions are compounds of type III (Schmitt et al., 1967) and VII (Fryer et al., 1964), which are under

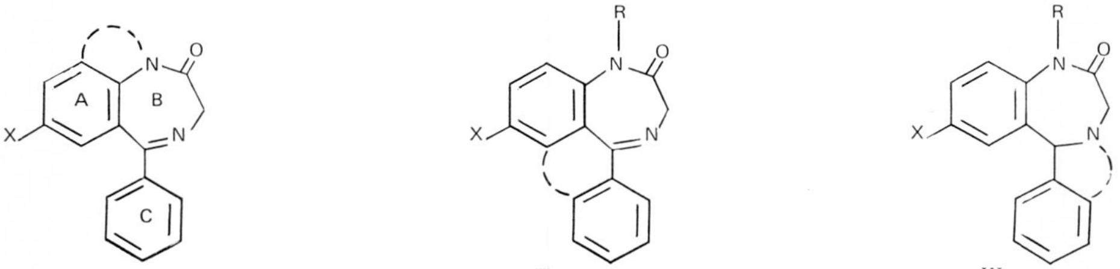

FIGURE 16

clinical study by Clin-Byla and Hoffmann-La Roche, respectively. The β- and γ-pyridyl derivatives corresponding to VII were also synthesized but proved to be less interesting.

Furthermore, compounds were prepared which linked two of the rings, as indicated in Fig. 16: (A- - - -B) (Härter and Liisberg, 1968), (A- - - -C) (Bell and Childress, 1968), (B- - - -C) (Müller and Zeller, 1966; Ott et al., 1968).

However, none of these products was of pharmacological or clinical value.

VII. COMPOUNDS OF CLINICAL INTEREST

At this time, the compounds shown in Fig. 17 are available to the practicing clinician. A good number of compounds are still under clinical study, as exemplified by the benzodiazepines shown in Fig. 18.

The compounds that are being investigated by us now represent only a fraction of the almost 30 compounds which have undergone preliminary clinical trials.

However, it is expected that these efforts will lead to additional, clinically useful benzodiazepines with specific properties as muscle relaxants, anticonvulsants, or, possibly, antidepressants. We still have great hopes and are continuing our studies mainly with these objectives in mind.

ACKNOWLEDGMENTS

I wish to acknowledge the dedicated work of my co-workers who contributed so greatly to the development of the chemistry of 1,4-benzodiazepines. In particular, I thank Mr. E. Reeder, Drs. G. Archer, G. Field, R. I. Fryer, W. Metlesics, G. Saucy, N. Steiger, and A. Stempel.

I also want to express my appreciation to Dr. L. O. Randall,

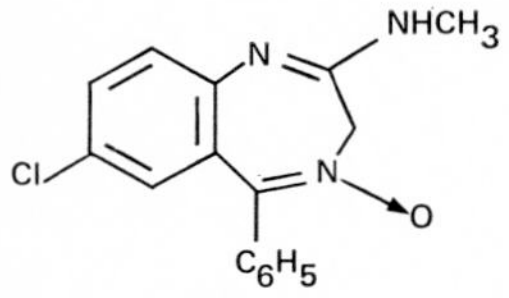

chlordiazepoxide
(LIBRIUM, 1960)

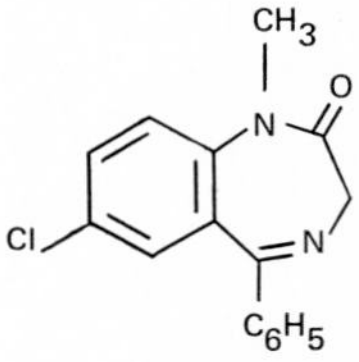

diazepam
(VALIUM, 1963)

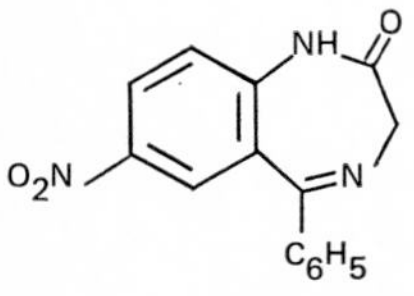

nitrazepam
(MOGADON, 1965)

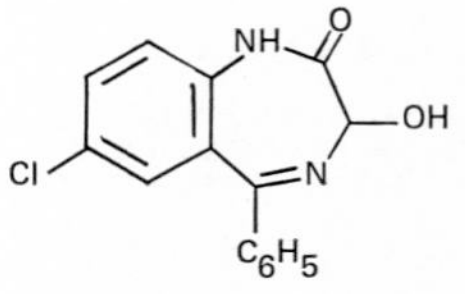

oxazepam
(SERAX, 1965)

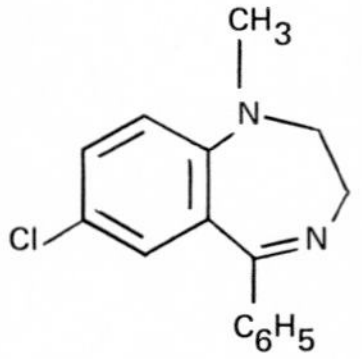

medazepam
(NOBRIUM, 1968)

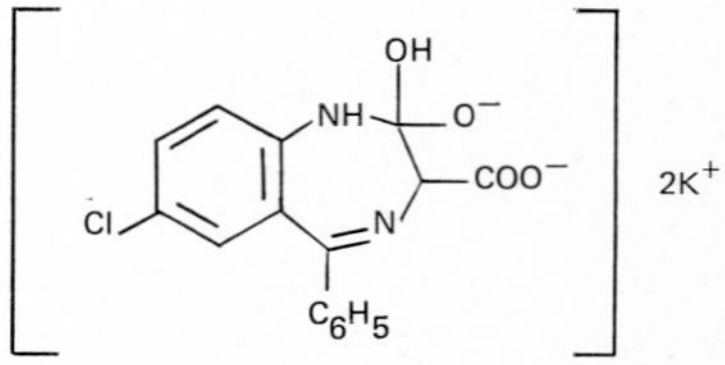

chlorazepate
(TRANZENE, 1969; TRANXILIUM, 1971)

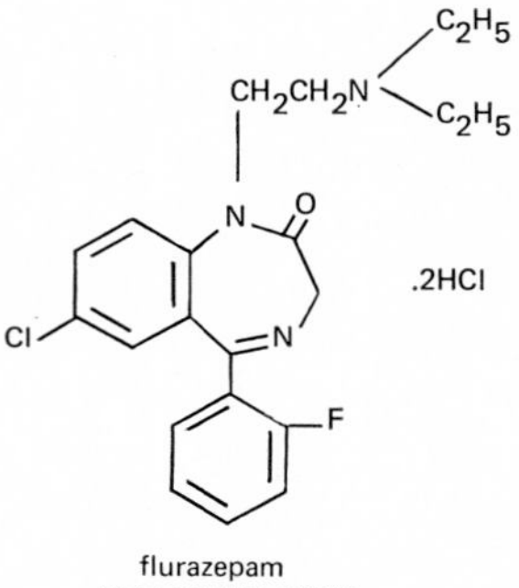

flurazepam
(DALMANE, 1970)

FIGURE 17

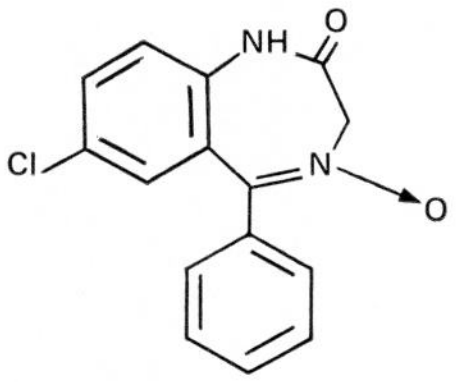

1. Demoxepam

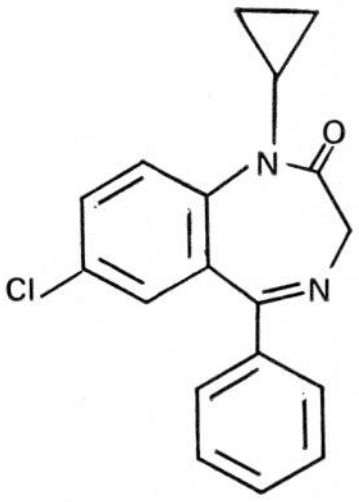

2. Prazepam

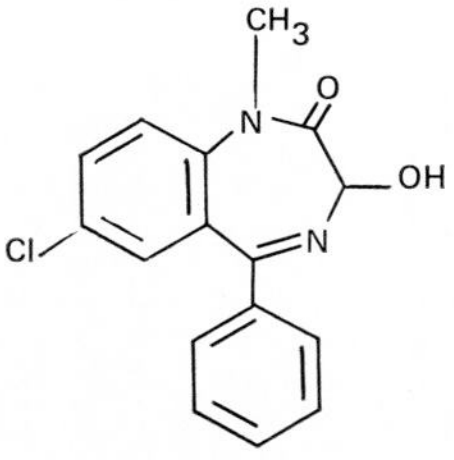

3. Temazepam

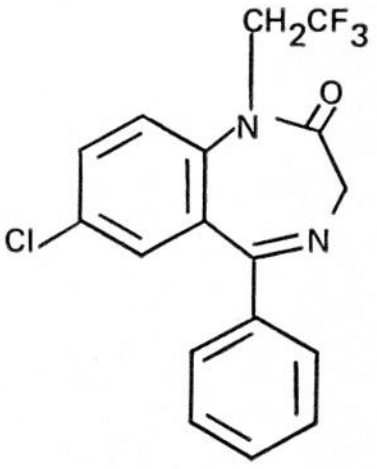

4. SCH 12041

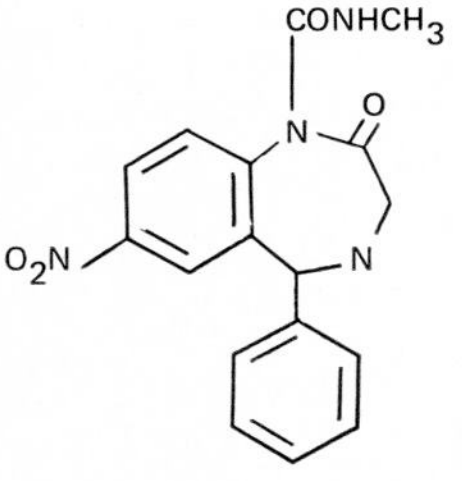

5. D-58SI

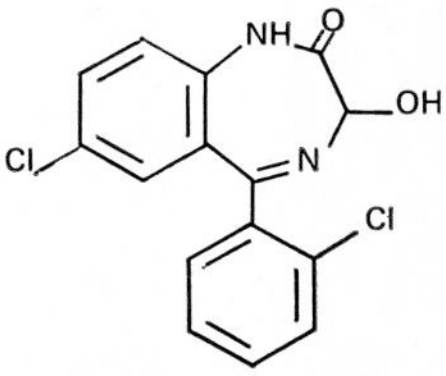

6. Lorazepam

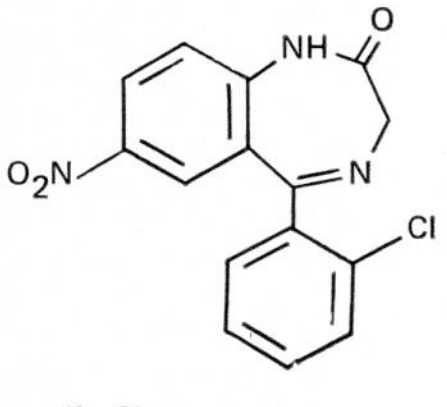

7. Clonazepam

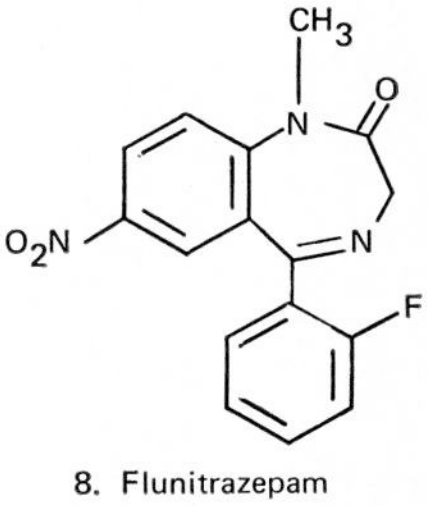

8. Flunitrazepam

FIGURE 18A

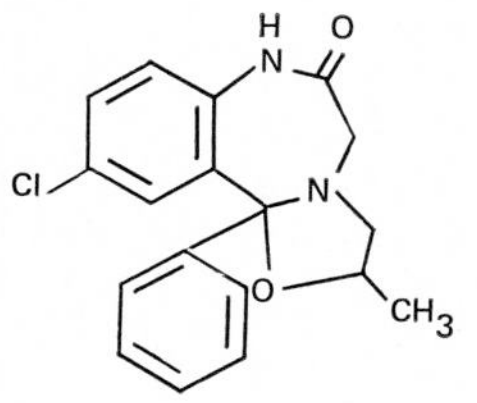

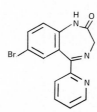

9. Oxazolazepam 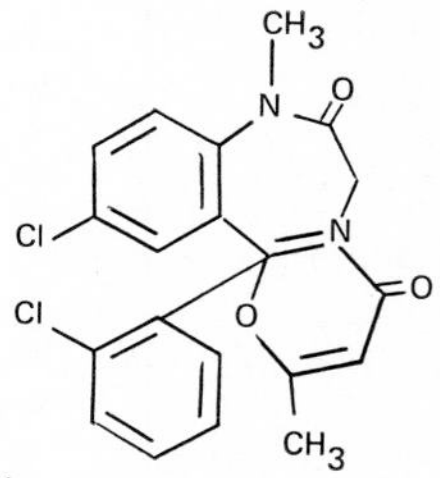   10. Ketazolam

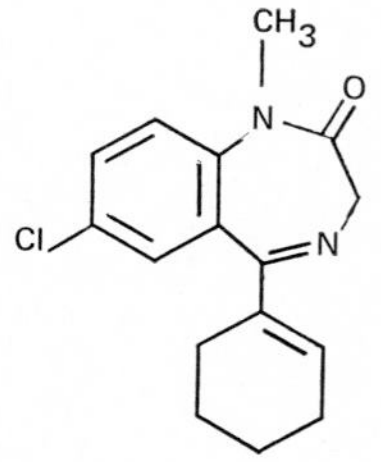

11. Tetrazepam   12. Bromazepam

FIGURE 18B

Director of Pharmacology, and his co-workers for their close cooperation which resulted in the efficient biological evaluation of these compounds.

REFERENCES

Archer, G. A., and Sternbach, L. H. (1964): Quinazolines and 1,4-benzodiazepines. XVI. Synthesis and transformations of 5-phenyl-1,4-benzodiazepine-2-thiones. *Journal of Organic Chemistry,* 29:231–233.

Auwers, K., and von Meyenburg, F. (1891): Ueber eine neue synthese von derivaten des isindazoles. *Berichte der Deutschen Chemischen Gesellschaft,* 24:2370–2388.

Bell, S. C., and Childress, S. J. (1962): A rearrangement of 5-aryl-1,3-dihydro-2H-1, 4-benzodiazepine-2-one 4-oxides. *Journal of Organic Chemistry,* 27:1691–1695.

Bell, S. C., and Childress, S. J. (1968): Fused ring compounds. *United States Patent* 3,417,101 (December 17, 1968).

Blazévić, N., and Kaifez, F. (1970): A new ring closure of 1,4-benzodiazepine. *Journal of Heterocyclic Chemistry,* 7:1173–1174.

Dziewonski, K., and Sternbach, L. (1933): Ueber die einwirkung von benzoylchlorid auf α-naphthylamin. *Extrait du Bulletin de l'Académie Polonaise des Sciences et des Lettres. Classe des Sciences Mathématiques et Naturelles. Série A: Sciences Mathématiques,* pp. 416–431. In: *Chemical Abstracts,* 28:2717[3] (1934).

Dziewonski, K., and Sternbach, L. (1935): Weitere studien ueber reaktionen zwischen benzoylchlorid und aromatischen aminen und ueber ihre produkte, verbindungen der chinazolinreihe. *Extrait du Bulletin de l'Académie Polonaise des Sciences et des Lettres. Classe des Sciences Mathématiques et Naturelles. Série A: Sciences Mathématiques*, pp. 333–348. In: *Chemical Abstracts*, 30:2971[3] (1936).

Fryer, R. I., Schmidt, R. A., and Sternbach, L. H. (1964): Quinazolines and 1, 4-benzodiazepines. XVII. Synthesis of 1,3-dihydro-5-pyridyl-2H-1,4-benzodiazepine derivatives. *Journal of Pharmaceutical Sciences*, 53:264–268.

Härter, H. P., and Liisberg,V S. (1968): 7-Chlor-5-phenyl-1,9-trimethylen-1, 4-benzodiazepin-2-on. *Acta Chemica Scandinavica*, 22:3332–3333.

Metlesics, W., Silverman, G., and Sternbach, L. H. (1963): Quinazolines and 1, 4-benzodiazepines. XII. Preparation and reactions of 2,3-dihydro-1H-1, 4-benzodiazepine 4-oxides. *Journal of Organic Chemistry*, 28:2459–2460.

Müller, M., and Zeller, P. (1966): Chinazoline und benzodiazepine. XXX. Die synthese von tetrahydro-isochino[2,1-*d*] [1,4]benzodiazepin-6(7H)-onen. *Helvetica Chemica Acta*, 49:1222–1226.

Ott, H., Hardtmann, G. E., Denzer, M., Frey, A. J., Gogerty, J. H., Leslie, G. H., and Trapold, J. H. (1968): Tetrahydroisoquino[2,1-*d*] [1,4]benzodiazepines. Synthesis and neuropharmacological activity. *Journal of Medicinal Chemistry*, 11:777–787.

Petersen, J. B., and Lakowitz, K. H. (1969): New methods for the preparation of 1,4-benzodiazepinones carbostyrils, and indolo [2,3-c]quinolones. *Acta Chemica Scandinavica*, 23:971–974.

Podesva, C. (1966): Benzodiazepine derivatives. *Netherlands Patent*, 6,500,446.

Saucy, G., and Sternbach, L. H. (1962): Chinazoline und 1,4-Benzodiazepine. VII. Trifluoromethyl-substituierte verbindungen. *Helvetica Chimica Acta*, 45:2226–2241.

Schmitt, J., Comoy, P., Suquet, M., Boitard, J., LeMeur, J., Basselier J. J., Brunaud, M., and Salle, J. (1967): Sûr un nouveau myorelaxant de la classe des benzodiazepines: Le tétrazepam. *Chemica Therapeutica*, 2:204-259.

Stempel, A., and Landgraf, F. W. (1962): Quinazolines and 1,4-benzodiazepines. IX. 2-Carbobenzoxyglycylamidobenzophenones and their conversion to 1,4-benzodiazepinones. *Journal of Organic Chemistry*, 27:4675–4676.

Stempel, A., Reeder, E., and Sternbach, L. H. (1965): Quinazolines and 1, 4-benzodiazepines. XXVII. Mechanism of ring enlargement of quinazoline-3-oxides with alkali to 1,4-benzodiazepin-2-one 4-oxides. *Journal of Organic Chemistry*, 30:4267–4271.

Sternbach, L. H. (1971): 1,4-Benzodiazepines. Chemistry and some aspects of the structure-activity relationship. *Angewandte Chemie* (International Edition), 10:34–43.

Sternbach, L. H., Archer, G., and Reeder, E. (1963): Quinazolines and benzodiazepines. XV. 7-Nitro- and 7-trifluoromethyl-2,3-dihydro-5-phenyl-1H-1,4-benzodiazepines and their transformations. *Journal of Organic Chemistry*, 28:3013–3016.

Sternbach, L. H., Fryer, R. I., Metlesics, W., Reeder, E., Sach, G., Saucy, G., and Stempel, A. (1962): Quinazolines and 1,4-benzodiazepines. VI. Halo-, methyl-, and methoxy substituted 1,3-dihydro-5-phenyl-2H-1,4-benzodiazepin-2-ones. *Journal of Organic Chemistry*, 27:3788–3796.

Sternbach, L. H., Kaiser, S., and Reeder, E. (1960): Quinazoline 3-oxide structure of compounds previously described in the literature as 3.1.4-benzoxadiazepines. *Journal of the American Chemical Society*, 82:475–480.

Sternbach, L. H., Randall, L. O., Banziger, R., and Lehr, H. (1967): Structure-activating relationships in the 1,4-benzodiazepine series. In: *Drugs Affecting the Central Nervous System*, Vol. 2, edited by A. Burger. Marcel Dekker, Inc., New York, pp. 237–264.

Sternbach, L. H., Randall, L. O., and Gustafson, S. R. (1964): 1,4-benzodiazepines (chlordiazepoxide and related compounds). In: *Psychopharmacological Agents*, Vol. 1, edited by M. Gordon. Academic Press, New York, pp. 137–224.

Sternbach, L. H., and Reeder, E. (1961*a*): Quinazolines and 1,4-benzodiazepines. II. The rearrangement of 6-chloro-2-chloromethyl-4-phenylquinazoline 3-oxide into 2-amino derivatives of 7-chloro-5-phenyl-3H-1,4-benzodiazepine 4-oxide. *Journal of Organic Chemistry*, 26:1111–1118.

Sternbach, L. H., and Reeder, E. (1961*b*): Quinazolines and 1,4-benzodiazepines. IV. Transformations of 7-chloro-2-methylamino-5-phenyl-3H-1,4-benzodiazepine 4-oxide. *Journal of Organic Chemistry*, 26:4936–4941.

Sternbach, L. H., Reeder, E., Keller, O., and Metlesics, W. (1961): Quinazolines and 1,4-benzodiazepines. III. Substituted 2-amino-5-phenyl-3H-1,4-benzodiazepine 4-oxides. *Journal of Organic Chemistry*, 26:4488–4497.

Yamamoto, H., Inaba, S., Hirohashi, T., and Ishizumi, K. (1968): Notiz ueber ein neues verfahren zur herstellung von 1,4-benzodiazepin-derivaten aus 2-aminomethyl–indol-derivaten. *Chemische Berichte*, 101:4245–4247.

The Benzodiazepines, Raven Press, New York © 1973

PHARMACOLOGICAL ACTIVITY OF SOME BENZODIAZEPINES AND THEIR METABOLITES

Lowell O. Randall and Beryl Kappell

Department of Pharmacology, Research Division, Hoffmann-La Roche Inc., Nutley, New Jersey

I. INTRODUCTION

Several reviews have correlated the chemical, pharmacological, and clinical activities of benzodiazepines (Sternbach et al., 1964; Svenson and Gordon, 1965; Zbinden and Randall, 1967; Randall and Schallek, 1968). This review will compare the pharmacological activity of certain marketed benzodiazepines and of some of their very potent derivatives with that of phenobarbital. It will delineate the unique safety margin of the benzodiazepines over that of phenobarbital in a number of animal species. The metabolites of chlordiazepoxide, diazepam, medazepam, flurazepam, and nitrazepam and their activity in a variety of tests and animal species will be discussed.

Historically, chlordiazepoxide hydrochloride (the active ingredient in LIBRIUM) was the first member of the benzodiazepine class which was synthesized in the middle 1950's by Sternbach and Reeder (1961). It was found to have potent sedative, muscle-relaxant, taming, and anticonvulsant activity (Randall et al., 1960). Initial clinical trials indicated that chlordiazepoxide had potent antianxiety effects in human subjects (Harris, 1960). Tobin and Lewis (1960) proposed that the antianxiety effects could be separated from the sedative effects which were equated with the occurrence of drowsiness and improvement of sleep pattern. The evaluation of a psychostimulant effect was predicated on a feeling of well-being, increase in social activity, drive, verbal productivity, and appetite, and reduction of anxious depression. The anticonvulsant properties of chlordiazepoxide hydrochloride were recognized by Kaim and Rosenstein (1960), who demonstrated an increase in threshold to pentylenetetrazol (METRAZOL) convulsions in human epileptics.

Diazepam (the active ingredient in VALIUM), another member of the benzodiazepine class, was found to have much stronger sedative, muscle-

27

relaxant, taming, and anticonvulsant properties (Randall et al., 1961). Oxazepam (the active ingredient in SERAX) was later introduced as a mild tranquilizer, followed by flurazepam hydrochloride (the active ingredient in DALMANE), a hypnotic.

Nitrazepam, a hypnotic, and medazepam hydrochloride, a mild tranquilizer, have been introduced in Europe.

Bromazepam, clonazepam, and flunitrazepam are undergoing intensive clinical investigation.

II. METHODS

Inclined screen. Male mice were given a maximum dose of 500 mg/kg p.o. and left on the screen at least 4 hr for observation of paralyzing effects severe enough to cause the mice to slide off the screen (Randall et al., 1960). On the basis of six mice per dose and at least two points between 100 and 0%, the ED_{50} was calculated.

Foot shock. The dose required to tame pairs of mice induced to fight by applying electrical shocks to their feet was determined according to the method of Tedeschi et al. (1959). Pairs of mice were treated 1 hr prior to the second shocking. At least three pairs of mice at three dose levels were used. The dose which blocked fighting in three out of three pairs was defined as 100% effective.

Unanesthetized cat. Normal cats were treated p.o. and observed for gross changes in behavior (Randall et al., 1960). The minimum effective dose (MED) was the lowest dose at which muscle relaxation was observed. Muscle relaxation was identified by the limpness in the legs when the cat was suspended by the scruff of the neck. At least three cats per dose at three dose levels were used.

Anti-pentylenetetrazol. This technique was carried out according to the method of Everett and Richards (1944). The ED_{50} was calculated as the dose which would prevent convulsions in 50% of the mice tested after administration of 125 mg/kg of pentylenetetrazol by the subcutaneous route. Groups of eight mice at a minimum of three dose levels were used.

Maximal electroshock. This test was performed according to the method described by Swinyard et al. (1952), with the following modifications. CF-1S male mice were used, and the shock was delivered in the same form and duration described above. However, 30 mA was delivered as the electroshock instead of 50 mA. One hr after receiving the compound, the animals were shocked in order to determine any anticonvulsant effects. Prevention of tonic hind limb extension was considered to be the end point of anticonvulsant activity. Eight mice were used per dose level at a minimum of three dose levels.

Acute toxicity. CF-1 mice of both sexes, weighing 17 to 25 g, were administered the compound orally, intraperitoneally, subcutaneously, and intravenously. The compound was suspended in 5% gum acacia solution for oral administration. Ten mice were used per dose level, and a minimum of three dose levels was employed for each toxicity determination. The dosed animals were observed 72 hr for mortality. LD_{50} levels were calculated by the method of Miller and Tainter (1944).

Hot plate test. This test was used to measure the analgesic effects of compounds on mice exposed to thermal pain (55.0 $\pm$ 0.5°C). Five CF-1 male mice, weighing 20 to 30 g, were used per dose level. The ED_{50} was the dose that decreased the threshold to thermal pain by 50% (Eddy et al., 1950).

Rotating rod test. This test was used to measure the effects of compounds on muscle tone and/or muscular coordination, particularly of muscle relaxants, sedatives, and stimulants in mice. Five male CF-1 mice, weighing 20 to 22 g, were used per dose level. The ED_{50} was the dose that caused 50% of the mice to fall off the rotating rod (Dunham and Miya, 1957).

Continuous avoidance (rats). Male albino rats (Charles River, CD) were employed, each weighing about 300 g at the start of the experiment. The subjects were trained to respond in the Sidman (1953) continuous-avoidance procedure, as modified by Heise and Boff (1962). Rats were placed in Skinner-type boxes containing two levers and a grid floor through which an electrical foot shock (0.6 mA at 350 V) could be delivered. Each response on the left lever (avoidance lever) postponed the onset of shock for 40 sec. When a rat paused longer than 40 sec between avoidance responses, it received a foot shock lasting a maximum of 5 sec, but the shock could be turned off by a response on the right (escape) lever. In the absence of avoidance responses, shocks were delivered every 20 sec.

The duration of each avoidance session was 5½ hr, and each rat was given one session per week. The avoidance procedure was automatically controlled through a series of transistorized logic circuits, and the subjects' responses were recorded on digital counters and cumulative recorders.

In the nondrugged (control) state, a trained rat avoided more than 90% of the possible shocks by pressing the avoidance lever at a steady rate of about three responses per minute. Each rat was dosed by the i.p. route at weekly intervals for 3 to 4 consecutive weeks. Following such a series of drug administrations, each rat was retested without drug in the avoidance procedure for a 5-hr session to demonstrate the reliability of control behavior.

Continuous avoidance (squirrel monkey). Male squirrel monkeys (*Samiri sciurea*), ranging in weight from 450 to 1,000 g, were used. Observations of these monkeys in their cages indicated that they were docile, active, and sociable toward other squirrel monkeys; however, handling by humans precipitated violent aggressive behavior. Their resistance to being handled provided a useful means of assessing taming effects typically produced by benzodiazepines.

The Skinner boxes and recording and controlling apparatus used for the experiments on monkeys were essentially the same as the equipment used in the rat studies, except for two modifications: (1) the monkey box was 2 in higher, and (2) the intensity of the shock delivered to the monkeys' feet was 2.5 mA.

The avoidance schedule used with monkeys was identical to the procedure employed in the experiments on rats. The schedule of drug administration described in the procedure for rats was also used in the tests on monkeys. However, in monkeys, all drugs were administered orally (by intubation into the stomach), and the interval between successive dosings was at least 2 weeks. The intervening week(s) served as nondrug controls. Methods used to determine significant drug effects on both rats and squirrel monkeys on the continuous avoidance schedule were the same as those described by Heise and Boff (1962).

Behavior check list (cynomolgus monkey). This procedure involved measuring "aggression" and "activity" in normally vicious cynomolgus monkeys, according to the method described by Randall et al. (1965). Two types of behavior were studied independently: (1) general motor activity, and (2) aggressive behavior directed at the experimenter. General activity of the monkeys was observed through a one-way window to prevent interaction between the experimenter and the monkey. "Aggression" was measured while the experimenter was in the room with the monkey by recording the presence or absence of symptoms of hostile behavior, e.g., baring teeth, attacking handler's glove. The monkeys' behavior was observed before and after drug administration. The "taming" effect of a drug was based on a maximal reduction in the "aggression," with only a minimal reduction in the general motor activity score. The occurrence of side effects such as ataxia or drowsiness was recorded. The monkeys were placed in the observation room in the same pairs once each week. Each monkey served as its own control.

III. PHARMACOLOGICAL ACTIVITY

A. Activity in mice

The pharmacological profile in mice for diazepam, chlordiazepoxide hydrochloride, medazepam, flurazepam, and bromazepam is illustrated in the top section of Fig. 1. The log doses in mg/kg, p.o., are plotted on the ordinate. The anti-pentylenetetrazol, rotarod, antifighting, antimaximal electroshock, hypnotic, and lethal effects are plotted on the abscissa. The data were obtained from publications by Zbinden and Randall (1967) and Randall and Schallek (1968).

For diazepam, the anti-pentylenetetrazol test was the most sensitive, followed by the rotarod, antifighting, and antimaximal electroshock. Chlordiazepoxide hydrochloride was less potent than diazepam in all tests. Medazepam resembled chlordiazepoxide hydrochloride in potency in most tests, but rotarod activity was greater. Flurazepam was equal to diazepam in all tests, except the antimaximal shock, in which it was weaker. Bromazepam was more potent then diazepam in the anti-pentylenetetrazol and rotarod tests, but similar in the antifighting and antimaximal electroshock tests. Bromazepam was the least toxic of this group. For all five compounds, there was a wide range between the LD_{50} and the ED_{50} in any given test.

In the bottom section of Fig. 1, diazepam is compared with three benzodiazepines, substituted with the nitro group in the 7-position, and with phenobarbital. Nitrazepam, clonazepam, and flunitrazepam show stronger anti-pentylenetetrazol, rotarod, and antifighting activities than diazepam. Clonazepam had only slight antimaximal shock activity. The lethal effects of these 7-nitro benzodiazepines were much less than diazepam, giving them an even greater safety margin. Mice treated with these compounds exhibited an unusual hyperactivity consisting of stereotyped movements. In marked contrast, phenobarbital was much weaker than the benzodiazepines in the tests for sedation but was equal in the antimaximal shock, and showed a much lower safety margin between the activity in the various tests and the hypnotic and lethal effects.

B. Activity in cats

The muscle-relaxant activity in cats shown in Table I was measured by observing the minimum doses which caused the cats to hang relaxed when suspended by the scruff of the neck. The eight benzodiazepines and

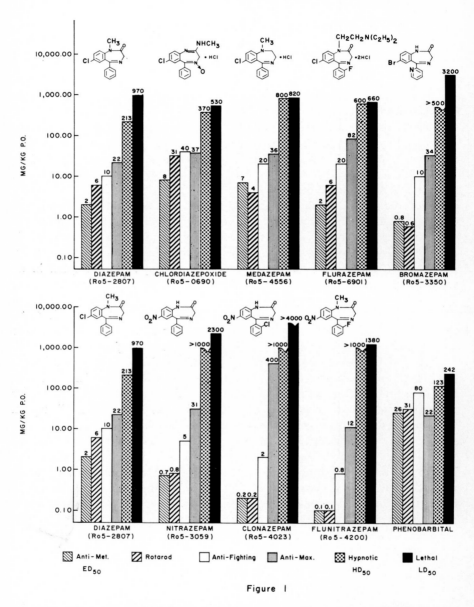

Figure I

phenobarbital were administered to two cats each at doses of 0.01 mg/kg p.o., and the doses were doubled daily until the muscle-relaxant effect was observed. The daily doubling of dosing was then continued until loss of

TABLE I. Muscle-relaxant activity in unanesthetized cats

Drug	7	2'	1-N	Misc.	Muscle relaxant MED	Loss of righting reflex (mg/kg)	Lethal p.o.	Time to death (days)
Chlordiazepoxide hydrochloride	Cl			2-NHCH$_3$ 4-N → 0	2	200	200	5
Diazepam	Cl		CH$_3$		0.2	500	500	1
Medazepam	Cl			2-H$_2$	4	400	500	1
Flurazepam	Cl	Fl	(CH$_2$)$_2$-N-(C$_2$H$_5$)$_2$		2	100	400	1 hr
						convulsions		
Nitrazepam	NO$_2$				0.1	200	200	3
Clonazepam	NO$_2$	Cl			0.05	1,000	>1,000	none
Flunitrazepam	NO$_2$	Fl	CH$_3$		0.02	200	200	3
Bromazepam	Br			5-pyridine	0.2	1,000	1,000	3
Phenobarbital					50	100	100	1

righting reflex occurred, at which dose cats usually died in 1 to 5 days from depression. Flurazepam was unusual in that death occurred from convulsions. Clonazepam did not cause death even at 1,000 mg/kg (the highest dose given).

The muscle-relaxant figure obtained in this experiment agreed closely with data published by Zbinden and Randall (1967). Diazepam was the most potent of the 7-chloro compounds, and had the highest safety margin between the effective muscle-relaxant dose and the dose which caused loss of righting reflex and death. The 7-nitro compounds were more potent than diazepam, and the safety margin was even greater. Clonazepam had the highest safety margin. Bromazepam was similar to diazepam in potency. The safety margin between muscle-relaxant and lethal doses of the benzodiazepines ranged from 100 to 20,000 whereas that of phenobarbital was only 2.

C. Activity in rats

The pharmacological activity in rats of the eight benzodiazepines, in comparison with that of phenobarbital, is given in Table II. The conditioned-avoidance data were taken from Zbinden and Randall (1967). The shock-rate increase measures the MED for modifying behavior, while the escape-failure dose measures the incapacitating dose. The rotarod measures the CNS-incapacitating dose on the rotating rod at 30 min, the time at which the peak effect occurs in rats.

In contrast to mice and cats, results in rats did not show the superiority of diazepam over the other 7-chloro derivatives. Nitrazepam, flunitrazepam, and bromazepam were more potent than diazepam. Clonazepam was equal to diazepam in the conditioned avoidance procedure but stronger in the rotarod test, and showed a higher safety margin than diazepam. In the conditioned avoidance test, phenobarbital was weaker than the benzodiazepines in shock-rate increase, but equal to diazepam and clonazepam in escape failure. Phenobarbital showed a safety margin of only 5, while the benzodiazepines had safety margins of 70 to 3,000.

D. Activity in monkeys

In Table III are given the pharmacological data in monkeys, taken from Zbinden and Randall (1967), for the eight benzodiazepines and phenobarbital. The taming effect was measured in vicious cynomolgus monkeys. In the conditioned avoidance experiments in squirrel monkeys, the shock increase measures the MED for modifying lever pressing, while the

TABLE II. Pharmacological activity in rats[a]

Drug	Substituent in benzodiazepine structure				Conditioned avoidance		Rotarod ED_{50}	LD_{50}
	7	2'	1-N	Misc.	Shock increase	Escape failure	(mg/kg; p.o.)	(mg/kg; p.o.)
					MED (mg/kg; i.p.)			
Chlordiazepoxide hydrochloride	Cl			2-NHCH$_3$ 4-N → 0	4	18	25	1,315
Diazepam	Cl		CH$_3$		10	67	25	710
Medazepam	Cl		CH$_3$	2-H$_2$	4	5	25	900
Flurazepam	Cl	Fl	(CH$_2$)$_2$-N-(C$_2$H$_5$)$_2$		10	28	75	1,300
Nitrazepam	NO$_2$				0.8	14	4	825
Clonazepam	NO$_2$	Cl			5	>60	2	>3,000
Flunitrazepam	NO$_2$	Fl	CH$_3$		0.3	32	5	485
Bromazepam	Br			5-pyridine	1	6	5	3,050
Phenobarbital					29	63	50	162

[a] See Table I for benzodiazepine structure.

TABLE III. *Pharmacological activity in monkeys*[a]

| Drug | Substituent in benzodiazepine structure | | | | Cynomolgus monkey | Squirrel monkey | | |
| | 7 | 2' | 1-N | Misc. | Taming | Conditioned avoidance | | |
						Shock increase MED (mg/kg; p.o.)	Escape failure	Safety ratio
Chlordiazepoxide hydrochloride	Cl			2-NHCH$_3$ 4-N → 0	1	1	29	29
Diazepam	Cl		CH$_3$		1	1	33	33
Medazepam	Cl		CH$_3$	2-H$_2$	5	60	>320	>5
Flurazepam	Cl	Fl	(CH$_2$)$_2$-N-(C$_2$H$_5$)$_2$		5	33	40	1
Nitrazepam	NO$_2$				0.1	1.6	9	6
Clonazepam	NO$_2$	Cl			2	1	320	320
Flunitrazepam	NO$_2$	Fl	CH$_3$		0.5	0.8	7	9
Bromazepam	Br			5-pyridine	1	2	8	3
Phenobarbital					20 (s.c.)	80	80	1

[a] See Table I for benzodiazepine structure.

escape failure measures the incapacitating dose. In monkeys, diazepam was similar in potency and safety margin to chlordiazepoxide, but stronger than medazepam and flurazepam. The 7-nitro compounds showed similar potency to diazepam. In contrast to diazepam, these 7-nitro compounds induced hyperactivity at the taming dose. Clonazepam was unusual in showing a very large safety margin in squirrel monkeys. Phenobarbital differed from all the benzodiazepines in showing weak activity in modifying monkey behavior, and there was no difference between the effective dose and the escape-failure dose.

IV. SPECIES VARIATION AMONG SOME BENZODIAZEPINES

Zbinden and Randall (1967) correlated the results of the screening tests in animals to the MED in human subjects of 22 benzodiazepines. The MED was determined in some patients by the calming and relaxing effect and in others by the dose producing drowsiness. The compounds were ranked in humans according to high potency of 0.25 to 3 mg/day, median potency of 3 to 15 mg/day, and low potency of 20 mg/day or higher. The best correlation of potency in humans was with the cat muscle-relaxant and the mouse anti-pentylenetetrazol tests. Scheckel and Boff (1968) used the same human and animal data, but proposed that the taming in monkeys correlated best with the human data.

The results in Table IV taken from Zbinden and Randall (1967), illustrate the separation of eight drugs into low potency (10 to 15 mg/day) and high potency (1.5 to 3 mg/day) in humans. The screening tests in cats, mice, rats, cynomolgus monkeys, and squirrel monkeys showed great variation in order of potency, and could only be separated roughly into classes of low and high potency. Medazepam was usually weaker than diazepam and the 7-nitro compounds (nitrazepam, clonazepam, and fluni-trazepam). Phenobarbital was less potent than the benzodiazepines and showed similar variability in sedative effects among species.

V. METABOLITES OF CHLORDIAZEPOXIDE HYDROCHLORIDE

Chlordiazepoxide hydrochloride is converted to the N-demethylated metabolite RO 5-0883 in man and other species (Schwartz and Postma, 1966) and also to the lactam RO 5-2092, (Koechlin et al., 1965). The major metabolite found in blood is RO 5-0883; RO 5-2092 appears on chronic administration of chlordiazepoxide hydrochloride. Both RO 5-2092 and the amino acid resulting from hydrolysis of the lactam (opened lactam) are urinary metabolites of chlordiazepoxide hydrochloride in man. In mice,

TABLE IV. *Species variation among benzodiazepines*[a]

Drug	Substituent in benzodiazepine structure				MED (mg/kg)					
	7	2'	1-N	Misc.	Cat muscle relaxant (p.o.)	Mouse anti-pent. (p.o.)	Cynomolgus monkey taming (p.o.)	Squirrel monkey cond. avoid. (p.o.)	Rat cond. avoid. (i.p.)	Man anti-anxiety (p.o.)
Chlordiazepoxide hydrochloride	Cl			2-NHCH_3 $4\text{-N} \rightarrow 0$	2	8	1	1	4	10
Diazepam	Cl		CH_3		0.2	2	1	1	10	10
Medazepam	Cl			2-H_2	4	7	5	60	4	15
Flurazepam	Cl	Fl	$(CH_2)_2\text{-N-}(C_2H_5)_2$		2	2	5	33	10	10
Nitrazepam	NO_2				0.1	0.7	0.1	1	0.8	10
Clonazepam	NO_2	Cl			0.05	0.2	2	1	5	3
Flunitrazepam	NO_2	Fl	CH_3		0.02	0.1	0.5	0.8	0.3	1.5
Bromazepam	Br			5-pyridine	0.2	0.7	1	2	1	3
Phenobarbital					50	26	20	80	29	100

[a] See Table I for benzodiazepine structure.

TABLE V. Metabolites of chlordiazepoxide hydrochloride

Drug	MED (Cat) Muscle relaxant	ED$_{50}$ (mg/kg, p.o.) Mouse Anti-pentylene-tetrazol	Anti-fighting	Rotarod	Antimaximal electroshock	Hot plate	Mouse LD$_{50}$ (mg/kg, p.o.)
Chlordiazepoxide hydrochloride	2	8	40	31	37	19	530
RO5-0883	1	6	80	36	31	37	2,150
RO5-2092	1	4	40	20	40	16	1,950
RO7-2932	>10	>800	>100	>200	>800	>200	>800
RO20-2646	>20	>800	>100		>800		

after oral administration of chlordiazepoxide hydrochloride, the anti-pentylenetetrazol activity appeared to parallel the brain levels of the N-demethylated metabolite RO 5-0883 rather than those of chlordiazepoxide hydrochloride or RO 5-2092 (Coutinho et al., 1969). In the monkey, the taming effects appeared to parallel the concentration of the parent compound in the brain rather than that of the metabolites (Coutinho et al., 1971).

The central nervous system effects of chlordiazepoxide hydrochloride and its major metabolites were described previously (Randall et al., 1965) and are summarized in Table V. The active metabolites RO 5-0883 and RO 5-2092 showed the same activity as chlordiazepoxide hydrochloride in muscle relaxation in cat, anticonvulsant effect in the anti-pentylenetetrazol test in mice, antiaggressive activity in the antifighting mouse test, muscle incoordination in the rotarod test, anticonvulsant activity in the electroshock test, and analgesic activity in the hot plate test.

The phenolic hydroxy-substituted compounds RO 7-2932 (4'-hydroxy) and RO 20-2646 (9-hydroxy) were not very active as sedatives or anticonvulsants.

In order to compare species variation among metabolites of chlordiazepoxide hydrochloride (Table VI), a sensitive test was chosen for each

TABLE VI. *Species variation among metabolites of chlordiazepoxide hydrochloride*

	MED (mg/kg)				
Drug	Cat muscle relaxant (p.o.)	Mouse anti-pentyl-enetetrazol (p.o.)	Cynomolgus monkey taming (p.o.)	Squirrel monkey conditioned avoidance (p.o.)	Rat conditioned avoidance (i.p.)
Chlordiazepoxide hydrochloride	2	8	1	1	4
RO 5-0883	1	6	10	20	20
RO 5-2092	1	4	2.5	15	>200

species. In the cat muscle-relaxant and mouse anti-pentylenetetrazol test, chlordiazepoxide hydrochloride and its metabolites RO 5-0883 and RO 5-2092 showed equal activity. In the monkey-taming and the monkey and rat conditioned-avoidance tests, the metabolites were less active. Thus the metabolites could account for the major activity of chlordiazepoxide hydrochloride in mice and cats but not in rats or monkeys. These results agree with the conclusions of Coutinho et al. (1971) that the metabolites in the brain of mice account for the anti-pentylenetetrazol activity, but that the parent compound accounts for the taming effect in monkeys.

VI. METABOLITES OF DIAZEPAM

Diazepam was shown to be metabolized in man and dog by N-demethylation to RO 5-2180, by hydroxylation in the 3-position to RO 5-5345, and by N-demethylation plus 3-hydroxylation to oxazepam (Schwartz et al., 1965). The rat not only utilizes the above pathways, but also hydroxylates in the 5-phenyl ring to form 4'-hydroxydiazepam (RO 7-3351), 4'-hydroxy RO 5-2180 (RO 7-2900), and 4'-hydroxy RO 5-5345 (Schwartz et al., 1967). Single oral doses of diazepam in man produced diazepam blood levels which declined in 24 hr, but repeated daily doses caused gradually rising levels (de Silva et al., 1966). The N-desmethyl metabolite RO 5-2180 was the major metabolite which appeared in measurable levels after 24 hr and gradually rose with daily doses and persisted in the blood longer than diazepam after discontinuing dosing.

After oral administration to mice, diazepam and its three major metabolites were measured in the brain by Coutinho et al. (1970). The anti-pentylenetetrazol activity in mice appeared to parallel the fall-off of the N-desmethyl product RO 5-2180. The oxazepam levels remained high for 24 hr, whereas the anti-pentylenetetrazol activity and the N-desmethyl concentration declined in 12 hr. After i.v. administration, diazepam levels in the brain of mice and rats were similar, and the rates of fall-off were similar (Marcucci et al., 1968, 1970a, b). The anti-pentylenetetrazol activity in rats was of short duration and paralleled the diazepam curves. The anti-pentylenetetrazol activity in mice was of long duration and paralleled the desmethyl metabolites; this long duration was attributed to the accumulation of oxazepam in the brain.

The pharmacological profile of diazepam and its metabolites (Randall et al., 1965) is shown in Table VII. The N-desmethyl metabolite RO 5-2180 is generally equal to or less than diazepam in activity as a muscle relaxant in cats, as a sedative and antiaggressive agent in mice, and has a lower toxicity. The 3-hydroxy product RO 5-5345 was similar in its profile to diazepam, except that the hot plate and toxicity figures were higher. Oxazepam, the major excretion product, showed less activity than diazepam as a muscle relaxant in cats and analgesic in mice, and had less antiaggressive effect and was less toxic, but it showed equal activity as a sedative in the anti-pentylenetetrazol and rotarod tests in mice. The phenolic 4'-hydroxy compound RO 7-2900 is not very active as a sedative and anticonvulsant.

When using sensitive screening tests in various species, the metabolites of diazepam show large variations (Table VIII). In the cat muscle-relaxant

TABLE VII. Metabolites of diazepam

Drug	MED	ED$_{50}$ (mg/kg, p.o.)					LD$_{50}$ (mg/kg, p.o.)
	Cat muscle relaxant	Mouse anti-pentylene-tetrazol	Mouse anti-fighting	Mouse rotarod	Mouse antimaximal electroshock	Mouse hot plate	Mouse
Diazepam	0.2	2	10	6	22	3	970
RO5-2180	1	1	20	4	19	75	2,950
RO5-5345	0.5	0.7	5	4	12	44	2,600
RO5-6789 Oxazepam	1	0.7	40	7	28	>200	>4,000
RO7-2900	>20	>800	>100	>200	>200	>200	>800

TABLE VIII. *Species variation among metabolites of diazepam*

Drug	MED (mg/kg)				
	Cat muscle relaxant (p.o.)	Mouse anti-pentyl-enetetrazol (p.o.)	Cynomolgus monkey taming (p.o.)	Squirrel monkey conditioned avoidance (p.o.)	Rat conditioned avoidance (i.p.)
Diazepam	0.2	2	1	1	10
RO 5-2180	1	1	0.1	5	7
RO 5-5345	0.5	0.7	2.5	20	15
Oxazepam RO 5-6789	1	0.7	>40	65	8

test, the metabolites were weaker than diazepam, but in the mouse anti-pentylenetetrazol test they were nearly equal to diazpam. In the cynomolgus monkey, the taming action of the N-desmethyl metabolite RO 5-2180 was stronger than diazepam, but the other metabolites were weaker. In the conditioned avoidance test, diazepam and its metabolites were equal in rats, but the metabolites were less active in squirrel monkeys. Thus the activity of diazepam could not be fully accounted for either by the parent compound or by the metabolites in the various species.

VII. METABOLITES OF MEDAZEPAM

Medazepam is rapidly and extensively metabolized in man, dog, and rat (Schwartz and Carbone, 1970). After oral administration, the blood in man contains medazepam and the metabolites desmethylmedazepam (RO 5-2925), diazepam, and N-desmethyldiazepam (RO 5-2180) (de Silva and Puglisi, 1970). The 3-hydroxy product oxazepam is excreted in the urine. The pharmacological properties of medazepam and its major metabolites have been described (Randall et al., 1970). The pharmacological profiles of medazepam and some of its metabolites are shown in Table IX. The N-desmethyl products RO 5-2925, RO 7-0220, and RO 5-2180 had activities similar to medazepam, with these exceptions: RO 5-2925 was slightly active in the cat, RO 7-0220 was less active in the rotarod and inactive in the hot plate, and RO 5-2180 was less active in the hot plate. RO 5-2180 is a common metabolite of medazepam and diazepam in man, dog, and rat. Diazepam is a metabolite of medazepam in man and rat; its status as a metabolite in the dog is not clear. Diazepam was more potent than medazepam in all tests represented here, but showed less muscle incoordination in the rotarod test.

TABLE IX. Metabolites of medazepam

Drug	MED Cat Muscle relaxant	ED$_{50}$ (mg/kg, p.o.) Mouse					Mouse LD$_{50}$ (mg/kg, p.o.)
		Anti-pentylene-tetrazol	Anti-fighting	Rotarod	Antimaximal electroshock	Hot plate	
Medazepam	4	7	20	4	36	12	820
RO 5-2925	20	4	10	12	19	8	1,340
RO 7-0220	4	12	40	122	50	>200	1,160
RO 5-2180	1	1	20	4	19	75	2,950
Diazepam RO 5-2807	0.2	2	10	6	22	3	970

TABLE X. Species variation among metabolites of medazepam

Drug	MED (mg/kg)				
	Cat muscle relaxant (p.o.)	Mouse anti-pentyl-enetetrazol (p.o.)	Cynomolgus monkey taming (p.o.)	Squirrel monkey conditioned avoidance (p.o.)	Rat conditioned avoidance (i.p.)
Medazepam RO 5-4556	4	7	5	60	4
RO 5-2925	20	4	—	>40	5
RO 7-0220	4	12	—	40	6
RO 5-2180	1	1	0.1	5	10
Diazepam RO 5-2807	0.2	2	1	1	10

There is great species variation among the metabolites of medazepam, when using a sensitive test for each species (Table X). In cats and mice, the N-desmethyl products RO 5-2925 and RO 7-0220 were less active, but the N-desmethyldiazepam RO 5-2180 was equal to medazepam. Diazepam, a major metabolite of medazepam, was more active than medazepam in cats but equal in mice. In the monkey tests, the N-desmethyl product RO 5-2180 and diazepam were more active than medazepam. In the rat conditioned-avoidance test, the activity of the various metabolites was nearly equal to medazepam. It appears that the major metabolites RO 5-2180 and diazepam could account for much of the activity of medazepam in various species.

VIII. METABOLITES OF FLURAZEPAM

The extensive metabolism of flurazepam in man and dog was established by Schwartz and Postma (1970). Flurazepam is absorbed in man and rapidly metabolized. The pharmacology of flurazepam was reported by Randall et al. (1969). Some pharmacological properties of flurazepam and its major metabolites are given in Table XI. The deethylated products RO 7-2431 and RO 7-1986 did not differ greatly from flurazepam, except that they were inactive in the hot plate test. The deaminated product RO 7-2750, which has an alcohol side chain on N_1, is the major metabolite of flurazepam in man. It showed stronger activity than flurazepam in the screening tests. The N_1-desalkyl metabolites RO 5-3367, with the N_1 side chain removed, and RO 7-5205, the 3-hydroxy product, were also more active than the parent compound. Oxidation of the ethanol side chain to N_1-acetic acid yields an inactive product, RO 7-5096. Thus it has not been

TABLE XI. *Metabolites of flurazepam hydrochloride*

Drug	MED Cat Muscle relaxant	ED$_{50}$ (mg/kg, p.o.) Mouse Anti-pentylene-tetrazol	Anti-fighting	Rotarod	Antimaximal electroshock	Hot plate	Mouse LD$_{50}$ (mg/kg, p.o.)
Flurazepam RO5-6901	2	2	20	6	82	32	660
RO 7-2431/1 N-CH$_2$CH$_2$NHC$_2$H$_5$ .2HCl	1	2	10	8	100	>200	740
RO 7-1986 N-CH$_2$CH$_2$NH$_2$	1	3	10	7	75	>200	1,500
RO 7-2750 N-CH$_2$CH$_2$OH	0.2	0.3	1	0.7	15	8	940
RO 5-3367 N-H	0.1	0.08	5	0.8	2	2	2,250
RO 7-5205 N-H 3-OH	0.4	0.5	2	3	18	>200	>4,000
RO 7-5096 N-CH$_2$COOH	>20	46	>100	70	800	>100	>800

Drug	MED Cat Muscle relaxant	ED$_{50}$ (mg/kg, p.o.) Mouse Anti-pentylene-tetrazol	Anti-fighting	Rotarod	Antimaximal electroshock	Hot plate	Mouse LD$_{50}$ (mg/kg, p.o.)
Nitrazepam RO 5-3059	0.1	0.7	5	0.8	31	6	2,300
RO 5-5191	10	3	100	15	177	35	>800
RO 5-3072	>20	>800	>100	27	600	200	>800
RO 5-3308	>20	>800	>100	>200	>800	200	>4,000
RO 7-5124	—	800	—	—	>800	>100	—

established which of the metabolites of flurazepam accounts for its pharmacological activity in various species.

IX. METABOLITES OF NITRAZEPAM

Nitrazepam was shown by Rieder (1965) to be rapidly absorbed in man, with maximum blood levels in 4 hr and a half-life of 7 hr. The major metabolites are the 7-amino metabolite (RO 5-3072) and the 7-acetylamino metabolite (RO 5-3308). The pharmacological profile of these metabolites and nitrazepam was described by Randall et al. (1965). The results in Table XII show that 7-amino compounds are not very active. The 3-hydroxylated metabolite of nitrazepam (RO 5-5191) retained weak sedative, muscle-relaxant, and anticonvulsant activity. Thus it is the parent compound nitrazepam rather than its metabolites which accounts for the pharmacological activity in various species.

X. SUMMARY

Screening tests for sedative, muscle-relaxant, taming, antiaggressive, and anticonvulsive activity in mice, rats, cats, cynomolgus monkeys, and squirrel monkeys show large variations in potency among eight benzodiazepines. They may be separated roughly into classes of low and high potency; all have exceptionally high safety margins.

Phenobarbital is primarily a sedative and a hypnotic in all species. The benzodiazepines and phenobarbital have equal antimaximal shock activity, but phenobarbital has a narrow margin of safety compared to the benzodiazepines.

The metabolites of chlordiazepoxide hydrochloride, diazepam, medazepam, flurazepam, and nitrazepam differ among rats, dogs, and man, and also show large species variation in the various screening tests.

It has not been established how much of the pharmacological activity of these benzodiazepines can be attributed to the metabolites in various species.

The greatest difference between the benzodiazepines and phenobarbital lies in the far superior safety margins of the benzodiazepines in all animal species tested.

ACKNOWLEDGMENTS

The author gratefully acknowledges the skilled technical assistance of D. Hane, H. Baruth, and E. Boff.

REFERENCES

Coutinho, C. B., Cheripko, J. A., and Carbone, J. J. (1969): Relationship between the duration of anticonvulsant activity of chlordiazepoxide and systemic levels of the parent compound and its major metabolites in mice. *Biochemical Pharmacology*, 18: 303–316.

Coutinho, C. B., Cheripko, J. A., and Carbone, J. J. (1970): Correlation between the duration of the anticonvulsant activity of diazepam and its physiological disposition in mice. *Biochemcal Pharmacology*, 19:363–379.

Coutinho, C. B., King, M., Carbone, J. J., Manning, J. E., Boff, E., and Crews, T. (1971): Chlordiazepoxide metabolism as related to reduction in aggressive behavior of cynomolgus primates. *Xenobiatica*, 1:287–301.

de Silva, J. A. F., Koechlin, B. A., and Bader, G. (1966): Blood level distribution patterns of diazepam and its major metabolite in man. *Journal of Pharmaceutical Sciences*, 55:692–702.

de Silva, J. A. F., and Puglisi, C. V. (1970): Determination of medazepam (NOBRIUM), diazepam (VALIUM) and their major biotransformation products in blood and urine by electron capture gas-liquid chromatography. *Analytical Chemistry*, 42:1725–1736.

Dunham, N. W., and Miya, T. S. (1957): A note on a simple apparatus for detecting neurological deficit in rats and mice. *Journal of the American Pharmaceutical Association, Scientific Edition*, 46:208–209.

Eddy, N. B., Touchberry, C. F., and Lieberman, J. E. (1950): Synthetic analgesics. I. Methadone isomers and derivatives. *Journal of Pharmacology and Experimental Therapeutics*, 98:121–137.

Everett, G. M., and Richards, R. K. (1955): Comparative anticonvulsant action of 3,5,5-trimethyloxazolidine-2,4-dione (TRIDIONE), DILANTIN and phenobarbital. *Journal of Pharmacology and Experimental Therapeutics*, 81:402–407.

Harris, T. H. (1960): Methaminodiazepoxide. *Journal of the American Medical Association*, 172:1162–1163.

Heise, G. A., and Boff, E. (1962): Continuous avoidance as a base-line for measuring behavioral effects of drugs. *Psychopharmacologia* (Berlin), 3:264–282.

Kaim, S. C., and Rosenstein, I. N. (1960): Anticonvulsant properties of a new psychotherapeutic drug. *Diseases of the Nervous System*, 21 (*Suppl.*):46–48.

Koechlin, B. A., Schwartz, M. A., Krol, G., and Oberhansli, W. (1965): The metabolic fate of C^{14}-labeled chlordiazepoxide in man, in the dog, and in the rat. *Journal of Pharmacology and Experimental Therapeutics*, 148:399–411.

Marcucci, F., Fanelli, R., Mussini, E., and Garattini, S. (1970a): Further studies on the long lasting antimetrazol activity of diazepam in mice. *Biochemical Pharmacology*, 19:1771–1776.

Marcucci, F., Guaitani, A., Kvetina, J., Mussini, E., and Garattini, S. (1968): Species differences in diazepam metabolism and anticonvulsant activity. *European Journal of Pharmacology*, 4:467–470.

Marcucci, F., Mussini, E., Fanelli, R., and Garattini, S. (1970b): Species differences in diazepam metabolism. I. Metabolism of diazepam metabolites. *Biochemical Pharmacology*, 19:1847–1851.

Miller, L. C., and Tainter, M. L. (1944): Estimation of the ED_{50} and its error by means of logarithmic-probit graph paper. *Proceedings of the Society for Experimental Biology and Medicine*, 57:261–264.

Randall, L. O., Heise, G. A., Schallek, W., Bagdon, R. E., Banziger, R. F., Boris, A., Moe, R. A., and Abrams, W. B. (1961): Pharmacological and clinical studies on VALIUM, a new psychotherapeutic agent of the benzodiazepine class. *Current Therapeutic Research*, 3:405–425.

Randall, L. O., and Schallek, W. (1968): Pharmacological activity of certain benzodiazepines. In: *Psychopharmacology: A Review of Progress 1957-1967*, edited by D. H. Efron, pp. 153–184. United States Public Health Service Publication Number 1836.

Randall, L. O., Schallek, W., Heise, G. A., Keith, E. F., and Bagdon, R. E. (1960): The psychosedative properties of methaminodiazepoxide. *Journal of Pharmacology and Experimental Therapeutics*, 129:163–171.

Randall, L. O., Schallek, W., Scheckel, C. L., Bagdon, R. E., and Rieder, J. (1965a): Zur Pharmakologie von MOGADON, einem Schlafmittel mit neuartigem Wirkungsmechanismus *Schweizerische Medizinische Wochenschrift*, 95:334–337.

Randall, L. O., Schallek, W., Scheckel, C. L., Stefko, P. L., Banziger, R. F., Pool, W., and Moe, R. A. (1969): Pharmacological studies on flurazepam hydrochloride (Ro 5-6901), a new psychotropic agent of the benzodiazepine class. *Archives Internationales de Pharmacodynamie et de Thérapie*, 178:216–241.

Randall, L. O., Scheckel, C. L., and Banziger, R. F. (1965b): Pharmacology of the metabolites of chlordiazepoxide and diazepam. *Current Therapeutic Research*, 7:590–606.

Randall, L. O., Scheckel, C. L., and Pool, W. (1970): Pharmacology of medazepam and metabolites. *Archives Internationales de Pharmacodynamie et de Thérapie*, 185:135–148.

Rieder, J. (1965): Methoden zur Bestimmung von 1,3-Dihydro-7-nitro-5-phenyl-2H-1,4-benzodiazepin-2-on und seinen Hauptmetaboliten in biologischen Proben und Ergebnisse von Versuchen ueber die Pharmakokinetik und den Metabolismus dieser Substanz bei Mensch und Ratte. *Arzneimittel-Forschung*, 15:1134–1148.

Scheckel, C. L., and Boff, E. (1968): The effects of drugs on conditioned avoidance and aggressive behavior. In: *Use of Nonhuman Primates in Drug Evaluation. A Symposium*, edited by H. Vagtborg, pp. 301–312. University of Texas Press, Austin.

Schwartz, M. A., Bommer, P., and Vane, F. M. (1967): Diazepam metabolites in the rat; characterization by high resolution mass spectrometry and nuclear magnetic resonance. *Archives of Biochemistry and Biophysics*, 121:508–516.

Schwartz, M. A., and Carbone, J. J. (1970): Metabolism of ^{14}C-medazepam hydrochloride in dog, rat and man. *Biochemical Pharmacology*, 19:343–361.

Schwartz, M. A., Koechlin, B. A., Postma, E., Palmer, S., and Krol, G. (1965): Metabolism of diazepam in rat, dog and man. *Journal of Pharmacology and Experimental Therapeutics*, 149:423–435.

Schwartz, M. A., and Postma, E. (1966): Metabolic N-demethylation of chlordiazepoxide. *Journal of Pharmaceutical Sciences*, 55:1358–1362.

Schwartz, M. A., and Postma, E. (1970): Metabolism of flurazepam, a benzodiazepine, in man and dog. *Journal of Pharmaceutical Sciences*, 59:1800–1806.

Sidman, M. (1953): Avoidance conditioning with brief shock and no exteroceptive warning signal. *Science*, 118:157–158.

Sternbach, L. H., Randall, L. O., and Gustafson, S. R. (1964): 1,4-Benzodiazepines (chlordiazepoxide and related compounds). In: *Psychopharmacological Agents*, Vol. 1, edited by M. Gordon, pp. 137–224. Academic Press, New York.

Sternbach, L. H., and Reeder, E. (1961): Quinazolines and 1,4-benzodiazepines. II. The rearrangement of 6-chloro-2-chloromethyl-4-phenylquinazoline 3-oxide into 2-amino derivatives of 7-chloro-5-phenyl-3H-1,4-benzodiazepine 4-oxide. *Journal of Organic Chemistry*, 26:1111–1118.

Svenson, S. E., and Gordon, L. E. (1965): Diazepam: A progress report. *Current Therapeutic Research*, 7:367–391.

Swinyard, E. A., Brown, W. C., and Goodman, L. S. (1952): Comparative assays of antiepileptic drugs in mice and rats. *Journal of Pharmacology and Experimental Therapeutics*, 106:319–330.

Tedeschi, R. E., Tedeschi, D. H., Mucha, A., Cook, L., Mattis, P. A., and Fellows, E. J. (1959): Effects of various centrally acting drugs on fighting behavior of mice. *Journal of Pharmacology and Experimental Therapeutics,* 125:28–34.

Tobin, J. M., and Lewis, N. D. C. (1960): New psychotherapeutic agent, chlordiazepoxide. Use in treatment of anxiety states and related symptoms. *Journal of the American Medical Association,* 174:1242–1249.

Zbinden, G., and Randall, L. O. (1967): Pharmacology of benzodiazepines: Laboratory and clinical correlations. In: *Advances in Pharmacology,* Vol. 5, edited by S. Garattini and P. A. Shore, pp. 213–291. Academic Press, New York.

PATHWAYS OF METABOLISM OF THE BENZODIAZEPINES

Morton A. Schwartz

Department of Biochemistry and Drug Metabolism, Hoffmann-La Roche Inc., Nutley, New Jersey

I. INTRODUCTION

This chapter is designed to illustrate the pathways by which the 1, 4-benzodiazepines are metabolized in experimental animals and man. The biotransformation of three members of this class, chlordiazepoxide, medazepam, and flurazepam, will be described in detail. These three drugs are in widespread clinical use, and each possesses structural features not shared by the other two (Fig. 1). Chlordiazepoxide is distinguished by a C_2-meth-

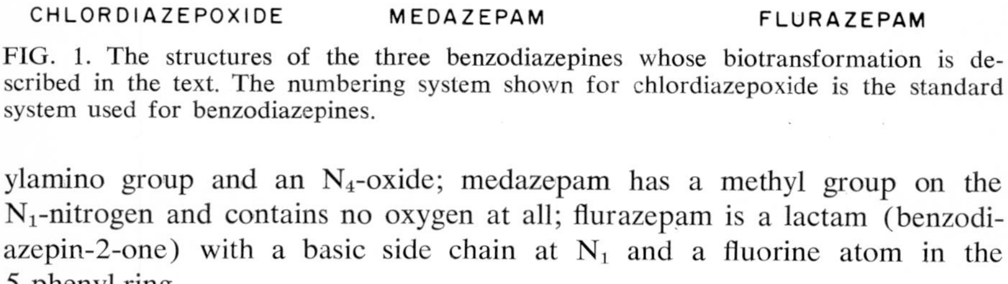

FIG. 1. The structures of the three benzodiazepines whose biotransformation is described in the text. The numbering system shown for chlordiazepoxide is the standard system used for benzodiazepines.

ylamino group and an N_4-oxide; medazepam has a methyl group on the N_1-nitrogen and contains no oxygen at all; flurazepam is a lactam (benzodiazepin-2-one) with a basic side chain at N_1 and a fluorine atom in the 5-phenyl ring.

The biotransformation of two additional benzodiazepin-2-ones in clinical use, diazepam and oxazepam, will be briefly noted since one is a metabolite of chlordiazepoxide and both are metabolites of medazepam. A more extensive discussion of diazepam metabolism will be presented later in this volume by Prof. Silvio Garattini. Furthermore, since the reductive pathways involved in the metabolism of nitrazepam will be described elsewhere in this volume by Prof. J. Rieder, the 7-nitrobenzodiazepines will not be discussed here. All the drugs mentioned will be designated by generic name, and metabolites will be given a name which is descriptive of their derivation from the parent drug; the reader is referred to the original papers for all chemical names.

It is appropriate here to mention briefly some techniques used in the characterization of benzodiazepine metabolites. The extraction of lipophilic metabolites into an organic solvent and the hydrolysis of hydrophilic metabolites by a crude mixture of β-glucuronidase and sulfatase to free the extractable lipophilic metabolite moiety have been extensively used. Any metabolite present in a form hydrolyzable by glusulase, presumably as a glucuronide and/or sulfate, is denoted as a conjugate. Once extracted, many metabolites were identified by a comparative chromatographic procedure which is based on the assertion that, if an isolated metabolite and a known reference compound cannot be separated, they are identical. Although a strictly unambiguous result is obtained only when a separation is achieved and the lack of identity is proved, this technique does yield a creditable identification if nonseparation is demonstrated in at least two solvent systems. This relatively simple method, however, requires (1) that the metabolite's structure be postulated both on the basis of the considerable literature available on the metabolism of foreign compounds (Williams, 1959; Parke, 1968) and on the developing knowledge of benzodiazepine biotransformation and (2) that an organic chemist synthesize the postulated metabolite.

If this relatively simple procedure cannot be used or proves unsuccessful in yielding an identification of the metabolite, then the structure must be determined by some physicochemical technique. Of particular usefulness has been the combination of thin-layer chromatography (TLC) for separation and isolation of metabolites with high-resolution mass spectrometry for their identification (Schwartz et al., 1967, 1968*a, b*). This combination has allowed for the isolation of the few micrograms of metabolite (which should be free of other metabolites but need not be completely pure) required for high-resolution mass spectral analysis. When successful, this analysis yields a molecular weight so precise that the empirical formula is attainable as well as a pattern of fragments of explicitly defined atomic content which can indicate the moiety in which the metabolic change has occurred. In some instances, this technique has been complemented by the use of nuclear magnetic resonance in order to locate more definitively the point of metabolic attack; examples will be cited below. The application of various spectroscopic techniques for the characterization of metabolites was discussed by Kuntzman et al. (1968) and has recently been reviewed by Bommer and Vane (1971).

II. CHLORDIAZEPOXIDE METABOLISM IN MAN, DOG, AND RAT

Chlordiazepoxide was the first benzodiazepine to be the subject of a metabolic study. While many aspects of its biotransformation are now known, there are still many metabolites which have not yet been identified. One important pathway in all species so far examined is shown in Fig. 2.

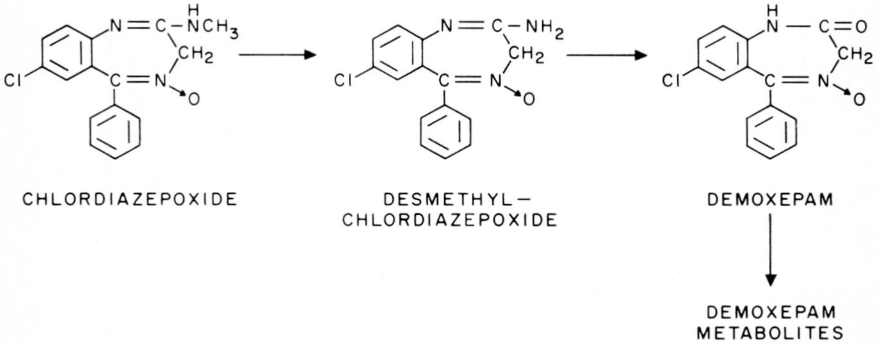

CHLORDIAZEPOXIDE   DESMETHYL–
CHLORDIAZEPOXIDE   DEMOXEPAM

DEMOXEPAM
METABOLITES

FIG. 2. Outline of chlordiazepoxide biotransformation in man and dog. The demoxepam metabolites are given in Fig. 3.

Koechlin and D'Arconte (1963) devised an assay for chlordiazepoxide in which the drug was first hydrolyzed to demoxepam (also designed Ro 5-2092 or "lactam"), which was then converted photochemically to a fluorescent product. Using this assay, they found that demoxepam itself was present in the plasma of chlordiazepoxide-treated human subjects and dogs. Subsequent studies with chlordiazepoxide-2-^{4}C (Koechlin et al., 1965) yielded evidence, obtained by comparative paper chromatography, that demoxepam and a hydrolysis product, "opened lactam," were urinary metabolites in man. While other metabolites were detected, these were the only ones which could be characterized at the time. The urinary metabolites of labeled chlordiazepoxide in the dog, but not in the rat, resembled those of man in extractability and paper chromatographic mobility.

The existence of the N-demethylated metabolite, desmethylchlordiazepoxide, was first demonstrated *in vitro* with slices, homogenate, and 9,000 $\times$ *g* supernatant of rat liver (Schwartz and Postma, 1966). The identification was made by comparative TLC. This oxidative demethylation of chlordiazepoxide was further shown to be stimulated by phenobarbital pretreatment of the rats; the 18% conversion to desmethylchlordiazepoxide on incubation of chlordiazepoxide with 9,000 $\times$ g supernatant of liver from control rats was increased to 56% by this pretreatment. The conversion of chlordiazepoxide or the demethylated metabolite to demoxepam, however, could not be demonstrated *in vitro*.

Desmethylchlordiazepoxide was also shown by TLC to be present in the blood of the rat, squirrel monkey, and man following chlordiazepoxide administration. A fluorometric determination of this metabolite was then developed which, on incorporation into the Koechlin and D'Arconte (1963) assay, provided the necessary specificity for determining the plasma levels of both chlordiazepoxide and desmethylchlordiazepoxide. Data obtained on using this new assay procedure provided indirect evidence that desmethylchlordiazepoxide was an intermediate in the biotransformation of chlordiazepoxide to demoxepam. In human subjects given chlordiazepoxide, plasma levels of demoxepam were never detected prior to the appearance of plasma desmethylchlordiazepoxide. Both metabolites, however, did accumulate in the plasma after chronic chlordiazepoxide administration.

Direct evidence for this pathway has been obtained from a pharmacokinetic study of chlordiazepoxide disposition in the dog (Kaplan et al., 1970). Chlordiazepoxide, desmethylchlordiazepoxide, and demoxepam were each separately administered intravenously, and the rates of elimination of the administered compounds and the rates of formation of metabolites were determined. Chlordiazepoxide was shown to be eliminated by the dog by virtually quantitative demethylation to desmethylchlordiazepoxide, which was itself almost completely biotransformed, with as much as 50% being converted to demoxepam. In addition, the rates of elimination of both metabolites were shown to be slower than that of chlordiazepoxide.

In a cross-over study in man, the elimination of chlordiazepoxide was shown to be two to four times faster than that of demoxepam; the chlordiazepoxide half-life ranged from 7 to 28 hr and the half-life of demoxepam ranged from 14 to 95 hr (Schwartz et al., 1971a). In addition, the administration of a single 20-mg oral dose of chlordiazepoxide resulted in persistent plasma levels of the desmethyl metabolite. Demoxepam, however, was not detectable in the plasma after this single dose. These findings suggest that in man, as in the dog, both metabolites are eliminated more slowly than the parent drug.

The same two chlordiazepoxide metabolites are also produced in the mouse. Coutinho et al. (1969) reported that both metabolites were present in the brain following a 20 mg/kg oral dose of chlordiazepoxide. From 0.5 to 6 hr, the major drug-related substance in the brain was desmethylchlordiazepoxide. Evidence that this metabolite contributed greatly to the antipentylenetetrazol (METRAZOL) activity of chlordiazepoxide was also obtained.

Kimmel and Walkenstein (1967) provided the first clue to the existence of a metabolic pathway which results in the reduction of the N-oxide function. They presented chromatographic evidence for a slight conversion of chlordiazepoxide to oxazepam in the chlordiazepoxide-2-^{14}C-treated dog. It was estimated that 1% of a 26 mg/kg oral dose of labeled drug was excreted as oxazepam in the urine and in the feces.

This pathway has been further elucidated by recent studies of the biotransformation of demoxepam-2-^{14}C in the dog (Schwartz et al., 1971*b*). As shown in Fig. 3, four distinct metabolic pathways are in operation. One, the hydrolysis to "opened lactam," had been suggested earlier (Koechlin et al., 1965). Two of the four pathways are concerned with the hy-

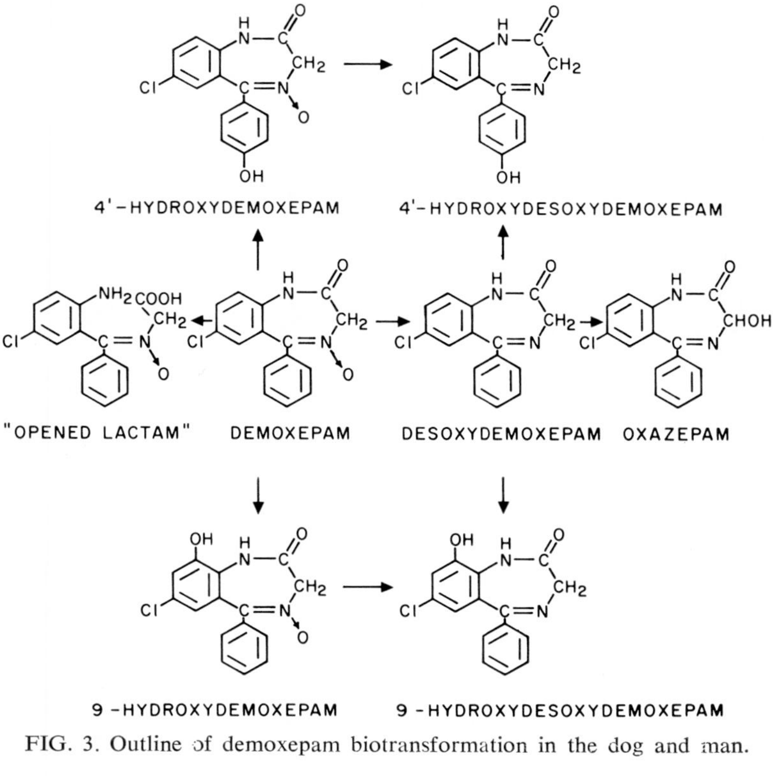

4'-HYDROXYDEMOXEPAM   4'-HYDROXYDESOXYDEMOXEPAM

"OPENED LACTAM"   DEMOXEPAM   DESOXYDEMOXEPAM   OXAZEPAM

9-HYDROXYDEMOXEPAM   9-HYDROXYDESOXYDEMOXEPAM

FIG. 3. Outline of demoxepam biotransformation in the dog and man.

droxylation of demoxepam to phenolic metabolites. The *para*-hydroxy metabolite, 4'-hydroxydemoxepam, was identified by comparative two-dimensional TLC. Accounting for 10% of the dose, it was the major metabolite identified in the urine and was excreted primarily as a conjugate. The other phenolic metabolite, 9-hydroxydemoxepam, was also excreted as a conjugate but was not so easily identified. Although high-resolution mass spectral analysis indicated that a phenolic hydroxyl function had been added to demozepam, nuclear magnetic resonance (NMR) was required to demonstrate that this new substituent was located at the C_9 position. The last pathway, the reduction of demoxepam, was established by the identification (by high-resolution mass spectrometry with confirmation by comparative two-dimensional TLC) of desoxydemoxepam as a fecal metabolite. In addition, small amounts of oxazepam were found in both the urine and feces.

Desoxydemoxepam, which may also be designated desmethyldiazepam, is both a metabolite of diazepam and a metabolic precursor of oxazepam (Schwartz and Postma, 1968; Kvetina et al., 1968; Marcucci et al., 1969). Therefore, the biotransformation of chlordiazepoxide to oxazepam may be postulated as: chlordiazepoxide → desmethylchlordiazepoxide → demoxepam → desoxydemoxepam → oxazepam.

Two additional N-desoxy compounds, 4'-hydroxydesoxydemoxepam and 9-hydroxydesoxydemoxepam, were identified as demoxepam metabolites in the dog. The fact that the N-desoxy metabolites (desoxydemoxepam and the two phenolic derivatives of desoxydemoxepam) predominated in the feces whereas the corresponding N-oxides (demoxepam and its two phenolic derivatives) predominated in the bile suggested that the intestinal tract was an important site of N-oxide reduction. It should be noted, however, that the reduction of demoxepam to desoxydemoxepam may also occur systemically, as has been reported for a number of other N-oxides (Bickel, 1969). As shown in Fig. 3, desoxydemoxepam, apart from forming oxazepam, may also act as an additional precursor of the N-desoxy phenols.

The biotransformation of single 20-mg oral doses of ^{14}C-demoxepam in man has been studied (Schwartz and Postma, 1972), and the pathways of Fig. 3 were found to apply also to man. Of all the metabolites found in the dog, only 4'-hydroxydesoxydemoxepam was not detected in human excreta. From the limited studies which have been completed in each species, it appears that the production of "opened lactam" and oxazepam may be more prominent in man, and the *para*-hydroxylation of the 5-phenyl ring may be preferred in the dog. Since desoxydemoxepam was also detected in human feces, the N-oxide reduction found in the dog may also occur in the intestinal tract of man. Although these studies of demoxepam metabolism provide additional information on the metabolites of chlordiazepoxide, further studies with chlordiazepoxide are needed to delineate more fully its biotransformation in man and dog.

The urinary metabolites of chlordiazepoxide in the rat were investigated because of the earlier finding (Koechlin et al., 1965) that these metabolites were chromatographically different from those excreted by the man and dog. Using TLC for metabolite separation and high-resolution mass spectrometry for the primary characterization, Schwartz et al. (1968a) showed that each of four metabolites of chlordiazepoxide-2-^{14}C excreted in rat urine was hydroxylated in the 5-phenyl ring (Fig. 4). Evidence that the hydroxy

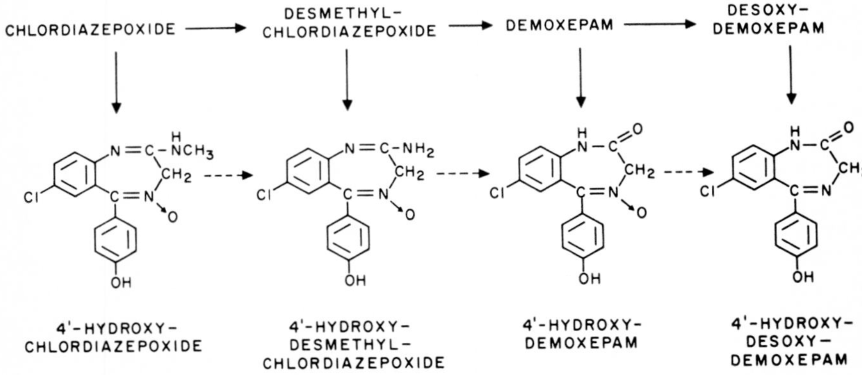

FIG. 4. Outline of chlordiazepoxide biotransformation in the rat. The pathways designated by the solid arrows are more likely to occur than those designated by broken arrows (see text).

group was added to the *para* position was obtained from comparative two-dimensional TLC; each of the metabolites migrated as its corresponding *para*-hydroxy reference compound in solvent systems capable of separating authentic 4'-hydroxydesoxydemoxepam from its 2'-hydroxy isomer.

Unconjugated 4'-hydroxychlordiazepoxide, conjugated 4'-hydroxydesoxydemoxepam, and both forms of the remaining two metabolites were excreted. As a group, these metabolites accounted for roughly 15% of the dose, or almost one-half of the total radioactivity excreted in the urine. The solid arrows in Fig. 4 show the more probable pathways: the formation of each phenol from a nonphenolic precursor rather than the conversion of one phenol to another. This is based on the consideration that since the phenols are excreted in the urine, they are less likely to be further metabolized than are the nonexcreted precursors. Two of these metabolites, 4'-hydroxyde-moxepam and 4'-hydroxydesoxydemoxepam, were subsequently shown (as noted above) to be metabolites of demoxepam in the dog.

Further studies were initiated to find out why demoxepam, which is a urinary metabolite of chlordiazepoxide in man and the dog, was not found in the urine of the chlordiazepoxide-treated rat. Unchanged demoxepam excreted in the urine of a rat given a 5 mg/kg oral dose of ^{14}C-demoxepam accounted for less than 2% of the dose (Schwartz and Postma, *unpublished data*). Moreover, the total excretion of 4'-hydroxydemoxepam, two-thirds of which was fecal, accounted for 20% of the dose, and the total excretion of 4'-hydroxydesoxydemoxepam, almost all of which was fecal, accounted for another 18% of the dose. This large excretion of *para*-hydroxylated metabolites suggests that unchanged demoxepam is not excreted because hydroxylation at the 4' position is very rapid in the rat.

III. MEDAZEPAM METABOLISM IN MAN, DOG, AND RAT

A considerable amount of information on the biotransformation of medazepam has recently been obtained (Schwartz and Carbone, 1970; Schwartz and Kolis, 1972). The pathways of metabolism, which are outlined in Fig. 5, have been elucidated primarily as a result of *in vitro* studies with fractions of rat and dog liver. Using medazepam-5-^{14}C as substrate and comparative two-dimensional TLC for metabolite identification and quantitation, the 9,000 $\times$ g supernatants of both rat and dog liver were shown to be capable of N-demethylation to form desmethylmedazepam and of 2-hydroxylation to form 2-hydroxymedazepam. A second 2-hydroxy derivative, 2-hydroxydesmethylmedazepam, while not actually found in the incubation media, was postulated as an intermediate on the basis of the structures of other metabolites which were identified. The reactions shown within the box were common to the 9,000 $\times$ g supernatants of both rat and dog liver.

A striking species difference in the further metabolism of the 2-hydroxy metabolites is evident from the data shown in Table I. Whereas the 9,000 $\times$ g supernatant of rat liver produced diazepam equivalent to 14 to 15% of the amount of medazepam incubated, the 9,000 $\times$ g supernatant of dog liver yielded less than 2% diazepam. Apparently, rat liver is capable of oxidizing the amino carbinols 2-hydroxymedazepam and 2-hydroxydesmethylmedazepam to the 2-keto metabolites diazepam and desmethyldiazepam, respectively. Desmethylmedazepam and both 2-keto metabolites predominated after the 30-min incubations. The two amino carbinols, however,

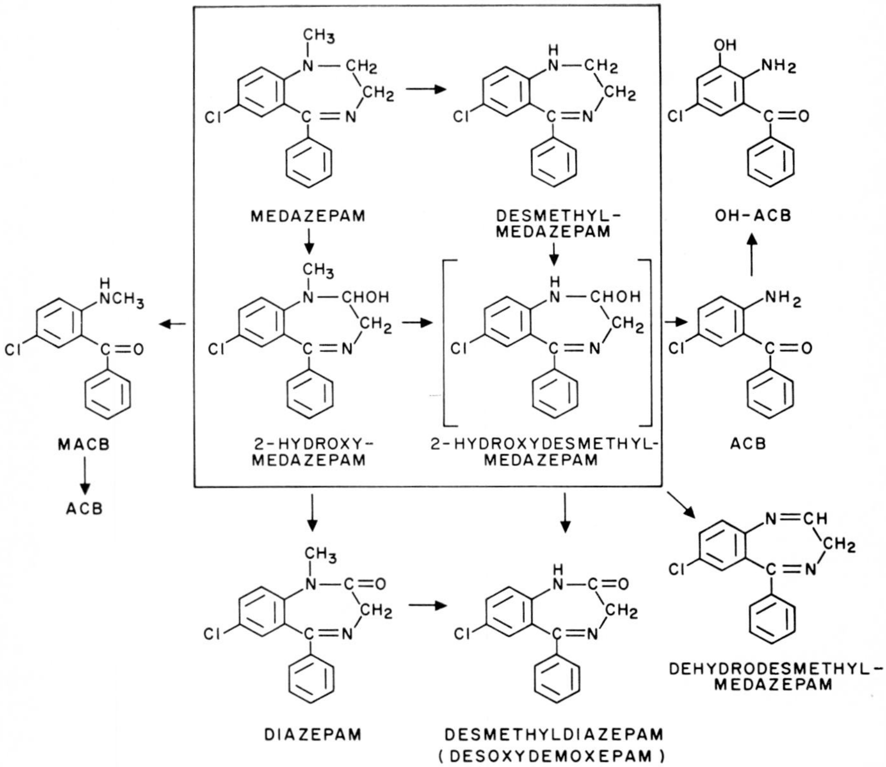

FIG. 5. Outline of medazepam biotransformation in man, dog, and rat. The reactions within the box were established *in vitro* with 9000 × g supernatants of rat and dog liver. The extent to which the various pathways operate in each species is discussed in the text.

did not appear to be appreciably oxidized in this manner by the 9,000 × g supernatant of dog liver; the metabolites which accumulated were desmethylmedazepam, 2-hydroxymedazepam, and dehydrodesmethylmedazepam. Since prior investigations had shown no appreciable difference in the amount of diazepam metabolized by 9,000 × g supernatants of rat and dog liver (Schwartz and Postma, 1968), the low levels of diazepam found in the dog liver supernatant indicate a lack of its formation rather than more extensive metabolism. It is concluded, therefore, that the dog liver enzymes preferentially dehydrate the postulated metabolite, 2-hydroxydesmethylmedazepam, which is formed from both desmethylmedazepam and 2-hydroxymedazepam.

TABLE I. Metabolites of medazepam produced in vitro.[a]

Enzyme source	Incubation time (min)	% formation:					
		Desmethyl-medazepam	2-Hydroxy-medazepam	Diazepam	Desmethyl-diazepam	Dehydro-desmethyl-medazepam	ACB
Rat liver:							
9000 × g supernatant	30[b]	14		14	16	6	<2
	30[c]	21	7	14	7	5	4
	30[c]	22	4	15	11	7	4
Dog liver:							
9000 × g supernatant	30[c]	12	17	1.5	1.5	11	5
	60[b]	13		<2	4	21	10

[a] Medazepam (0.3 to 0.5 μmole) was incubated aerobically at 37°C with each designated liver fraction.
[b] Data from Schwartz and Carbone (1970).
[c] Data from Schwartz and Kolis (1972).

The 2-hydroxy metabolites also displayed a chemical instability; 2-hydroxymedazepam and dehydrodesmethylmedazepam (which is most probably hydrated to 2-hydroxydesmethylmedazepam prior to hydrolysis) are each cleaved to the respective benzophenones, 2-methylamino-5-chlorobenzophenone (MACB) and 2-amino-5-chlorobenzophenone (ACB), on incubation at pH 7.4 in the absence of liver enzymes. In Fig. 5, ACB is also shown arising from MACB; it is presumed that N-demethylation of MACB would occur readily in the 9,000 $\times$ g liver supernatant. The third benzophenone, 2-amino-3-benzoyl-5-chlorophenol (OH-ACB), was not found in the *in vitro* experiments. It was excreted, however, as a conjugated urinary metabolite of medazepam in both the man and dog. The fact that ACB was its precursor was demonstrated in a dog given ^{3}H-labeled ACB.

The oxidation of 2-hydroxymedazepam to diazepam in rat liver has been found to be mainly a microsomal reaction. This reaction, therefore, differs from those reported by Gillette (1959), Toki et al. (1963), and Kuntzman et al. (1968), in which a soluble enzyme, apparently alcohol dehydrogenase, was responsible for the *in vitro* oxidation of several different alcohols. The fact that 2-hydroxymedazepam is an amino carbinol may be responsible for this difference. It is of interest that the oxidation of medazepam to diazepam is analogous to the oxidation of nicotine to cotinine, in which an amino carbinol is also an intermediate (Hucker et al., 1960).

The *in vivo* metabolism of medazepam has been studied in rat, dog, and man. In the rat, the major plasma metabolites (desmethylmedazepam, diazepam, and desmethyldiazepam) were also the major metabolites produced *in vitro* by the 9,000 $\times$ g liver supernatant. One would expect from this that the metabolites of diazepam known to be formed in the rat would also be present after medazepam administration. It is therefore necessary at this point to discuss briefly the biotransformation of diazepam (Fig. 6).

The metabolism of diazepam to oxazepam *via* separate pathways, involving desmethyldiazepam and N-methyloxazepam as intermediates, was first suggested by the *in vivo* studies of Ruelius et al. (1965), and Schwartz et al. (1965). These pathways were subsequently demonstrated in the rat and dog by *in vitro* experiments with 9,000 $\times$ g liver supernatants (Schwartz and Postma, 1968), and in the rat and mouse by liver perfusion experiments (Kvetina et al., 1968) and liver microsome incubations (Marcucci et al., 1969). The conjugated metabolites of ^{3}H-diazepam in the rat intestinal contents were identified by high-resolution mass spectrometry and NMR spectroscopy as 4'-hydroxydiazepam, 4'-hydroxydesmethyldiazepam, N-methyloxazepam, and 4'-hydroxy-N-methyloxazepam (Schwartz et al.,

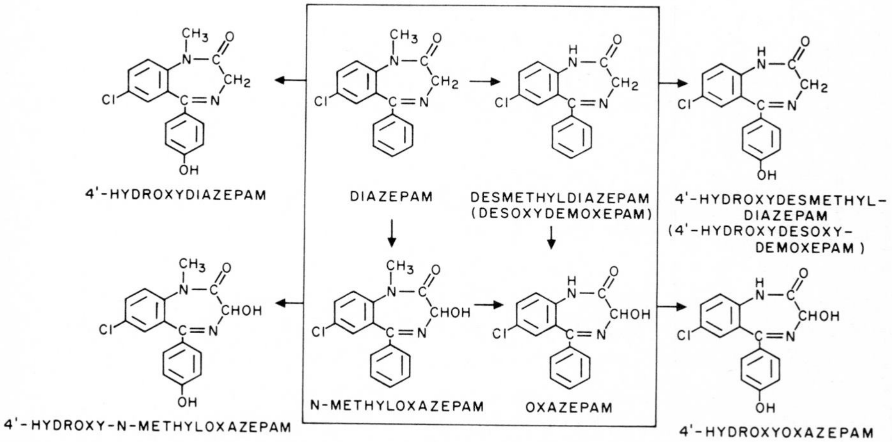

FIG. 6. Outline of diazepam biotransformation in man, dog, rat, and mouse. The reactions within the box are common to all of the above species. The phenolic metabolites outside the box are predominant in the rat and are excreted as conjugates.

1967). No oxazepam was detected, but this is not surprising since Marcucci et al. (1970) have recently shown that oxazepam is rapidly eliminated by the rat (much faster than by the mouse). This rapid elimination is known to proceed via the formation of several metabolites (Walkenstein et al., 1964), of which 4′-hydroxyoxazepam has recently been reported (Sisenwine et al., 1970) to be the major one.

Returning to medazepam metabolism in the rat, three phenolic metabolites of diazepam, 4′-hydroxydiazepam, 4′-hydroxy-N-methyldiazepam, and 4′-hydroxydesmethyldiazepam, were found to be present as conjugated metabolites of medazepam in the intestinal contents and urine of the rat (Schwartz and Carbone, 1970). This is additional evidence for the thesis that the conversion of medazepam to diazepam is an important pathway in the rat.

The metabolites of medazepam formed in substantial amounts by the dog liver 9,000 × *g* supernatant—2-hydroxymedazepam, desmethylmedazepam, dehydrodesmethylmedazepam, and ACB—were also prominent plasma metabolites in a dog given a 10 mg/kg i.v. dose of ^{14}C-medazepam HCl (Schwartz and Kolis, 1972). However, the major metabolite in the plasma 1 hr after the dose, was desmethyldiazepam, which *in vitro* had accounted for less than 5% of the medazepam incubated. In Fig. 7, the plasma levels of diazepam and desmethyldiazepam found in a dog after 10 mg/kg of i.v. medazepam hydrochloride (equivalent to 8.8 mg/kg of

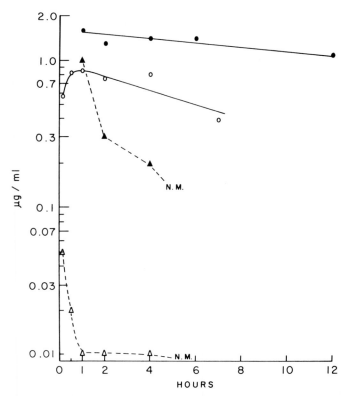

FIG. 7. Comparison of the plasma levels of diazepam and desmethyldiazepam in a dog given a 10 mg/kg i.v. dose of ^{14}C-medazepam HCl with the blood levels of the same compounds in a dog given a 10 mg/kg i.v. dose of ^{3}H-diazepam. After medazepam: plasma diazepam (Δ - - Δ) and desmethyldiazepam (O—O) in μg/ml (Schwartz and Kolis, 1972). After diazepam: blood diazepam (▲ - - ▲) and desmethyldiazepam (●—●) in μg/ml (Schwartz et al., 1965). N.M. = not measurable.

medazepam base) are compared to the blood levels of each seen after a 10 mg/kg i.v. dose of diazepam. While the diazepam levels declined rapidly following the administration of either drug, the plasma levels of desmethyldiazepam after medazepam administration were, during the first 4 hr, more than one-half the levels in blood following diazepam administration. It is clear that some diazepam and substantial amounts of desmethyldiazepam were produced from medazepam in the dog. However, the magnitude of diazepam formation and the extent to which desmethyldiazepam was formed from diazepam or from other precursors is not known.

Preliminary evidence obtained from two human subjects treated with

[14]C-medazepam hydrochloride suggested that diazepam was also a medazepam metabolite in man. The TLC of blood extracts not only provided a tentative identification of desmethylmedazepam, diazepam, and desmethyldiazepam but also indicated that the apparent desmethyldiazepam levels were increasing with time (Rieder and Rentsch, 1968; Schwartz and Carbone, 1970). These tentative results have been confirmed by de Silva and Puglisi (1970), who developed a gas-liquid chromatographic (GLC) assay for medazepam and its metabolites. They found that the chronic administration of medazepam led to increasing blood levels of desmethyldiazepam which became the dominant metabolite with time (Fig. 8). The accumulation of blood diazepam, medazepam, and desmethylmedazepam was either minimal or undetectable. With respect to the human urinary metabolites of medazepam, small amounts of the diazepam metabolites, N-methyloxazepam and oxazepam were found in the urine (Schwartz and Carbone, 1970; de Silva and Puglisi, 1970).

In conclusion, the conversion of medazepam to diazepam is significant in the rat and man and, possibly, in the dog. The production of desmethyldiazepam as an *in vivo* metabolite of medazepam, however, has been estab-

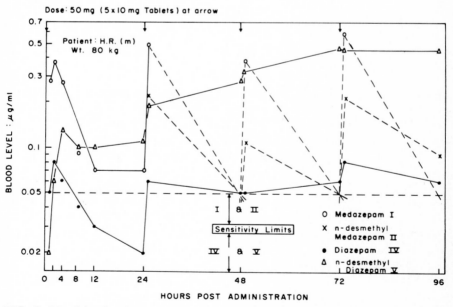

FIG. 8. Blood levels of medazepam and its major metabolites in man following the chronic administration of medazepam. Reprinted from de Silva and Puglisi, *Analytical Chemistry*, 42:1725–1736 (1970) by permission of the American Chemical Society.

lished in all three species. It is of interest that blood desmethyldiazepam, which accumulated on repeated diazepam administration in man (de Silva et al., 1966; van der Kleijn, 1971), also accumulated on chronic medazepam administration (de Silva and Puglisi, 1970).

IV. FLURAZEPAM METABOLISM IN MAN AND DOG

Since flurazepam has a diethylaminoethyl side chain at the N_1 position, one would expect its biotransformation to reflect the presence of this vulnerable moiety. This expectation was realized in a study of the urinary metabolites of flurazepam, using TLC for metabolite isolation and high-resolution mass spectrometry for identification (Schwartz et al., 1968*b*). Five metabolites, in addition to unchanged drug, were identified in the urine of a dog chronically dosed with 40 mg/kg of flurazepam hydrochloride. Four of these metabolites, monodesethylflurazepam, didesethylflurazepam, flurazepam N_1-ethanol, and N_1-desalkyl-3-hydroxyflurazepam, are included in the outline of flurazepam metabolism shown in Fig. 9. An additional urinary metabolite was isolated but was not completely identified. While its mass spectrum indicated that it was a phenolic derivative of N_1-desalkyl-flurazepam, the point of attachment of the hydroxyl function was not defined. All three of the hydroxylated metabolites were excreted as conjugates, and mono- and didesethylflurazepam were apparently excreted in both conjugated and nonconjugated forms.

Although this initial study did provide information on the metabolism of flurazepam, the major urinary metabolite in the dog was not detected. A subsequent study (Schwartz and Postma, 1970) with labeled drug flurazepam-5-^{14}C hydrochloride revealed that the major portion of the urinary radioactivity was extractable only from acidified urine. Once this was realized, the major metabolite in this fraction was readily identified by comparative TLC as the acetic acid analog of flurazepam N_1-ethanol. This urinary metabolite, flurazepam N_1-acetic acid, accounted for almost one-half of the total radioactivity excreted in the urine, or 15% of the oral 2 mg/kg dose administered. With respect to the other urinary metabolites, conjugated N_1-desalkyl-3-hydroxyflurazepam and flurazepam N_1-ethanol accounted for only 1 to 2% of the dose and less than 0.5% of the dose, respectively. These findings suggest that flurazepam is successively de-ethylated and deaminated to yield the postulated acetaldehyde derivative which is oxidized to flurazepam N_1-acetic acid. After a single dose, the reduction of this postulated aldehyde to flurazepam N_1-ethanol does not appear to be significant in the dog.

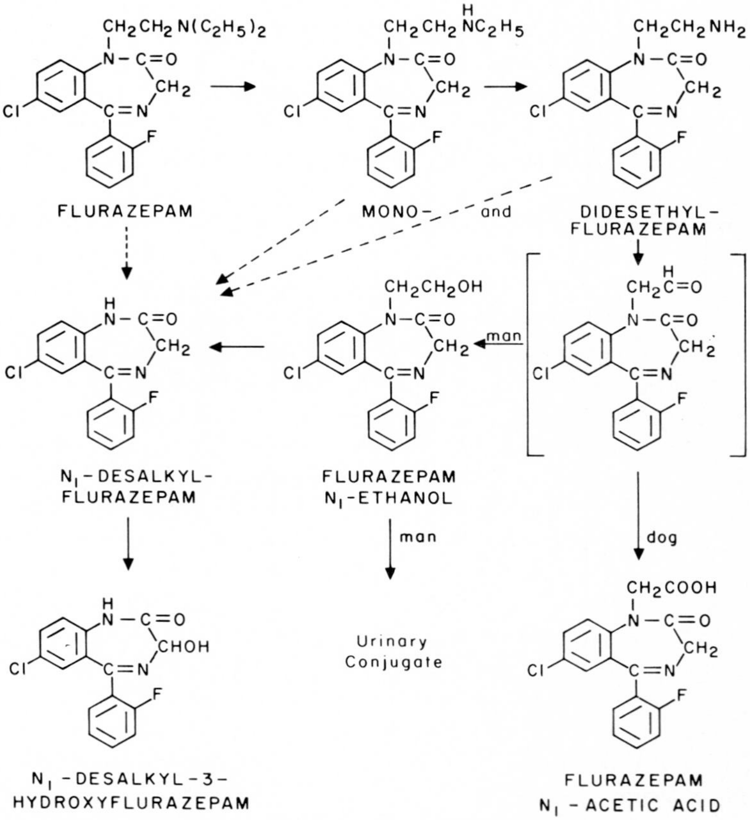

FIG. 9. Postulated pathways of flurazepam biotransformation in man and dog. The pathways labeled "man and dog" depict the major reactions which decide the fate of the postulated aldehyde intermediate in each species.

The formation of N_1-desalkylflurazepam from flurazepam N_1-ethanol is shown in Fig. 9 as a definite pathway. This was established in a dog given flurazepam N_1-ethanol hydrochloride (Schwartz and Postma, 1970). Not only was N_1-desalkylflurazepam detected in the plasma, but conjugated N_1-desalkyl-3-hydroxyflurazepam was found in the urine together with flurazepam N_1-acetic acid. These findings, which have recently been confirmed by de Silva and Strojny (1971), show that the dog can both dealkylate the N_1-ethanol moiety (and hydroxylate the resulting N_1-desalkylflurazepam) and oxide this moiety to form the analogous acid. Since there is no reason to conclude that N_1-desalkylflurazepam is formed from flurazepam exclusively via the N_1-ethanol metabolite, pathways leading from other possible precursors are shown as broken arrows in Fig. 9.

Finally, it is necessary to point out that roughly 50% of the ^{14}C administered intravenously to a dog as labeled flurazepam hydrochloride was excreted in the feces. These fecal metabolites, presumably secreted into the GI tract via the bile, have not been investigated. Therefore, a final estimate of the quantitative importance of the formation of the N_1-acetic acid metabolite must await the identification of these unknown metabolites.

The major urinary metabolite excreted by two human subjects, each given a 28-mg oral dose of ^{14}C-flurazepam hydrochloride, was the conjugated N_1-ethanol; this metabolite accounted for approximately one-quarter of the dose (Schwartz and Postma, 1970). Minor amounts of the mono- and didesethyl metabolites and of N_1-desalkyl-3-hydroxyflurazepam were also excreted in the urine. These results have also been confirmed by de Silva and Strojny (1971), who developed an assay for flurazepam and its metabolites, in which the respective benzophenones produced by acid hydrolysis were separated by TLC and measured either spectrophotometrically or spectrofluorometrically. In fact, they reported that the excretion of conjugated flurazepam N_1-ethanol in man, after a 90-mg oral dose of flurazepam hydrochloride, approached 50% of the dose. It thus appears that the dog and man both form the postulated aldehyde intermediate and that the dog preferentially oxidizes it to the N_1-acetic acid whereas man reduces it to the N_1-ethanol and excretes this alcohol as a urinary conjugate (Fig. 9).

Although the excretion of the 3-hydroxy derivative of flurazepam was also monitored, no detectable amounts were found in the urine of man or dog. Apparently, the metabolic alteration of the N_1-side chain was preferred over 3-hydroxylation, thus indicating that N_1-desalkylflurazepam was the most likely immediate precursor of N_1-desalkyl-3-hydroxyflurazepam in both species. Since conjugated N_1-desalkyl-3-hydroxyflurazepam did not appear in the urine of one subject until the second day and was then eliminated relatively slowly (half-life of 42 hr), Schwartz and Postma (1970) suggested that this 3-hydroxylation was probably a relatively slow metabolic reaction. Support for this hypothesis was obtained by de Silva and Strojny (1971), who recently reported that the blood levels of N_1-desalkyl-flurazepam in man do appear to decline more slowly than those of unchanged flurazepam or flurazepam N_1-ethanol. Further human studies are required, however, to definitely establish whether N_1-desalkylflurazepam is formed faster than it is further metabolized.

V. CONCLUDING REMARKS

As a general rule, the pharmacologically active benzodiazepines are lipophilic compounds which are eliminated from the body almost entirely by biotransformation. The chlordiazepoxide metabolite demoxepam is an exception in that it possesses psychotropic activity in animals and man (Zbinden and Randall, 1967), and yet is excreted unchanged to an appreciable extent by man and dog. Demoxepam is not atypical, however, in being an active compound. Since, as discussed elsewhere in this volume by Dr. L. O. Randall, demoxepam is one of the many metabolites mentioned in this chapter which exhibit activity, it is evident that any detailed pharmacological study of the benzodiazepines must take into consideration a probable contribution by the metabolites to the activity of the parent drug.

There are numerous positions on the benzodiazepine molecule which are vulnerable to metabolic attack. The N-dealkylation of secondary and tertiary amines, which is a well-established microsomal reaction (McMahon, 1966), is definitely in operation at the N_1 position; *e.g.*, the N-demethylation of medazepam and diazepam and the dealkylation of the N_1-ethanol moiety of flurazepam N_1-ethanol. Prazepam, a benzodiazepine in which the N_1-methyl group of diazepam has been replaced by a cyclopropylmethyl group, is also metabolized by N-dealkylation (DiCarlo et al., 1970).

These N_1-dealkylations appear to occur readily in man, dog, and rat. In man and dog, the N_1-desalkyl metabolites are formed faster than they are further metabolized. It is of particular interest that in man this has resulted in the accumulation of plasma demoxepam after chronic chlordiazepoxide administration (Schwartz and Postma, 1966), the accumulation of blood desmethyldiazepam after chronic diazepam treatment (de Silva et al., 1966; van der Kleijn, 1971), the accumulation of blood desmethyldiazepam and, to a lesser extent, of desmethylmedazepam, after multiple doses of medazepam (de Silva and Puglisi, 1970), and the apparent slower disappearance of blood N_1-desalkylflurazepam than of intact drug after single doses of flurazepam (de Silva and Strojny, 1971). Desmethyldiazepam is also a prazepam metabolite in man (DiCarlo et al., 1970), and one would expect it to accumulate on chronic prazepam administration.

All the benzodiazepines of current interest are lactams (benzodiazepin-2-ones), except for chlordiazepoxide and medazepam. These two compounds, however, contain metabolically vulnerable moieties at the C_2 position, and both are metabolized to active lactam metabolites—chlordiazepoxide to demoxepam and medazepam to diazepam and desmethyldiazepam. Of course, further studies would be required to establish conclusively the thesis that the lactam structure is an absolute requirement for activity.

Hydroxylation of benzodiazepin-2-ones at the C_3 position is an important metabolic pathway in man, dog, rat, and mouse. The fact that no nonlactams have been shown to be hydroxylated at this position, while certainly not conclusive, suggests that a C_2 carbonyl function may be required for the C_3-hydroxylation to proceed to a significant extent. Also, the presence of an N_4-oxide (as in demoxepam) appears to block this reaction. The accumulation of N_1-desalkyl-2-one metabolites in the blood (discussed above) indicates that, in man, 3-hydroxylation is a relatively slow metabolic reaction.

The formation of an N-oxide at the N_4 position has not as yet been reported. An N-oxide at this position, however, has been shown to be reduced in the GI tract. This reaction, which may also occur systemically, is an essential one in the pathway leading from chlordiazepoxide to oxazepam.

Aromatic hydroxylation at the 4' or *para* position is a prominent reaction in the rat. Even the 3-hydroxylated products, oxazepam and N-methyloxazepam, are hydroxylated at the 4' position before conjugation and excretion. To date, this metabolic reaction appears to be of a decreasing order of importance in man and dog. Of further interest, the slight excretion of C_9-hydroxylated metabolites of demoxepam in man and dog demonstrated that phenol formation is not limited to the 4' position.

It must be emphasized that the metabolites discussed above were those which were readily detected, isolated, and identified. They were either lipophilic metabolites or their conjugates and consequently were considered to be possible contributors to the central nervous system activity of the parent drugs. However, a significant portion of the excreted metabolites of each drug was not identified, and this fraction undoubtedly contains additional metabolites of biochemical and pharmacological interest.

VI. SUMMARY

The metabolism of chlordiazepoxide, medazepam, and flurazepam is described in order to illustrate the metabolic pathways by which the benzodiazepine class of drugs are biotransformed. A striking characteristic of these three benzodiazepines, and of others as well, is their biotransformation to pharmacologically active metabolites. The general metabolic reactions operative in man, dog, and rat include the N-dealkylation of vulnerable groups at the N_1 position and on side chain nitrogens, the conversion of the nonlactam drugs, chlordiazepoxide and medazepam, to lactam (benzodiazepin-2-one) metabolites, and the oxidation of benzodiazepin-2-ones to 3-hydroxy metabolites. Another pathway which appears to be especially important in the rat is the formation of phenolic metabolites. Both the 3-hydroxy and the phenolic metabolites are excreted primarily as conjugates of glucuronic and/or sulfuric acid. The evidence to date suggests that N-desalkylbenzodiazepin-2-one metabolites are eliminated more slowly in man than are the parent drugs. Therefore, these metabolites may contribute significantly to the therapeutic effect of the drug.

REFERENCES

Bickel, M. H. (1969): The pharmacology and biochemistry of N-oxides. *Pharmacological Reviews*, 21:325–355.

Bommer, P., and Vane, F. M. (1971): The application of various spectroscopies to the identification of drug metabolites. In: *Handbook of Experimental Pharmacology*, Vol. 28, Part 2, edited by B. B. Brodie and J. R. Gillette. Springer-Verlag, Berlin, pp. 209–225.

Coutinho, C. B., Cheripko, J. A., and Carbone, J. J. (1969): Relationship between the duration of anticonvulsant activity of chlordiazepoxide and systemic levels of the parent compound and its major metabolites in mice. *Biochemical Pharmacology*, 18: 303–316.

de Silva, J. A. F., Koechlin, B. A., and Bader, G. (1966): Blood level distribution patterns of diazepam and its major metabolite in man. *Journal of Pharmaceutical Sciences*, 55:692–702.

de Silva, J. A. F., and Puglisi, C. V. (1970): Determination of medazepam (NOBRIUM), diazepam (VALIUM), and their major biotransformation products in blood and urine by electron capture gas-liquid chromatography. *Analytical Chemistry*, 42: 1725–1736.

de Silva, J. A. F., and Strojny, N. (1971): Determination of flurazepam and its major biotransformation products in blood and urine by spectrophotofluorometry and by spectrophotometry. *Journal of Pharmaceutical Sciences*, 60:1303–1314.

DiCarlo, F. J., Viau, J. P., Epps, J. E., and Haynes, L. J. (1970): Prazepam metabolism by man. *Clinical Pharmacology and Therapeutics*, 11:890–897.

Gillette, J. R. (1959): Side chain oxidation of alkyl substituted ring compounds. I. Enzymatic oxidation of p-nitrotoluene. *Journal of Biological Chemistry*, 234:139–143.

Hucker, H. B., Gillette, J. R., and Brodie, B. B. (1960): Enzymatic pathway for the formation of cotinine, a major metabolite of nicotine in rabbit liver. *Journal of Pharmacology and Experimental Therapeutics*, 129:94–100.

Kaplan, S. A., Lewis, M., Schwartz, M. A., Postma, E., Cotler, S., Abruzzo, C. W., Lee, T. L., and Weinfeld, R. E. (1970): Pharmacokinetic model for chlordiazepoxide HCl in the dog. *Journal of Pharmaceutical Sciences*, 59:1569–1574.

Kimmel, H. B., and Walkenstein, S. S. (1967): Oxazepam excretion by chlordiazepoxide-^{14}C-dosed dogs. *Journal of Pharmaceutical Sciences*, 56:538–539.

Koechlin, B. A., and D'Arconte, L. (1963): Determination of chlordiazepoxide (LIBRIUM) and of a metabolite of lactam character in plasma of humans, dogs, and rats by a specific spectrofluorometric micro method. *Analytical Biochemistry*, 5:195–207.

Koechlin, B. A., Schwartz, M. A., Krol, G., and Oberhansli, W. (1965): The metabolic fate of ^{14}C-labeled chlordiazepoxide in man, in the dog, and in the rat. *Journal of Pharmacology and Experimental Therapeutics*, 148:399–411.

Kuntzman, R., Sernatinger, E., Tsai, I., and Klutch, A. (1968): New methodology for studies of drug metabolism. In: *Importance of Fundamental Principles in Drug Evaluation,* edited by D. H. Tedeschi and R. E. Tedeschi. Raven Press, New York, pp. 87–103.
Kvetina, J., Marcucci, F., and Fanelli, R. (1968): Metabolism of diazepam in isolated perfused liver of rat and mouse. *Journal of Pharmacy and Pharmacology,* 20:807–808.
Marcucci, F., Fanelli, R., Mussini, E., and Garattini, S. (1969): The metabolism of diazepam by liver microsomal enzymes of rats and mice. *European Journal of Pharmacology,* 7:307–313.
Marcucci, F., Mussini, E., Fanelli, R., and Garattini, S. (1970): Species differences in diazepam metabolism. I. Metabolism of diazepam metabolites. *Biochemical Pharmacology,* 19:1847–1851.
McMahon, R. E. (1966): Microsomal dealkylation of drugs. *Journal of Pharmaceutical Sciences,* 55:457–466.
Parke, D. V. (1968): *The Biochemistry of Foreign Compounds.* Pergamon Press, Oxford.
Rieder, J., and Rentsch, G. (1968): Metabolismus und pharmakokinetik des neuen psychopharmakons 7-chlor-2-3-dihydro-1-methyl-5-phenyl-1H-1,4-benzodiazepin (Ro 5-4556) beim menschen. *Arzneimittel-Forschung,* 18:1545–1556.
Ruelius, H. W., Lee, J. M., and Alburn, H. E. (1965): Metabolism of diazepam in dogs: Transformation to oxazepam. *Archives of Biochemistry and Biophysics,* 111:376–380.
Schwartz, M. A., Bommer, P., and Vane, F. M. (1967): Diazepam metabolites in the rat: Characterization by high-resolution mass spectrometry and nuclear magnetic resonance. *Archives of Biochemistry and Biophysics,* 121:508–516.
Schwartz, M. A., and Carbone, J. J. (1970): Metabolism of ^{14}C-medazepam hydrochloride in dog, rat and man. *Biochemical Pharmacology,* 19:343–361.
Schwartz, M. A., Koechlin, B. A., Postma, E., Palmer, S., and Krol, G. (1965): Metabolism of diazepam in rat, dog, and man. *Journal of Pharmacology and Experimental Therapeutics,* 149:423–435.
Schwartz, M. A., and Kolis, S. J. (1972): Pathways of medazepam metabolism in the dog and rat. *Journal of Pharmacology and Experimental Therapeutics,* 180:180–188.
Schwartz, M. A., and Postma, E. (1966): Metabolic N-demethylation of chlordiazepoxide. *Journal of Pharmaceutical Sciences,* 55:1358–1362.
Schwartz, M. A., and Postma, E. (1968): Metabolism of diazepam *in vitro. Biochemical Pharmacology,* 17:2443–2449.
Schwartz, M. A., and Postma, E. (1970): Metabolism of flurazepam, a benzodiazepine, in man and dog. *Journal of Pharmaceutical Sciences,* 59:1800–1806.
Schwartz, M. A., and Postma, E. (1972): Metabolites of demoxepam, a chlordiazepoxide metabolite, in man. *Journal of Pharmaceutical Sciences,* 61:123–125.
Schwartz, M. A., Postma, E., and Gaut, Z. (1971a): The biological half-life of chlordiazepoxide and its metabolite, demoxepam, in man. *Journal of Pharmaceutical Sciences,* 60:1500–1503.
Schwartz, M. A., Postma, E., and Kolis, S. J. (1971b): Metabolism of demoxepam, a chlordiazepoxide metabolite, in the dog. *Journal of Pharmaceutical Sciences,* 60:438–444.
Schwartz, M. A., Vane, F. M., and Postma, E. (1968a): Chlordiazepoxide metabolites in the rat. Characterization by high resolution mass spectrometry. *Biochemical Pharmacology,* 17:965–974.
Schwartz, M. A., Vane, F. M., and Postma, E. (1968b): Urinary metabolites of 7-chloro-1-(2-diethylaminoethyl)-5-(2-fluorophenyl)-1,3-dihydro-2H-1,4-benzodiazepin-2-one dihydrochloride. *Journal of Medicinal Chemistry,* 11:770–774.

Sisenwine, S. F., Tio, C. O., and Ruelius, H. W. (1970): The metabolic transformation of oxazepam in man, pig, and rat. *The Pharmacologist,* 12:272.

Toki, K., Toki, S., and Tsukamoto, H. (1963): Metabolism of drugs. XXXVIII. Enzymatic oxidation of methylhexabital. 2. Reversible oxidation of 3-hydroxy-methylhexabital. *Journal of Biochemistry* (Tokyo), 53:43–49.

van der Kleijn, E. (1971): Pharmacokinetics of distribution and metabolism of ataractic drugs and an evaluation of the site of antianxiety activity. *Annals of the New York Academy of Sciences,* 179:115–125.

Walkenstein, S. S., Wiser, R., Gudmundsen, C. H., Kimmel, H. B., and Corradino, R. A. (1964): Absorption, metabolism, and excretion of oxazepam and its succinate half-ester. *Journal of Pharmaceutical Sciences,* 53:1181–1186.

Williams, R. T. (1959): *Detoxication Mechanisms.* John Wiley and Sons, Inc., New York.

Zbinden, G., and Randall, L. O. (1967): Pharmacology of benzodiazepines: Laboratory and clinical correlations. In: *Advances in Pharmacology,* Vol. 5, edited by S. Garattini and P. A. Shore. Academic Press, New York, pp. 213–291.

The Benzodiazepines, Raven Press, New York © 1973

METABOLIC STUDIES ON BENZODIAZEPINES IN VARIOUS ANIMAL SPECIES

Silvio Garattini, Emilio Mussini, Franca Marcucci, and Amalia Guaitani

Istituto di Ricerche Farmacologiche "Mario Negri," Milan, Italy

I. INTRODUCTION

Diazepam is a widely used drug of the benzodiazepine group, which has powerful anticonvulsant, muscle-relaxant, and tranquilizing activities (Zbinden and Randall, 1967). Several studies *in vitro* and *in vivo,* including some carried out in this laboratory, have been devoted to the understanding of the metabolic fate of diazepam. Figure 1 represents a summary of these investi-

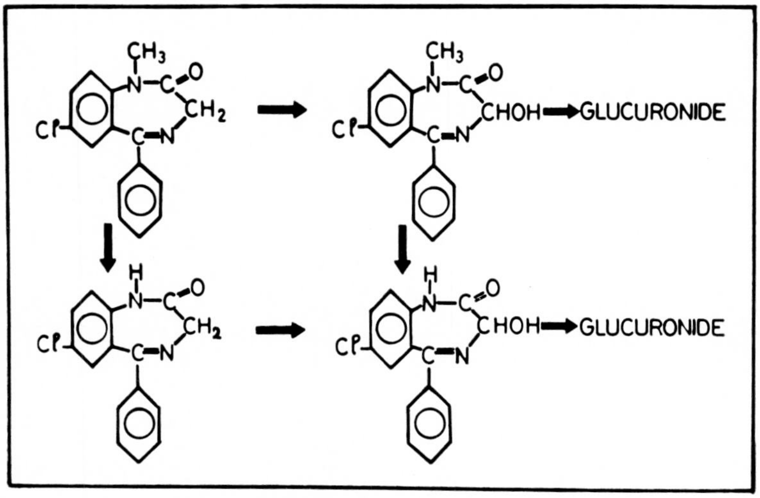

FIG. 1. Metabolic fate of diazepam.

gations (Schwartz et al., 1965; Schwartz and Postma, 1968; Květina et al., 1968; Marcucci et al., 1970). It is evident that diazepam undergoes a process of N_1-demethylation to form N_1-demethyldiazepam, and a process of C_3-hy-

75

droxylation to form N_1-methyloxazepam. These two metabolites are then C_3-hydroxylated and N_1-demethylated, respectively, to form a common metabolite known as oxazepam (Marcucci et al., 1969, 1970c; Mussini et al., 1971). The two hydroxylated metabolites of diazepam, that is, N_1-methyloxazepam and oxazepam, are then conjugated with glucuronic acid to form the respective glucuronides. It has also been established that, under given conditions, there is a hydroxylation in the *para* position of the phenyl ring (Jommi et al., 1964; Schwartz et al., 1967). These hydroxylated products may also then be conjugated with glucuronic acid (Schwartz et al., 1967). The degree of these metabolic events depends on a number of conditions and particularly on the species used.

II. IN VITRO METABOLISM

An example of the importance of the animal species in the metabolism of diazepam is reported in Table I, which summarizes the N_1-demethylation

TABLE I. *In vitro metabolism of diazepam by liver microsomes of several animal species*

Animal species	% of N_1-demethylation	% of C_3-hydroxylation
Mouse	22	15
Rat	2	60
Guinea pig	60	<0.1
Rabbit	3	5
Cat	0.5	10
Dog	15	13

and the C_3-hydroxylation obtained with liver microsome preparations from mice, rats, guinea pigs, rabbits, cats, and dogs (Marcucci et al., 1969; Mussini et al., 1971). Under the same experimental conditions, a study was also made of the C_3-hydroxylation of N-demethyldiazepam and the N_1-demethylation of N-methyloxazepam (see Table II). The results obtained indicate

TABLE II. *In vitro metabolism of N-demethyldiazepam* (DDZ) *and N-methyloxazepam* (MOX) *by liver microsomes of several animal species*

Animal species	DDZ % of C_3-hydroxylation	MOX % of N_1-demethylation
Mouse	2.6	37
Rat	4.5	4
Guinea pig	<0.1	7
Rabbit	<0.1	2
Cat	<0.1	1.5
Dog	<0.1	<0.1

a different pattern, depending on the various animal species used. For instance, in mice, rabbits, and dogs, the two metabolic steps occur to about the same extent when diazepam is used as a substrate, whereas in rats and cats, hydroxylation is predominant. Conversely, guinea pigs show primarily the capacity to N-demethylate diazepam. When N-demethyldiazepam was employed as a substrate for liver microsomes, hydroxylation occurred in mice and rats, but only to a small degree in the other animal species. N-methyloxazepam was demethylated in decreasing order by mice, guinea pigs, rats, rabbits, cats, and dogs (Garattini et al., 1972).

In most cases, the metabolites which are formed account for the disappearance of the substrate. However, for rabbits and cats (Garattini et al., 1972), other metabolites may also be present which have not yet been identified.

The presence of unknown metabolites was evident for all the tested animal species when liver microsomes were obtained from mice, rats, and guinea pigs, and pretreated with phenobarbital, a known inducer of microsomal enzymes. Table III summarizes these findings (Marcucci et al., 1970b;

TABLE III. *Effect of phenobarbital on in vitro metabolism of diazepam and of its metabolites by liver microsomes of three animal species*

| | | Increase by phenobarbital induction | |
| | | C_3-hydroxylation % | N_1-demethylation % |
Substrate	Species		
Diazepam	Rat	+233	+323
	Mouse	+109	0
	Guinea pig	0	+426
N-demethyl-diazepam	Rat	+647	—
	Mouse	+382	—
	Guinea pig	0	—
N-methyl-oxazepam	Rat	—	+700
	Mouse	—	−560*
	Guinea pig	—	+460

* 30% total recovery.

Mussini et al., 1972a) and indicates that phenobarbital increases N-demethylation in guinea pigs and rats, while it increases hydroxylation in rats and mice.

III. IN VIVO METABOLISM (BLOOD LEVELS)

Diazepam is metabolized *in vivo* quite differently, depending on the animal species used. Table IV shows the levels of diazepam and its metabolites,

TABLE IV. Levels of diazepam and of its metabolites in blood of rats, mice, and guinea pigs after administration of diazepam (5 mg/kg, i.v.)

Time after DZ	Rat (μg/ml ± S.E.)			Mouse (μg/ml ± S.E.)			Guinea pig (μg/ml ± S.E.)		
	DZ	DDZ	OX	DZ	DDZ	OX	DZ	DDZ	OX
1 min	2.33 ±0.10	—	—	1.36 ±0.11	0.05 ±0.002	—	1.70 ±0.10	0.07 ±0.01	—
5 min	1.32 ±0.11	0.10 ±0.01	—	0.90 ±0.05	0.77 ±0.03	—	0.64 ±0.04	0.20 ±0.03	—
30 min	0.28 ±0.03	—	—	0.10 ±0.007	1.15 ±0.04	0.14 ±0.01	0.28 ±0.05	0.26 ±0.06	—
1 hr	0.09 ±0.005	—	—	0.05 ±0.002	1.02 ±0.04	0.13 ±0.02	0.16 ±0.04	0.37 ±0.07	—
3 hr	0.04 ±0.005	—	—	0.01 ±0.002	0.83 ±0.01	0.22 ±0.02	0.04 ±0.01	0.20 ±0.04	—
5 hr	<0.01	—	—	<0.01	0.14 ±0.002	0.21 ±0.02	0.02 ±0.004	0.08 ±0.006	—
10 hr	—	—	—	<0.01	0.02 ±0.002	0.09 ±0.002	<0.01	0.03 ±0.002	—
20 hr	—	—	—	—	<0.01	<0.01	<0.01	0.02 ±0.004	—
40 hr	—	—	—	—	—	<0.01	<0.01	<0.01	—

DZ = diazepam.
DDZ = N-demethyldiazepam.
OX = oxazepam.

TABLE V. *Levels (± S.E.) of N-demethyldiazepam (DDZ) and its metabolite oxazepam (OX) in blood of rats, mice, and guinea pigs after administration of N-demethyldiazepam (5 mg/kg, i.v.)*

Time after DDZ	Rat (μg/ml) DDZ	OX	Mouse (μg/ml) DDZ	OX	Guinea pig (μg/ml) DDZ	OX
1 min	5.12 ± 0.18	—	3.98 ± 0.12	—	1.19 ± 0.11	—
5 min	3.29 ± 0.04	—	1.76 ± 0.09	<0.05	0.96 ± 0.4	—
30 min	1.58 ± 0.14	—	1.31 ± 0.03	0.10 ± 0	0.94 ± 0.02	—
1 hr	0.098 ± 0.003	—	0.47 ± 0.03	0.13 ± 0.002	0.18 ± 0.01	—
3 hr	0.007 ± 0.0005	—	0.12 ± 0.01	0.29 ± 0.01	0.14 ± 0.01	—
5 hr	0.005 ± 0.0005	—	<0.005	0.21 ± 0.0002	0.10 ± 0.02	—
10 hr	<0.005	—	<0.005	0.13 ± 0.007		

with relation to time, in the blood of rats, mice, and guinea pigs, after intravenous administration of 5 mg/kg of diazepam. The peak level of diazepam and the rate of its disappearance are about the same in the three animal species considered. However, although only traces of the metabolites can be detected in rats, guinea pigs show a sustained level of N-demethyldiazepam, lasting for about 10 hr. In mice, after the disappearance of diazepam, there is a peak of N-demethyldiazepam, followed by the presence of oxazepam, so that the total presence of diazepam and its metabolites lasts for about 10 hr.

It should be emphasized that the metabolites of diazepam are also metabolized *in vivo* in a different manner, depending on the animal species, as summarized in Tables V and VI. Again, the peak levels and the rates of disappearance of N-demethyldiazepam and N-methyloxazepam are comparable in mice, rats, and guinea pigs, while the accumulation of the metabolite

TABLE VI. *Levels (±S.E.) of N-methyloxazepam (MOX) and of its metabolite oxazepam (OX) in blood of rats, mice, and guinea pigs after administration of N-methyloxazepam (5 mg/kg, i.v.)*

Time after MOX	Rat (μg/ml)		Mouse (μg/ml)		Guinea pig (μg/ml)	
	MOX	OX	MOX	OX	MOX	OX
1 min	5.70 ±0.04	—	3.54 ±0.23	0.17 ±0.01		
5 min	1.74 ±0.02	<0.05	1.39 ±0.25	0.40 ±0.02	0.95 ±0.03	0.03 ±0.005
30 min	1.28 ±0.05	0.21 ±0.01	0.84 ±0.04	0.56 ±0.03	0.69 ±0.01	0.02 ±0.001
1 hr	0.24 ±0.03	0.09 ±0.01	0.15 ±0.02	0.72 ±0.05	0.64 ±0.03	0.02 ±0.001
3 hr	<0.09	—	<0.09	0.14 ±0.003	0.08 ±0.01	0.02 ±0.001
5 hr	—	—	<0.09	0.13 ±0.01	—	0.02 ±0.003
10 hr	—	—	—	0.13 ±0.01	—	<0.02

oxazepam in blood has an important value only in mice. This probably reflects not only a different rate of oxazepam formation but also a different rate of its disappearance. In fact, when oxazepam was injected, its presence in blood was more lasting in mice and in guinea pigs than in rats (see Table VII).

IV. ENTEROHEPATIC CIRCULATION

Although the hydroxylated metabolites of diazepam are not detectable in

TABLE VII. *Levels ($\pm$S.E.) of oxazepam (OX) in blood of rats, mice, and guinea pigs after administration of this drug (5 mg/kg, i.v.)*

Time after oxazepam	Rat OX (μg/ml)	Mouse OX (μg/ml)	Guinea pig OX (μg/ml)
1 min	1.63 ± 0.09	3.16 ± 0.05	
5 min	1.50 ± 0.04	2.08 ± 0.05	1.81 ± 0.12
30 min	0.55 ± 0.02	1.51 ± 0.05	1.52 ± 0.05
1 hr	0.23 ± 0.01	1.10 ± 0.02	0.76 ± 0.18
3 hr	0.06 ± 0.002	0.71 ± 0.05	0.21 ± 0.007
5 hr	<0.05	0.47 ± 0.02	0.09 ± 0.01
10 hr	—	0.20 ± 0.01	0.07 ± 0.01

the blood of guinea pigs, they certainly must be formed, since they are excreted as glucuronides in the bile. Table VIII compares the biliary excretion of diazepam and its metabolites in mice, rats, and guinea pigs.

It is evident that the excretion of hydroxylated conjugated metabolites, independent of the compound administered, is always higher in mice than in guinea pigs and higher in guinea pigs than in rats. Particularly in the case of oxazepam, there is a striking correlation between the percentage of the glucuronide excreted and the persistence of oxazepam in blood in the animal species considered (Marcucci et al., 1970c). This suggested the presence of an enterohepatic circulation. In fact, when the bile of guinea pigs treated with oxazepam was injected intraduodenally in untreated guinea pigs, there was a rapid apppearance of oxazepam in the blood. Table IX also shows that guinea pigs, bearing a biliary fistula, have lower blood levels of oxazepam after intravenous administration of oxazepam than sham-operated guinea pigs. It is, however, not yet established whether oxazepam glucuronide is hydrolyzed by the intestinal flora or the intestinal wall, or if it is absorbed as such and then broken down in the blood. In this connection, it should be mentioned that rats and mice are able to hydrolyze oxazepam hemisuccinate injected intravenously (Mussini et al., 1972b).

The differences observed in the various animal species with regard to the biliary excretion of the hydroxylated glucuronides formed during diazepam metabolism may reflect diverse rates of glucuronide formation and/or variations in the threshold of the urinary as compared to the biliary excretion of glucuronides (Smith and Williams, 1966).

V. STORAGE IN ADIPOSE TISSUE

Diazepam and its metabolites, being lipophylic compounds, tend to accumulate in fat (Marcucci et al., 1968). Table X shows that the accumula-

TABLE VIII. *Biliary excretion of conjugated hydroxylated benzodiazepines in the rat, guinea pig, and mouse*

% of the administered dose ± S.E.

Drug administered (5 mg/kg, i.v.)	N-Methyloxazepam			Oxazepam		
	Rat	Guinea pig	Mouse	Rat	Guinea pig	Mouse
Diazepam	—	—	—	—	3.0 ± 0.8	13.1 ± 1.9
N-demethyldiazepam	—	—	—	—	4.0 ± 0.7	33.4 ± 3.9
N-methyloxazepam	8.5 ± 0.2	20.8 ± 1.1	9.9 ± 1	traces	8.1 ± 0.8	23.0 ± 0.8
Oxazepam	—	—	—	5.3 ± 0.4	34.7 ± 3.4	49.7 ± 5.3

The above figures represent the average of four determinations of the conjugated forms present in the bile collected during 3 hr after drug administration, calculated as N-methyloxazepam or oxazepam.

No diazepam or N-demethyldiazepam was found in the bile either as the free or conjugated form, and no free N-methyl-oxazepam or oxazepam was found after the administration of the four benzodiazepines.

This method permits detection of amounts higher than 0.15% of the administered benzodiazepines.

TABLE IX. *Oxazepam (OX) blood levels of sham-operated and bile-fistula guinea pigs given this drug (5 mg/kg i.v.)*

Time after oxazepam	Sham-operated OX (μg/ml)	Bile-fistula* OX (μg/ml)
30 min	0.99	0.39
60 min	0.86	0.34
90 min	0.69	0.31
150 min	0.50	0.26

* Oxazepam recovered from bile collected over a period of 150 min, after incubation with β-glucuronidase, was 1.92 mg.
No free oxazepam was present in the bile.

tion in epididymal adipose tissue of the benzodiazepines considered is similar for both nonhydroxylated (diazepam and N-demethyldiazepam) and hydroxylated (N-methyloxazepam and oxazepam) compounds. The ratio between adipose tissue and blood concentrations reveals higher values for the former in the case of all the benzodiazepines, and this proportion tends to increase with time. This finding may be of importance in explaining the different brain levels of benzodiazepines in animals of varying body weight. Preliminary experiments carried out in mice made obese by the administration of aurothioglucose indicate that diazepam given at the same dose in milligrams per kilogram is present in the brain of obese mice at a level which is about

TABLE X. *Accumulation of several benzodiazepines in the epididymal adipose tissue of mice, and ratio between adipose tissue and blood drug levels*

Drug injected (5 mg/kg, i.v.)	Time after administration	Benzodiazepine level in adipose tissue (μg/g $\pm$ S.E.)	Drug ratio: adipose tissue/blood
DZ	5 min	3.67 $\pm$ 0.09	4.07
	30 min	4.86 $\pm$ 0.12	4.86
	5 hr	0.06 $\pm$ 0.007	>6
DDZ	5 min	5.51 $\pm$ 0.07	3.13
	30 min	12.50 $\pm$ 0.62	9.54
	5 hr	0.23 $\pm$ 0.01	>45
MOX	5 min	6.20 $\pm$ 0.41	4.46
	30 min	8.60 $\pm$ 0.32	10.23
	5 hr	0.21 $\pm$ 0.01	>2.33
OX	5 min	3.50 $\pm$ 0.04	1.68
	30 min	7.35 $\pm$ 0.12	4.86
	5 hr	1.38 $\pm$ 0.01	2.93

DZ = diazepam.
DDZ = N-demethyldiazepam.
MOX = N-methyloxazepam.
OX = oxazepam.

TABLE XI. Brain levels of several benzodiazepines after their administration (5 mg/kg, i.v.) to mice, rats, and guinea pigs, and ratio between brain and blood drug levels

Drug	Time after administration	Benzodiazepine levels in brain (μg/g ± S.E.)			Drug level ratio: brain/blood		
		Mouse	Rat	Guinea pig	Mouse	Rat	Guinea pig
Diazepam	1 min	7.04 ±0.37	4.28 ±0.14		5.17	1.83	
	5 min	3.06 ±0.12	3.64 ±0.15	6.28 ±0.30	3.40	2.75	9.81
	30 min	0.74 ±0.09	1.01 ±0.02	1.13 ±0.06	7.40	3.60	4.03
	1 hr	0.13 ±0.007	0.42 ±0.02	0.45 ±0.02	2.60	4.66	2.81
	3 hr	<0.01	0.23 ±0.01	0.13 ±0.02	—	5.75	3.25
	5 hr	<0.01	0.08 ±0.01	0.08 ±0.008	—	>8	4.00
	10 hr	<0.01	<0.01	0.01 ±0.002	—	—	—
	20 hr	<0.01	—	0.01 ±0.002	—	—	—
N-demethyldiazepam	1 min	8.16 ±0.21	7.82 ±0.37		2.05	1.52	
	5 min	6.52 ±0.12	6.35 ±0.09	5.23 ±0.13	3.70	1.93	4.39
	30 min	5.55 ±0.05	3.01 ±0.07	5.12 ±0.04	4.24	1.90	5.33
	1 hr	4.52 ±0.18	0.59 ±0.07	2.20 ±0.11	9.61	6.02	2.34
	3 hr	0.62 ±0.01	0.02 ±0.001	1.06 ±0.05	5.16	3.00	5.88
	5 hr	<0.01	0.01 ±0.0005	0.76 ±0.06	—	3.00	5.42
	10 hr	<0.01	<0.01	0.50 ±0.05	—	—	5.00
	20 hr	—	—	0.06 ±0.001	—	—	2.35

TABLE XI. (Continued)

Drug	Time after administration	Benzodiazepine levels in brain (µg/g ± S.E.)			Drug level ratio: brain/blood		
		Mouse	Rat	Guinea pig	Mouse	Rat	Guinea pig
N-methyloxazepam	1 min	6.98 ±0.54	9.40 ±0.26	6.06 ±0.20	1.97	1.64	6.37
	5 min	5.25 ±0.63	8.04 ±0.34	5.97 ±0.21	3.77	4.62	8.65
	30 min	2.63 ±0.20	1.20 ±0.19	3.96 ±0.16	3.13	0.93	6.18
	1 hr	0.44 ±0.01	0.52 ±0.02	0.78 ±0.03	2.93	2.16	6.18
	3 hr	0.21 ±0.02	<0.03	0.07 ±0.007	>2.33	—	9.75
	5 hr	<0.03	—	0.05 ±0.004	—	—	>3.5
	10 hr	<0.03	—	0.039 ±0.003	—	—	>2.5
	20 hr	—	—		—	—	>1.9
Oxazepam	1 min	3.59 ±0.05	3.06 ±0.05		1.13	1.87	
	5 min	14.30 ±0.17	4.50 ±0.03	3.47 ±0.47	6.87	3.00	1.91
	30 min	6.84 ±0.12	2.90 ±0.02	3.21 ±0.08	4.52	5.27	2.11
	1 hr	5.21 ±0.03	1.43 ±0.05	2.84 ±0.08	4.73	6.21	3.73
	3 hr	3.12 ±0.01	0.20 ±0.0005	1.37 ±0.08	4.39	3.33	6.52
	5 hr	2.78 ±0.02	<0.03	0.80 ±0.04	5.91	—	8.88
	10 hr	2.25 ±0.02	—	0.48 ±0.06	11.25	—	6.80
	20 hr	0.11 ±0.01	—	0.15 ±0.07	>5.5	—	3.00

50% of that found in normal animals. These data may have implications in establishing the dosage and schedule of treatment in obese patients.

VI. BRAIN LEVELS

Diazepam and its metabolites enter the brain very rapidly from the blood stream. Thus, even 1 min after intravenous administration, the concentrations in the brain are always higher than those in the blood (Marcucci et al., 1970a, 1971b). This may explain why diazepam acts so rapidly in interrupting an epileptic attack in man (Gastaut et al., 1965).

Table XI summarizes the brain levels of diazepam and its metabolites (after a single i.v. dose of 5 mg/kg), together with the ratios between brain and blood concentrations in mice, rats, and guinea pigs. It is evident that the presence of the injected benzodiazepines in the brain is similar for the three species considered. However, oxazepam is present in the brains of mice and guinea pigs for a longer time than in the brains of rats.

It should be noted that when diazepam is injected, the presence of the total benzodiazepines, diazepam, and the metabolites formed in the brain lasts longer in mice and guinea pigs than in rats. The distribution in the brain is uneven, as indicated by preliminary observations in rats, showing that diazepam accumulates more in the hemispheres than in the olfactory bulbs. These findings are expanded by Morselli et al. (1972), who found that in cats there is a positive correlation between early regional concentration and regional blood flow.

VII. RELATIONSHIP BETWEEN BRAIN CONCENTRATIONS AND ANTICONVULSANT ACTIVITY

The differences observed in the brain levels of diazepam and its metabolites for the various animal species also seem to have a pharmacological significance. Figure 2 indicates the duration of the anti-pentylenetetrazol activity occurring in mice, guinea pigs, and rats after a single administration of diazepam. This effect lasts for about 20 hr in mice, 10 hr in guinea pigs, and 3 hr in rats. It is also evident from Fig. 2 that the duration of the anticonvulsant activity does not correlate directly with the presence in brain of diazepam or of one of the metabolites, with the exception of rats, but that it is more precisely reflected by the sequential presence in the brain of diazepam, N-demethyldiazepam, and oxazepam. Hence, the hypothesis that both diazepam and its known metabolites may exert an anti-pentylenetetrazol effect in those

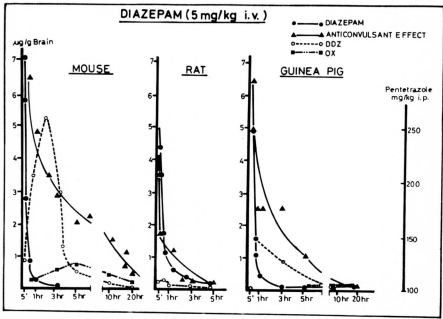

FIG. 2. Correlation between anti-pentylenetetrazol activity and brain levels of diazepam and its metabolites after i.v. administration of diazepam to different animal species.

animal species in which these metabolites are formed. Table XII confirms, indeed, that when N-demethyldiazepam, N-methyloxazepam, or oxazepam is

TABLE XII. *Comparison of the anti-pentylenetetrazol activity of diazepam (DZ) and that of N-demethyldiazepam (DDZ), N-methyloxazepam (MOX), and oxazepam (OX) after i.v. administration to mice*

Drug	Time (min)	Anti-pentylenetetrazol activity (ED_{50} mg/kg, i.v., and 95% fiducial limits)	LD_{50} (mg/kg, i.p. and 95% fiducial limits)*
DZ	5	0.280 (0.233–0.336)	355 (294–382)
	180	0.680 (0.486–0.952)	
DDZ	5	0.198 (0.177–0.221)	290 (261–322)
	180	0.640 (0.427–0.960)	
MOX	5	0.155 (0.119–0.202)	310 (263–366)
	180	0.680 (0.557–0.830)	
OX	5	0.342 (0.263–0.445)	>1500
	180	0.690 (0.460–1.035)	

* The LD_{50} was calculated 72 hr after drug administration to groups of 60 mice for each drug.

administered intravenously, one compound at a time, the anti-pentylenetetrazol activity of each of these metabolites of diazepam is comparable to that of the parent drug. It should also be noted that the ratio between toxicity and anti-pentylenetetrazol activity is more specific to oxazepam than to the other benzodiazepines, including diazepam. But, in order to prove that the metabolites of diazepam, and particularly oxazepam, are responsible for the long-lasting anti-pentylenetetrazol activity of diazepam in mice, it was also necessary to demonstrate that the "active" brain concentrations of oxazepam were similar when measured at appropriate times after the administration of either diazepam or oxazepam.

Table XIII provides this evidence and shows that the active levels (50%

TABLE XIII. *Brain levels of oxazepam (OX) and its anti-pentylenetetrazol activity in mice*

Drug administered (mg/kg, i.v.)	Time after treatment (hr)	Brain level of OX (μg/g $\pm$ S.E.)	Anti-pentylenetetrazol activity (% protection)*
Diazepam 5	20	0.17 $\pm$ 0.01	87
Oxazepam 5	18	0.15 $\pm$ 0.02	80
Oxazepam 5	24	0.11 $\pm$ 0.01	60
Oxazepam 1	15	0.06 $\pm$ 0.005	0

* The dose of 120 mg/kg, i.p., of pentylenetetrazol was lethal to all the control mice treated only with the solvent.

inhibition of pentylenetetrazol-induced convulsions) of oxazepam are around 0.1 μg/g of brain. This is remarkable, considering that this concentration of oxazepam antagonizes the convulsant effect of pentylenetetrazol present at

TABLE XIV. *Brain levels of oxazepam required to sustain an anticonvulsant activity (ED_{50}, i.v.) in different animal species*

Species	Time*	ED_{50}, mg/kg, i.v. (and 95% fiducial limits)	Brain levels (ng/g $\pm$ S.E.)
Rat	5	0.470 (0.348–0.635)	624 $\pm$ 18
	30	0.833 (0.712–0.975)	618 $\pm$ 12
	180	>10.000	
Mouse	5	0.342 (0.263–0.445)	96 $\pm$ 2
	30	0.380 (0.336–0.429)	133 $\pm$ 13
	180	0.690 (0.460–1.035)	94 $\pm$ 1
	720	4.850 (3.819–6.160)	117 $\pm$ 12
Guinea pig	5	0.465 (0.388–0.558)	275 $\pm$ 7
	30	0.520 (0.354–0.764)	242 $\pm$ 13
	180	0.818 (0.649–1.031)	118 $\pm$ 12
	720	4.400 (3.520–5.500)	91 $\pm$ 4

* Minutes between oxazepam and pentylenetetrazol administration (120 mg/kg, i.p.).

about 100 μg/g of brain (Marcucci et al., 1971*a*).

However, the "active" anticonvulsant brain levels of oxazepam are not the same for the three animal species considered. Table XIV shows that, in rats, the active brain concentration of oxazepam with respect to pentylenetetrazol convulsions is about six times higher than that required to exert the same effect in mice, although the brain concentrations of pentylenetetrazol are about the same in the two animal species (Marcucci et al., 1971*a*).

In guinea pigs, the active brain concentration of oxazepam varied, depending on the time after its administration. This may indicate that oxazepam, soon after its administration, is also stored in sites which are not involved in exerting the anti-pentylenetetrazol activity, and that the rate of removal of oxazepam is faster from the "aspecific" than from the "specific" sites. It is of interest that, after a long period, the brain concentration of oxazepam in guinea pigs is also around 0.1 μg/g, that is, the same as in mice.

VIII. BRAIN CONCENTRATIONS AND ANTICONVULSANT ACTIVITY IN NEWBORNS

Age is also an important factor affecting the metabolism of benzodiazepines and the related pharmacological activity.

Table XV shows that in newborn rats the ED_{50} of diazepam effective against pentylenetetrazol convulsions is about 15 times less than in adults. At these doses, the brain levels of diazepam were around 0.2 μg/g, as opposed to 0.82 μg/g in adults. It should be noted, however, that in the brain of adult rats, 0.06 μg/g of N-demethyldiazepam was also present. The levels of brain pentylenetetrazol were comparable in both experimental conditions. It appears, therefore, that diazepam is metabolized in newborn rats at a somewhat slower rate. The fact that the metabolism of diazepam by liver microsomes in the newborn is considerably lower than in adult rats may account for this phenomenon. However, other factors which may contribute to the differences observed should not be excluded. Thus, the fact that newborn rats are very poor in myelin may be relevant, since the differences between the activity of diazepam in newborns and adults was much less marked when animals which are born fully myelinized, such as guinea pigs, were used (see Table XV).

IX. STRUCTURE-ACTIVITY RELATIONSHIP

The problem of structure-activity relationship has been discussed in detail by Sternbach et al. (1968) and by Zbinden and Randall (1967).

TABLE XV. Brain levels of benzodiazepines and pentylenetetrazol after administration of diazepam at the ED_{50} dose against pentylenetetrazol in newborn and adult rats and guinea pigs

Animal	Age*	Time**	Pentylenetetrazol ED_{100}, i.p. (mg/kg)	Diazepam ED_{50}, s.c. (mg/kg and 95% fiducial limits)	Brain levels ($\mu g/g \pm$ S.E.)†			
					DZ	DDZ	OX	PT
Rat	n.b.	60	160	0.058 (0.040–0.084)	0.207 ±0.02	<0.02	<0.02	105.81 ±2.30
Rat	a	60	82	0.890 (0.751–1.055)	0.082 ±0.008	0.062 ±0.01	<0.02	96.04 ±2.07
Guinea pig	n.b.	60	78	0.212 (0.159–0.282)	0.092 ±0.008	0.186 ±0.005	0.064 ±0.005	104.16 ±5.28
Guinea pig	a	60	80	0.393 (0.328–0.472)	0.196 ±0.010	0.355 ±0.017	0.027 ±0.002	106.08 ±4.9

* n.b. = newborn, a = adult.
** Minutes between diazepam and pentylenetetrazol administration.
† DZ = diazepam; DDZ = N-demethyldiazepam; OX = oxazepam; PT = pentylenetetrazol.

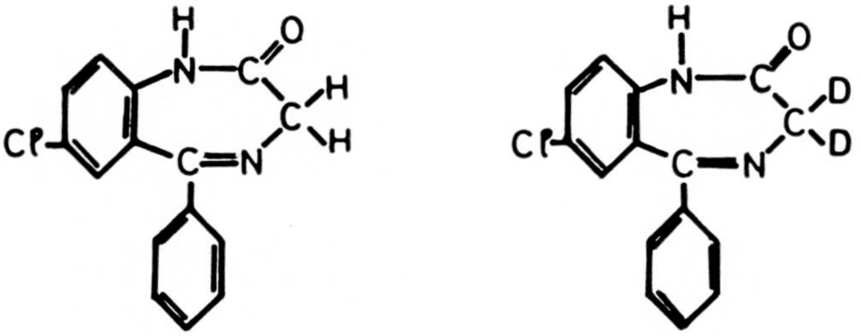

FIG. 3. Chemical structures of N-demethyldiazepam (1) and its deuterated analog (2).

Here it is interesting to report how changes in the structure of the benzodiazepines may result in a different metabolism of the particular compound and also, therefore, in a modification of its pharmacological activity.

In the structure of N-demethyldiazepam, the two atoms of hydrogen in position 3 may be replaced, by a suitable procedure (McMillan and Pattison, 1965), with two atoms of deuterium (see Fig. 3). This deuterated N-demethyldiazepam was much less hydroxylated (about 15% with respect to the nondeuterated compound) by liver microsomes of mice. Table XVI shows that, *in vivo,* the deuterated compound is also less metabolized than the hydrogenated N-demethyldiazepam. In fact, the brain levels of oxazepam were much lower in the case of the deuterated compound than in that of the hydrogenated compound. As a result, the duration of the anticonvulsant activity was much shorter with the deuterated compound than with the hydrogenated N-demethyldiazepam. These findings confirm again that the long-lasting anticonvulsant effect exerted by N-demethyldiazepam in

TABLE XVI. *Brain levels of N-demethyldiazepam (DDZ) and its metabolite oxazepam (OX) after administration (5 mg/kg, i.v.) of normal and deuterated (DDZ$_{D2}$) N-demethyldiazepam to mice*

Drug administered	Time after treatment (hr)	Benzodiazepine brain level (μg/g $\pm$ S.E.)	
		DDZ	OX
DDZ	1	4.52 $\pm$ 0.18	2.37 $\pm$ 0.02
DDZ$_{D2}$	1	5.66 $\pm$ 0.25	0.21 $\pm$ 0.01
DDZ	3	0.62 $\pm$ 0.01	3.93 $\pm$ 0.07
DDZ$_{D2}$	3	3.50 $\pm$ 0.20	0.38 $\pm$ 0.01
DDZ	10	<0.002	0.75 $\pm$ 0.02
DDZ$_{D2}$	10	0.74 $\pm$ 0.04	0.09 $\pm$ 0.007
DDZ	20	<0.002	0.09 $\pm$ 0.01
DDZ$_{D2}$	20	0.09 $\pm$ 0.01	0.04 $\pm$ 0.02

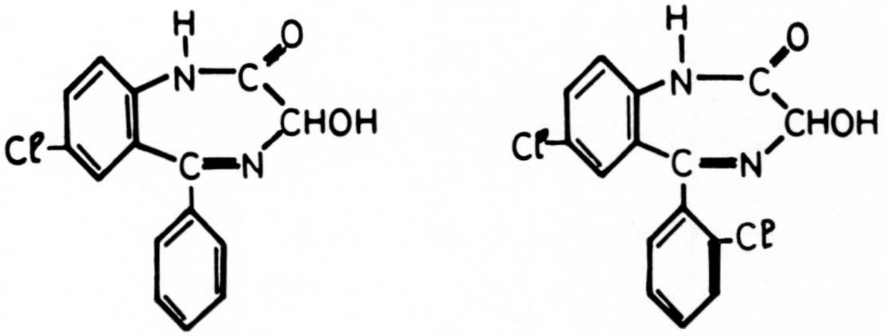

FIG. 4. Chemical structures of oxazepam (1) and lorazepam (2).

mice depends on the formation and accumulation of oxazepam in the brain.

Another example of a modification of the chemical structure, which also results in changes in metabolism and pharmacological activity, is represented by oxazepam and its derivative, *o*-chlorooxazepam (lorazepam) (Fig. 4). The introduction of chlorine in the *o* position of the phenyl ring causes a marked increase in pharmacological activity. This increase is due not only to a different disposition but, probably, also to an increased intrinsic activity. Table XVII shows, in fact, that the ED_{50} against pentylenetetrazol is achieved at a brain concentration which is about three times less for *o*-chlorooxazepam than for oxazepam.

TABLE XVII. *Brain levels of lorazepam and oxazepam after administration of the ED_{50} dose against pentylenetetrazol to mice*

Drug	Time*	ED_{50}, mg/kg, i.v. (and 95% fiducial limits)	Brain level (ng/g ± S.E.)
Lorazepam	30	0.031 (0.035–0.027)	19 ± 1
	180	0.255 (0.337–0.193)	27 ± 2
	720	0.640 (0.768–0.533)	31 ± 2
Oxazepam	30	0.380 (0.429–0.336)	98 ± 2
	180	0.690 (1.035–0.460)	94 ± 1
	720	4.850 (6.160–3.819)	95 ± 6

* Minutes between drug (lorazepam or oxazepam) and pentylenetetrazol administration (120 mg/kg, i.p.).

X. RELATIONSHIP BETWEEN BRAIN LEVELS OF BENZODIAZEPINES AND PHARMACOLOGICAL ACTIVITY

The correlations discussed above concern only the anti-pentylenetetrazol activity exerted by diazepam. It should be stressed, however, that this does

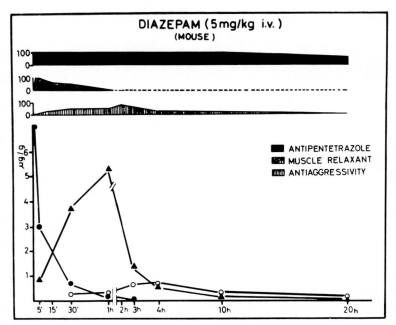

FIG. 5. Correlation between pharmacological activities and brain levels of diazepam and its metabolites after i.v. administration of diazepam (5 mg/kg) to mice.

not necessarily apply to other pharmacological effects. Thus, for instance, Fig. 5 reports brain levels of diazepam and its metabolites in mice, and their anti-pentylenetetrazol, muscle-relaxant, and antiaggressive activity. It was found that, whereas the duration of anti-pentylenetetrazol activity depends on the presence of diazepam, N-demethyldiazepam, and oxazepam in the brain, the muscle-relaxant activity seems to be related more to the level of diazepam, and the antiaggressive activity to the level of N-demethyl-diazepam.

Figure 6 shows that the anti-pentylenetetrazol activity of oxazepam correlates with its brain levels, while its antiaggressive effect seems to be independent of its levels in brain or blood. Further studies are therefore needed to relate each pharmacological effect of a given benzodiazepine to its corresponding level in the whole brain or a specific part of it.

XI. STUDIES IN MAN

The findings reported above may help in understanding some of the factors responsible for the variability in blood levels which can be found

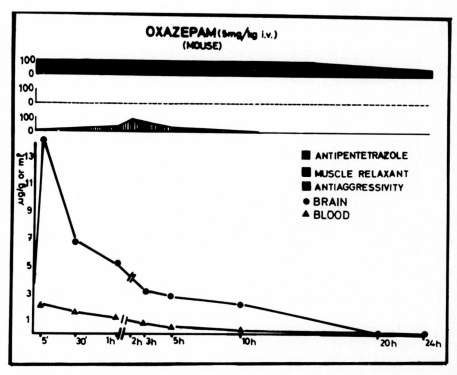

FIG. 6. Correlation between pharmacological activities and blood and brain levels of oxazepam after i.v. administration (5 mg/kg) to mice.

in man after the administration of a single oral dose (15 mg) of diazepam. Figure 7 shows that the blood levels at 3 hr after diazepam administration may differ by a factor of more than 20. The different levels may be of clinical importance, since they seem to correlate, at least after an acute administration, with the degree of drowsiness shown by the subjects, as indicated in Table XVIII.

TABLE XVIII. *Correlation between blood diazepam peak and drowsiness*

DZ blood peak (μg/ml)	No. cases/No. cases showing drowsiness	% cases showing drowsiness
<0.10	11/2	18
0.10–0.15	11/4	36
0.15–0.20	7/3	43
0.20–0.30	7/7	100
>0.30	5/5	100

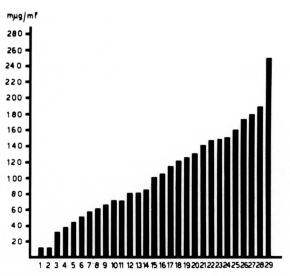

FIG. 7. Variability in diazepam blood levels 3 hr after its oral administration (15 mg) to men.

With regard to the metabolism of diazepam in man, in agreement with other authors (De Silva et al., 1966; van der Kleijn, 1971), it was found that repeated administration of this drug results in a blood level of N-demethyldiazepam which is even higher than the level of diazepam (see Fig. 8). The reason for this is that N-demethyldiazepam persists in blood longer than diazepam (see Table XIX).

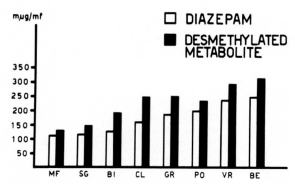

FIG. 8. Human blood levels of diazepam and its demethylated metabolite after chronic diazepam oral administration (15 mg/day for 48 days).

TABLE XIX. *N-demethyldiazepam (DDZ) blood levels (µg/ml) after a single oral administration of DDZ (20 mg)*

Case	Time after administration (hr)						
	0	1	2	4	8	24	48
A	0	0.54	0.56	0.74	0.48	0.47	0.40
B	0	0.0	0.23	0.25	0.21	0.18	0.05
C	0	0.16	0.18	n.c.	0.18	0.06	0.02
D	0	0.52	0.62	n.c.	0.85	0.42	n.c.
E	0	0.25	0.28	n.c.	0.27	0.16	n.c.
Mean	0	0.294 ±0.104	0.374 ±0.090	n.c.	0.398 ±0.125	0.258 ±0.079	n.c.

n.c. = blood sampling not carried out.

XII. CONCLUSIONS

It is believed that the investigations described above, which were carried out on the metabolism of diazepam and its metabolites, may have some importance in designing new chemical structures and in interpreting the pharmacological results obtained in various animal species. Furthermore, these studies may throw some light on the mechanism of action of the benzodiazepines and, hopefully, give an indication for improving the use of these drugs in human therapy.

REFERENCES

De Silva, J. A. F., Koechlin, B. A., and Bader, G. (1966): Blood level distribution patterns of diazepam and its major metabolite in man. *Journal of Pharmaceutical Science,* 55:692–702.
Garattini, S., Marcucci, F., and Mussini, E. (1972): Benzodiazepine metabolism "in vitro." *Drug Metabolism Reviews (in press).*
Gastaut, H., Naquet, R., Poiré, R., and Tassinari, C. A. (1965): Treatment of status epilepticus with diazepam (VALIUM). *Epilepsia (Amsterdam),* 6:167–182.
Jommi, G., Manitto, P., and Silanos, M. A. (1964): Metabolism of diazepam in rabbits. *Archives of Biochemistry and Biophysics,* 108:562–568.
Květina, J., Marcucci, F., and Fanelli, R. (1968): Metabolism of diazepam in isolated perfused liver of rat and mouse. *Journal of Pharmacy and Pharmacology,* 20:807–808.
McMillan, F. H., and Pattison, I. (1965): Substituted 1,4-benzodiazepines. French Patent 1, 394, 287 (*Chemical Abstracts,* 63:8387).
Marcucci, F., Airoldi, M. L., Mussini, E., and Garattini, S. (1971a): Brain levels of metrazole determined with a new gas chromatographic procedure. *European Journal of Pharmacology,* 16:219–221.
Marcucci, F., Guaitani, A., Fanelli, R., Mussini, E., and Garattini, S. (1971b): Metabolism and anticonvulsant activity of diazepam in guinea pigs. *European Journal of Pharmacology,* 20:1711–1713.

Marcucci, F., Fanelli, R., Mussini, E., and Garattini, S. (1970a): Further studies on species difference in diazepam metabolism. *European Journal of Pharmacology*, 9: 253–256.

Marcucci, F., Fanelli, R., Mussini, E., and Garattini, S. (1970b): Effect of phenobarbital on the "in vitro" metabolism of diazepam in several animal species. *Biochemical Pharmacology*, 19:1771–1776.

Marcucci, F., Mussini, E., Fanelli, R., and Garattini, S. (1970c): Species differences in diazepam metabolism. I. Metabolism of diazepam metabolites. *Biochemical Pharmacology*, 19:1847–1851.

Marcucci, F., Fanelli, R., Mussini, E., and Garattini, S. (1969): The metabolism of diazepam by liver microsomal enzymes of rats and mice. *European Journal of Pharmacology*, 7:307–313.

Marcucci, F., Fanelli, R., Frova, M., and Morselli, P. L. (1968): Levels of diazepam in adipose tissue of rats, mice and man. *European Journal of Pharmacology*, 4:464–466.

Morselli, P. L., Cassano, G. B., Placidi, G. F., Muscettola, G. B., and Rizzo, M. (1972): Kinetics of the distribution of C^{14}-diazepam and its metabolites in various areas of cat brain. *This Symposium*.

Mussini, E., Marcucci, F., Fanelli, R., and Garattini, S. (1972a): Metabolism of diazepam metabolites in guinea pigs. *Chemico-Biological Interactions* (*in press*).

Mussini, E., Marcucci, F., Fanelli, R., Guaitani, A., and Garattini, S. (1972b): Analytical and pharmacokinetic studies on the optic isomers of oxazepam succinate half-ester. *Biochemical Pharmacology*, 21:127–129.

Mussini, E., Marcucci, F., Fanelli, R., and Garattini, S. (1971): Metabolism of diazepam and its metabolites by guinea pig liver microsomes. *Biochemical Pharmacology*, 20:2529–2531.

Schwartz, M. A., and Postma, E. (1968): Metabolism of diazepam "in vitro." *Biochemical Pharmacology*, 17:2443–2449.

Schwartz, M. A., Bommer, P., and Vane, F. M. (1967): Diazepam metabolites in the rat: Characterization by high-resolution mass spectrometry and nuclear magnetic resonance. *Archives of Biochemistry and Biophysics*, 121:508–516.

Schwartz, M. A., Koechlin, B. A., Postma, E., Palmer, S., and Krol, G. (1965): Metabolism of diazepam in rat, dog and man. *Journal of Pharmacology and Experimental Therapeutics*, 149:423–435.

Sisenwine, S. F., Tio, C. O., and Relius, H. W. (1970): The metabolic transformation of oxazepam in man, pig and rat. *Pharmacologist*, 12:272.

Smith, L. R., and Williams, R. T. (1966): Implication of the conjugation of drugs and other exogenous compounds. In: *Glucuronic Acid Free and Combined*, edited by G. J. Dutton, Academic, New York, pp. 457–491.

Sternbach, L. H., Randall, L. O., Banziger, R., and Lehr, H. (1968): Structure-activity relationships in the 1,4-benzodiazepine series. In: *Drugs Affecting the Central Nervous System*, edited by A. Burger, Marcel Dekker, New York, pp. 237–264.

van der Kleijn, E. (1971): Pharmacokinetics of distribution and metabolism of ataractic drugs and an evaluation of the site of antianxiety activity. *Annals of the New York Academy of Sciences*, 179:115–125.

Zbinden, G., and Randall, L. O. (1967): Pharmacology of benzodiazepines: Laboratory and clinical correlations. In: *Advances in Pharmacology*, Vol. 5, edited by S. Garattini and P. A. Shore, Academic, New York, pp. 213–291.

The Benzodiazepines, Raven Press, New York © 1973

PHARMACOKINETICS AND METABOLISM OF THE HYPNOTIC NITRAZEPAM *

Jürg Rieder and Gernot Wendt

Department of Experimental Medicine, F. Hoffmann-La Roche & Co., Ltd., Basel, Switzerland

I. INTRODUCTION

A part of the information included in this report represents a synthesis of data contained in three other papers by the same authors (Rieder, 1972*a*, *b*; Rieder and Wendt, 1973), which will be published shortly. These papers give all the details of methods, experimental design, direct analytical results, and their evaluation, which cannot be presented here owing to lack of space. Therefore, only an outline of the methods used is given, followed by a survey of the results obtained.

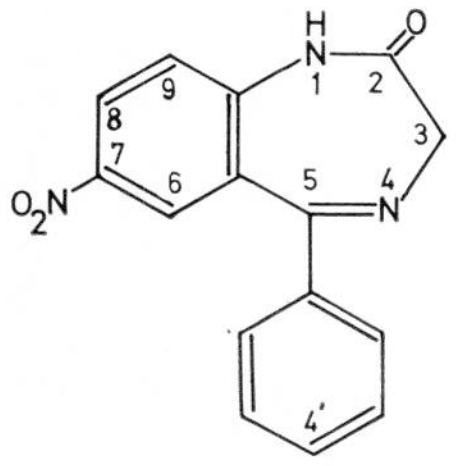

Nitrazepam

II. METHODS OF DETERMINING THE UNCHANGED NITRAZEPAM AND ITS MAIN METABOLITES

The first method available to determine nitrazepam was an absorption photometric technique devised by Rieder (1965), which was based on the reduction of nitrazepam by sodium hydrosulfite to the corresponding amine

---

* Generic name of the active principle of the preparation MOGADON® (Roche). Chemical name: 1,3-dihydro-7-nitro-5-phenyl-2H-1,4-benzodiazepine-2-one.

and a subsequent Bratton-Marshall reaction. The method permits the concentrations of unchanged nitrazepam in the human plasma to be measured
up to 12 hr after an intravenous injection of 30 mg of the drug. Even
after such a large dose, the resulting plasma levels were close to the sensitivity limit of the procedure, and their determination was therefore endang-
ered by relatively large technical variations. This, unfortunately, led us to
derive from our findings, obtained in four subjects, an erroneous average
elimination half-life of 6.2 hr for the unaltered nitrazepam. As will be
shown in this report, the correct value is three to four times higher.

The unsatisfactory sensitivity of our first method prompted us to continue our pharmacokinetic studies on nitrazepam with the aid of a radioactive preparation labeled with ^{14}C in position 2. This kind of labeling,
which was the only one possible at that time, has the disadvantage that the
tracer is lost if the benzodiazepine ring is cleaved during metabolism. Only
several years later, thanks to progress in methods of synthesis achieved in
the meantime, could another radioactive nitrazepam labeled with ^{14}C in
the metabolically stable position 5 be prepared by our chemists and used
for biological experiments.

During the last two years, we have also developed a highly sensitive
fluorometric method for measuring nonradioactive unchanged nitrazepam
and the sum of its two main metabolites, II and III, in biological samples.
This technique is based on a selective extraction of the compounds and on
coupling them, after reduction of the nitro group with metallic zinc, with
ortho-phthalaldehyde. It allows the plasma levels of the drug in man to be
followed for at least 72 hr after oral administration of a single therapeutic
dose.

III. METABOLISM OF NITRAZEPAM

Figure 1 shows the metabolites of nitrazepam known at present to be
formed in man and in the rat. With the exception of substance IV, which
was first described by Beyer and Sadée (1969), and substance X, which is
still hypothetical, the other compounds listed have been proved by us to
be biotransformation products of the drug appearing in the urine. They
have been isolated by various procedures of extraction, column chromatography, and thin-layer chromatography, and their chemical structures have
been elucidated by chemical reactions, comparison with synthetized reference substances, mass spectrometry, nuclear magnetic resonance (NMR)

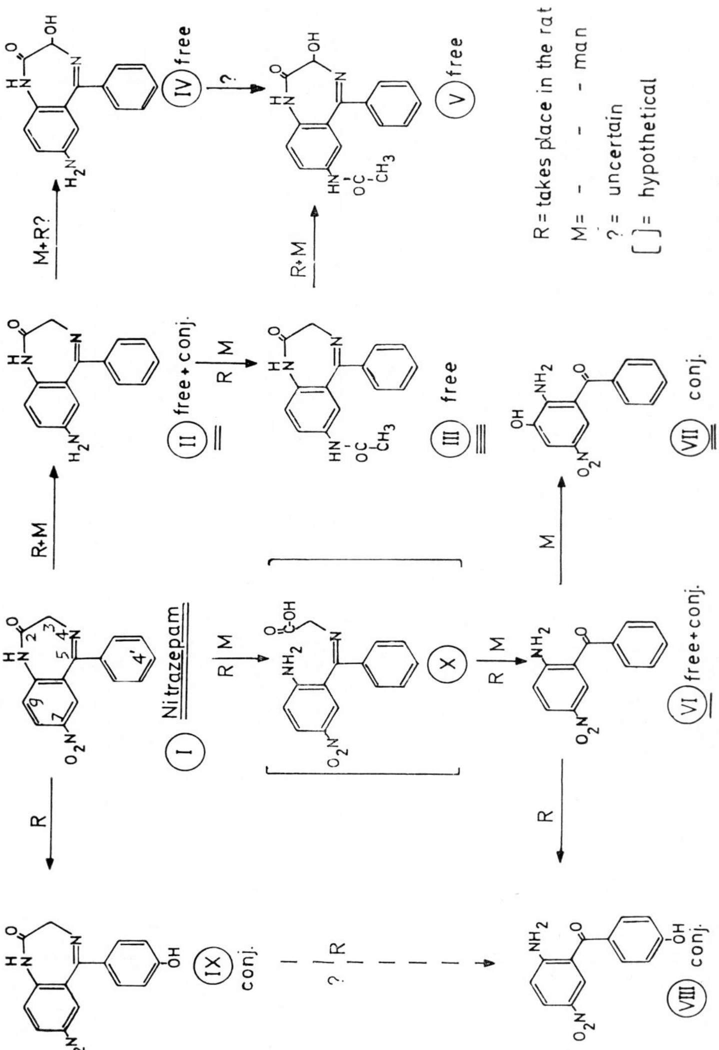

FIG. 1. Proposed metabolic pathways of nitrazepam in man and in rat.

spectrometry, and, in the cases of II and III, also by ultraviolet (UV) and infrared (IR) spectrometry.

From this figure it can be seen that, in man and in the rat, the main line of metabolization leads by reduction of the nitro group to the corresponding amine II and—by acetylation of the latter—to the 7-acetamido derivative III, which is the prevailing metabolite. A small proportion of II and III is hydroxylated in position 3, yielding compounds IV and V.

A secondary line of metabolization consists of the cleavage of the benzodiazepine ring, with the formation of the benzophenone derivatives VI, VII, and VIII. There is some evidence that the acid X is formed during the generation of the benzophenones, which can be called an opened lactam. The end product of this line is, in man, the 2-amino-3-hydroxy-5-nitro-benzophenone VII and, in the rat, the 2-amino-5-nitro-4′-hydroxy-benzophenone VIII. Part of this latter substance may possibly derive from the 4′-hydroxylated nitrazepam-IX which we have found in the urine of the rat, but not in man. The phenolic substances VII, VIII, and IX are excreted almost exclusively; II and VI only to a minor part in conjugated form. The number of lines under the numbers in Fig. 1 indicate the quantitative importance of the respective substances.

IV. PHARMACOKINETICS OF NITRAZEPAM IN MAN

A. Absorption data

1. The absorption quota (i.e., the percentage of the dose absorbed) of unchanged nitrazepam, which by definition is identical with the physiological availability of the drug, was determined in six healthy adult volunteers (Nos. 34–39) who first received an intravenous injection of 10 mg of nitrazepam base dissolved in 2 ml of glycerolformal, and 14 days later an equal oral dose of the drug in the form of two commercially produced tablets of MOGADON® (Roche), ingested with some water on an empty stomach. Plasma samples from blood, taken at selected times between the 2nd and the 120th hr after each administration, were examined for non-metabolized nitrazepam by means of the highly sensitive fluorimetric method developed by Rieder (1972a). The absorption quota was calculated according to the method of "the corresponding areas" of Dost and Gladtke (1963); see also footnote to Table I.

The results obtained are indicated in the second line of Table I. These figures show that, despite considerable interindividual variation, ranging from 53 to 94% of the dose, as a rule a large proportion of nitrazepam is absorbed in unchanged form (an average of 78% in the six subjects tested). One reason for an occasionally reduced amount of the unaltered preparation passing into the blood plasma is probably the formation of 2-amino-5-

TABLE 1. *Absorption quota of unchanged nitrazepam in six healthy humans after oral intake of 10 mg of the drug in the form of two tablets of MOGADON® (Roche)*

Subjects (No., sex, age, and weight)	34 fem. 22 years 56 kg	35 male 29 years 70 kg	36 male 26 years 60 kg	37 male 25 years 68 kg	38 male 18 years 60 kg	39 male 28 years 64 kg	Average
Absorption quota in % of the dose*	53	94	83	66	81	91	78
Invasion constant, k_i	2.199	2.201	2.187	2.219	2.125	2.056	2.164
Plasma levels (μg/ml) at 2 hr after med.	108	68	94	75	82	79	84
Distribution coeff. Δ' after 10 mg, i.v.	1.50	2.76	2.00	1.85	1.80	1.80	1.95

* Values corrected (according to a procedure described by Rieder 1972b) for differences existing in the elimination rates of the same subjects after i.v. and oral administration (see Table IVA).

nitro-benzophenone, caused by prolonged exposure to strong hydrochloric acid in the stomach, which we have proved to occur in certain cases.

2. The invasion constant of unchanged nitrazepam has been determined in the same six subjects (Nos. 34–39), according to the method described by Dost and Gladtke (1963), utilizing their formula No. 24.21. The values obtained on the assumption that the time of maximal plasma levels t^{max} was 2 hr in each case are listed in the second line of Table I. The figures show that there is little interindividual variation and also indicate that the unchanged nitrazepam is absorbed very rapidly from the gastrointestinal tract.

3. The time (t^{max}) and the value of maximal concentration (c^{max}) of the unchanged nitrazepam in the plasma could not be determined accurately in the six subjects mentioned, but it can be deduced from our data that t^{max} was reached approximately 2 hr after ingestion of 10 mg of the drug. The maximal plasma levels c^{max} measured at this time are shown in the third line of Table I. They ranged from 68 to 108 ng/ml, with a mean of 84.

B. Data concerning the distribution of the drug in the body

1. The distribution coefficient (Δ') of unchanged nitrazepam, determined in the six subjects (Nos. 34–39) on the basis of the plasma levels measured during the interval between the 12th and the 72nd hr after intravenous injection of 10 mg of the drug, was found to vary from 1.50 to 2.76 liters/kg body weight, with a mean of 1.95, as shown in the last line of Table I. These values indicate that by far the largest fraction of the total quantity of the compound present in the human body is distributed outside the intravasal compartment and must accumulate temporarily in certain tissues.

2. The constant of distribution (k_d) was approximately calculated in three of our subjects (Nos. 35, 37, and 39), and values of 5.0, 2.4, and 3.0 were obtained. They are an expression of a relatively high rate of passage of the drug from the blood plasma into the extravasal compartment.

3. The distribution of nitrazepam labeled with ^{14}C in position 5, in the body of the mouse, as examined by whole-body autoradiography confirmed the conclusion drawn under B. 3, for humans. Owing to lack of space, however, we will not discuss these findings here.

4. The passage of unchanged nitrazepam into the cerebrospinal fluid in man could not yet be determined exactly, because we were unable to obtain fluid samples taken immediately before the administration of the

drug in order to measure blank values. This would have required repeated punctures within short intervals. At present, we can only infer from our findings that nitrazepam passes into the cerebrospinal fluid, where, however, it only reaches lower concentrations than those existing simultaneously in the plasma. This view is in accordance with the data given in the following paragraph.

5. The binding of unchanged nitrazepam to the proteins of human plasma was examined by means of an equilibrium-dialysis technique at pH 7.4 and 37°C, using a mixture of equal volumes of plasma from 11 healthy adults. Nitrazepam labeled with ^{14}C in position 5 and checked for purity by thin-layer chromatography was added *in vitro* in 10 concentrations graded from 10 to 10,000 ng/ml, and the measurements were made by liquid scintillation counting.

It was found that within the indicated limits of plasma levels of total nitrazepam—which comprise and exceed by far the range of those resulting in man from single or daily repeated oral doses of 5 to 20 mg of the drug, that is, 10 to 200 ng/ml—the percentage of the non-protein-bound fraction, although it rises in principle with the total concentration of the drug, remains almost constant between 13.2 and 14.2%. Table II shows the binding data for a series of "round" concentrations of total nitrazepam, as calculated from the regression equation of the 10 experimentally determined pairs of values (log C_{total} versus $C_{unbound}$). Further data and graphic presentations of the correlations between total, protein-bound, and unbound nitrazepam in mixed human plasma will be given by Rieder (1972*b*).

6. The passage of nitrazepam and its metabolites through the human placenta into the fetus was examined in six mothers at the time of delivery. Three to six hr prior to birth, a single oral dose of 10 mg of nitrazepam, labeled to the very low activity of 10 μC with ^{14}C in position 2, was administered. The radioactive preparation was used to measure the sum of the unchanged substance and the metabolites of the drug, since the passage through the placental barrier is a toxicological problem comprising all these compounds. At the time of birth, the total radioactivity of the blood plasma of the mother and of the child, the latter obtained from the blood contained in the umbilical veins, was then measured and calculated as ng-equivalents of unchanged nitrazepam/ml.

The last column of Table III shows the values for the newborn as a percentage of those of the mother. The figures suggest that, after the administration of nitrazepam to pregnant women, the drug and its metabolites reach fairly rapidly about the same concentration in the fetal blood as is present

TABLE II. Binding of unchanged nitrazepam to the total proteins of human plasma (mixture of equal volumes of plasma obtained from 11 healthy adults)

	5	10	20	40	60	80	100	150	200	300
C_{total} = ng of total nitrazepam/ml of plasma	5	10	20	40	60	80	100	150	200	300
$C_{unbound}$ = ng of unbound nitrazepam/ml of plasma	0.66	1.33	2.67	5.40	8.13	10.87	13.61	20.50	27.42	41.30
% unbound = $C_{unbound}$ in % of C_{total}	13.20	13.30	13.35	13.50	13.55	13.59	13.61	13.67	13.71	13.77

Method: equilibrium dialysis at pH 7.4 and +37°C; analysis by scintillometry using pure 5-^{14}C-nitrazepam. In order to show the concentrations and percentages of the unbound nitrazepam valid for "round" values of the total concentration of the drug in the plasma within the range occurring under therapeutical conditions, the former two kinds of values were calculated from the regression equation of 10 directly determined pairs of values (log C_{total} versus log $C_{unbound}$).

TABLE III. *Passage of nitrazepam and its metabolites into the fetal blood plasma at the time of birth, after a single oral dose of 10 mg of 2-^{14}C-nitrazepam*

Time after administration (hr)	Total radioactivity (ng-equivalents of unchanged nitrazepam/ml)		
	Plasma of mothers	Plasma of babies	Values for the babies in % of those of their mothers
3	25.1	10.8	43.0
3½	46.8	22.2	47.4
3½	31.4	15.8	50.3
4	41.2	29.2	70.9
5	43.8	39.0	89.0
6	41.6	32.2	77.4

in the blood of the mother. This means that the placenta is permeable to the compounds examined in both directions.

7. To investigate the passage of nitrazepam and its metabolites into the mother's milk, it was again necessary to determine the passage into the milk not only of the unchanged drug, but also of its biotransformation products. Furthermore, we had to ascertain that even very low concentrations, possibly occurring in the milk, would still be measured. This could be accomplished only by using a radioactive tracer. Since the secretion and the composition of the milk change during the first days of its production, the preparation had to be given repeatedly.

Five women received daily, at 8 p.m., from the 1st to the 5th day following parturition, an oral dose of 5 mg of nitrazepam labeled to the activity of 10 μC with ^{14}C in position 2. On each following day, a sample of venous blood was taken from the mother at 9 a.m., and her milk was collected at the same time and at 5:45 p.m., that is, about 2 hr before the next dose. The babies were, of course, fed milk from other nontreated women.

Figure 2 shows the curves of the total radioactivity in the plasma of the five mothers. An initial accumulation up to a plateau, at values corresponding presumably to 0.15 to 0.2 μg-equivalents of nitrazepam/ml can be seen.

Figure 3 gives the respective curves of the total radioactivity in the milk. As in the plasma, the values rose initially to attain, within a few days, plateau levels corresponding to about 0.05 to 0.1 μg-equivalents of nitrazepam/ml. This concentration range thus approximates the lower limit of the respective range in the mother's plasma. One can calculate that, during the steady state of the secretion into the milk, the amount of drug and/or its metabolites ingested by the baby in 100 ml of milk from a mother receiving daily doses of 5 mg of nitrazepam would be of the order of 5 to 10 μg-equivalents of

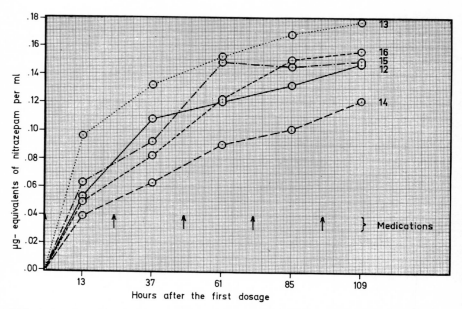

FIG. 2. Total radioactivity (calculated as μg-equivalents of nitrazepam) in the plasma of five mothers after oral medication of 5 mg of 2-14C-nitrazepam on five succeeding days. The drug was given at 8 p.m. on the 1st to 5th day after parturition, and blood samples were taken at 9 a.m. on each following day.

the unaltered compound. It seems unlikely that these very small quantities could produce any pharmacological or toxic effects in the child, even if one were to consider that the excretion of the compounds in question during the first weeks of life may possibly occur somewhat more slowly than in adults. Neither this excretion rate in the newborn nor the composition of the material entering into the milk from doses of nitrazepam given to the mother has yet been examined.

C. Plasma levels and the elimination of nitrazepam and its metabolites following administration of a single dose of the drug

Table IV shows the concentrations of unchanged nitrazepam, measured by our fluorimetric method in the plasma of our subjects Nos. 34–39. The values without parentheses were obtained after intravenous injection, those within parentheses after oral administration of a single 10-mg dose of the drug.

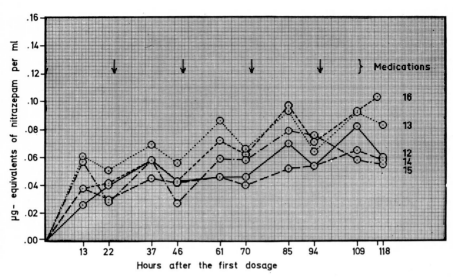

FIG. 3. Total radioactivity (calculated as μg-equivalents of nitrazepam) in the milk of five mothers after oral medication of 5 mg of 2-^{14}C-nitrazepam each on five succeeding days. The drug was given at 8 p.m. on the 1st to 5th day after parturition, and the milk was taken at 9 a.m. and 5:45 p.m. on the 2nd to 6th day.

TABLE IV. *Plasma levels of unchanged nitrazepam in six healthy adult humans follow-ing a single dose of 10 mg of the drug*

Subjects (sex, age, weight)	Hours after administration						
	2	4	8	12	24	48	72
No. 34 f/22 yr/56 kg	124 (108)	101 (83)	82 (55)	73 (47)	60 (38)	26 (20)	11 (11)
No. 35 m/29 yr/70 kg	61 (68)	56* (55)	46 (48)	39 (42)	23 (31)	11* (17)	6 (8)
No. 36 m/26 yr/60 kg	92 (94)	84 (77)	64 (66)	56 (60)	38 (47)	19 (26)	8 (12)
No. 37 m/25 yr/68 kg	102 (75)	90 (71)	68 (50)	60 (45)	36 (34)	20 (18)	10 (10)
No. 38 m/18 yr/60 kg	86 (82)	78 (73)	65 (71)	60 (64)	39 (36)	17 (18)	7 (10)
No. 39 m/28 yr/64 kg	86 (79)	78 (79)	65 (73)	56 (60)	31 (38)	15 (20)	5 (8)

The drug was administered as follows:

1. Dissolved in 2 ml of glycerol formal as an i.v. injection (figures without paren-theses).

2. Given 14 days later orally in the form of two tablets of MOGADON® (Roche) (figures within parentheses). Values indicated in ng/ml of plasma.

* Value interpolated graphically because of analytical failure.

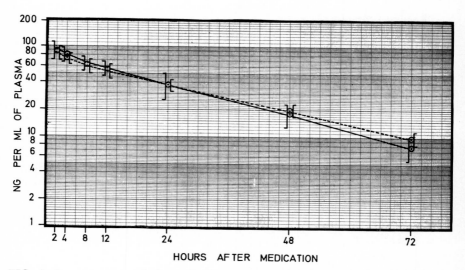

FIG. 4. Average curves of plasma levels ± S.D. of unchanged nitrazepam in subjects Nos. 34–39, after a single dose of 10 mg of the drug, administered orally in a tablet (broken line) or dissolved in 2 ml of glycerol formal, as an intravenous injection (solid line).

Figure 4 serves as a further illustration. It shows the average plasma-level curves of the unchanged nitrazepam and the respective standard deviations obtained in our six subjects after administration of 10 mg of the drug intravenously (solid line), and orally, in tablet form (broken line).

The elimination of the unchanged nitrazepam from the plasma after intravenous administration occurs in two phases. In phase I, which lasts about 8 to 12 hr, the processes of pure elimination by metabolization and excretion are combined with those of distribution of the drug among the different compartments of the body. The semilogarithmic plasma-level curves are, therefore, slightly bent during this phase, indicating nonexponential elimination kinetics. A constant or half-life of elimination cannot, therefore, be calculated for this interval.

In the succeeding phase II, beginning about 8 to 12 hr after intravenous and oral administration and lasting at least up to the 72nd hr following medication, the equilibrium of distribution and elimination is obtained. Thus, a semilogarithmic plot of the plasma levels against time yields a straight line, a sign of first-order kinetics of elimination. As shown in Table IVA, in the six subjects mentioned, the elimination half-lives, determined for the period

TABLE IVA. *Half-lives (t$_{1/2}$) of elimination of unchanged nitrazepam from the plasma of six subjects (Nos. 34–39), as determined in the interval between the 12th and the 72nd hr after a single intravenous or oral dose of 10 mg of the drug (values indicated in hr)*

Subjects No., sex, age, and weight	*34* fem. 22 years 56 kg	*35* male 29 years 70 kg	*36* male 26 years 60 kg	*37* male 25 years 68 kg	*38* male 18 years 60 kg	*39* male 28 years 64 kg	Average
After 10 mg, i.v.	21.4	22.6	21.6	24.0	19.4	17.8	21.1
After 10 mg, oral	28.1	25.2	25.8	27.4	23.0	21.2	25.1

between the 12th and the 72nd hr varied after intravenous injection of 10 mg of the drug from 17.8 to 24.0 hr, with an average of 21.1, and after oral administration of the same dose from 21.2 to 28.1 hr, with a mean of 25.1. The unexpected finding that in all six persons the elimination half-life during phase II was markedly longer (on an average by 4 hr) after oral dosage than after intravenous injection cannot yet be explained conclusively. It may possibly be connected with the different circulation conditions of the blood carrying the compound after the two routes of medication.

D. Plasma levels and the elimination of nitrazepam after repeated administration

Four healthy subjects (Nos. 40–43) each received a total of 26 oral doses of 5 mg of nitrazepam, the dosage intervals being 34 hr between the 1st and the 2nd dose, then 24 hr from the 2nd to the 25th dose, and 62 hr between the 25th and the final one. The plasma levels of the unchanged drug and also of the sum of its two main metabolites, II and III, were measured first every day, later every 2 days, and finally every 3 days, each time 5 min preceding the following dose. Furthermore, the elimination half-life was determined from samples taken in close succession after the first and the last doses of this long-term experiment.

Figures 5 and 6 show, as examples, single curves resulting from semilogarithmic plots of the respective plasma levels against time, as obtained in two of the subjects. In Fig. 5, the curve of the unaltered nitrazepam, marked by a solid line, runs above that of the two metabolites. Figure 6 shows that the reverse is also possible.

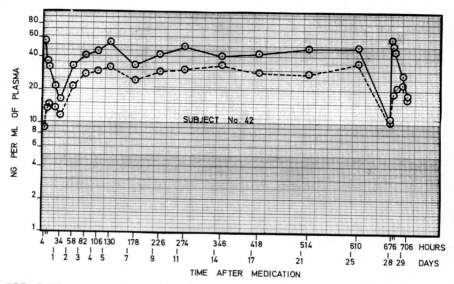

FIG. 5. Plasma levels of unchanged nitrazepam (solid line) and the sum of its main metabolites, the 7-amino and the 7-acetamido derivatives (broken line), in a healthy adult during repeated oral administration of the drug.

Dosage: 5 mg of nitrazepam was given in the form of tablets of MOGADON® (Roche) every 24 hr for 24 days and, in addition, 34 hr preceding and 62 hr following the period of regular daily administration.

Sampling: Plasma samples were obtained 5 min before and 4, 8, 12, 24, and 34 hr after the first and the last of the doses mentioned, as well as 5 min before 11 of the 24 daily doses given in the meantime.

The diagram given in Fig. 7, which shows the average curves for all four persons examined, may serve to explain the main results of these assays. If elimination after the initial and the terminal dose is disregarded, it can be seen that the unchanged nitrazepam accumulated in the plasma between the 1st and about the 4th day, then reached an approximate plateau which persisted with little variation up to the end of the daily treatment.

In order to indicate for each person in single figures the minimum concentrations (c^{min}) per dosage interval of the unaltered drug and of the sum of its metabolites II and III, the averages of the eight respective values determined from the 5th to the 25th day of the experiment were calculated. The results are indicated in the first and second lines of Table V.

It can be seen that there was little interindividual variation of the unaltered drug, but somewhat wider variation of the sum of its two main

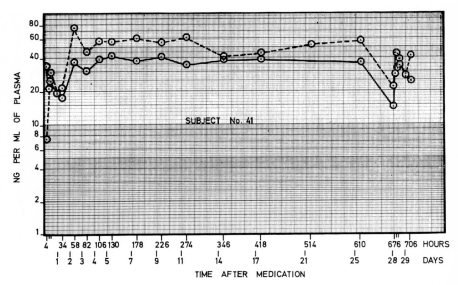

FIG. 6. Plasma levels of unchanged nitrazepam (solid line) and the sum of its main metabolites, the 7-amino and the 7-acetamido derivatives (broken line), in a healthy adult during repeated oral administration of the drug.

Dosage and sampling: As in Fig. 5.

metabolites, II and III. The average of this sum was 30.7 ng/ml and that of the unchanged nitrazepam was 38.6.

Further information furnished by the same experiments concerns the question of whether repeatedly administered nitrazepam influences its own elimination rate. The existence of such an influence in man can be checked by comparing the elimination half-life of the drug at the beginning and at the end of a prolonged daily treatment. The last two lines of Table V shows the elimination half-lives of each of the four subjects, Nos. 40–43, determined after the first dose and 2 days following 24 daily oral doses of 5 mg of nitrazepam. (The interval of 2 days before the second determination was introduced in order to lower the initial concentrations of the unchanged nitrazepam, as well as of its metabolites, to approximately those present at time t^{max} after the first dose.)

Intraindividual differences were observed between the two times of determination, but these differences have no systematic direction and were found to be statistically not significant. This means that, under the conditions of the experiment, no signs of a change in the elimination rate of the

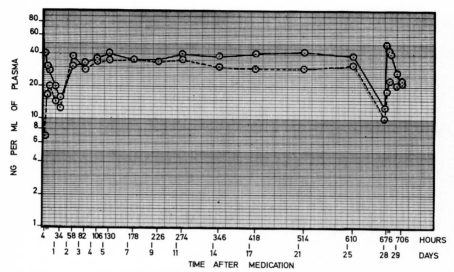

FIG. 7. Average plasma levels of unchanged nitrazepam (solid line) and of the sum of its main metabolites, the 7-amino and the 7-acetamido derivative (broken line), in four healthy adults during repeated oral administration of the drug.
Dosage and sampling: As in Fig. 5.

unchanged nitrazepam due to acceleration or inhibition by its own effects were observed. This finding is in agreement with the clinical experience that in man there is no reduction in the effect of the drug during prolonged administration at a constant dosage and that it is not necessary to increase the dosage in order to obtain constant effects.

TABLE V. *Average of minimal plasma levels of c^{min} of unchanged nitrazepam and of the sum of metabolites II and III, during the steady state reached within 24 days of medication with 5 mg of the drug given orally every evening, and half-life of elimination $t_{1/2}$ of the unchanged nitrazepam after the first and last doses of the treatment*

Subjects No., sex, age, and weight	40 male 35 years 74 kg	41 male 37 years 71 kg	42 fem. 26 years 51 kg	43 male 35 years 73 kg	Average
c^{min} of I (ng/ml)	36.6	38.0	44.4	35.2	38.6
c^{min} of II + III (ng/ml)	18.2	52.9	30.2	21.4	30.7
$t_{1/2}$ of I at the beginning and at the end of a 24-day treatment (hr)	28.5 29.1	30.5 30.8	22.5 17.4	22.0 23.5	—

E. Proportions of the unchanged nitrazepam and its metabolites in human plasma

In one of our early experiments, two healthy adults (Nos. 44 and 45) received a single oral dose of 5 mg of nitrazepam labeled with ^{14}C in position 2, with an activity of 100 μC. Blood samples were taken 6, 12, 24, 48, 72, 96, and 120 hr afterwards. The plasma separated from these samples was extracted three times at pH 10 by a 2:1 mixture of methylene chloride and ethyl acetate, which takes up all the known nonconjugated metabolites of nitrazepam. The compounds contained in the pooled extracts were fractionated by thin-layer chromatography in the system benzene plus *n*-propanol plus 25% aqueous ammonia (80:20:1), thus separating the total extractable radioactive material into the unchanged nitrazepam, its metabolites II and III, and a residual part composed of several unidentified further transformation products. The respective zones of the chromatograms were eluted and the radioactivity of the eluates measured by liquid-scintillation counting.

Figure 8 shows the proportions of the fractions mentioned, calculated as μg-equivalents of the unchanged drug and indicated in percentage of the total radioactivity of the extract. It will be seen that the unchanged drug amounted to about 70% at the 6th hr and then dropped at an increasing rate to about 35% at the 120th hr after medication. Metabolite II, the amino derivative, constituted about 20% of the total during the first 3 days; after this its share decreased to about 15%. Metabolite III, the acetamido derivative, remained (with the exception of only one point) between 1 and 4% during the entire observation time. The sum of the remaining extractable radioactive material, marked by dotted curves, was about 10 to 15% during the 1st 48 hr, but its proportion rose afterwards at an increasing rate to about 40 to 50% at the 120th hr after administration.

The composition of this "residual part" could not be examined any further. On the basis of our results obtained in other experiments with nitrazepam labeled in position 5, one can assume that this fraction was at least partly formed by 3-hydroxy derivatives of the metabolites II and III.

That the relative rise of the dotted curves does not mean that there is also an absolute accumulation of the respective materials in the blood can be inferred from Figs. 9 and 10, which show the curves of the total radioactivity measured in the plasma of eight adult persons after an intravenous injection of 5 mg of nitrazepam labeled with ^{14}C in position 2, with an activity of 50 μC. The values, calculated as ng-equivalents of unchanged nitrazepam/

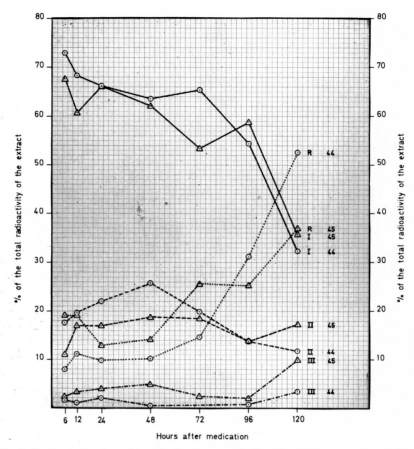

FIG. 8. Proportions of unchanged nitrazepam (I), 7-amino (II), and 7-acetamido derivative (III), and residual radioactivity (R) found in extracts from the plasma of two persons (Nos. 44 and 45), after a single oral dose of 5 mg of $2\text{-}^{14}C\text{-}$nitrazepam.

ml, decreased, with elimination half-lives between 37.6 and 56.1 hr from 32.0 to 55.4 ng-equivalents at the 6th hr to 5.6 to 11.5 ng-equivalents at the 122nd hr after administration.

F. Data concerning the excretion of nitrazepam and its metabolites

1. The cumulative renal excretion of the drug and its transformation products was examined in an earlier experiment in six adult persons (Nos. 44–49) with normal liver and kidney functions, following a single oral

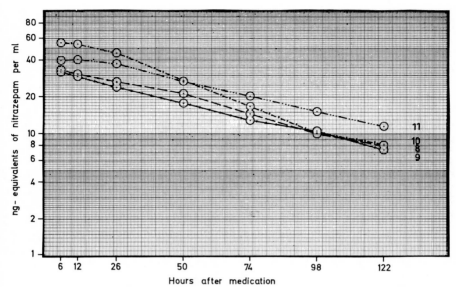

FIG. 9. Total radioactivity (calculated as ng-equivalents of unchanged nitrazepam) in the plasma of four adult subjects with normal liver and kidney function (Nos. 24–27) after an i.v. dose of 5 mg of 2-^{14}C-nitrazepam.

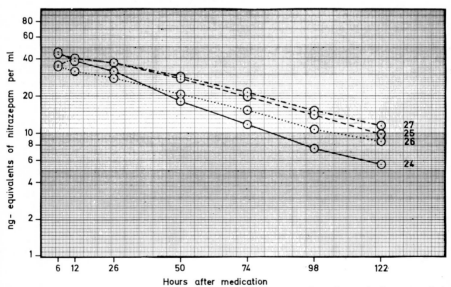

FIG. 10. Total radioactivity (calculated as ng-equivalents of unchanged nitrazepam) in the plasma of four adult subjects with normal liver and kidney function (Nos. 8–11) after an i.v. dose of 5 mg of 2-^{14}C-nitrazepam.

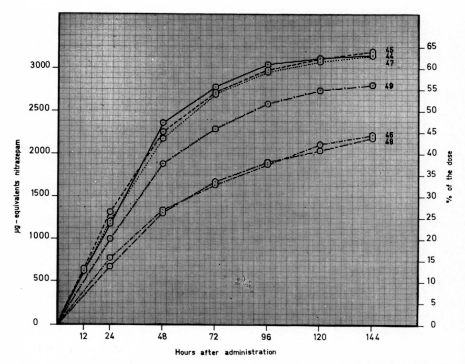

FIG. 11. Total excretion of nitrazepam in the human urine. Cumulative curves of the total radioactive material recovered from the urine of six adults (Nos. 44–49) after a single oral dose of 5 mg of 2-^{14}C-nitrazepam. Values calculated in μg-equivalents of unchanged nitrazepam and in percent of the dose.

administration of 5 mg of nitrazepam labeled with ^{14}C in position 2, with an activity of 100 μC.

Figure 11 shows that at the 144th hr after dosage, when the curves were still rising slowly, 45 to 65% of the total radioactivity introduced with the dose had been excreted in the urine. These figures are too low, considering the amount of free and conjugated benzophenones contained in the urine, probably because these compounds lost the tracer when the diazepine ring was cleaved.

Figure 12 shows the corresponding curves of two other healthy adults (Nos. 50 and 51) who ingested a single dose of 10 mg of nitrazepam, labeled with ^{14}C in position 5, with an activity of 50 μC and, marked with triangles, a further curve of one of them (subject No. 50) was obtained 3 months

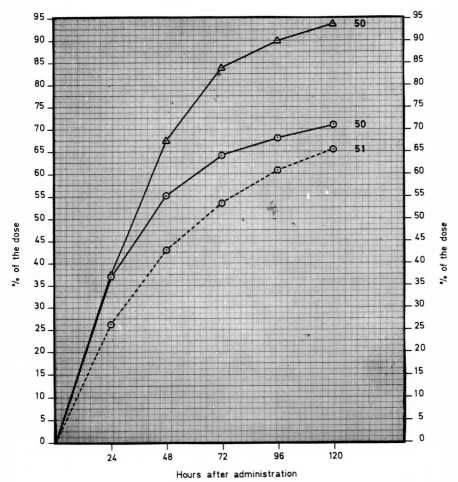

FIG. 12. Total excretion of nitrazepam in the human urine. Cumulative curves of the total radioactive material recovered from the urine of two adults (Nos. 50 and 51) after a single oral dose (⊙) of 10 mg of 5-^{14}C-nitrazepam, and after an equal intravenous dose (△) in subject 50 only. Values calculated in percent of the dose.

after he had received an equal dose of the same preparation by intravenous injection.

In the cases in which benzophenone metabolites could not escape the measurement, it was found that at the 120th hr after administration (i.e., 24 hr earlier than the end of the first experiment), 65 and 71%, respectively, of the dose had been excreted in the urine after oral dosage, but 93% after intravenous medication. The lower percentage after oral administration is explained by incomplete absorption.

2. The cumulative excretion of nitrazepam or its metabolites in the feces could be followed in four adult healthy subjects (Nos. 17–20) with the aid of nitrazepam labeled with ^{14}C in position 2 only, thus again presenting the risk that the figures obtained would be slightly too low due to benzophenones escaping detection.

The curves in Fig. 13 show that the percentage of the dose excreted in the feces during the first 144 hr after administration of 30 mg of the drug,

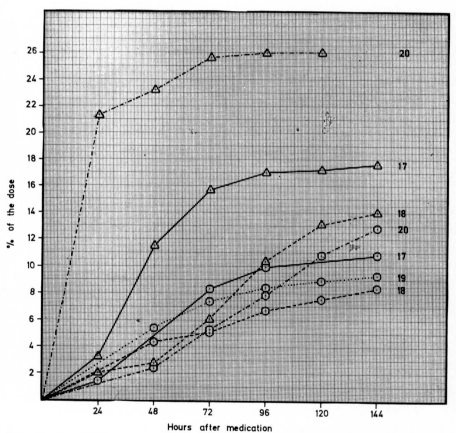

FIG. 13. Cumulative curves of the total radioactive material excreted in the feces of four adults (Nos. 17–20) after oral (△) and i.v. (⊙) application of a 30-mg dose of 2-^{14}C-nitrazepam.

with an activity of 194 μC, reached 8 to 13% of the dose when the preparation was injected intravenously, but 14 to 20% when it was given orally. The higher values in the latter case are due to incomplete absorption. After an intravenous dose of only 4 mg in another person, the portion recovered in the feces during the same interval was reduced to only 2% of the dose.

The total data available at present on the excretion of nitrazepam in man give rise to the assumption that the drug and its metabolites are almost completely eliminated from the body in the course of approximately 2 weeks following cessation of treatment with therapeutic doses.

3. The proportions of the various excretion products of nitrazepam appearing in human urine are shown in Fig. 14. The fractionation of the extractable and, therefore, unconjugated radioactive compounds present in the blood plasma of two subjects (Nos. 44 and 45) has already been reported in Fig. 8. Figure 14 shows, for the same two persons, the results of a corre-

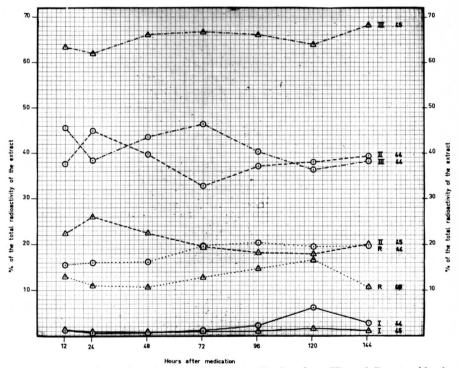

FIG. 14. Proportions of unchanged nitrazepam (I), 7-amino (II), and 7-acetamido derivative (III), and residual radioactivity (R) extracted directly from the urine of two persons after a single oral dose of 5 mg of 2-^{14}C-nitrazepam.

sponding fractionation of the total radioactive material contained in the urine, excreted during the first 144 hr after ingestion of 5 mg of nitrazepam labeled with ^{14}C in position 2, with an activity of 100 μC.

From the nearly horizontal course of all the relevant curves, it can be inferred that the proportions of all fractions that were measured separately, and that do not include the benzophenones possibly present, have not changed considerably between the 12th and the 120th hr after administration. In addition, it can be seen that nitrazepam is excreted in the urine of man almost exclusively in the form of its metabolites. Among these, the 7-amino derivative II and the 7-acetamido derivative III are the main products of biotransformation. As can be seen from the figure, however, there are inter-individual differences in the capacity for acetylating the 7-amino metabolite of nitrazepam. The same phenomenon is well known in other drugs such as isoniazid or sulfonamides.

The single components of the remaining radioactive material contained in the extract examined, the sum of which is represented in the figure by dotted curves, could not be separated in these experiments. Furthermore, the conjugated metabolites present in the urine are not shown in this diagram. Table VI, therefore, gives the proportions of unconjugated and conjugated excretion products of nitrazepam, determined in the pooled urine excreted by a healthy adult man in the first 24 hr after a single oral dose, and 3 months later, after a single intravenous dose of 10 mg of the drug labeled with ^{14}C in position 5, with an activity of 50 μC. Attention is drawn to the following peculiarities in these data.

Unchanged free nitrazepam appeared in the urine examined only in the very small amount of 0.2 and 0.7%, respectively, of the dose. The main excretion product of the drug after both oral and intravenous administration was the unconjugated 7-acetamido derivative III, which represented 4.8 and 10.5% of the dose, respectively. Its precursor, the 7-amino derivative II, amounted after oral dosage to 1.8% of the dose in free form, whereas it was not found as a conjugate. After intravenous injection, it reached 5.3% of the dose in free form and 1.7% as a conjugate. The sum of the two benzophenones VI and VII, after oral medication, made up 1.9% in free and 7.2% in conjugated form. The respective figures, after intravenous administration, were only 0.7 and 1.9% of the dose. The 3-hydroxylated derivatives IV and V, in free form, and the metabolite II, as a conjugate, could be measured only after intravenous dosage. Their amounts possibly present in the urine examined, after oral medication, fell below the sensitivity limit of the analytical procedures.

TABLE VI. Proportions of nitrazepam and its metabolites found in the urine of a healthy adult man in the first 24 hr after oral and i.v. administration of 10 mg of the drug labeled with ^{14}C in position 5, indicated in % of the dose

		I	II	IV	III	V	VI	VII	sums of comp. extr. at pH 10
after oral admin. total act.=37.0%	in free form	0.2	1.8	--	4.8	--	1.5	0.4	11.6
	as conjugates	0.5	--	--	--	--	1.9	5.3	10.8
	sums +14.6% unknown acidic radioact. comp.	0.7	1.8	--	4.8	--	3.4	5.7	22.4
after i.v. admin. total act.=37.5%	in free form	0.7	5.3	0.4	10.5	0.5	0.6	0.1	20.9
	as conjugates	--	1.7	--	--	--	1.1	0.8	10.5
	sums +6.1% unknown acidic radioact. comp.	0.7	7.0	0.4	10.5	0.5	1.7	0.9	31.4

The sums of all the conjugated fractions were, with 10.8 and 10.5% of the dose, almost the same after both routes of administration, whereas the fraction of the unconjugated compounds, extractable at pH 10 into the organic solvent mixture, was about twice as large after intravenous injection as after oral intake of the same dose, probably because a larger amount of acidic derivatives of nitrazepam remained unextracted at pH 10 following oral dosage. These acidic compounds have not yet been identified, but, hypothetically, we can expect them to consist mainly of the free "opened lactam" X, which is probably formed as an intermediary product in the generation of benzophenones. This opinion would be consistent with our finding, also demonstrated in this table, that the fraction of the nonconjugated plus the conjugated 2-amino-5-nitrobenzophenone VI and its 3-hydroxylated derivative VII present in the urine of the first 24 hr amounted to 2.6% of the dose after intravenous injection, but to 9.1% of the dose, that is, 3.5 times more, after oral administration.

It therefore seems very likely that the benzophenones are predominantly formed in the stomach during prolonged exposure to hydrochloric acid. In our opinion, they must be considered as degradation products of nitrazepam rather than as real metabolites. The acid hydrolysis of nitrazepam to the respective benzophenone is also known to occur *in vitro,* and our preliminary trials furnished indications that the "opened lactam" is produced as a precursor under these extracorporeal conditions as well. Its probable occurrence *in vivo* is further suggested by the fact that an analogous "opened lactam" has been identified in the urine of man by Koechlin et al. (1965) as a metabolite of chlordiazepoxide, a benzodiazepine of similar structure to nitrazepam. The fact that benzophenones are formed relatively extensively after oral intake of the drug but not after its intravenous injection is the main reason why the proportions of all other metabolites found in the urine are lower after oral intake than after intravenous administration.

V. SUMMARY

A fairly complete picture of the metabolism and the pharmacokinetics of nitrazepam in man is presented. It has been obtained partly with the aid of preparations of the drug labeled in position 2 or 5 with ^{14}C and partly by using a new fluorimetric method. The details concerning the analytical procedures and other complementary information are not included in this chapter owing to lack of space, but they will soon be published elsewhere.

The main results communicated above are the following:

With the exception of some minor, polar and predominantly acidic metabolites, shown to exist but not identified up to now, the main lines of biotransformation in man and, for comparison, in the rat have been clarified and are displayed in Fig. 1.

The absorption quota of unchanged nitrazepam and thus the physiological availability of the drug after oral administration of 10 mg of the drug was found to vary in six subjects from 53 to 94% of the dose, with an average of 78%.

The invasion constant k_i was determined in six subjects, with values of 2.056 to 2.219, indicating rapid absorption of the drug from the gastrointestinal tract.

Peak plasma levels of unchanged nitrazepam were reached in six persons about 2 hr after a single oral dose of 10 mg, taken in the form of two tablets of MOGADON® (Roche).

The distribution constant k_d was determined in three subjects with values of between 2.4 and 5.0, indicating a relatively high rate of passage of the drug from the plasma into the extravasal compartment of the body.

The distribution coefficient Δ' of the unchanged nitrazepam, determined in six subjects following intravenous injection of 10 mg of the drug, varied from 1.50 to 2.76. This means that by far the largest portion of the total amount of unaltered nitrazepam present in the body is localized in the extravasal compartment.

Concerning the passage of the compound into the cerebrospinal fluid, it can be said at present only that the concentrations attained in the fluid are lower than those existing simultaneously in plasma.

Within the range of the plasma levels stabilizing into a steady state during a prolonged daily oral administration of therapeutic doses, roughly 86 to 87% of the unchanged nitrazepam present in the plasma was found to be bound to the total proteins of a pooled sample obtained from 10 healthy adults.

The placental barrier can be passed in either direction by unchanged nitrazepam and/or its metabolites, so that the same concentrations as those present in the plasma of a mother receiving therapeutic oral doses of the drug are reached fairly rapidly in the plasma of the fetus.

Unchanged nitrazepam passes into the milk of mothers ingesting daily doses of 5 mg of nitrazepam to the extent that 100 ml of the milk contains only about 5 to 10 μg-equivalents of the drug unaltered and metabolized.

The plasma levels of unchanged nitrazepam measured in six subjects at stated times after administration of a single oral or intravenous dose of

10 mg of the drug are indicated in Table IV. Figure 4 shows the corresponding average curves, together with the standard deviations.

The plasma levels of unchanged nitrazepam and of the sum of its two main metabolites, II and III, resulting from prolonged treatment with daily oral doses of 5 mg of the drug, were determined in four subjects. The curves of minimum concentrations present immediately before a succeeding dose are shown in Figs. 5 through 7. The average plateau levels of these values were 38.6 ng/ml for the unchanged drug and 30.7 ng/ml for the sum of metabolites II and III.

The elimination rate of unaltered nitrazepam in man was not changed significantly by its own effects during a daily oral treatment with 5 mg of the drug for 24 days.

The proportions of unchanged nitrazepam and its metabolites in the plasma were examined in two subjects. The respective curves are shown in Fig. 8.

The cumulative excretion of nitrazepam plus its metabolites in urine was followed in eight adults, that in the feces in four persons. Under the most favorable conditions of measurement that could be applied, the total renal excretion determined during the first 120 hr was 65 to 75% after an oral administration of 10 mg of nitrazepam, and 93% after intravenous injection of the same dose. Only a small fraction of the dose, varying considerably with the circumstances, was excreted in the feces.

The proportions of the various excretion products of nitrazepam appearing in the urine were determined in two persons, restricted to the unconjugated compounds, and in one person, including the unconjugated and the conjugated metabolites. The relevant information is given in Fig. 14 and in Table VI.

REFERENCES

Beyer, K. H., and Sadée, W. (1969): Spektrophotometrische Bestimmung von 5-Phenyl-1.4-benzodiazepin-Derivaten und Untersuchungen über den Metabolismus des Nitrazepams. *Arzneimittel Forschung*, 19:1929.

Dost, F. H., and Gladtke, E. (1963): Ein Verfahren zur Ermittlung der Bilanz einverleibter Stoffe ohne Kenntnis der Ausscheidung. *Zeitschrift für klinische Chemie*, 1:14–18.

Koechlin, B. A., Schwartz, M. A., Krol, G., and Oberhänsli, W. (1965): The metabolic fate of ^{14}C-labelled chlordiazepoxide in man, in the dog and in the rat. *Journal of Pharmacology and Experimental Therapeutics*, 148:399.

Rieder, J. (1965): Methoden zur Bestimmung von 1,3-Dihydro-7-nitro-5-phenyl-2H-1,4-benzodiazepin-2-on und seinen Hauptmetaboliten in biologischen Proben und Ergebnisse von Versuchen über die Pharmakokinetik und den Metabolismus dieser Substanz bei Mensch und Ratte, *Arzneimittel Forschung*, 15:1134–1148.

Rieder, J. (1972*a*): A fluorimetric method for determining nitrazepam and the sum of its main metabolites in plasma and urine. *Arzneimittel Forschung.*

Rieder, J. (1972*b*): Plasma levels and derived pharmacokinetic characteristics of unchanged nitrazepam in man. *Arzneimittel Forschung.*

Rieder, J., and Wendt, G. (1973): Metabolism of nitrazepam in man and in the rat. (*in preparation*).

The Benzodiazepines, Raven Press, New York © 1973

KINETICS OF THE DISTRIBUTION OF ^{14}C-DIAZEPAM AND ITS METABOLITES IN VARIOUS AREAS OF CAT BRAIN

P. L. Morselli, G. B. Cassano*, G. F. Placidi*, G. B. Muscettola, and M. Rizzo

*Istituto di Ricerche Farmacologiche "Mario Negri," Milan, and *Clinical Psichiatrica, Università di Pisa, Pisa, Italy*

I. INTRODUCTION

It has been previously suggested (Placidi and Cassano, 1968; Cassano and Placidi, 1969; Cassano, Gliozzi, and Ghetti, 1969) that changes in the motor-behavior pattern may be more closely related to the kinetics of distribution of a given drug in discrete areas of the central nervous system (CNS) than to the drug concentration in the whole brain.

The distribution of diazepam (DZ) and its metabolites in the CNS has been extensively studied in mice, rats, guinea pigs, and dogs, both with autoradiographic and analytical techniques.

As other benzodiazepines do in mice (Placidi and Cassano, 1968), ^{14}C-DZ accumulates very rapidly in the brain where it localizes first in the grey matter structures, whereas later the radioactivity is mainly concentrated in the white matter (van der Kleijn, 1969).

With regard to the metabolism of the drug, it is known that the rate of formation of the various metabolites and their brain uptake may differ consistently depending on the various animal species involved (Marcucci, Fanelli, Mussini, and Garattini, 1970a, 1970b; Marcucci, Mussini, Fanelli, and Garattini, 1970c; Marcucci, Guaitani, Fanelli, Mussini, and Garattini, 1971; Schwartz, Koechlin, Postma, Palmer, and Krol, 1965). However, no data are at present available on the metabolism and distribution of DZ in the cat, a species frequently used for neurophysiological studies. This chapter will summarize data obtained using both autoradiographic and chemical analysis on the distribution of ^{14}C-DZ and its metabolites in brain and other tissues of the cat.

129

II. MATERIAL AND METHODS

DZ labeled with ^{14}C in position 2 (33 μC/mg) was kindly obtained by courtesy of Hoffmann-La Roche, Basel, Switzerland. Its chemical purity was checked by means of thin-layer chromatography ($CHCl_3$-acetone: 9:1; v/v), followed by scanning on a Packard Radiochromatoscanner-7201, and found to be 98%.

Two series of experiments were run on a total of 20 cats. In the first experiment, four adult cats received 1 mg/kg (33 μg/kg) of ^{14}C-DZ, by i.v. injection, and were then killed by exsanguination, under light ether anesthesia 7, 60, 180, and 360 min after drug administration. The brains were quickly removed, frozen in dry ice, and used for whole-brain autoradiography. In the second experiment, 16 cats were injected intravenously with 1 mg/kg (33 μC/kg) of ^{14}C-DZ and sacrificed 1, 5, 30, 60, and 240 min after drug administration. In this case, half of the brain was used for autoradiography and, from the remaining part, different areas such as white and grey matter, olfactory bulbs, hippocampus, thalamus, hypothalamus, mesencephalon, pons, bulb, and cerebellum were removed and used for chemical analysis. Plasma, myocardium, skeletal muscle (anterior legs), and adipose tissue were also collected for analysis.

A. Autoradiography

The brains used for the autoradiography were frozen on dry ice, embedded in carboxymethylcellulose, and fixed on the object board of a Leitz Sleigh Microtome. Serial coronal sections, 40 μ thick, were taken at different levels. This procedure was performed in a refrigerated room at about $-15°C$ in order to prevent the leaking of labeling. The dried sections were pressed against X-ray films (Gevaert), and exposed for about 3 months, according to Ullberg (1954).

B. Chemical analysis

The various brain areas and tissues were homogenized with a Potter homogenator in 3 ml of EtOH 95%. The tubes and the homogenizer were rinsed with another 2 plus 2 ml of EtOH 95%. The homogenates and the combined rinsing fluids were pooled and centrifuged for 10 min at 5°C and 2,000 rpm in a refrigerated Mistral-2L centrifuge.

The supernatants were then transferred to glass-stoppered test tubes and 0.5 ml of the supernatant was assayed for total radioactivity; 5.5 ml of the remaining portions of the tissue homogenate supernatants were then brought to dryness at 60°C under a gentle stream of nitrogen. The dry residues were then redissolved in 100 μl of EtOH 95%, and 1 μl of the ethanol solution was again assayed for total radioactivity. The rest was applied to thin-layer chromatography plates (20 × 5 cm precoated with silica gel F 254-Merck A.G.), together with cold carriers of DZ, N-demethyldiazepam, N-methyloxazepam, and oxazepam.

Occasionally 1.5 μl of the ethanol solution was also analyzed by gas-liquid chromatography (GLC), according to the technique described by Garattini, Marcucci, and Mussini (1969) in order to compare the results obtained by radiochemical analysis with those by gas chromatography. The thin-layer chromatography plates were then developed in a CHCl$_3$-acetone (8:2; v/v) system. The benzodiazepines were visualized by U.V. light and the radioactive zones checked by means of a Packard Radiochromatoscanner-7201. The radioactive and the nonradioactive zones were then scraped and eluted with 2 ml of 95% EtOH. One ml of this solution was then transferred into counting vials and assayed for radioactivity.

C. Assay of radioactivity

Assay of radioactivity was performed with a Beckman LS-250 liquid scintillation counter. The scintillation solution consisted of 2-′(4-t-butyl-phenyl) 5-(4-biphenyl)-1,3,4-oxidiazole (butyl-PBD) (7 g/liter) in toluene-absolute EtOH (14:1; v/v). The efficiency of the counting system was 92%. The efficiency of the total procedure of extraction, spotting, developing, scraping, and counting was 75 ± 1%.

III. RESULTS

A. Autoradiography

The autoradiograms obtained from the brains of cats killed 1 min after i.v. injection of ^{14}C-DZ showed that the radioactivity was mainly confined to the grey matter; the highest uptake was found in such sites as thalamus, colliculi, geniculate bodies, midbrain nuclei, cerebral and cerebellar cortex. At this time, the radioactivity in the white matter was low (Fig. 1*a*).

Five to 7 min after i.v. injection, the highest concentration of radioactivity was still in the cell-rich areas such as thalamus, mesencephalon and cortex. The leptomeninges and the walls of small blood vessels, which penetrate from the meninges into the subpial tissues, were highly radioactive.

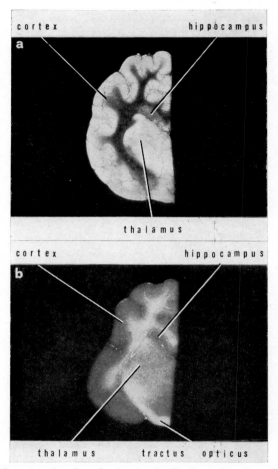

FIG. 1. Autoradiograms showing the brain distribution of radioactivity (*light areas*) 1 min (*a*) and 1 hr (*b*) after i.v. injection of 1 mg/kg (33 μC/kg) of ^{14}C-DZ in the cat. Corenal sections. Note the low accumulation of the radioactivity in the hippocampus, when compared with some other grey structures, such as cortex and thalamus.

As time progresses, the concentration of the labeled material tends to become more uniform throughout the whole brain, and the autoradiograms obtained 30 min after drug administration did not show marked differences between grey and white matter.

One hr after the i.v. injection of ^{14}C-DZ, the radioactivity was much higher in the white than in the grey matter (Fig. 1*b*, 2*b*, *d*, 3). A fairly high amount of radioactivity was also present in the leptomeninges and the

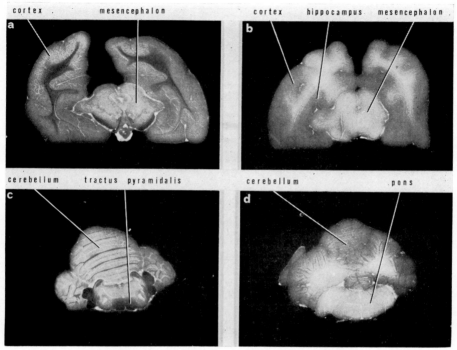

FIG. 2. Autoradiograms showing the brain distribution of radioactivity (*light areas*) 7 min (*a,c*) and 1 hr (*b,d*) after i.v. injection of 1 mg/kg (33 μC/kg) of ^{14}C-DZ in the cat. Coronal sections.

subpial blood vessels up to one hour after the injection.

Four hr after drug administration the difference between grey and white matter was still evident.

B. Radiochemical analysis

1. Brain areas

The radiochemical analysis performed on the various brain areas, at different times after ^{14}C-DZ intravenous injection, confirmed the distribution pattern observed autoradiographically.

The radioactivity present 1 min after drug administration was practically all due to unmetabolized DZ (Figs. 4 and 5).

The initial distribution pattern appears to depend mainly upon vascular-

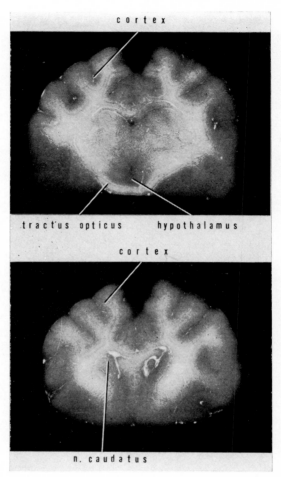

FIG. 3. Autoradiograms showing the brain distribution of radioactivity (*light areas*) 1 hr after i.v. injection of 1 mg/kg (33 μC/kg) of ^{14}C-DZ in the cat. Coronal sections at two different levels.

ity and blood flow, since the highest DZ concentrations were found in the cell-rich areas of the grey matter, such as thalamus, hypophysis, mesencephalon, and cortical grey matter (Table I). The relatively low drug level in the hippocampus should be noted. DZ levels in the subcortical white matter were, at this time (1 min), about 56 to 30% of those found in the grey matter.

A significant positive correlation was found to be present between DZ concentrations in the various areas 1 min after drug administration and the

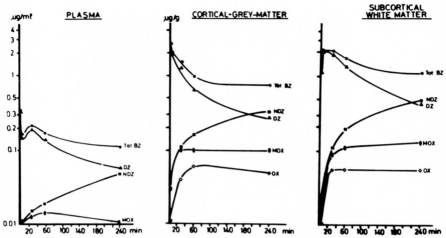

FIG. 4. Kinetics of distribution and metabolism of DZ in plasma and brain of the cat. (Tot BZ) total benzodiazepine; (diazepam) DZ; (NDZ) N-demethyldiazepam; (MOX) N-methyloxazepam, and (OX) oxazepam concentrations are expressed as μg/ml (plasma) or μg/g (tissue) on a log scale.

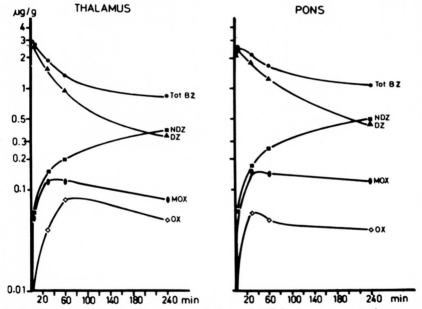

FIG. 5. Kinetics of distribution and metabolism of DZ in two representative areas of cat brain (thalamus and pons). (Tot BZ) total benzodiazepine; (diazepam) DZ; (NDZ) N-demethyldiazepam; (MOX) N-methyloxazepam, and (OX) oxazepam concentrations are expressed as μg/g tissue on a log scale.

TABLE I. *Diazepam concentrations in various brain areas of the cat 1 min after i.v. administration of ^{14}C-diazepam*

Brain area	Diazepam ($\mu g/g \pm$ S.E.)
Frontal grey matter	2.68 ± 0.33
Occipital grey matter	2.60 ± 0.69
Temporal grey matter	2.51 ± 0.25
Parietal grey matter	2.50 ± 0.47
Olfactory bulb	2.46 ± 0.37
Cerebellar grey matter	2.41 ± 0.17
Thalamus	2.95 ± 0.36
Hippocampus	1.88 ± 0.21
Hypothalamus	2.08 ± 0.24
Mesencephalon	2.65 ± 0.19
Pons	2.24 ± 0.17
Bulb	2.13 ± 0.21
Frontal white matter	1.04 ± 0.14
Temporal white matter	1.48 ± 0.44
Occipital white matter	1.07 ± 0.08
Cerebellar white matter	1.62 ± 0.24

Each figure is the average of 4 determinations.

relative cerebral blood flow as illustrated in Fig. 6.

DZ concentrations, 5 and 30 min after drug administration, were more or less uniformly distributed between grey and white structures (Table II),

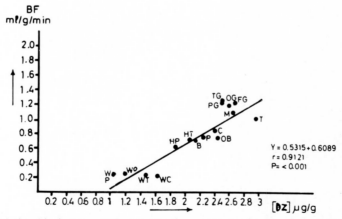

FIG. 6. Relationship between DZ concentrations $\mu g/g$ 1 min after i.v. administration of 1 mg/kg of ^{14}C-DZ and the blood flow (ml/g/min) in the various areas of cat brain (Kety, 1965; Reivich, Jehle, Sokoloff, and Kety, 1969). FG = frontal grey; OG = occipital grey; TG = temporal grey; PG = parietal grey; M = mesencephalon; T = thalamus; C = cerebellum grey; OB = optic bulb; P = pons; B = bulb; HT = hypothalamus; HP = hippocampus; WC = cerebellar white; WT = temporal white; WO = occipital white; WP = parietal white.

TABLE II. Diazepam levels in various brain areas and tissues of the cat after i.v. administration of ^{14}C-diazepam (1 mg/kg)

Area	Diazepam levels (μg/g ± S.E.)				
	(1 min)	(5 min)	(30 min)	(60 min)	(240 min)
Cortical grey matter	2.60 ± 0.16	1.92 ± 0.26	1.25 ± 0.11	0.64 ± 0.08	0.28 ± 0.05
White matter	1.11 ± 0.11	2.17 ± 0.32	1.93 ± 0.04	1.34 ± 0.18	0.45 ± 0.03
Thalamus	2.95 ± 0.36	2.66 ± 0.49	1.57 ± 0.09	0.96 ± 0.07	0.34 ± 0.04
Hypothalamus	2.08 ± 0.24	1.90 ± 0.40	1.09 ± 0.12	0.73 ± 0.03	0.33 ± 0.05
Hippocampus	1.88 ± 0.21	2.18 ± 0.40	1.55 ± 0.07	0.96 ± 0.08	0.32 ± 0.02
Mesencephalon	2.65 ± 0.19	2.71 ± 0.56	1.68 ± 0.06	0.98 ± 0.05	0.37 ± 0.02
Pons	2.24 ± 0.17	2.42 ± 0.41	1.80 ± 0.01	1.24 ± 0.13	0.41 ± 0.02
Bulb	2.13 ± 0.21	2.70 ± 0.30	1.69 ± 0.07	1.19 ± 0.15	0.41 ± 0.02
Cerebellum	2.41 ± 0.17	2.15 ± 0.42	1.21 ± 0.09	0.74 ± 0.05	0.29 ± 0.01
Adipose tissue	0.42 ± 0.10	2.24 ± 0.72	2.47 ± 0.92	2.88 ± 0.67	3.34 ± 0.20
Myocardium	3.08 ± 0.31	1.81 ± 0.20	1.54 ± 0.34	1.32 ± 0.14	0.38 ± 0.05
Skeletal muscle	0.61 ± 0.11	0.82 ± 0.14	1.01 ± 0.03	0.53 ± 0.08	0.34 ± 0.01
Plasma	0.34 ± 0.09	0.15 ± 0.01	0.18 ± 0.03	0.13 ± 0.01	0.05 ± 0.01

Each figure is the average of three determinations.

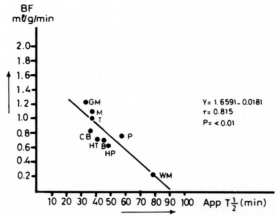

FIG. 7. Relationship between DZ apparent half-life (T½ min) and blood flow (ml/g/min) in the various areas of cat brain (Kety, 1965; Reivich, Jehle, Sokoloff, and Kety, 1969). GM = cortical grey matter; M = mesencephalon; T = thalamus; P = pons; CB = cerebellum; HT = hypothalamus; B = bulb; HP = hippocampus; WM = subcortical white matter.

whereas at later times (60 and 240 min), a clear shift in favor of the white-matter-rich areas was evident, with drug levels in the subcortical white matter about twice those in cortical grey matter.

It may be of interest that the DZ disappearance rate in the various areas reflected the relative blood flow (Fig. 7). Marked differences were, in

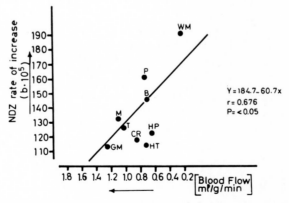

FIG. 8. Relationship between N-demethyldiazepam (NDZ) rate of increase (b × 10⁵) in the various brain areas of the cat and regional blood flow (ml/g/min) (Kety, 1965; Reivich, Jehle, Sokoloff, and Kety, 1969). GM = cortical grey matter; CR = cerebellum; T = thalamus; M = mesencephalon; HT = hypothalamus; HP = hippocampus; B = bulb; P = pons; WM = subcortical white matter.

fact, observed between cortical grey and subcortical white matter, with an apparent drug half-life of 34 and 78 min respectively (Fig. 7).

As far as the metabolites are concerned, N-demethyldiazepam, present only in traces at 1 and 5 min after DZ administration, was found in considerable amounts after longer time intervals, with levels in the white-matter-rich areas constantly exceeding those in the grey ones (Figs. 4 and 5).

It should be noted that N-demethyldiazepam was still accumulating 4 hr after drug administration, and also that the rate of increase of N-demethyldiazepam concentrations was different for the various brain areas and appeared to be inversely related to the relative blood flow (Fig. 8). This may suggest that the compound has higher affinity for or binds more extensively with the lipoid-rich white matter areas than it does with the cell-rich grey areas.

N-methyloxazepam and oxazepam were found to be present in lower concentrations. Their levels were virtually constant in the various areas from 30 to 240 min and no remarkable differences were evident between grey and white matter concentrations of these two metabolites (Figs. 4 and 5).

Preliminary observations on the concentration of DZ and its metabolites in various areas of cat brain, after chronic administration of DZ, tend to confirm these findings. As shown in Fig. 9, N-demethyldiazepam appears to

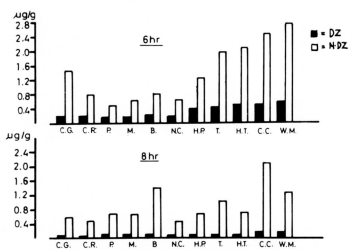

FIG. 9. DZ and N-demethyldiazepam (NDZ) concentration in various brain areas 6 and 8 hr after the last administration of a repeated treatment with DZ (0.5 mg/kg b.i.d. × 4 days). CG = cortical grey; CR = cerebellum; P = pons; M = mesencephalon; B = bulb; NC = nucleus caudatus; HP = hippocampus; T = thalamus; HT = hypothalamus; CC = corpus callosum; WH = subcortical white matter.

be the main metabolite present in the cat brain. As in acute administration, the white matter structures showed the highest levels of N-demethyldiazepam, with maximal concentration in the corpus callosum. N-methyloxazepam and oxazepam were present only in very small concentrations and they were, again, quite uniformly distributed in the various brain areas.

2. *Peripheral tissues*

As previously observed in other species, such as rats and mice, as well as in humans (Marcucci, Fanelli, Frova, and Morselli, 1968), DZ and N-demethyldiazepam showed a strong tendency to accumulate in adipose tissue (Table II; Fig. 10). After 240 min, DZ concentrations were, in fact, about 5 to 6 times higher than those in the CNS, while N-demethyldiazepam levels were of the same magnitude as those observed in the CNS.

N-methyloxazepam also tended to accumulate in adipose tissue, while oxazepam did not.

A different pattern of uptake was also evident for the myocardium and skeletal muscles, as illustrated in Fig. 10 and Table II.

Peak levels for DZ were, in fact, attained in the myocardium 1 min after drug administration, while in the skeletal muscles, maximum concentrations were reached only after 30 min.

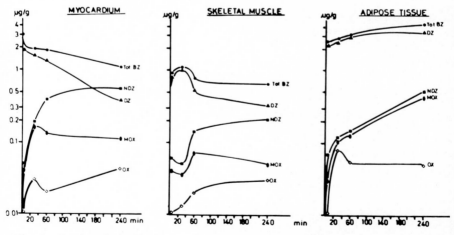

FIG. 10. Kinetics of distribution and metabolism of DZ in peripheral tissues of the cat. (Tot BZ) total benzodiazepine; (diazepam) DZ; (NDZ) N-demethyldiazepam; (MOX) N-methyloxazepam; and (OX) oxazepam concentrations expressed as $\mu g/g$ of tissue on a log scale.

The rate of accumulation of N-demethyldiazepam was also significantly different in the two muscle tissues, where concentrations of the compound, at 240 min after DZ administration, were twofold higher in the myocardium than in the skeletal muscles.

IV. DISCUSSION

In cats, DZ undergoes metabolic degradation with the formation of N-demethyldiazepam, N-methyloxazepam, and oxazepam.

The data presented suggest that DZ penetration into cat brain is not uniform and that the lipoid-rich white matter is more slowly penetrated than the cell-rich grey matter.

Equilibrium between grey and white structures is achieved within 30 min but after that time higher levels of DZ were found in the white matter.

The early distribution pattern of DZ in various brain areas and the more rapid fall of its concentrations in the grey structures are indicative of the importance of vascularity. However, blood flow does not seem to be the only factor involved. A different affinity for the various tissue consitituents of the different parts of the brain may exist, possibly as indicated by the different rates of accumulation of the metabolites.

DZ did not seem to follow the same pattern of distribution of compounds, such as chlorpromazine, imipramine, amitriptyline, and the butyrophenones which show a high affinity for the hippocampus and the caudate nucleus (Cassano, Sjöstrand, and Hansson, 1965a, b; Cassano and Hansson, 1966; Janssen and Allewijn, 1969).

Its distribution pattern is instead more similar to that of the barbiturates and diphenylhydantoin (Roth and Barlow, 1965; Firemark, Barlow, and Roth, 1963) especially considering the late accumulation in the white structures.

Among the metabolites, N-demethyldiazepam seemed to play a major role after both an acute or a chronic DZ administration. The compound is, in fact, retained for an extended period, with levels in the white matter always exceeding those in the grey.

At variance with what has been observed in other animal species, N-methyloxazepam and oxazepam represent only a minor fraction of the total amount of benzodiazepines present in the cat brain.

These data are in good agreement with some recent findings of Mussini and Marcucci (1971), who observed *in vitro* that liver microsomes metabolize DZ more according to the pathway of N-demethylation than C_3-hydroxylation.

It is tempting to speculate that the rapid sedative and anticonvulsant effect of DZ may depend on its widespread depressant effect, presumably correlated with an action of the drug on the nerve cell bodies, as indicated by the early and high accumulation of DZ in grey matter.

The persistence of DZ and its metabolites in white matter, observed at longer periods after administration, may be useful in indicating the possible action of the drug on the nerve fibers, such as that recently observed by Barnes and Moolenaer (1971) on the optic tract. DZ and N-demethyldiazepam could, in fact, interfere with membrane permeability affecting the nerve action potentials.

The lower concentration of benzodiazepines in unmyelinated brains of fetuses in respect to those of the mothers (Cassano and Placidi, 1971; van der Kleijn, 1969) indicates the importance of white matter in the accumulation and retention of such drugs in the brain.

Finally, the data reported here stress once again the necessity of evaluating the penetration and localization of drugs in relation to specific brain structures rather than to the brain as a whole.

ACKNOWLEDGMENTS

This work was supported by a grant from the National Institutes of Health (No. P01 GM 18376-01 PTR) for the work carried out in Milan, and by contracts from the Italian National Research Council (CNR) (Nos. 69.1661.115.1052 and 70.01821.04.115.1052) for the work carried out in Pisa.

We wish to thank Mrs. S. M. Standen for the statistical evaluations, and Mr. C. Pantarotto, Miss E. Riva, and Mr. M. Brogiotti for their technical assistance.

REFERENCES

Barnes, C. D., and Moolenaar, G. M. (1971): Effects of diazepam and picrotoxin on the visual system. *Neuropharmacology*, 10:193–201.

Cassano, G. B., Gliozzi, E., and Ghetti, B. (1969): The relationship between dynamic features of the brain distribution of C^{14}-chlordiazepoxide and motor behaviour changes in mice. In: *The Present Status of Psychotropic Drugs: Pharmacological and Clinical Aspects*, Proc. 6th C.I.N.P. pp. 342–346. Congress, Tarragona, edited by A. Cerletti and F. J. Bové. Excerpta Medica, Amsterdam.

Cassano, G. B., and Hansson, E. (1966): Autoradiographic distribution studies in mice with ^{14}C-imipramine. *International Journal of Neuropsychiatry*, 2:269–278.

Cassano, G. B., and Placidi, G. F. (1969): Penetration and distribution of neuropharmacological agents in the brain. *Pharmakopsychiatrie Neuropsychopharmakologie*, 2:160–175.

Cassano, G. B., and Placidi, G. F. (1971): Unpublished results.
Cassano, G. B., Sjöstrand, S. E., and Hansson, E. (1965a): Distribution of ¹⁴C-labelled
amitriptyline in the cat brain. *Psychopharmacologia*, 8:12–22.
Cassano, G. B., Sjöstrand, S. E., and Hansson, E. (1965b): Distribution of ³⁵S-chlor-
promazine in cat brain. *Archives Internationales de Pharmacodynamie et de Thérapie*,
156:48–58.
Firemark, H., Barlow, C. F., and Roth, L. J. (1963): The entry, accumulation and
binding of diphenylhydantoin-2-C¹⁴ in brain. Studies on adult, immature and hyper-
capnic cats. *International Journal of Neuropharmacology*, 2:25–38.
Garattini, S., Marcucci, F., and Mussini, E. (1969): Gas–chromatographic analysis of
benzodiazepines. In: *Gas-Chromatography in Biology and Medicine,* edited by R.
Porter, pp. 161–172. Churchill, London.
Janssen, P. A. J., and Allewijn, F. T. N. (1969): The distribution of the butyrophenones,
haloperidol, trifluperidol, moperone and clofluperol in rats and its relationship with
their neuroleptic activity. *Arzneimittel Forschung*, 19:199–208.
Kety, S. S. (1965): Measurement of regional circulation in the brain and other organs
by isotopic techniques. In: *Isotopes in Experimental Pharmacology,* edited by L. J.
Roth, pp. 211–218. University of Chicago Press, Chicago.
Marcucci, F., Fanelli, R., Frova, M., and Morselli, P. L. (1968): Levels of diazepam
in adipose tissue of rats, mice and man. *European Journal of Pharmacology,*
4:464–466.
Marcucci, F., Fanelli, R., Mussini, E., and Garattini, S. (1970a): Further studies on
species differences in diazepam metabolism. *European Journal of Pharmacology,*
9:253–256.
Marcucci, F., Fanelli, R., Mussini, E., and Garattini, S. (1970b): Further studies on
the long-lasting antimetrazol activity of diazepam in mice. *European Journal of Pharmacology,* 11:115–116.
Marcucci, F., Guaitani, A., Fanelli, R., Mussini, E., and Garattini, S. (1971): Metabolism and anticonvulsant activity of diazepam in guinea pigs. *Biochemical Pharmacology,* 20:1711–1713.
Marcucci, F., and Mussini, E. (1968): A metabolic explanation for differences between
species of the anticonvulsant activity of diazepam. *British Journal of Pharmacology,*
34:667P.
Marcucci, F., Mussini, E., Fanelli, R., and Garattini, S. (1970c): Species difference
in diazepam metabolism. I. Metabolism of diazepam metabolites. *Biochemical Pharmacology,* 19:1847–1851.
Mussini, E., and Marcucci, F. (1971): Unpublished results.
Placidi, G. F., and Cassano, G. B. (1968): Distribution and metabolism of ¹⁴C-labelled-
chlordiazepoxide in mice. *International Journal of Neuropharmacology*, 7:383–389.
Reivich, M., Jehle, J., Sokoloff, L., and Kety, S. S. (1969): Measurement of regional
cerebral blood flow with antipyrine-¹⁴C in awake cats. *Journal of Applied Physiology,*
27:296–300.
Roth, L. J., and Barlow, C. F., (1965): Autoradiography of drugs in the brain. In:
Isotopes in Experimental Pharmacology, edited by L. J. Roth, pp. 49–62. University of Chicago Press, Chicago.
Schwartz, M. A., Koechlin, B. A., Postma, E., Palmer, S., and Krol, G. (1965): Metabolism of diazepam in rat, dog, and man. *Journal of Pharmacology and Experimental
Therapeutics,* 149:423–435.
Ullberg, S. (1954): Studies on the distribution and fate of ³⁵S-labelled benzyl-penicillin
in the body. *Acta Radiologica, Suppl.* 118.
van der Kleijn, E. (1969): Kinetics of distribution and metabolism of diazepam and
chlordiazepoxide in mice. *Archives Internationales des Pharmacodynamie et Thérapie,*
178:193–215.

KINETICS OF GENERAL AND REGIONAL BRAIN DISTRIBUTION OF CLOSELY RELATED 7-CHLORO-1,4-BENZODIAZEPINES

Eppo van der Kleijn, Pieter J. M. Guelen, Trudy C. M. Beelen, Nico V. M. Rijntjes, and Ton L. B. Zuidgeest

Sint Radboud Hospital Pharmacy, Department of Clinical Pharmacy, University of Nijmegen, The Netherlands

I. INTRODUCTION

The group of 7-chloro- and 7-nitro-benzodiazepines is known to display many shades of neuropharmacological activity, depending on the dose, method of administration, and time course of dosing. Physicochemical properties can differ greatly among the various compounds. Lipophylicity, for example, has been shown to be related to observed rates of distribution and of biotransformation (Cassano et al., 1967; Lien and Hansch, 1968; van der Kleijn, 1969*b*).

There are many clinical data available which indicate the usefulness of individual compounds as central muscle relaxants, ataractics, hypnotics, or as anticonvulsants, depending upon conditions. With few exceptions, the metabolism of this group of compounds is a stepwise degradation via pharmacologically active substances.

Neurophysiological and pharmacological data differentiate the benzodiazepines from barbiturates and neuroleptics. The limbic system has been recognized as the substrate for the antianxiety activity (Schallek et al., 1964; Randall et al., 1970), whereas the brainstem is involved in central depression (Przybyla and Wang, 1968).

In most instances, laboratory animals show higher rates of distribution and elimination than humans. It is therefore very difficult in pharmacokinetic studies to compare the physiological conditions in animals with those in humans after a single dose. It has been the purpose of these studies to investigate the distribution of the benzodiazepines in animals, by mimicking the conditions in which they are used most frequently in humans, and to gain information on the rates of distribution and metabolism during chronic treatment.

In this investigation, we were concerned with the relation between the drug concentration in plasma and the regional concentration in the target organ, and with the possible relevance of this concentration with regard to

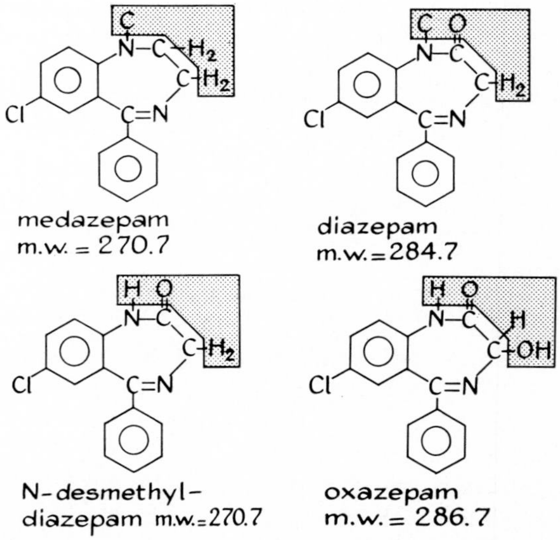

FIG. 1. Structural formulae of the 1,4-benzodiazepines studied for their kinetics of distribution and metabolism. The compounds are closely related since they are either precursor or metabolite of one another. Physicochemical characteristics are controlled mainly by the side groups in the diazepine ring.

the pharmacological effect. All the compounds studied were closely related structurally since they were either precursor or product of one another.

II. MATERIALS AND METHODS

Medazepam-5-[14]C (specific activity: 23 μC/mg), desmethyldiazepam-2-[14]C (specific activity: 25 μC/mg), and oxydesmethyldiazepam-2-[14]C (specific activity: 20 μC/mg) were used in animal and *in vitro* experiments. Diazepam (VALIUM) was used in the form of commercial tablets of 10 mg in human studies. Male Swiss mice, weighing approximately 20 g, and pregnant female mice after 18 days of gestation, weighing approximately 40 g, were used for the bolus injection studies. Squirrel monkeys of both sexes, weighing approximately 500 g, were used in the infusion studies.

A. Chemical analyses

Resolution of the parent compound from its metabolites, in the animal experiments, was carried out by thin-layer chromatography, using silica gel plates No. F 254 (Merck, Darmstadt, Germany) and chloroform-absolute ethanol (90:10) over 10 cm, followed by cyclohexane-diethylamide-benzene (80:15:5) over 20 cm, as the mobile phases. Spots and bands were qualitatively detected by quenching of U.V. light and quantitatively assayed by scanning with a Berthold-Desaga Scanner, Model 12-2 LB 2031, or by scraping off the silica gel from the plates, followed by liquid scintillation spectrometry, counting with Instagel (Packard) in a Packard Tri-carb Liquid Scintillation Spectrometer, Model 3380 AAA.

B. Studies in mice

Mice were injected intravenously in the tail vein, as indicated in Table I. The compounds studied were solubilized by dissolving them in acetone

TABLE I. *Doses of the compounds administered to mice for autoradiographic studies*

Compounds	^{14}C	Cold
Medazepam	200 μC/kg	8.43 mg/kg
Desmethyldiazepam	453 μC/kg	18.12 mg/kg
Oxazepam	400 μC/kg	20.0 mg/kg

containing polysorbate 80 (Tween 80, Atlas Chemical Company). After evaporation of the acetone, physiological salt solution was added to make up a solution to be administered on the basis of 10 ml/kg. The final concentration of polysorbate 80 was 2%.

The animals were sacrificed at various periods (see Figs. 2 through 4, 5, 7, and 9) by immersing them in isopenthane cooled in dry ice at about —80°. Whole-body autoradiography (Ullberg, 1954) and tissue-homogenate analyses were performed as described earlier (van der Kleijn, 1969b).

C. Studies in squirrel monkeys with desmethyldiazepam

Infusion studies were designed in order to simulate chronic treatment and to achieve steady-state conditions at plasma concentrations comparable with therapeutic conditions found in humans. Solutions of the drug solubilized in 2% polysorbate 80 in physiological salt solution were made as described under point B above. Infusion into the femoral vein was carried out by means of a Cenco pulsating pump with a rate of 0.9 ml/hr. The dose rates were 0.128, 0.25, and 0.27 mg desmethyldiazepam-2-^{14}C/hr, respectively, for each of three monkeys.

Infusion was continued, in two cases, for approximately 21 hr. In one case, infusion was continued for 53 hr. After that period, the animal was sacrificed. Whole-body autoradiography and tissue-homogenate analyses were performed as described previously (van der Kleijn, 1971).

D. Human study

A psychiatric patient (dB), under the guidance of Dr. G. T. J. A. van Lier of St. Servatius Hospital at Venray, The Netherlands, received three tablets of 10 mg diazepam, and plasma samples were taken over a period of 12 hr following the intake. Subsequently, treatment with three times 10 mg of diazepam daily was continued over 14 days. After the last morning dose on the 15th day, the treatment was stopped and plasma samples were taken during the subsequent 4 days.

The analyses of diazepam and desmethyldiazepam in plasma were performed using the method originally reported by Marcucci et al. (1968; van der Kleijn, 1969b; van der Kleijn et al., 1971a).

III. RESULTS

A. Autoradiography

1. Studies in mice

Medazepam is the most lipophylic compound in the series of benzodiazepines studied in this report. Its distribution in the gray matter structures of brain and in body fat, after intravenous injection, is extremely rapid. Nevertheless, the drug is also rapidly eliminated. Although the ^{14}C label is located in a position also present in the pharmacologically active metabolites (Rieder and Rentsch, 1968; Schwartz and Carbone, 1970), which is known to have a much lower elimination rate, the radioactivity after 30 min is present predominantly in excretory organs such as liver, intestine, and kidney. After this period, radioactivity is almost absent in the central nervous system as well as in the spinal cord.

The general and regional brain distribution of medazepam over short periods has very much in common with that of the other 7-chloro-benzodiazepines. Similarly, it shows a high affinity for the myocardium. The differences are reflected by the kinetic behavior: the rate of elimination is much faster than that of diazepam (van der Kleijn, 1969a), desmethyldiazepam, and oxydesmethyldiazepam. Differences in the accumulation in and sustained release from the myelinated structures in brain, spinal cord, and peripheral nerves are less obvious than for the other compounds (Figs. 2 and 3).

The autoradiography in mice of diazepam, as such, through the N-^{14}CH$_3$ compound and of diazepam, together with its accumulating metabolites, as the 2-^{14}C compound, has been reported previously (van der Kleijn, 1969a; van der Kleijn et al., 1971a). Although diazepam is rather rapidly eliminated, its metabolites desmethyldiazepam and, to a smaller extent, oxydesmethyl-diazepam (oxazepam) show a lower elimination rate.

The distribution of desmethyldiazepam-2-^{14}C in mice is in accordance with what was expected from previous studies with diazepam-2-^{14}C (van der Kleijn, 1969a). The compound shows a considerably slower elimination from the central nervous system than both medazepam and diazepam. The uptake in body fat is rather slow and reaches a considerably lower maximum level than it does after injection of the previous drugs (Figs. 4 and 5). The accumulation of radioactivity in the white matter structures, however, is more pronounced (Fig. 6). The general distribution and the metabolic aspects will be reported elsewhere.

N-desmethyl-3-hydroxydiazepam (oxazepam): Figure 7 shows the distribution pattern of oxazepam-2-^{14}C in pregnant mice. Noteworthy, in respect to its distribution in the brain, is the slow uptake into the gray matter structures, compared to the previously mentioned compounds. Radioactivity accumulates in the fiber tracts and other white matter structures in the central nervous system and peripheral nerves. The rate of uptake into the body fat is slow and rather limited when compared with medazepam. The rate and extent of placental transfer is small when compared to that of diazepam (van der Kleijn, 1969c).

A more detailed report on the kinetics of distribution and metabolism in mice and monkeys is in preparation.

2. N-desmethyldiazepam studies in squirrel monkeys

Desmethyldiazepam-2-^{14}C has been studied after linear intravenous infusion; it reaches maximum levels about four to six times the levels that

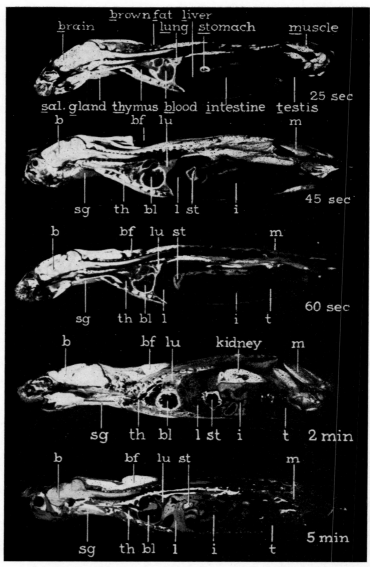

FIG. 2. Autoradiograms showing the distribution of medazepam-5-^{14}C in mice during the first 5 min following bolus intravenous administration. White areas correspond to the presence of radioactivity. Note the rapid uptake of the compound in the brain, the myocardium, and the somewhat slower but extensive uptake by body fat. High concentrations are found in the adrenal cortex.

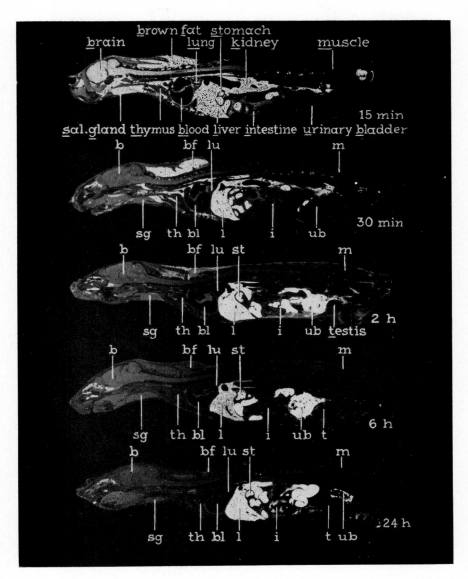

FIG. 3. Autoradiogram showing the distribution of medazepam-5-^{14}C in mice during the period of 15 min to 24 hr. Note the high extent of uptake of radioactivity in body fat, the rather rapid elimination from the brain, and the large size of the bile bladder at the 2- and 24-hr period. The radioactive material is being recycled in the GI tract for a long period.

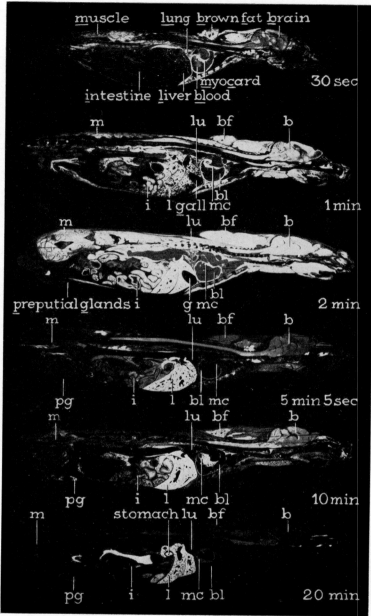

FIG. 4. Autoradiograms showing the distribution of desmethyldiazepam in mice during the period from 30 sec to 20 min. Notice that the uptake in the brain is lower than in the case of medazepam (Figs. 2 and 3). The uptake by body fat is also slower and the elimination is rather rapid. After 5 min, the distribution becomes even. Later an accumulation in the white matter structures can be observed. The affinity for the myocardium (14) noted with medazepam is also valid for desmethyldiazepam.

Radioactivity in the intestines is due to biliary excretion, primarily of the metabolites.

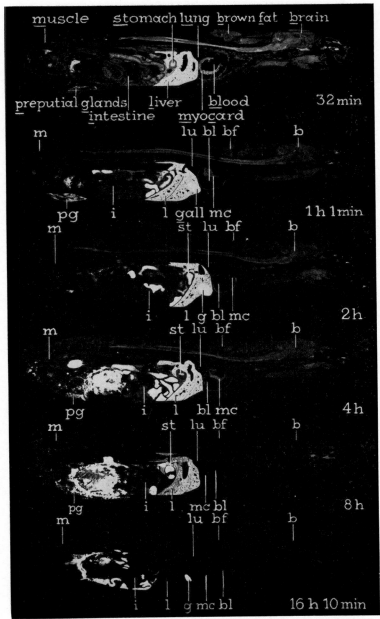

FIG. 5. Autoradiograms showing the distribution of desmethyldiazepam-2-^{14}C in mice from 32 min to more than 16 hr. Although intravenously injected, a small amount of the original drug is present in the bile and intestine. Radioactivity in the central nervous system is localized predominantly in the white matter: corpus callosum, medulla cerebelli, fiber connections in the brainstem, peripheral nerves, and the spinal cord. The high affinity for the myocardium can be observed over a long period. The blood concentration has fallen rapidly to relatively low levels. The high uptake by the preputial glands is unusual for the benzodiazepines and cannot be explained.

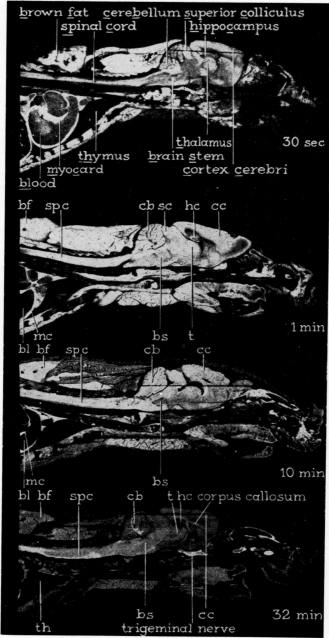

FIG. 6. Details of the distribution of desmethyldiazepam, illustrating the rapid uptake and change in regional distribution in the brain during the first 32 min. The rapid decrease in blood concentration can be observed within the myocardium. Further explanation is given in Fig. 5.

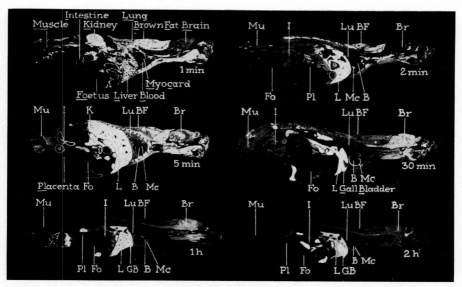

FIG. 7. Autoradiograms showing the distribution of oxydesmethyldiazepam-2-^{14}C (oxazepam) in pregnant mice in the period from 1 min to 2 hr. A much slower accumulation of radioactivity in the brain than with the previously mentioned compounds is observed. The uptake by body fat is limited and the drug is rapidly eliminated.

Again a gradual accumulation in the white matter structures can be observed. The drug is excreted by the gallbladder which appears to be larger than normal. No direct explanation for this enlargement is available. It has also been observed after administration of medazepam (Fig. 3). The rate and extent of placental transfer is relatively small. The highest concentrations in the fetus are found in the GI tract and myocardium.

would be achieved after a chronic treatment with desmethyldiazepam, e.g., with 10 mg, three times a day in humans. Figure 8 gives details of the distribution pattern in the brain at approximately 45% of the theoretical steady state.

The drug accumulates to a large extent in the white matter structures of the central as well as of the peripheral nervous system: the fiber tracts of the cerebellum to the spinal cord and to the brainstem and those in the thalamus. Radioactivity is present in the white matter of the cortex and also in the peripheral nerves, including, among others, the optic tract and the oculomotor nerve. The concentration of the drug in the gray matter structures of the brain is lower but is still higher than in the blood of, e.g., the vena cerebri magna. In the thalamus, the highest concentrations are found in the laminae. In the hippocampal region, the highest concentrations are in the stratum granulare of the dentate gyrus, the fimbria, and the stratum pyramidale.

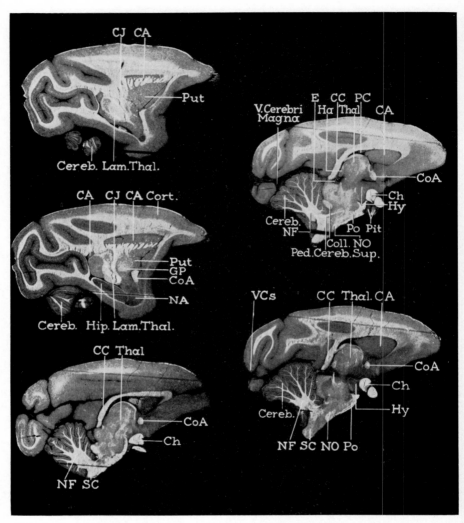

FIG. 8. Regional brain distribution of desmethyldiazepam-2-^{14}C in the squirrel mon-key, 21 hr after linear intravenous infusion at a rate of 0.250 mg/hr. The plasma con-centration, as can be seen in Fig. 9, lies well in the range that can also be achieved therapeutically (see vena cerebri magna). At this time, about 45% of the theoretical steady state has been achieved. Notice the high concentration of radioactivity in such white matter structures as the internal capsule, the laminae thalami, nuclei and medulla cerebelli, the fiber connections of the cerebellum with brainstem, thalamus and with the spinal cord. Relatively low concentrations are found in the gray matter caudate nucleus, hypothalamus, hippocampus, cortex cerebri, and the molecular layer of the cerebellum.

B. Biochemical analyses

1. Squirrel monkey

The heparanized samples were analyzed for: total radioactivity of the plasma; total radioactivity of the organic solvent residue after 2 $\times$ extraction

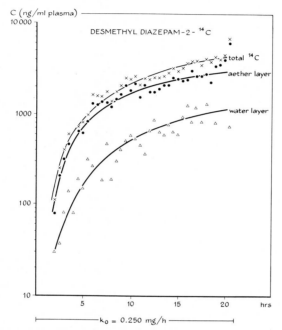

FIG. 9. Linear intravenous infusion into the femoral vein of the squirrel monkey for 21 hr. Log plasma concentration versus time for total ^{14}C, radioactivity in ether layer and in the aqueous layer remaining after extraction. The theoretical maximum value of the plasma concentration (C_{ss}: Table II) is found by the best fit of the curve of the residuals: log (C_{ss} — C_t) versus time. This allows the calculation of the plasma half-life ($t_{1/2}$) and, subsequently, the apparent volume of distribution (V_{ss}). This value is of importance for the design of optimal dosage regimens. Linear infusion in the squirrel monkey can mimic the repeated intake of the drug when the dosage interval is short compared to $t_{1/2}$.

Abbreviations: CI, Capsula Interna; CA, Nucleus Caudatus; Cereb., Cerebellum; Lam. Thal., Laminae Thalami; Put, Putamen; V., Vena; E, Epiphysis; Ha, Habenula; CC, Corpus Callosum; Thal., Thalamus; PC, Plexus Choroideus; CoA, Commis sura Anterior; Hy, Hypothalamus; Pit, Pituitary (anterior and posterior lobe); Po, Pons; Coll., Colliculus; NO, Nervus Opticus; Ped. Cereb. Sup., Pedunculus Cerebelli Superior; SC, Spinal Cord; NF, Nucleus Fastigii.

TABLE II. *Pharmacokinetic parameters calculated from the three experiments with linear intravenous infusion with desmethyl-diazepam*

Squirrel monkey	k_0	BW	Sex	C_{ss} ^{14}C	C^1_{ss} DD	C^2_{ss} M	$t_{1/2}$ ^{14}C	$t_{1/2}$ DD	$t_{1/2}$ M	k_{12}	V^1_{ss}	MC (1)	MC (2)	ΔBW DD	ΔMC (1)	ΔMC (2)
Exp. 1	0.128	0.50	♂	4.0	2.0	2.0	44	30	72	0.0231	2.76	0.064	0.064	6.144	0.142	0.142
Exp. 2	0.250	0.45	♂	8.5	4.5	4.0	45	30	55	0.0231	2.42	0.056	0.063	5.376	0.124	0.120
Exp. 3	0.270	0.60	♀	6.5	5.5	1.0	45	34	60	0.0203	2.42	0.049	0.027	4.033	0.082	0.045

k_0 = linear infusion rate (mg/hr); BW = body weight (kg); C_{ss} = concentration at the steady state (μgr/ml); DD = desmethyldiazepam; M = metabolites; $t_{1/2}$ = biological halflife (hrs); k_{12} = first order rate constant (hrs^{-1}); V_{ss} = apparent distribution volume at the steady state (1); 1 = compartment of the parent drug; 2 = compartment of the metabolites; MC = metabolic clearance (1/hr); ΔBW = relative apparent distribution volume (1/kg); ΔMC = relative metabolic clearance (1/hr/kg).

with ether; and radioactivity of the remaining aqueous phase. Analysis of the ether extract by TLC failed to resolve radioactive compounds other than the parent desmethyldiazepam.

Figure 9 shows the plasma concentration-time relationship in experiment 3 for total radioactivity, the aqueous phase, and the other extract. Plotting log $(C_{ss} - C_t)$ versus t permits the calculation of the biological halflife $(t_{1/2})$. Knowing the rate of infusion, it is possible to calculate a variety of pharmaco-kinetic parameters (van der Kleijn et al., 1971b). Kinetic parameters obtained from the three experiments are given in Table II.

2. Human study

Figure 10 gives an example of the plasma-concentration pattern after a single oral dose of 30 mg diazepam (A), and following a 2-week period of multiple dosing with three times 10 mg diazepam daily (B).

The large differences in rates of elimination, maximum plasma concentration between A and B, and the presence of a considerable concentration and slow elimination of desmethyldiazepam should be noted (van der Kleijn et al., 1971a).

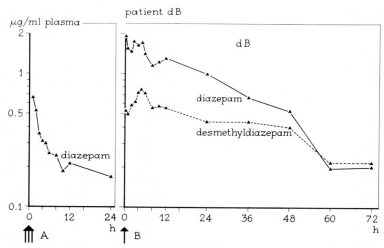

FIG. 10. Log of plasma concentration versus time for diazepam and desmethyldiaze-pam in patient dB after a single oral dose of three commercial tablets of 10 mg diaze-pam (VALIUM) (A), and following a subchronic treatment of 3 times 10 mg diaze-pam daily (B). Notice the rather rapid elimination of the drug after the single dose and the accumulation, decrease of the rate of elimination and the presence of desmethyl-diazepam as metabolite with pharmacological properties following the subchronic treat-ment.

IV. DISCUSSION

Biochemical pharmacology and pharmacokinetics are based principally on the existence of a definable relation between the concentration of the drug in the medium available for analysis and the concentration in the so-called biophase, i.e., the concentration that is responsible for the ultimate, measurable effect. This philosophy identifies the site where the mechanism of the effect is located with the site of accumulation of the drug.

Physicochemical properties such as the lipophylicity of drugs play an important role in the kinetics of distribution, metabolism, binding to macro-molecules, and partitioning with lipoid media, such as body fat. The pK_a values of the compounds are too low to cause differences in the degree of dissociation at the physiological pH. Lipophylicity, expressed as the partition coefficient between an immiscible lipoid layer and water (buffer), or as the R_f value in partition chromatography (van der Kleijn, 1969c), decreases in the following sequence: medazepam, diazepam, desmethyldiazepam, and oxazepam.

A. Penetration into the central nervous system

If one compares the various initial rates of penetration into the gray matter of the central nervous system and into body fat, one can notice the same sequence as from higher to lower lipophylic nature.

However, if one focuses on the accumulation of radioactivity in the white matter structures and peripheral nerves, then the inverse relation can be observed. It cannot be concluded whether the opposing rates of accumulation and elimination or the specific affinity for myelin plays the predominant role.

The similarity in the protein-binding capacity is surprising, being about 96% for all four compounds at therapeutic levels (Rieder and Rentsch, 1968; van der Kleijn, 1969c; van der Kleijn et al., 1972). This observation favors the explanation that the rapid elimination of medazepam prevents sustained accumulation in the white matter after a single dose.

B. Placental transfer

In accordance with what has been reported earlier, the rate and extent of placental transfer after a single dose in pregnant mice is rather small (van der Kleijn, 1969a). Nevertheless, the highest relative concentrations in the fetus are found in the myocardium and the contents of the intestines, although the concentrations remain much lower than in the corresponding maternal tissues. This conclusion, however, is not necessarily applicable to multiple dosing.

C. Biliary excretion

All four compounds show an extensive transfer of radioactivity to the gallbladder and subsequently to the small and large intestines. Biochemical analyses of the contents of the intestines of the squirrel monkeys reveal that the main fraction appears to be unconjugated oxazepam, although a small amount of desmethyldiazepam can be identified. It has not been investigated whether the free oxazepam originates from a conjugate that is hydrolyzed either in the intestine or during the procedure, or whether it represents a hepatic-enteric cycling for oxazepam as such. It is our impression that the sizes of the gallbladders of the mice which have been treated with medazepam and oxazepam for periods longer than 1 hr are extraordinarily large. This observation can be correlated with that of Randall et al. (1970), showing an influence on the gallbladder after high doses of medazepam.

D. Regional brain distribution

The aim of these and similar studies is to detect body regions that accumulate high concentrations of the compounds in an attempt to resolve their site of pharmacological activity.

The experiments after the bolus injection of medazepam, desmethyldiazepam, and oxazepam in mice show, initially, a high uptake in the gray matter structures that are known to be densely vascularized and rapidly perfused. The highest concentrations are found in the spinal cord, pons, colliculi, and cerebral and cerebellar cortex. At later periods, the white matter dominates. After 30 min, the highest concentrations of desmethyldiazepam and oxazepam are found in the corpus callosum, the medulla of the cerebellum, the white matter of the spinal cord, and the fiber tracts through the medulla oblongata and the brainstem. This distribution pattern is essentially similar to that of diazepam in mice (van der Kleijn, 1969a) and in rhesus monkey fetus (van der Kleijn and Wijffels, 1971), although rate differences among the compounds can be observed.

E. Relevance of plasma concentration for pharmacological effect

It is subject to speculation what the relevance of this distribution pattern can be for the different shades of pharmacological activity. The initial effect that can be observed after intravenous injection in mice is muscle relaxation and hypnosis. A rather rapid recovery within 10 to 20 min has been observed. As has been described earlier (van der Kleijn, 1971) distinction should be made between the short-lasting pharmacological properties and the longer-lasting but less defined pharmacological effects following chronic or subchronic treatment. It is for these latter and as yet unexplored effects that accumulation in the white matter structures may be of importance. No further progress in relation to the mechanisms has been made. An essentially similar distribution pattern has been observed after linear infusion for the anticonvulsive drug diphenylhydantoin (DPH) (van der Kleijn et al., 1972). Although the tissue-plasma distribution coefficients for both white and gray matter are considerably lower than in the case of diazepam and desmethyldiazepam, the distribution in the brain is essentially the same.

Data are available that DPH inhibits peripheral nerve conductance (Spaans, 1971) and that it is therapeutically active in trigeminal neuralgia. These data are not available for the benzodiazepines. The difficulties in defining antianxiety activity play a large role in the uncertainty about plasma or tissue concentration-activity relationships. It remains to be investigated what the actual significance of the accumulation reported here is in relation to the different shades of activity observed both after acute and after chronic administration. It is obvious that there exists quite a difference between the rate of elimination of, e.g., diazepam after a single dose and after muliple dosing in humans (van der Kleijn et al., 1971a). Moreover, the additive combination of the pharmacologically active metabolite desmethyldiazepam has to be taken into account after repeated intake.

Although a relation between a minimum plasma concentration and effect may be established, the time course of the effect may be protracted manyfold because of the apparent extension of the biological halflife and increased plasma concentration following subchronic treatment, compared with a single dose. Until now, no such extension of a well-defined and measurable effect in humans has been found. Since several body regions can accumulate the compound or metabolite—the intestines, the cortex of adrenals, the myelinated structures, and, for the more lipophylic benzodiazepines, also the body fat—there is a need for studies with combinations of drugs that can alter the rate of reabsorption, distribution, or elimination and consequently the actual intensity and time course of pharmacological activity.

ACKNOWLEDGMENTS

The labeled medazepam and desmethyldiazepam were synthesized by Dr. Hans Kaegi and supplied by Hoffmann-La Roche Inc., Nutley, New

Jersey. Oxazepam-2-^{14}C was obtained from Dr. Hans Ruelius of Wyeth Inc., Radnor, Pennsylvania.

This research was supported in part by the Dutch organizations FUNGO Z. W. O., and the Dr. Saal van Zwanenberg Foundation.

The biotechnical assistance of Mr. Th. Arts and M. G. T. J. Grutters of the Central Animal Laboratory and the preparation of the figures by Mr. J. Konings of the Medical Illustrations Department is gratefully acknowledged.

We thank Dr. A. W. Schwartz for reviewing the manuscript.

REFERENCES

Cassano, G. B., Ghetti, B., Gliozzi, E., and Hansson, E. (1967): Autoradiographic distribution study of "shortacting" and "longacting" barbiturates ^{35}S-thiopentone and ^{14}C-phenobarbitone. *British Journal of Anaesthesia*, 39:11–20.

Lien, E. J., and Hansch, C. (1968): Correlation of ratios of drug metabolism by microsomal subfractions with partition coefficients. *Journal of Pharmaceutical Sciences*, 57:1027–1028.

Marcucci, F., Fanelli, R., and Mussini, E. (1968): A method of gaschromatographic determination of benzodiazepines. *Journal of Chromatography*, 37:318–320.

Przybyla, A. C., and Wang, S. C. (1968): Locus of central depressant action of diazepam. *Journal of Pharmacology and Experimental Therapeutics*, 163:439.

Randall, L. O., Schallek, W., Scheckel, C., Banziger, R. U., and Moe, R. A. (1968): Zur Pharmakologie des neuen Psychopharmakons 7-Chlor-2,3-dihydro-1-methyl-5-phenyl-1H-1,4-benzodiazepin. *Arzneimittel Forschung*, 18:1542–1545.

Rieder, J., and Rentsch, G. (1968): Metabolismus and Pharmacokinetik des neuen Psychopharmakons 7-Chlor-2,3-dihydro-1-methyl-5-phenyl-1H-1,4-benzodiazepin. *Arzneimittel Forschung*, 18:1545–1551.

Schallek, W., Zabransky, F., and Kuehn, A. (1964): Effect of benzodiazepines on central nervous system of cat. *Archives Internationales de Pharmacodynamie*, 149:467–483.

Schwartz, M. A., and Carbone, J. J. (1970): Metabolism of ^{14}C medazepam hydrochloride in dog, rat, and man. *Biochemical Pharmacology*, 19:343–361.

Spaans, F. (1970): Epilepsy and folic acid: A double blind study on the effect of folic acid administration in epileptic patients (*in Dutch*). *Dissertation*, University of Amsterdam. Drukkerij Zeldenthuis, Amsterdam.

Ullberg, S. (1954): Studies on the distribution and fate of ^{35}S-labeled benzylpenicillin in the body. *Acta Radiologica*, Suppl. 118.

van der Kleijn, E. (1969a): Kinetics of distribution and metabolism of diazepam and chlordiazepoxide in mice. *Archives Internationales de Pharmacodynamie*, 178:193–215.

van der Kleijn, E. (1969b): Pharmacokinetics of ataractic drugs. *Dissertation*, University of Nijmegen. Ste. Catharine Press, Bruges.

van der Kleijn, E. (1969c): Protein binding and lipophylic nature of ataractics of the meprobamate group and diazepam group. *Archives Internationales de Pharmacodynamie*, 179:225–250.

van der Kleijn, E. (1971): Pharmacokinetics of distribution and metabolism of ataractic drugs and an evaluation of the site of anti-anxiety activity. *Annals of the New York Academy of Sciences*, 179:115–125.

van der Kleijn, E., Beelen, G. C., and Frederick, M. A. (1971a): Determination of tranquillizers by GLC in biological fluids. *Clinica Chimica Acta*, 34:345–356.

van der Kleijn, E., Rijntjes, N. V. M., Guelen, P. J. M., and Wijffels, C. C. G. (1972): Systemic and brain distribution of diphenylhydantoin in the squirrel monkey. In: *Antiepileptic Drugs,* edited by D. M. Woodbury, J. K. Penry, and R. P. Schmidt. Raven Press, New York, page 124.

van der Kleijn, E., Van Rossum, J. M., Muskens, A. T. J. M., and Rijntjes, N. V. M. (1971*b*): Pharmacokinetics of diazepam in dogs, mice, and humans. *Acta Pharmacologica,* 29:109-129.

van der Kleijn, E., and Wijffels, C. C. G. (1971): Wholebody and regional brain distribution of diazepam in newborn rhesus monkeys. *Archives Internationales de Pharmacodynamie,* 192:255–264.

The Benzodiazepines, Raven Press, New York © 1973

COMPARATIVE PHARMACOKINETICS OF OXAZEPAM AND NORTRIPTYLINE AFTER SINGLE ORAL DOSES IN MAN

Jörgen Vessman, Balzar Alexanderson, Folke Sjöqvist, Bengt Strindberg, and Anders Sundwall

Research Department, Kabi Group, Stockholm, Department of Clinical Pharmacology, University of Linköping, Linköping, Department of Pharmacology and Toxicology, University of Uppsala, Uppsala, Sweden

I. INTRODUCTION

Recent studies suggest that there are great inter-individual differences in the rate of metabolism of certain psychotropic drugs in man. Thus, patients treated with standard doses of desmethylimipramine (Hammer et al., 1967), nortriptyline (Hammer and Sjöqvist, 1967), or chlorpromazine (Curry et al., 1968) show markedly different steady-state plasma concentrations. For the two antidepressants, 20- to 30-fold inter-individual differences have been observed (Curry et al., 1968). Experimental investigations in twins and families reveal that the steady-state plasma concentration of nortriptyline is mainly determined by genetic factors (Alexanderson et al., 1969), and the mode of inheritance is likely to be polygenic (Asberg et al., 1971); however, concomitant treatment with barbiturates, for example, (Alexanderson et al., 1969) tends to lower the steady-state plasma concentration. It has been suggested (Sjöqvist et al., 1971) that the rate of hydroxylation of nortriptyline to a large extent determines the individual steady-state plasma concentration, as 10-hydroxynortriptyline is the main metabolite of NT formed in man. Pharmacokinetic cross-over studies have been performed with desmethylimipramine and nortriptyline because they are both highly distributed extravascularly and are metabolized in two steps, an initial hydroxylation followed by conjugation. These studies showed a highly significant correlation within individuals between the kinetics (steady-state plasma concentration and plasma half-life) of nortriptyline and desmethylimipramine (Hammer et al., 1969; Alexanderson, *unpublished observations*).

Oxazepam is a drug which is almost exclusively metabolized by conjugation with glucuronic acid in man (Walkenstein et al., 1964). The purpose of

165

the present study was to compare indices of hydroxylation (nortriptyline) and glucuronidation (oxazepam) in the same subjects.

II. METHODS

A. Subjects

All experiments were performed in 12 volunteers working in the respective research groups. All were healthy, as determined by history, physical examination, electrocardiogram, and routine laboratory analyses, including hemogram, sedimentation rate, and screening for protein, glucose, and bacteria in urine. Details about sex, age, and weight are given in Table I. The

TABLE I. *Plasma half-lives and clearances of nortriptyline (NT) and oxazepam (OX) calculated from single oral dose data in 12 healthy subjects*

Subjects	Sex	Age	Weight (kg)	Plasma half-lives (hr)*		Plasma clearances (liter/hr/kg)**	
				NT	OX	NT	OX ($\times\ 10^{-1}$)
BaA	M	29	79	34.2	11.4	0.43	1.93
GuA	M	27	83	18.2	13.9	0.94	0.90
InB	F	40	51	20.2	21.2	0.67	0.51
OlB	M	32	61	19.0	15.9	1.09	0.75
BeC	M	30	71	38.5	12.5	0.43	0.44
InF	F	26	58	25.0	15.6	0.35	0.50
MaG	M	24	67	24.7	14.8	0.64	0.73
MaL	F	33	67	30.4	11.8	0.67	0.75
SeM	F	30	55	37.2	19.4	0.66	0.28
RoO	M	55	83	27.4	8.8	0.68	0.44
AnR	M	28	80	35.0	16.6	0.42	0.84
HeS	M	32	74	17.9	16.1	1.72	0.68

* Determined from the terminal apparent monoexponential part of the curve.
** Calculated as the dose divided by the total area under the plasma concentration curve assuming complete availability of the administered dose.

subjects had not taken any drugs for at least 3 weeks before the study and none had been treated with drugs regularly for 6 months.

B. Oxazepam

A preliminary study was performed in four healthy volunteers (BeC, ScM, RoO, and HeS), regarding absorption, metabolism, and excretion of oxazepam after an oral dose of 15 mg (as the commercial tablet SOBRIL). All experiments were performed in the morning on an empty stomach. Five hundred ml of an orange drink was given during the first hour in order to facilitate urine collection 6 hr after drug administration.

Oxazepam in serum, urine, and feces was analyzed with a gas chromatographic procedure, which is based on electron-capture detection of the hydrolysis product, a benzophenone. Lorazepam was used as the internal standard (de Silva et al., 1969; Knowles et al., 1971). Lorazepam differs from oxazepam only by an additional chlorine atom in the 2′ position.

Oxazepam and the internal standard were extracted into methylene chloride and then back to 12 N sulfuric acid. After dilution of the acid to 8 N, the compounds were hydrolyzed to the corresponding benzophenones, which were subsequently extracted into heptane and determined by electron-capture gas chromatography on a column with 10% OV-17. Recovery and precision was 100% ± 2.8 from both serum and urine. Analysis of oxazepam concentrations in serum and plasma samples obtained at the same time gave equivalent values.

Oxazepam glucuronide in serum or urine was first hydrolyzed to oxazepam by β-glucuronidase and then determined as the benzophenone, as described above. In urine and feces, the glucuronide was also analyzed as the aminochlorobenzophenone, following acid hydrolysis. The same results were obtained with both methods when urine was analyzed.

C. Nortriptyline

Nortriptyline hydrochloride was administered by mouth in a dose of 1 mg/kg body weight. The drug was given in the morning on an empty stomach and food intake was not allowed until 4 hr after the administration. The drug was dispensed as gelatin capsules at the Military Pharmacy, Karolinska Hospital, Stockholm. All doses were weighed for each individual, with an accuracy of ± 1 mg.

Before nortriptyline administration, a blood specimen was drawn and used as a plasma blank. Eight to 12 blood samples were obtained at different time points (usually 1, 2, 4, 5, 8, 12, 16, 24, 32, 48, 72, and 96 hr) after administration of a single oral dose, in order to map out the increase and decline of the nortriptyline concentration in plasma.

The blood samples were drawn in heparinized tubes and centrifuged at 2,000 rpm (650 × g) for 10 min. Six ml of plasma was removed and mixed with 1.5 ml of 0.1 N HCl. The plasma samples were stored at −20°C until assayed for nortriptyline, according to the method of Hammer and Brodie (1967), slightly modified as further described by Sjöqvist et al., (1969). The total concentration of nortriptyline in plasma was determined in terms of free base. The plasma samples were processed in triplicates.

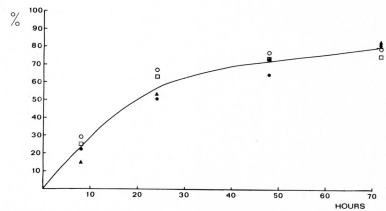

FIG. 1. Cumulative urinary excretion of oxazepam-glucuronide in man (percent of ad-ministered dose). Fifteen mg of oxazepam (tablets SOBRIL) was given orally (on an empty stomach) to healthy volunteers.

The interval between the oxazepam and nortriptyline experiments was at least 4 weeks. In 8 of the 13 volunteers nortriptyline was given first and already 3 months before oxazepam. This study is described in detail by Alexanderson (1972). The nortriptyline concentration was determined as described above. In the remaining four subjects, oxazepam was given as the first drug, and nortriptyline was given 4 weeks later. In these subjects (BeC, ScM, RoO, and HeS), the nortriptyline concentration in the plasma was deter-mined by use of gas chromatography-mass spectrometry (Borga et al., 1971). There is a good correlation between this method and that based upon *in vitro* ^{3}H-acetylation labeling previously used (Borga et al., 1971).

III. RESULTS AND COMMENTS

The cumulative urinary excretion of oxazepam glucuronide over 72 hr is shown in Fig. 1. Within this time, about 80% of the dose is found in the urine as the glucuronide.

TABLE II. *Excretion of oxazepam (ox.) and its glucuronide (gluc.) in urine and feces following oral administration to four human volunteers*

Time (hr)	Urine		Feces	
	ox. %	gluc. %	ox. %	gluc.
0– 6	0.3	22.7		
0–24	0.8	57.4		
0–48	1.2	72.2		
0–72	1.4	79.6	6.8	—

TABLE III. *Serum concentrations of oxazepam (ox.) and oxazepam-glucuronide (gluc.) after oral administration of 15 mg to human volunteers (ng/ml)*

Time (hr)	BeC		HeS		RoO		ScM	
	ox.	gluc.	ox.	gluc.	ox.	gluc.	ox.	gluc.
1	87	58	30	19	29	14	45	21
2	180	136	123	86	98	111	130	41
4	186	357	111	119	119	162	190	22
8	90	147	84	101	101	150	145	60
24	33	49	28	38	38	28	96	27

Table II shows the amount of oxazepam and its glucuronide present in urine and feces. Approximately 1% of the dose is excreted unchanged in the urine. In two of the four subjects, feces was collected over 72 hr. Ten and 3.4% respectively, of the dose was recovered unchanged in feces; no glucuronide could be detected. Thus, about 85% of the given dose could be accounted for in urine and feces, indicating that the compound is almost completely absorbed from the gastrointestinal tract.

Serum concentrations of oxazepam and oxazepam glucuronide in the four subjects are shown in Table III. Peak concentrations of oxazepam vary from

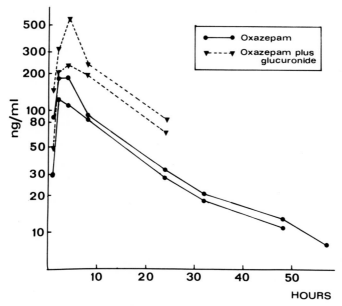

FIG. 2. Serum concentrations of oxazepam and oxazepam-glucuronide from the same experiment as in Fig. 1.

119 to 190 ng/ml and of the glucuronide from 60 to 357 ng/ml. The proportion of the glucuronide in the serum samples is usually between 20 and 60%. The absorption seems to be relatively slow with peak concentrations after 2 to 4 hr (Fig. 2).

The plasma half-lives of oxazepam (equal to the serum half-lives; *see* Section II) and nortriptyline were calculated from the terminal, apparent monoexponential part of the curves (β-slopes). All data are presented in Table I. The half-lives of oxazepam in four subjects are shown in Fig. 3.

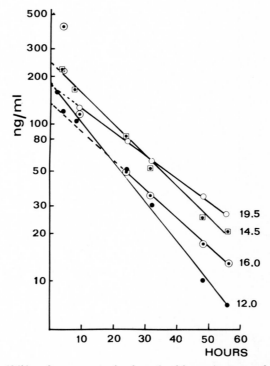

FIG. 3. Serum half-life of oxazepam in four healthy volunteers, following oral administration of 15 mg in the form of tablets SOBRIL on an empty stomach.

Figure 4 shows a diagram where the half-life of oxazepam has been plotted against the half-life of nortriptyline in the same subject. No correlation was found between the half-lives of the two drugs ($r = -0.46; p > 0.05$).

Nortriptyline and oxazepam are both almost completely metabolized (Walkenstein et al., 1964; Sjöqvist et al., 1971) before renal excretion. In order to search further for a possible relationship between the kinetics of

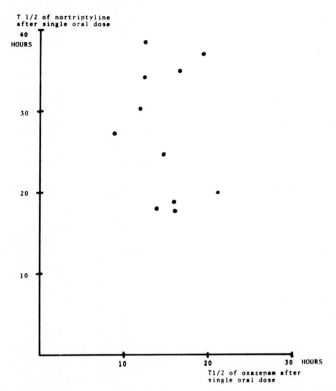

FIG. 4. Relationship between plasma half-lives of nortriptyline and oxazepam in 12 healthy subjects.

oxazepam and nortriptyline, their plasma clearances were compared (Table I). The plasma clearance of a drug can be obtained as the product of its apparent volume of distribution and the elimination rate constant (Wagner et al., 1965). The respective apparent volumes of distribution were calculated as the dose divided by the total area under the plasma level curve, assuming complete availability of the dose administered.

It is known from studies in animals and man that tricyclic antidepressants are easily absorbed from the intestine (Bernhard and Beer, 1962; Herrman, 1963; Yates et al., 1963). Oxazepam seems to be almost completely absorbed since approximately 85% of the given dose could be accounted for in urine and feces. It would be desirable to compare plasma clearances of the two drugs after i.v. administration. However, intravenous injection of tricyclic antidepressants may be dangerous because of the risks of cardiac side effects.

No relationship was found between the clearances of oxazepam and nortriptyline ($r = -0.10$; $p > 0.05$) in this study, and the overall data indicate that there is no correlation between a subject's ability to oxidize nortriptyline and to conjugate oxazepam.

IV. SUMMARY

Twelve healthy volunteers were given oxazepam (15 mg) and nortriptyline (1 mg/kg), orally, on two different occasions, and the plasma concentrations were determined for several days after drug administration. Oxazepam and its glucuronide were analyzed in serum, plasma, urine, and feces by gas chromatography. Nortriptyline was quantitatively analyzed by either ^{3}H-acetylation *in vitro* or by use of gas chromatography-mass spectrometry. The plasma half-lives and clearances of the two drugs were determined in each subject.

About 80% of the oxazepam dose could be accounted for in urine as its glucuronide. Approximately 1% of the dose is excreted unchanged in the urine and less than 10% in feces. Only a small percent of a given dose of nortriptyline is excreted unchanged in the urine; most of the drug is hydroxylated.

No relationship was found between plasma half-lives or clearances of oxazepam and nortriptyline, indicating no correlation between a subject's ability to oxidize nortriptyline and to conjugate oxazepam.

ACKNOWLEDGMENTS

We wish to thank Mrs. Ingegerd Bertling, Miss Grethel Freij, and Miss Margareta Lind for skillful technical assistance.

This study was partly supported by the Swedish Medical Research Council (B72-14X-1021-07A and B72-14P-3589-01), by U.S. Public Health Service grant GM 13978-06, and by funds from Karolinska Institutet.

REFERENCES

Alexanderson, B. (1972): Pharmacokinetics of nortriptyline in man after single and multiple oral doses: The predictability of steady-state plasma concentrations from single-dose plasma-level data. *European Journal of Clinical Pharmacology*, 4:82–91.

Alexanderson, B., Price Evans, D. A., and Sjöqvist, F. (1969): Steady-state plasma levels of nortriptyline in twins: Influence of genetic factors and drug therapy. *British Medical Journal*, 4:764–768.

Åsberg, M., Price Evans, D. A., and Sjöqvist, F. (1971): Genetic control of nortriptyline kinetics in man. A study of the relatives of propositi with high plasma concentrations. *Journal of Medical Genetics*, 8:129–135.

Bernhard, K., and Beer, H. (1962): Aktivitäten der Expikationskohlensäure, des ZNS und anderer Organe nach Gaben von ^{14}C-signiertem N-(γ-Dimethyl-aminopropyl)-iminodibenzyl (Psychopharmakon Tofranil) an Rattan und Hunde. *Helvetica Physiologica et Pharmacologica Acta*, 20:114–121.

Borgå, O., Palmér, L., Linnarsson, A., and Holmstedt, B. (1972): Quantitative determination of nortriptyline and desmethylnortriptyline in human plasma by combined gas chromatography-mass spectrometry. *Analytical Letters*, 4:837–849.

Curry, S. H., Marshall, J. H. L., and William, A. (1968): Plasma levels of chlorpromazine and some of its relatively non-polar metabolites in psychiatric patients. *Life Sciences*, 7:9–17.

Hammer, W., and Brodie, B. B. (1967): Application of isotope derivative technique to assay of secondary amines: Estimation of desipramine by acetylation with H^3-acetic anhydride. *Journal of Pharmacology and Experimental Therapeutics*, 157: 503–508.

Hammer, W., Ideström, C.-M., and Sjöqvist, F. (1967): Chemical control of antidepressant drug therapy. In: *Proceedings of the First International Symposium on Antidepressant Drugs*, edited by S. Garattini and M. N. G. Dukes, Milan 1966. *Excerpta Medica International Congress Series* No. 122, Amsterdam, pp. 301–310.

Hammer, W., Martens, S., and Sjöqvist, F. (1969): A comparative study of the rate of metabolism of desmethylimipramine, nortriptyline, and oxyphenylbutazone in man. *Clinical Pharmacology and Therapeutics*, 10:44–49.

Hammer, W., and Sjöqvist, F. (1967): Plasma levels of monomethylated tricyclic antidepressants during treatment with imipramine-like compounds. *Life Sciences*, 6: 1895–1903.

Herrman, B. (1963): Quantitative Methoden zur Untersuchung des Stoffwechsels von Tofranil., *Helvetica Physiologica et Pharmacologica Acta*, 21:402–408.

Knowles, J. A., Comer, W. H., and Ruelius, H. W. (1971): Disposition of 7-chloro-5-(o-chlorophenyl)-1,3-dihydro-3-hydroxy-2H-1,4 9-benzodiazepin-2-one (Lorazepam) in humans. Determination of the drug by electron capture gas chromatography. *Arzneimittelforschung*, 21:1055–1059.

de Silva, J. A. F., Schwartz, M. A., Stefanivic, V., Kaplan, J., and D'Arconte, L. (1964): Determination of diazepam (VALIUM) in blood by gas-liquid chromatography. *Analytical Chemistry*, 36:2099–2105.

Sjöqvist, F., Hammer, W., Borgå, O., and Azarnoff, D. L. (1969): Pharmacological significance of the plasma level of monomethylated tricyclic antidepressants. In: *Proceedings Collegium Internationale Neuropsychopharmacologicum* (Tarragona, Spain, April, 1968). *Excerpta Medica International Congress Series* No. 180, Amsterdam, pp. 128–136.

Sjöqvist, F., Alexanderson, B., Åsberg, M., Bertilsson, L., Borgå, O., Hamberger, B., and Tuck, D. (1971): Pharmacokinetics and biological effects of nortriptyline in man. *Acta Pharmacologica et Toxicologica*, 29: (*Suppl. 3*): 255–280.

Wagner, J. G., Northam, J. I., Alway, C. D., and Carpenter, O. S. (1965): Blood levels of drug at equilibrium state after multiple dosing. *Nature*, 207:1301–1302.

Walkenstein, S. S., Wiser, R., Gudmundsen, C. H., Kimmel, H. B., and Corradino, R. A. (1964): Absorption, metabolism, and excretion of oxazepam and its succinate half-ester. *Journal of Pharmaceutical Sciences*, 55:1181–1186.

Yates, M. C., Todrick, A., and Tait, C. A. (1963): Aspects of the clinical chemistry of desmethylimipramine in man. *Journal of Pharmacy and Pharmacology*, 15: 432–439.

DISCUSSION SUMMARY

A. Breckenridge

Department of Clinical Pharmacology, Royal Postgraduate Medical School, London, United Kingdom

When asked about extrahepatic drug metabolism of the benzodiazepines, Schwartz said there was little information available. He was also asked to comment on the possibility that chlordiazepoxide might stimulate its own metabolism but could not confirm that this happened in any animal species studied.

Rieder said there was no evidence for nitrazepam inducing its own metabolism in man.

Discussion on Garattini's presentation concerned the correlation of plasma and brain levels of diazepam with pharmacological effects. Garattini indicated some problems associated with this; not only are there active metabolites in plasma and tissues, but their pharmacological effects differ from those of the parent drug. The nonpredictability of *in vivo* metabolism from *in vitro* metabolism and their differing patterns in different species made interpretation of plasma levels extremely difficult.

Rieder was asked to comment on the small range of values for the plasma half-life of nitrazepam found in man, which contrasted with the marked individual variation in steady-state plasma concentrations of diazepam in animals shown by Garattini. Rieder supposed this might be due to the relatively small number of subjects studied. Morselli commented on the variation in nitrazepam absorption. Some patients in whom the total percentage of drug absorbed was small had a higher 2-hr plasma concentration than patients whose total percentage absorbed was higher and whose 2-hr plasma concentration was relatively low. There was also considerable discussion on the possibility of gut bacteria metabolizing nitrazepam.

Morselli's and van der Kleijn's papers on the distribution of diazepam, especially within the brain, were discussed together. The changing pattern of radioactivity from gray to white matter, with time, was noticeable in both studies. The action of diazepam or its metabolites within white matter was presumably to alter neuronal conduction. The report of variations in the pattern of plasma levels of diazepam and demethyldiazepam in different subjects was of considerable interest, with some patients having a higher plasma level of the metabolite on chronic dosing. Morselli's data indicated that the same was true of their relative distribution within the brain.

Sundvall's paper (Vessman et al.) was discussed in terms of the concept of clearance of drugs from the body, stressing how differences in apparent volume of distribution were potentially as important as differences in plasma half-lives.

175

The Benzodiazepines, Raven Press, New York © 1973

SUBCELLULAR ACTIONS OF BENZODIAZEPINES

Walter B. Essman

Queens College of the City University of New York, Flushing, New York

I. INTRODUCTION

The central action of benzodiazepines has been given limited selective review (Randall, 1961, 1968) regarding their pharmacological properties and behavioral effects, but few consistent data have been provided concerning the effects of these compounds upon central nervous system metabolism. One of the several problems encountered in the study of the central action of benzodiazepines concerns the relationship between the uptake and deposition of such molecules and the specific locus where their central effects occur and through which their functions are modulated. This series of studies investigated several neurochemical changes associated with the action of two benzodiazepine compounds. Previous behavioral indications and studies of time course for central uptake, following parenteral administration, served as a basis for consideration of their action in terms of central synaptic localization and function.

Preliminary investigations in our laboratory suggested that the guinea pig was maximally sedated 60 min following the parenteral injection of chlordiazepoxide (CDP) in doses up to 10 mg/kg. It was further established that at these times maximum drug recovery (as measured from brain tissue by low-temperature spectrofluorescence assay following heptane-alcohol and ethyl acetate extraction from regional brain-tissue homogenates) was from the cerebral cortex, which represents a substantial proportion of the total brain weight of the guinea pig. On the basis of these preliminary findings and with the operating premise that CDP and related benzodiazepines probably exert profound central synaptic actions in bringing about their other physiological and behavioral effects, several indexes of such cerebral synaptic function were utilized as measures of CDP action, including electrolyte concentration and distribution, ribonucleic acid (RNA) content, and acetylcholine (ACh) levels.

Specific concern was given to the effects of two benzodiazepine compounds, CDP and SCH 12041, upon synaptosomal elements and subcellular organelles having possible significance for the functional integrity of the synapse. Surprisingly little has been reported concerning synaptic and sub-synaptic indexes of excitation processes that is useful to characterize the functional status of those systems within which such events may be operative. However, several relationships have emerged which, although not necessarily derived directly from synaptic sites, may serve as useful measures of synaptic status; these include the potassium to sodium ratio, magnesium concentration, and the relationship of both of these measures to the content and disposition of RNA and ACh. Cerebral electrolyte concentrations and ratios have been reasonably well established as indexes of excitability and excitability change (McIlwain, 1963); moreover, the sites and changes in nucleic acid binding of synaptic amines that are influenced by changes in ionic strength have been given further recent attention as a possible model for synaptic excitability coding (Essman, 1970a; Essman et al., 1971). ACh is one of the synaptic molecules that has significance not only as a putative transmitter but also, possibly, as a trophic agent, with which the former interrelationships appear intimately concerned.

Several other points support the rationale underlying the choice of the indexes utilized in the present studies. The K^+/Na^+ ratios reported for freshly isolated synaptosomes (Whittaker, 1969) were reasonably consistent (1.63) with those previously reported (McIlwain, 1963) for slices of fresh guinea pig cerebral cortex (1.86). There has also been some indication that the respiration of synaptosomes may be stimulated by K^+ (Whittaker, 1969) and that such cations may influence synaptic excitability through activation of phosphohydrolases (DeRobertis, 1970). Also, among these cations, Mg^{++} has been shown to block synaptic transmission as well as the release of ACh and to provide for a sedative-like effect, depending upon its cerebral elevation, approximating anesthesia. This change in the status of cerebral tissue Mg^{++} content has been shown to be associated with psychoactive agents for which central effects are consistent with such electrolyte changes (Essman, 1970b).

There also appears to be some agreement as to which benzodiazepine compounds lead to reduced regional norepinephrine turnover in the brain (Taylor and Laverty, *this Volume*; Fuxe et al., *this Volume*). These observations may perhaps be related to findings that pharmacologically induced decreases in norepinephrine levels lead to increases in both ACh concentration and the activity of cholinacetylase, the primary determinant of the nerve ending biosynthesis of ACh (Mandell, 1970).

In the present studies, the major focus was directed at consideration of those synaptic events with which the action of two benzodiazepine compounds was concerned. More specifically, all experiments involved the isolation of subcellular and subsynaptic fractions for which specific assays and molecular localizations were carried out.

II. SUBCELLULAR EFFECTS OF CDP

In initial experiments, six replicate subcellular studies were carried out with groups of animals (each consisting of four male guinea pigs, weighing approximately 350 g apiece) given 2, 4, 6, 8, or 10 mg/kg of CDP or an equivalent volume of 0.9% saline, i.p. All animals were killed by stunning and guillotining. The whole brain was removed from the skull, and the cerebral cortex, scraped of underlying white matter comprised chiefly of myelin, was weighed, pooled, and dispersed in an Aldridge-type homogenizer to yield a 10% (w/v) homogenate in 0.32 M sucrose. Through a series of differential and sucrose density-gradient centrifugation steps (Gray and Whittaker, 1962), three fractions were isolated, concentrated, and then prepared for assay of the constituents associated with each. The relative purity of the three fractions—myelin, synaptosomes, and mitochondria—was established

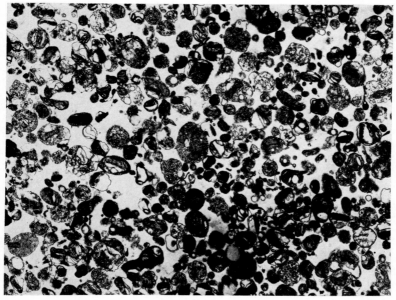

FIG. 1. Representative electron micrograph of a section through a fixed, pelleted fraction (B) of synaptosomes from guinea pig cerebral cortex (uranyl acetate-lead citrate stain). × 17,500

through assay for the activity of standard identifying enzyme markers (Whittaker, 1969) and through electron microscopy of an aliquot of the synaptosomal
fraction. The latter, fixed in 2.5% gluteraldehyle in phosphate buffer (pH
7.4), was dehydrated following treatment with osmium tetroxide, and was
then imbedded in Epon from which sections were cut and stained with uranyl
acetate-lead citrate. The sections were viewed in a Phillips 300 electron
microscope.

Figure 1, an electron micrograph of a synaptosome fraction derived
from fractionation of guinea pig cerebral cortex, shows a reasonably well-
distributed number of synaptosomes encapsulating their vesicular and intraterminal mitochondrial contents. Also shown are the minor contaminants of
this fraction—myelin fragments, mitochondria, and the more prevalent membrane fragments. This typically representative picture of the synaptosomal
fractions, upon which the present experiments were focused, provides reasonably good assurance of homogeneity as well as purity in these fractions.

Electrolyte determinations for each of the subcellular fractions obtained
(Na^+, K^+ and Mg^{++}) were made with appropriate sucrose dilutions and
measured in a Perkin-Elmer Atomic Absorption Spectrometer using the appropriate hollow cathode lamp. For all ion separations, gel filtration through
a strongly hyposmotic buffer allowed for a correction for macromolecular
binding of small ions, as compared with subcellular compartmentalized ions.

The results are summarized in Table I. In addition to the saline-treated
control, the effects of all doses of CDP administered are shown. There appeared to be a reduction in the K^+ concentration associated with the myelin

TABLE I. *Mean ($\pm\sigma$) concentration ($\mu At/mg$) of electrolytes associated with subcellular fractions of guinea pig cerebral cortex*

Treatment	Myelin			Synaptosomes			Mitochondria		
	Na	K	Mg	Na	K	Mg	Na	K	Mg
Saline	8.5	10.4	0.5	3.8	10.9	1.1	1.5	1.5	0.2
(0.9%)	(1.2)	(0.3)	(0.2)	(0.3)	(0.4)	(0.3)	(0.1)	(0.3)	(0.1)
CDP	10.0	7.2	0.5	4.6	8.2	0.6	1.1	1.1	0.2
(2.0 mg/kg)	(0.7)	(0.3)	(0.2)	(0.2)	(0.3)	(0.2)	(0.1)	(0.4)	(0.1)
(4.0 mg/kg)	6.7	7.1	0.5	4.9	10.6	0.8	1.3	1.1	0.2
	(1.1)	(0.3)	(0.2)	(0.3)	(1.4)	(0.3)	(0.8)	(0.4)	(0.1)
(6.0 mg/kg)	9.7	7.8	0.3	3.0	8.5	0.5	1.5	1.0	0.2
	(0.8)	(1.3)	(0.1)	(0.3)	(0.8)	(0.2)	(0.7)	(0.3)	(0.1)
(8.0 mg/kg)	5.0	8.0	0.5	5.5	11.7	0.8	1.1	0.7	0.4
	(1.6)	(1.8)	(0.2)	(0.2)	(2.0)	(0.3)	(0.6)	(0.2)	(0.1)
(10.0 mg/kg)	6.7	7.5	0.7	4.7	11.1	0.8	1.1	0.6	0.1
	(1.1)	(0.7)	(0.2)	(0.3)	(0.4)	(0.3)	(0.6)	(0.2)	(0.1)

fraction as a result of CDP treatment, but this effect was not clearly dose related. Reductions in mitochondria-associated K^+ levels were also observed, but only for the highest doses of CDP (8 and 10 mg/kg). Except for one dose level (6 mg/kg of CDP), there was an elevation in synaptosomal Na^+ levels, but again this increase was not dose related. It is interesting to note that the K^+/Na^+ ratio decreased, as calculated across all doses for the synaptosomal fraction, except at the 6 mg/kg dose of CDP where the ratio was unchanged from control conditions (2.8). Ratios at other doses, while decreased from control levels, showed no dose-dependent change.

These findings were considered in relation to parallel determinations, made under identical experimental conditions from the same tissue fractions, for levels of fraction-associated RNA. The data obtained which are summarized in Table IV were individually intercorrelated with the individual K^+/Na^+ ratios obtained for the same animals summarized in Table I. The intercorrelation between the synaptosomal K^+/Na^+ ratio and the corresponding level of synaptosomal RNA as a result of treatment with CDP was $+0.83$ ($p < 0.05$). Data from the control condition applied to this same intercorrelational analysis did not yield a statistically meaningful value ($r = 0.18$; $p > 0.50$). It therefore seems apparent that a significant drug-related, dose-independent relationship between K^+/Na^+ ratio and RNA level is indicated in a synaptosomal fraction from guinea pig cerebral cortex obtained 60 min after CDP treatment. This relationship suggests that, as the K^+/Na^+ ratio becomes reduced by 60 min after CDP administration, there is a corresponding increase in RNA level associated with those fractions.

III. TEMPORAL CHARACTERISTICS OF CDP EFFECTS

In view of the possibility that the time course (60 min) utilized in the same series of experiments might produce drug-related results that were distinctive from those obtained with a longer injection-fractionation interval, a second series of experiments was performed where tissue was obtained 90 min after injection of 4 mg/kg of CDP. Fractionation of scraped guinea pig cerebral cortex was carried out as previously described, and myelin, synaptosomal, and mitochondrial fractions were obtained. In addition, a crude synaptosomal fraction (P_2B), obtained under the same experimental conditions, was subfractionated, utilizing a sucrose density-gradient that provided for the isolation of seven distinct synaptosomal components (Whittaker et al., 1964). Six guinea pigs were used to obtain each set of synaptosomal and subsynaptosomal fractions for saline or CDP treatment conditions, and the experimental procedures were replicated six times. The electrolyte content

TABLE II. *Mean (±σ) concentration (μAt/mg) of electrolytes associated with subcellular fractions of guinea pig cerebral cortex*

Subcellular fraction	Saline (0.9%)			(4.0 mg/kg)		
	Na	K	Mg	Na	K	Mg
Myelin	4.0 (1.4)	5.5 (0.1)	0.5 (0.1)	5.0 (2.1)	5.7 (3.0)	0.8 (0.3)
Synaptosomes	2.1 (0.3)	4.9 (0.4)	0.8 (0.1)	2.8 (1.1)	5.1 (2.1)	1.1 (0.1)
Mitochondria	1.4 (0.8)	3.0 (1.2)	0.7 (0.1)	2.1 (0.2)	2.2 (0.4)	0.6 (0.4)
Soluble cytoplasm	15.5 (5.1)	23.7 (8.2)	0.5 (0.2)	11.8 (4.3)	7.6 (4.1)	4.1 (0.2)
Synaptic vesicles	3.5 (1.0)	4.3 (0.2)	0.4 (0.2)	3.0 (1.2)	5.0 (0.7)	0.4 (0.4)
ESM: I	2.3 (0.5)	2.3 (0.1)	0.7 (0.4)	3.4 (2.0)	3.1 (0.7)	0.8 (0.5)
ESM: II	1.1 (0.2)	1.4 (0.1)	0.6 (0.4)	1.4 (0.4)	2.0 (0.2)	6.2 (0.4)
ESM: III	2.1 (0.8)	3.0 (1.1)	0.7 (0.3)	1.4 (0.2)	2.4 (0.3)	0.6 (0.4)
Incompletely disrupted synaptosomes	0.8 (0.2)	1.1 (0.2)	0.6 (0.4)	1.1 (0.7)	1.3 (0.1)	0.6 (0.4)
Intraterminal mitochondria	0.9 (0.4)	0.7 (0.1)	0.4 (0.2)	1.1 (0.7)	0.5 (0.1)	0.4 (0.2)

measured in each fraction 90 min after drug or saline treatment has been summarized in Table II. One interesting observation is the significant increase (37%) in the level of synaptosomal Mg^{++}; this change occurred more consistently at 90 min after CDP treatment than at 60 min.

The K^+/Na^+ ratio for the synaptosome fraction, measured 90 min following administration of 4 mg/kg of CDP, was also consistent with the previous observations where drug treatment resulted in a decrease from control values (2.3) to a ratio of 1.8. Some contribution to this change may be reflected by the highly significant decrease in K^+, observed for the soluble cytoplasm of the synaptosome. Corresponding to this change in synaptosomal cytoplasm was an eightfold increase in Mg^{++} concentration. The intermediate-size, external synaptic membrane particles (II) constituted one fraction in which a drug-related increase of more than 10-fold occurred in the Mg^{++} concentration. It is of further interest that two of the subsynaptosomal fractions showed no associated electrolyte changes related to the drug effect; these were the synaptic vesicles and the intraterminal mitochondria.

Previous consideration was given to the interrelationship between decreased K^+/Na^+ ratios and changes in RNA concentration for the same subcellular fractions. It was apparent from these findings that the RNA content was significantly increased following CDP treatment as a function of the decreases in the K^+/Na^+ ratio. Since an elevation in synaptosomal Mg^{++} was observed 90 min—but not 60 min—after CDP treatment, it was proposed, in accord with previous theoretical and empirical considerations (Ess-

TABLE III. *Mean ($\pm\sigma$) concentration of RNA ($\mu g/mg$ protein) associated with subcellular fractions of guinea pig cerebral cortex 90 min after drug treatment*

Subcellular fraction	Treatment condition	
	Saline (0.9%)	CDP (4.0 mg/kg)
Myelin	0.78 (0.17)	1.02 (0.26)
Synaptosomes	0.35 (0.06)	0.75 (0.05)[a]
Mitochondria	0.26 (0.12)	0.29 (0.06)
Soluble cytoplasm	0.37 (0.03)	0.28 (0.04)
Synaptic vesicles	0.72 (0.01)	0.72 (0.09)
ESM: I	1.18 (0.19)	1.06 (0.24)
ESM: II	0.57 (0.25)	1.30 (0.08)[a]
ESM: III	0.77 (0.21)	0.86 (0.24)
Incompletely disrupted synaptosomes	0.85 (0.10)	0.36 (0.01)[a]
Intraterminal mitochondria	0.48 (0.07)	0.57 (0.21)

[a] $p < 0.01$.

man, 1970a; Essman et al., 1971), that synaptosomal RNA content could be related to changes in the Mg^{++} level of the presynaptic nerve ending. A correlation between Mg^{++} content associated with those subcellular fractions summarized in Table II and the RNA content of the same fractions summarized in Table III yielded a value of 0.69 ($p > 0.05$) for the saline-treated animals, as compared with 0.92 ($p < 0.02$) for the samples from CDP-treated animals. A comparison of differences between transformed correlation coefficients indicated that they were significant ($p < 0.02$).

Considered independently of the electrolyte data, it is apparent from Table IV, which shows RNA levels in the three major subcellular fractions obtained 60 min after drug treatment, that the synaptosome fraction at three of the five dose levels given resulted in statistically significant increases in RNA content. This change was not apparent for RNA associated with either myelin or mitochondria. The basis for this effect may, perhaps, be considered further in light of the RNA data presented in Table III, where there was again a significant increase (114%) in synaptosomal RNA for 90 min between treatment and assessment. It seems apparent that the sub-synaptosomal fraction in which an increase occurred consisted of intermediate fragments of external synaptic membrane (II); this 128% increase can be contrasted with the 55% decrease in RNA content of incompletely disrupted synaptosomes derived from drug-treated animals. This decrease might reflect changes in cytoplasmic RNA which was also decreased as a consequence of drug treatment and which would appear to represent the major constituent contained within incompletely disrupted presynaptic nerve endings.

TABLE IV. *Mean (±σ) concentration of RNA (µg/mg protein) associated with various subcellular fractions of guinea pig cerebral cortex*

Treatment	Subcellular fraction		
	Myelin	Synaptosomes	Mitochondrial
Saline (0.9%)	1.27	0.33	0.85
	(0.40)	(0.04)	(0.05)
CDP			
(2.0 mg/kg)	1.32	0.49[a]	0.86
	(0.08)	(0.16)	(0.14)
(4.0 mg/kg)	1.21	0.44[a]	0.79
	(0.10)	(0.08)	(0.14)
(6.0 mg/kg)	0.88	0.32	0.81
	(0.08)	(0.14)	(0.16)
(8.0 mg/kg)	1.14	0.51[a]	0.83
	(0.30)	(0.17)	(0.25)
(10.0 mg/kg)	1.04	0.39	0.78
	(0.08)	(0.13)	(0.17)

[a] $p < 0.01$.

IV. SUBCELLULAR ACTIONS OF SCH 12041

Another general series of experiments utilized a different benzodiazepine compound, SCH 12041 (7-chloro-1,3-dihydro-5-phenyl-1-(2,2,2-trifluoro-ethyl)-2H-1,4-benzodiazepin-2-one). The structural differences between this molecule and CDP may be seen in Fig. 2. A preliminary clinical report in

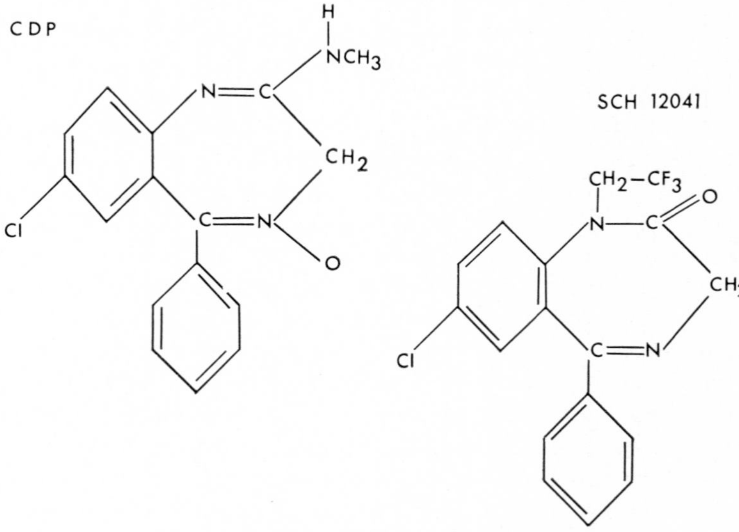

FIG. 2. Chlordiazepoxide and SCH 12041.

TABLE V. Mean (±σ) *electrolyte concentration* (μAt/mg) *associated with subcellular fractions of guinea pig cerebral cortex following in vivo treatment with SCH 12041*

Treatment	Subcellular fraction								
	Myelin			Synaptosomes			Mitochondria		
	Na	K	Mg	Na	K	Mg	Na	K	Mg
Control (DMSO + 0.9% saline)	3.6 (1.1)	4.7 (1.1)	1.7 (0.9)	2.8 (0.6)	8.0 (1.3)	5.1 (1.2)	7.1 (2.1)	4.7 (1.2)	0.8 (0.4)
SCH 12041									
(2.0 mg/kg)	4.1 (0.9)	6.7 (2.1)	2.4 (1.0)	3.9 (0.9)	10.9 (1.2)	5.7 (1.0)	7.2 (1.4)	4.8 (1.1)	1.1 (0.3)
(4.0 mg/kg)	3.1 (1.1)	4.9 (1.6)	1.6 (0.6)	2.0 (0.8)	6.9 (1.4)	4.5 (1.3)	4.8 (2.2)	3.9 (0.9)	0.9 (0.5)
(6.0 mg/kg)	4.9 (1.3)	8.6[a] (1.7)	2.1 (0.8)	3.4[a] (1.2)	11.6[a] (2.6)	6.6 (1.2)	10.0[a] (2.7)	8.0[a] (2.3)	2.0 (0.5)

[a] $p < 0.01$.

which the therapeutic properties of this compound were considered indicated some antianxiety activity but with no appropriate comparison with control treatment; there was also evidence of appreciable somnolence (Gallant and Bishop, 1971).

The compound was administered i.p. to groups of four male guinea pigs in doses ranging from 2 to 6 mg/kg; the drug was administered in a vehicle of DMSO and saline, which was used as a control treatment for another group of animals. Each series was replicated six times; Table V summarizes the data for all the experiments. The electrolyte concentrations associated with the subcellular fractions of guinea pig cerebral cortex were measured as previously described, 90 min following treatment. These data indicate that only the highest dose of SCH 12041 (6 mg/kg) exerted any significant effect upon the electrolyte content in subcellular fractions. This effect was observed for K^+ associated with the myelin fraction, which was significantly increased (83%) and for Na^+ and K^+ associated with synaptosomes, both of which were significantly increased; Na^+ and K^+ levels associated with mitochondria were both increased, as compared with control values. It is apparent that the nature of the subcellular electrolyte change associated with the action of this benzodiazepine compound differs from that previously observed with CDP, perhaps because of differences in the drug or—more likely—as a result of differences in uptake, distribution, or synaptic effect. It seemed evident that one consideration appropriate in regard to the present data was the nature of the subcellular distribution of CDP at the particular time when its subcellular effects were predominantly in evidence.

Some consideration was given to the issue of the subcellular distribution of CDP following its administration to guinea pigs, utilizing a low-temperature fluorescence method of localization in subcellular fractions of cerebral cortex prepared as previously indicated. With in vitro controls for redistribution effects following homogenization, it was apparent that in vivo distribution of the drug occurred in association with the soluble neuronal elements and myelin. There was no evidence of drug distribution to either mitochondria or synaptic vesicles. A correlation between the obtained CDP concentrations for those subcellular fractions with which it was associated and the electrolytes and RNA content for the same fractions yielded statistically significant intercorrelations for K^+ ($r = -0.80$; $p < 0.05$) and Mg^{++} ($r = +0.73$; $p < 0.05$), suggesting, particularly in view of the latter relationship, that potential contributions to cholinergic effects of CDP might be sought at the subcellular level.

V. SUBCELLULAR CHOLINERGIC ACTIONS OF BENZODIAZEPINES

Some final consideration was given to the effect of CDP and SCH 12041 upon levels of ACh in synaptosomes derived from guinea pig cerebral cortex, as measured 90 min after the i.p. administration of 2, 4, 6, 8, or 10 mg/kg of CDP, or 2, 4, and 8 mg/kg of SCH 12041. In addition, another group of animals was treated with 0.9% saline and served as a control for each set of determinations in this experiment. Four animals were used in each treatment group and each series was replicated three times. A synaptosome fraction was prepared, as previously described, and ACh was extracted and its concentration estimated against known standards, utilizing a guinea pig ileum as a biological assay system for these determinations (Chang and Gaddum, 1933). The results, summarized in Fig. 3, indicate that, as compared with saline-treated controls, all doses of CDP as well as of SCH 12041 were effective in significantly elevating synaptosomal ACh levels. These findings may again be related to our previous observations that excitability changes in the nerve ending, as inferred from either decreased K^+/Na^+ ratios or increased Mg^{++} in the synaptosome, may be related to cholinergic synaptosomes in much the same way as previous neuropharmacological evidence has supported the hypothesis that increased nerve-ending ACh content pro-

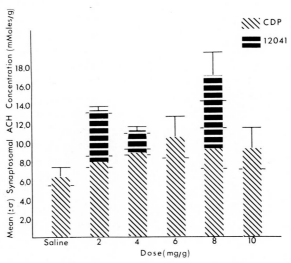

FIG. 3. Mean ($\pm\sigma$) concentration of ACh in synaptosomes from guinea pig cerebral cortex, obtained 90 min following i.p. injection of saline (0.9%), CDP, or SCH 12041.

vides for excitability decrements. Perhaps one clear example of this may be found in that anesthetics act to increase brain ACh levels. It may well be that those benzodiazepines utilized in the present experiments, at their respective doses and uptake times, may approximate sedative effects comparable to the CNS-excitability reduction properties of anesthetic agents. Changes in brain Mg^{++} level may be related to sedative effects associated with this condition, its synaptic effects being related to cholinergic transmission (Shitov, 1965) and to its potential role in increasing O-methylation of epinephrine (LaBrosse et al., 1958). Some previous consideration has been given to the effects of elevated magnesium level on brain 5-hydroxytryptamine content and metabolism (Essman, 1970b); these findings are also consistent with the psychosedative properties of compounds which lead to increased brain-Mg^{++} content.

VI. SUMMARY

In the foregoing studies, several findings have suggested that the two benzodiazepine compounds utilized exercise their effects at the presynaptic nerve endings of guinea pig cerebral cortex. Differences in electrolyte changes associated with such subcellular constituents and organelles have been shown to be related, both in terms of dose and time following treatment, to correlated changes in the concentration of RNA within the synaptosomal compartment and also to ACh stored within cholinergic nerve endings. Since the major storage pool of ACh in cholinergic nerve endings appears to be the synaptic vesicles, it was interesting to note that no detectable amounts of CDP were associated with synaptic vesicles and that electrolyte as well as RNA changes did not occur at this site following CDP treatment. It would, therefore, seem appropriate to consider that the cholinergic effects of CDP and SCH 12041, as observed in these experiments, probably reside within a much more complex framework, possibly related to membrane storage pools of this amine within the nerve endings. Although there are several bases upon which the action of benzodiazepines may be related indirectly to changes in other brain biogenic amines mediated through cholinergic effects, it seems more appropriate at this point to deal simply with the empirical data rather than to indulge in such speculation. It does seem appropriate, however, to point to those observations which obviously suggest the interactions of electrolytes, nucleic acids, and biogenic amines at the nerve ending, in response to the uptake and specific synaptic deposition of benzodiazepines.

ACKNOWLEDGMENTS

The research summarized in this chapter has been supported in part by Biomedical Sciences Support grant 5-SO5-FR-07064-05, from the National Institute of Mental Health, N.I.H.

I wish to thank Dr. Robert Taber and the Schering Corporation (Bloomfield, N. J.) for generously providing the SCH 12041 used in this investigation.

REFERENCES

Chang, H. C., and Gaddum, J. H. (1933): Choline esters in tissue extracts. *Journal of Physiology,* 79:255-285.

DeRobertis, E., and Rodriguez De'Lores Arnaiz, G. (1969): Structural components of the synaptic region. In: *Handbook of Neurochemistry,* Vol. 2, edited by A. Lajtha. Plenum Press, New York, p. 365.

Essman, W. B. (1970a): Some neurochemical correlates of altered memory consolidation. *Transactions of the New York Academy of Sciences,* 32:948-973.

Essman, W. B. (1970b): Central nervous system metabolism, drug action, and higher functions. In: *Drugs and Cerebral Function,* edited by W. L. Smith. Charles C Thomas, Springfield, Ill., pp. 151-175.

Essman, W. B., Bittman, R., and Heldman, E. (1971): Molecular and synaptic modulation of RNA 5-HT interactions. In: *Abstracts of Papers, Fourth Annual Winter Conference on Brain Research.* Snowmass-at-Aspen, Col.

Fuxe, K., Corrodi, H., Lidbrink, P., and Olson, L. (1973): The effects of benzodiazepines and barbiturates on central monoamine neurons. In: *The Benzodiazepines,* edited by S. Garattini, E. Mussini, and L. O. Randall. Raven Press, New York.

Gallant, D. M., and Bishop, M. P. (1971): SCH 12041. New antianxiety agents. *Current Therapeutic Research,* 13:107.

LaBrosse, E. H., Axelrod, J., and Kety, S. S. (1958): O-Methylation, the principle route of metabolism of epinephrine in man. *Science,* 128:593.

Mandell, A. J. (1970): Drug induced alterations in brain biosynthetic enzyme activity— A model for adaptation to the environment by the central nervous system. In: *Biochemistry of Brain and Behavior,* edited by R. E. Bowman and S. P. Datta. Plenum Press, New York, pp. 97-121.

McIlwain, H. (1963): *Chemical Exploration of the Brain.* Elsevier, Amsterdam.

Randall, L. O. (1961): Pharmacology of chlordiazepoxide (LIBRIUM). *Diseases of the Nervous System,* 22:7.

Randall, L. O., and Schallek, W. (1968): Pharmacological activity of certain benzodiazepines. In: *Psychopharmacology: A Review of Progress, 1957-1967,* edited by D. H. Efron. U. S. Government Printing Office, Washington, D.C., pp. 153-184.

Shitov, Ye. Ye. (1965): Vliyaniye sul'fata magniya v kombinatsü s trankvilizatorami na bioelektricheskuya i kholinesteraznuyu aktivnost'golovnogo mozga. (Rus.) (The effect of magnesium sulfate combined with tranquilizers on the bioelectric and cholinesterase activity of the brain.) *Farmakologiia i Toksikoligiia,* 28:13.

Taylor, K. M., and Laverty, R. (1973): The interaction of chlordiazepoxide, diazepam, and nitrazepam with catecholamines and histamine in regions of the rat brain. In: *The Benzodiazepines,* edited by S. Garattini, E. Mussini, and L. O. Randall. Raven Press, New York.

Whittaker, V. P. (1969): The synaptosome. In: *Handbook of Neurochemistry*, Vol. 2, edited by A. Lajtha. Plenum Press, New York, p. 327.

Whittaker, V. P., Michaelson, I. A., and Kirkland, R. J. A. (1964): The separation of synaptic vesicles from disrupted nerve-ending particles ('synaptosomes'). *Biochemical Journal*, 90:239.

The Benzodiazepines, Raven Press, New York © 1973

THE INTERACTION OF CHLORDIAZEPOXIDE, DIAZEPAM, AND NITRAZEPAM WITH CATECHOLAMINES AND HISTAMINE IN REGIONS OF THE RAT BRAIN

K. M. Taylor and R. Laverty

Department of Pharmacology, University of Otago Medical School, Dunedin, New Zealand, and Department of Pharmacology and Experimental Therapeutics, The Johns Hopkins University School of Medicine, Baltimore, Maryland

I. INTRODUCTION

Although many features of the pharmacology of the benzodiazepine drugs have been well studied (Sternbach et al., 1964), there is little information on the effect of these drugs on the synthesis, storage, and metabolism of putative neurotransmitters in the brain. This is surprising as many drugs that affect behavior do have effects on the uptake, storage, release, and metabolism of neurotransmitters, especially the biogenic amines in the brain. Perhaps the lack of interest is due to the fact that the more powerful tranquilizers, such as chlorpromazine, thioridazine, and haloperidol, have no effect on the endogenous levels of biogenic amines in the brain of animals (Andén et al. 1964; Laverty and Sharman, 1965), and also to the initial studies of Moe et al. (1962), who found that chlordiazepoxide had no effect on the whole brain levels of norepinephrine in the rabbit. More recently Corrodi et al. (1967) showed that a high dose of chlordiazepoxide had no effect on whole rat brain dopamine and norepinephrine levels.

However, static tissue levels, especially whole brain levels, do not necessarily indicate the effect of drugs on neurotransmitter function. Chlorpromazine and haloperidol have no effect on endogenous catecholamine levels, yet both increase the turnover rate of dopamine in the corpus striatum to compensate for the suggested receptor blockade (Andén et al., 1964; Laverty and Sharman, 1965). Therefore, we have made a detailed study of the effect of the benzodiazepine drugs on the catecholamine and histamine systems in regions of the rat brain. Some of these results have been reported earlier (Taylor and Laverty, 1969a).

191

II. EXPERIMENTAL

In all experiments, male Sprague-Dawley rats (180 to 220 g) received a subcutaneous dose of 10 mg/kg of chlordiazepoxide (LIBRIUM, Roche), diazepam (VALIUM, Roche), or nitrazepam (MOGADON, Roche). Diazepam and nitrazepam at this dose level caused sedation and ataxia, whereas chlordiazepoxide was without these effects. Behavioral observations have shown that chlordiazepoxide is 4 to 10 times less active than diazepam (Sternback et al., 1964). In a conflict situation in rats, approximately equipotent subcutaneous doses of chlordiazepoxide, diazepam, and nitrazepam were 20, 2, and 1 mg/kg, respectively (Laverty, 1969). Because of their insolubility in water, the drugs were dissolved in dimethylsulfoxide (DMSO). Control animals received an equivalent volume of DMSO. Recently it has been suggested that the solvent for the benzodiazepine drugs will affect their activity as measured by the loss of righting reflex (Crankshaw and Raper, 1971).

III. ENDOGENOUS CATECHOLAMINE LEVELS AND SYNAPTOSOMAL UPTAKE

The concentration of endogenous NE and DA in regions of the rat brain was determined after separation on a Dowex-50W cation-exchange column (Taylor and Laverty, 1969b), using the trihydroxyindole method of Laverty and Taylor (1968). Chlordiazepoxide, diazepam, and nitrazepam have no significant effect on the concentration of NE in all regions of the rat brain or on the DA concentration in the corpus striatum (Table I).

Many classes of drugs affect the uptake of catecholamines into catecholamine neurons (Horn et al., 1971). The effect of chlordiazepoxide and diazepam on the uptake of ^{3}H-norepinephrine into synaptosomes prepared from the hypothalamus and of ^{3}H-dopamine into synaptosomes prepared from the corpus striatum was evaluated using the method of Horn et al. (1969). In both preparations, chlordiazepoxide and diazepam proved to be very weak inhibitors of synaptosomal uptake (Table II). In the hypothalamus, both drugs were about 1,000 times less potent than the most active drug tested, desmethylimipramine, and about 100 times less potent than chlorpromazine. In the corpus striatum, chlordiazepoxide and diazepam were about 500 times less potent than benztropine, the most active drug tested, and about seven times less potent than chlorpromazine. Besides reuptake, catecholamines are thought to be inactivated after release by the enzymes monoamine oxidase and catechol-O-methyltransferase. In an *in*

TABLE I. *The effect of chlordiazepoxide, diazepam, and nitrazepam on rat brain regional catecholamine levels*

Drug	Norepinephrine					Dopamine
	Thalamus - hypothalamus-midbrain	Corpus striatum	Cerebral cortex	Cerebellum	Medulla-pons	Corpus striatum
Control	0.60 ± 0.04	0.39 ± 0.04	0.20 ± 0.01	0.27 ± 0.02	0.46 ± 0.03	3.92 ± 0.51
Chlordiazepoxide	0.67 ± 0.04	0.37 ± 0.04	0.19 ± 0.01	0.27 ± 0.03	0.46 ± 0.03	3.26 ± 0.27
Diazepam	0.63 ± 0.02	0.30 ± 0.02	0.20 ± 0.01	0.29 ± 0.02	0.48 ± 0.02	3.77 ± 0.22
Nitrazepam	0.58 ± 0.03	0.30 ± 0.02	0.19 ± 0.01	0.24 ± 0.04	0.46 ± 0.02	3.79 ± 0.50

All drugs were given subcutaneously 10 mg/kg for 5 hr. Levels are expressed as the mean $\pm$ S.E.M. (μg/g) of four experiments.

TABLE II. *Relative potency for inhibition of ^{3}H-catecholamine uptake by chlordiazepoxide and diazepam*

Drug	Relative potency	
	Corpus striatum	Hypothalamus
Benztropine	8,333	370
Desmethylimipramine	20	20,000
Chlorpromazine	147	2,000
Diazepam	21	56
Chlordiazepoxide	17	26

The values are the means of three independent determinations of ID_{50} values. Each determination was carried out in quadruplicate with three concentrations of inhibitor. The term "relative potency" is defined as the reciprocal of $ID_{50} \times 10^{-3}$, i.e. $1 = 1$ mM.

vitro study, it was found that chlordiazepoxide had no effect on rat liver monoamine oxidase activity using the method of Hendley and Snyder (1968), and rat brain catechol-O-methyltransferase activity using the method of Axelrod et al. (1959). These studies suggest that chlordiazepoxide does not interfere with the metabolism of released catecholamines.

IV. METABOLISM OF ^{3}H-DOPAMINE

The endogenous stores of catecholamines were labeled by administering 10 μC of ^{3}H-dopamine [ring-^{3}H(G); specific activity, 280 mC/mmole] by intraventricular injection. Rats received drugs or the vehicle 1 hr before the intraventricular injection and were killed 4 hr after ^{3}H-dopamine administration to allow for the metabolism of excess exogenous labeled dopamine. The regional level of labeled catecholamines and metabolites were determined after separation by ion-exchange chromatography (Taylor and Laverty, 1969*b*).

Pretreatment with chlordiazepoxide, diazepam, or nitrazepam reduced the disappearance rate of ^{3}H-norepinephrine formed from ^{3}H-dopamine. This was indicated by the significantly higher levels of ^{3}H-norepinephrine in all brain regions except the medulla-pons (Table III). This effect was more evident after pretreatment with chlordiazepoxide or diazepam than after nitrazepam. The benzodiazepine drugs had no effect on the level of other labeled metabolites measured in this study. These were normetanephrine, 3-methoxytyramine, 3,4-dihydroxymandelic acid, 3,4-dihydroxyphenylacetic acid, vanillylmandelic acid, and homovanillic acid.

The level of ^{3}H-catecholamines in the brain regions after ^{3}H-dopamine administration is the resultant of catecholamine synthesis, storage, reuptake, and release mechanisms. As these drugs do not affect the en-

TABLE III. *The effect of benzodiazepine drugs on the metabolism of* ^{3}*H-dopamine*

	% Increase		
Region affected and drug used	Total ^{3}H	^{3}H-NE	^{3}H-Neutral alcohol metabolites
Thalamus-hypothalamus-midbrain			
Chlordiazepoxide	61	41	62
Diazepam	42	37	32
Nitrazepam	19	31	18
Cerebral cortex			
Chlordiazepoxide	23	35	24
Diazepam	29	47	18
Nitrazepam	23	25	26
Cerebellum			
Chlordiazepoxide	98	45	45
Diazepam	15	40	10
Nitrazepam	47	32	32
Medulla-pons			
Chlordiazepoxide	0	4	3
Diazepam	5	8	0
Nitrazepam	—11	—8	—8
Corpus striatum		^{3}H-DA	
Chlordiazepoxide	26	128	22
Diazepam	28	132	14
Nitrazepam	—1	10	—7

The effect of chlordiazepoxide, diazepam, or nitrazepam (all 10 mg/kg, s.c., 5 hr before death) on the rat brain regional levels of tritiated catecholamines and metabolites was measured 4 hr after the intraventricular injection of 10 μC of ^{3}H-dopamine. Values are expressed as the mean percentage increase over control values. There were 10 to 12 animals in each group.

dogenous catecholamine levels or the uptake of catecholamines, the increased ^{3}H-catecholamine levels cannot be due to increased storage or reuptake. This suggests that the increased ^{3}H-catecholamine levels are due to decreased synthesis or decreased release of catecholamines. This would result in the same general effect of decreased turnover of catecholamines which would cause an increased intraneuronal retention time for labeled norepinephrine and dopamine.

The possible decrease in catecholamine turnover could be supported by the increase in neutral alcohol metabolites in the regions where increased ^{3}H-norepinephrine levels were found. Neutral alcohols are thought to be the products of physiologically released catecholamines and are removed from brain tissue by a simple diffusion process (Taylor and Laverty, 1969c). The increased levels of labeled alcohols could indicate reduced dilution by unlabeled alcohols formed from endogenous unlabeled catecholamines.

To confirm this theory, that the minor tranquilizers reduce catechol-amine turnover in the rat brain, the specific activities of the ^{3}H-neutral alcohols must be determined. At present this is not possible as there are no sensitive specific methods to measure the endogenous levels of all alcohol metabolites.

However, these results have been confirmed by Corrodi et al. (1971) who found with histochemical methods that chlordiazepoxide and diazepam reduced norepinephrine turnover in noradrenergic nerve terminals of the cerebral cortex and cerebellum. They also found that these drugs decreased dopamine turnover in the ascending dopamine neurons to the telencephalon.

Recently, Persson and Waldeck (1971) reported that in mice highly sedative doses of pentobarbital reduced norepinephrine turnover without affecting endogenous catecholamine levels. However, in our study, if the sedative effect of the benzodiazepine drugs was related to the decreased catecholamine turnover, the potency order should be nitrazepam, diazepam, and chlordiazepoxide last, as it has the least sedative properties.

V. ELECTRO-FOOTSHOCK STRESS AND THE METABOLISM OF ^{3}H-DOPAMINE

Rats were pretreated with chlordiazepoxide, diazepam, nitrazepam, or the vehicle 2 and 4 hr before death. They also received an intraventricular injection of 10 μC of ^{3}H-dopamine 4 hr before death, and the stressed group received electro-footshock treatment (1-sec electro-footshock every 10 sec) for 1 hr before death.

In all regions of the brain, electroshock stress significantly reduced the level of both endogenous norepinephrine and ^{3}H-norepinephrine formed from ^{3}H-dopamine (Fig. 1). This was in agreement with previous studies which showed that electro-footshock treatment generally increases norepi-nephrine turnover in all regions of the rat brain (Bliss et al., 1968; Taylor and Laverty, 1969b; Rosecrans, 1969).

Chlordiazepoxide, diazepam, and nitrazepam maintained both the norepinephrine and ^{3}H-norepinephrine levels nearer the control value and in most regions significantly higher than the stress-induced depleted levels (Fig. 1). This suggests that the benzodiazepine drugs prevent the increase in norepinephrine turnover by electro-footshock treatment. This effect was max-imal in the cortex, cerebellum and thalamus-hypothalamus-midbrain, the regions in which these drugs affected the metabolism of catecholamines in nonstressed rats, and least in the brainstem where catecholamine metabolism was generally unaffected by chlordiazepoxide, diazepam, and ni-

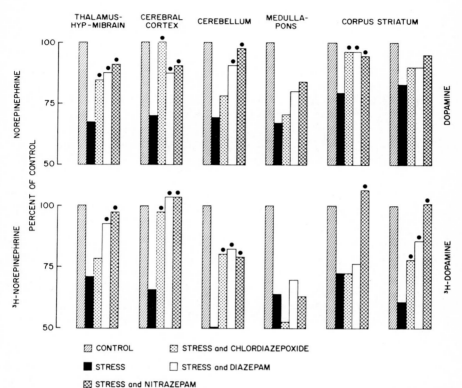

FIG. 1. The effect of electro-footshock stress (1 hr before death) and benzodiazepine drugs (two doses of 10 mg/kg s.c., 2 and 4 hr before death) on endogenous norepinephrine, dopamine, ^{3}H-norepinephrine, and ^{3}H-dopamine in regions of the rat brain 4 hr after the intraventricular injection of 10 μC of ^{3}H-dopamine. Each value is the mean ± S.E.M. of four rats. Significance of the difference between the stress group and groups pretreated with benzodiazepine drugs before stress: $p < 0.01$.

trazepam. The regional antagonistic effect of these drugs on the increased norepinephrine turnover resulting from footshock treatment supports the suggestion that these drugs can decrease catecholamine turnover in some regions of the rat brain.

The biochemical and histochemical results of Corrodi et al. (1971) also indicate that chlordiazepoxide, diazepam, and nitrazepam can completely counteract the increase in norepinephrine turnover in certain brain areas induced by immobilization stress.

Pretreatment with the benzodiazepine drugs also affected rat behavior during the electro-footshock treatment. The treated rats responded to the

1-sec shock, but in the 10-sec period between shocks they were ataxic as compared to the agitated control animals. Thus, both their behavioral and biochemical responses to the shock treatment were reduced by drug pretreatment.

VI. RESTRAINT AND COLD EXPOSURE AND ENDOGENOUS HISTAMINE LEVELS

Histamine satisfies several criteria for describing central neurotransmitters, suggesting that it might have a synaptic function in the brain (Green, 1970; Taylor and Snyder, 1971a; Snyder and Taylor, 1972). Recently we have developed a highly sensitive enzymatic-isotopic assay for histamine, histidine, histamine methyltransferase, and histidine decarboxylase activity in brain tissue (Snyder and Taylor, 1971; Taylor and Snyder, 1972). Using these methods, we found that rat brain histamine, especially in the hypothalamus, had a rapid turnover and this was affected by applying restraint and cold exposure to the animals before death (Taylor and Snyder, 1971b). These stressful procedures lowered hypothalamic histamine to 40% of unstressed animals.

To evaluate the effect of chlordiazepoxide and diazepam on hypothalamic histamine, rats were pretreated with these drugs 1 hr before being placed in restraint boxes in a cold room at 4°C for 2 hr. At the end of this period, the animals were sacrificed and the hypothalamus was assayed for histamine, histidine, and the activities of histamine methyltransferase and histidine decarboxylase. The benzodiazepine drugs tested had no effect on any of these parameters in unstressed animals, but they did prevent the reduction in hypothalamic histamine by cold exposure and restraint (Table IV) without affecting the activity of histidine decarboxylase or histamine methyltransferase. Restraint and cold exposure have also been shown to cause ulceration in the stomach of rats, and this action may be related to excessive histamine release (Senay and Levine, 1967; Radwan and West, 1971). Haot et al. (1964) showed that chlordiazepoxide pretreatment prevented the appearance of gastric ulcers in rats subjected to cold exposure and restraint.

VII. DISCUSSION

The biochemical study of the effects of drugs on putative neurotransmitters has been a useful tool in elucidating the mode of action of psychopharmacologic drugs. This study shows that the benzodiazepine drugs have little

TABLE IV. The effect of restraint and cold exposure (2 hr before death) and benzodiazepine drugs (10 mg/kg, 3 hr before death) on endogenous histamine, histidine, histidine decarboxylase, and histamine methyltransferase activity in the rat hypothalamus

Treatment	Histamine (ng/g)	Histidine (μg/g)	Histidine decarboxylase (nmole/g protein/hr)	Histamine methyltransferase (μmole/g protein/hr)
Control	205 ± 7	27.5 ± 0.8	30.5 ± 1.6	0.48 ± 0.04
Chlordiazepoxide	201 ± 13	26.7 ± 1.0	30.4 ± 2.0	0.46 ± 0.06
Diazepam	196 ± 15	26.8 ± 1.1	29.8 ± 1.9	0.51 ± 0.04
Restraint and cold exposure	141 ± 11*	27.4 ± 0.9	33.4 ± 2.1	0.46 ± 0.05
+ Chlordiazepoxide	188 ± 13	28.2 ± 1.2	31.9 ± 2.2	0.47 ± 0.07
+ Diazepam	194 ± 10	27.2 ± 0.7	30.4 ± 1.7	0.51 ± 0.04

Each value is the mean ± S.E.M. of eight determinations.
* Significance of difference between mean treated and mean control values: $p < 0.01$.

effect on catecholamine levels, metabolism, and reuptake, which is in contrast to other groups of centrally active drugs such as phenothiazines, tricyclic antidepressants, monoamine oxidase inhibitors, rauwolfia alkaloids, and amphetamines.

However, these drugs do reduce norepinephrine turnover in some regions of the rat brain. A histochemical study using more sophisticated dissection procedures (Corrodi et al., 1971) confirmed these results, and suggested that the minor tranquilizers decrease the neuronal activity of the ascending norepinephrine neurons from the locus coeruleus.

Chlordiazepoxide, diazepam, and nitrazepam also prevented the depletion of norepinephrine in specific areas of the brain by electro-footshock stress and by immobilization (Corrodi et al., 1971). Corrodi et al. (1971) indicated that meprobamate and barbiturates, such as phenobarbital, can also block stress-induced activation of central noradrenergic neurons, suggesting that this effect may be a common property of drugs belonging to the group of minor tranquilizers.

Therefore, it is tempting to suggest that part of the pharmacological effects of these drugs is due to a reduction in central noradrenergic neurotransmission in specific areas of the brain. However, from a study of this type it is very difficult to relate a biochemical finding directly to a behavioral effect of a drug. Perhaps this is exemplified by the fact that chlordiazepoxide and diazepam also prevent the reduction in hypothalamic histamine in rats after restraint and cold exposure. This could suggest that the antagonistic action of these drugs on these biochemical effects of stressful procedures may be an indirect effect involving other biochemical or physiological processes.

VIII. SUMMARY

The effect of chlordiazepoxide, diazepam, and nitrazepam on histamine and catecholamine systems in regions of the rat brain has been studied using fluorometric, radioisotopic, and enzymatic-isotopic techniques. These benzodiazepine drugs decrease norepinephrine turnover in the thalamus-hypothalamus-midbrain, cortex, and cerebellum regions and dopamine turnover in the corpus striatum. They also antagonize the increased catecholamine turnover in the above regions after electro-footshock stress. Chlordiazepoxide and diazepam are weak inhibitors of norepinephrine uptake into hypothalamic synaptosomes and dopamine uptake into striatal synaptosomes. They prevent the decrease in hypothalamic histamine elicited by restraint and cold exposure.

ACKNOWLEDGMENTS

The authors would like to thank Dr. Solomon H. Snyder for his expert advice.

This work was supported by U.S. Public Health Service grants MH-18501, NS-07275, and GM-16492 and by a grant from the Golden Kiwi Medical Research Distribution Committee, New Zealand.

REFERENCES

Andén, N. E., Roos, B. E., and Werdinius, B. (1964): Effects of chlorpromazine, halo-peridol, and reserpine on the levels of phenolic acids in rabbit corpus striatum. *Life Sciences,* 3:149–158.

Axelrod, J., Albers, W., and Clemente, C. D. (1959): Distribution of catechol-O-methyl-transferase in the nervous system and other tissues. *Journal of Neurochemistry,* 5: 68–72.

Bliss, E. J., Ailion, J., and Zwanziger, J. (1968): Metabolism of norepinephrine, seroto-nin, and dopamine in rat brain with stress. *Journal of Pharmacology and Experimental Therapeutics,* 164:122–131.

Corrodi, H., Fuxe, K., and Hökfelt, T. (1967): The effect of some psychoactive drugs on central monoamine neurons. *European Journal of Pharmacology,* 1:363–368.

Corrodi, H., Fuxe, K., Lidbrink, P., and Olson, L. (1971): Minor tranquilizers, stress, and central catecholamine neurons. *Brain Research,* 29:2–16.

Crankshaw, D. P., and Raper, C. (1971): The effect of solvents on the potency of chlordiazepoxide, diazepam, medazepam, and nitrazepam. *Journal of Pharmacy and Pharmacology,* 23:313–321.

Green, J. P. (1970): Histamine. In: *Handbook of Neurochemistry,* Vol. 4, edited by A. Lathja. Plenum Press, New York, pp. 221–250.

Haot, J., Djahanguiri, B., and Richelle, M. (1964): Action protéctrice du chlordiace-poxide sur l'ulcère de contrainte chez le rat. *Archives Internationales de Pharmacodynamie et de Thérapie,* 148:557–559.

Hendley, E. D., and Snyder, S. H. (1968): A simple and sensitive fluorescence assay for monoamine oxidase and diamine oxidase. *Journal of Pharmacology and Experimental Therapeutics,* 163:386–392.

Horn, A. S., Coyle, J. T., and Snyder, S. H. (1971): Catecholamine uptake by synapto-somes from rat brain. *Molecular Pharmacology,* 7:66–80.

Laverty, R. (1969): Effect of psychoactive drugs in an operant conflict situation in rats. *Proceedings of the University of Otago Medical School,* 47:57–58.

Laverty, R., and Sharman, D. F. (1965): Modification by drugs of the metabolism of 3,4-dihydroxyphenylethylamine, noradrenaline and 5-hydroxytryptamine in the brain. *British Journal of Pharmacology and Chemotherapy,* 24:759–772.

Laverty, R., and Taylor, K. M. (1968): The fluorometric assay of catecholamine and related compounds: Improvements and extensions to the hydroxyindole technique. *Analytical Biochemistry,* 22:269–278.

Moe, R. A., Bagdon, R. E., and Zbinden, G. (1962): The effect of tranquilizers on myocardial metabolism. *Angiology,* 13:4–12.

Persson, T., and Waldeck, B. (1971): A reduced rate of turnover of brain noradrena-line during pentobarbitone anaesthesia. *Journal of Pharmacy and Pharmacology,* 23: 377–378.

Radwan, A. G., and West, G. B. (1971): Effect of aminoguanidine, chlorpromazine, and NSD-1055 on gastric secretion and ulceration in the Shay rat. *British Journal of Pharmacology and Chemotherapy*, 41:167–169.

Rosecrans, J. A. (1969): Brain amine changes in stressed and normal rats pretreated with various drugs. *Archives Internationales de Pharmacodynamie et de Thérapie*, 180:400–470.

Senay, E. C., and Levine, R. J. (1967): Synergism between cold and restraint for rapid production of stress ulcers in rats. *Proceedings of the Society for Experimental Biology and Medicine*, 124:1221–1223.

Snyder, S. H., and Taylor, K. M. (1972): Assay of biogenic amines and their deaminating enzymes in animal tissues. In: *Methods in Neurochemistry*, edited by R. Rodnight and N. Marks. Plenum Press, New York (*in press*).

Snyder, S. H., and Taylor, K. M. (1972): Histamine in the brain: A neurotransmitter? In: *Perspectives in Neuropharmacology: A Tribute to Julius Axelrod*, edited by S. H. Snyder. Oxford University Press, New York (*in press*).

Sternbach, L. H., Randall, L. O., and Gustafson, S. R. (1964): Chlordiazepoxide and related compounds. In: *Psychopharmacological Agents, Medical Chemistry*, Vol. 4-1, edited by M. Gordon. Academic Press, New York, pp. 137–224.

Taylor, K. M., and Laverty, R. (1969a): The effect of chlordiazepoxide, diazepam, and nitrazepam on catecholamine metabolism in regions of the rat brain. *European Journal of Pharmacology*, 8:296–301.

Taylor, K. M., and Laverty, R. (1969b): The metabolism of tritiated dopamine in regions of the rat brain *in vivo*: The separation of catecholamines and their metabolites. *Journal of Neurochemistry*, 16:1361–1366.

Taylor, K. M., and Laverty, R. (1969c): The metabolism of tritiated dopamine in regions of the rat brain *in vivo*: The significance of the neutral metabolites of catecholamines. *Journal of Neurochemistry*, 16:1367–1376.

Taylor, K. M., and Snyder, S. H. (1971a): Histamine in rat brain: Sensitive assay of endogenous levels, formation *in vivo*, and lowering by inhibitors of histidine decarboxylase. *Journal of Pharmacology and Experimental Therapeutics*, 179:619-633.

Taylor, K. M., and Snyder, S. H. (1971b): Brain histamine: Rapid apparent turnover altered by restraint and cold stress. *Science*, 172:1037–1039.

Taylor, K. M., and Snyder, S. H. (1972): Isotopic microassay of histamine, histidine decarboxylase and histamine methyltransferase in brain tissue. *Journal of Neurochemistry*, 19:1343–1358.

The Benzodiazepines, Raven Press, New York © 1973

THE EFFECTS OF BENZODIAZEPINES, MEPROBAMATE, AND BARBITURATES ON CENTRAL MONOAMINE NEURONS

Peter Lidbrink*, Hans Corrodi**, Kjell Fuxe*, and Lars Olson*

*Department of Histology, Karolinska Institute, Stockholm,
**Department of Pharmacology, University of Göteborg,
and Research Laboratories of Hässle, Göteborg, Sweden

I. INTRODUCTION

For many years, chlordiazepoxide (LIBRIUM), diazepam (VALIUM), and meprobamate have been used in the treatment of psychiatric disorders in man, especially in "neurosis," and are usually known as antianxiety drugs. However, little is known about their neurochemical basis of action. Thus, their action on putative transmitters in the nervous system is as yet poorly understood. A few studies have been performed on their effects on central catecholamine (CA) turnover (Corrodi et al., 1968; Taylor, 1969; Taylor and Laverty, 1969; Chase et al., 1970; Fuxe et al., 1970b; Corrodi et al., 1971). The results of our own work are summarized here, and new findings are also presented. It has been considered of special interest to compare the effects of the minor tranquilizers with those of the barbiturates on the central monoamine neurons.

II. MATERIAL AND METHODS

A. Turnover studies

Monoamine turnover was evaluated with the help of amine-synthesis inhibitors in male Sprague-Dawley rats (150 to 200 g b. wt.) (see Andén et al., 1969; Fuxe et al., 1970a). Alpha-methyl-tyrosine methylester (H44/68) was used as a tyrosine-hydroxylase inhibitor (Andén et al., 1966a; Corrodi and Hansson, 1966); FLA63 [bis(4-methyl-1-homopiperazinylthiocarbonyl)disulphide] as a dopamine-β-hydroxylase inhibitor (Svensson and Waldeck, 1969; Corrodi et al., 1970a, b; Florvall and Corrodi, 1970); and α-propyl dopacetamide (H22/54) as a tryptophan-hydroxylase inhibitor. The rate of decline of the amine stores after amine-synthesis inhibition gives an indication of the amine turnover (Costa and Neff, 1970; Andén et al., 1969). Supramaximal doses of the amine-synthesis inhibitors were used in order to avoid the influence of drug-inhibitor interactions, such as changes in the metabolism and distribution of the inhibitor. In studies on noradrenaline (NA) turnover, both H44/68 and FLA63 can be used. These inhibitors are two chemically different substances. If similar results are obtained with the two inhibitors, drug-inhibitor interactions are unlikely to account for the changes found. Usually, however, the results themselves exclude such interactions because the dopamine (DA), NA and 5-hydroxytryptamine (5-HT) turnovers are often affected differently.

No method available to measure turnover can give the true turnover rate (see also Persson and Waldeck, 1970). We have, therefore, only been interested in detecting *changes* in amine turnover and have not described our results on turnover in absolute terms. The appropriate time intervals after synthesis inhibition used in our studies were obtained by investigating the amine-depletion curves in the original works (Andén et al., 1966*a, c*; Corrodi and Hansson, 1966; Corrodi et al., 1970*a*). Animals were thus sacrificed 2 and 4 hr after H44/68, 4 hr after FLA63, and 3 hr after H22/54. Differences in amine depletion between drug-treated and control animals at these time intervals are regarded as representing changes in amine turnover.

In view of the antianxiety action of minor tranquilizers and barbiturates, it was of interest to evaluate the effects of these drugs on stress-induced changes in monoamine turnover. Immobilization stress was induced by wrapping rats tightly in metal nets. This type of stress increases NA turnover and decreases DA turnover (Corrodi et al., 1968, 1971), and the effects of minor tranquilizers and barbiturates on the stress-induced changes in monoamine turnover were studied.

B. Analytical procedures

The monoamine stores were analyzed both histochemically and biochemically. By combining the two approaches, quantitative information can be obtained at the same time that discrete changes in turnover are detected in various terminal systems. The biochemical analysis of dopamine was made according to Carlsson and Waldeck (1958), and Carlsson and Lindqvist (1962), that of noradrenaline according to Bertler et al. (1958), and that of 5-hydroxytryptamine (5-HT) according to Bertler (1961). The histochemical analysis was performed with the Falck-Hillarp fluorescence technique (Falck et al., 1962; Hillarp et al., 1966; Corrodi and Jonsson, 1967; Fuxe and Jonsson, 1967). The cortical NA nerve terminals were visualized with the help of the smear technique described by Olson and Ungerstedt (1970*a*), which is a rapid and reliable way of demonstrating these terminals. Monoamine nerve terminals in other parts of the brain were studied in freeze-dried material (Olson and Ungerstedt, 1970*b*). It is known that a change in fluorescence intensity represents a change in amine levels (Olson et al., 1968; Jonsson, 1969). Furthermore, the concentration-fluorescence relationship seems to be linear even up to normal levels in cortical NA nerve terminals (Lidbrink and Jonsson 1971), central DA, and 5-HT nerve terminals (Fuxe, *unpublished data*), which facilitates the histochemical evaluation of changes along the entire course of amine depletion. It could be argued that changes in fluorescence could be due to interference by the drug with fluorophor formation. However, such a possibility can be excluded because the fluorescence intensity is not changed by the drug alone and because it is changed following drug and H44/68 treatment only in certain terminal systems. Also, the fact that the biochemical and histochemical results agree with one another rules out such effects.

Evaluations of fluorescence intensities were made on coded slides using a semiquantitative scale from 0 to 3+. When it could not be determined whether the intensity of a certain preparation belonged to one grade or the next, half a plus was added to the lower grade. So far, the histochemical and biochemical results have been in good agreement. Thus, the selective effects on NA turnover in the cortex cerebri (*see below*) observed histochemically were not found in the whole brain analysis but could be confirmed in a biochemical analysis of the cortex cerebri alone.

C. Functional studies

In order to understand how minor tranquilizers and barbiturates can interact with monoaminergic mechanisms, the effects of these drugs have been studied in three different functional models, which are highly dependent on DA, NA, and 5-HT receptor activity, respectively.

1. Studies in rats with a chronic unilateral lesion of the nigro-neostriatal DA pathway (see Andén et al., 1966*b*). The lesions were similar to but larger than those described by Hökfelt and Ungerstedt (1969). Without drug treatment, these rats are straight but, when treated with drugs which increase DA release (amphetamine; see Carlsson et al., 1966) or which directly stimulate the receptors (apomorphine; Ernst, 1967; Andén et al., 1967), the rats will rotate toward the denervated side. DA receptor-blocking agents, on the other hand, will cause marked asymmetry of the rat toward the innervated side. The effects of chlordiazepoxide, diazepam, and phenobarbital have been studied alone or in combination wtih amphetamine or apomorphine.

Locomotion: In order to evaluate effects on DA mechanisms, the influence on locomotion was studied with or without treatment with amphetamine or apomorphine. Locomotion was also induced with dopa combined with a peripheral dopa-decarboxylase inhibitor (RO 4-4602) given to reserpine pretreated animals. Locomotion is highly dependent on activity in the nigro-neostriatal DA neurons (see Carlsson, 1966; Randrup and Munkvad, 1968; Andén, 1970; Fuxe and Ungerstedt 1970), especially stereotyped behavior. Exploratory behavior with rearing activity is also dependent on NA mechanisms (Fuxe and Ungerstedt, 1970; Corrodi et al., 1970b; Randrup and Munkvad, 1970). Locomotion was measured in an activity box (Animex, Farad, Sweden).

2. *Effects on the flexor hindlimb reflex of the acutely spinalized rat.* This reflex is highly dependent on NA receptor activity (Carlsson et al., 1963; Andén et al., 1966d, e). Dopa combined with a monoamine oxidase inhibitor induces a dose-dependent increase in flexor activity. Similar results have been obtained with clonidine, a NA receptor stimulating agent (Andén et al., 1970). The effects of chlordiazepoxide, diazepam, and phenobarbital on this reflex were studied with or without treatment with dopa or clonidine.

3. *Effect on the extensor hindlimb reflex of the acutely spinalized rat.* This reflex is highly dependent on 5-HT receptor activity (Andén et al., 1964; Andén, 1968; Meek et al., 1970). 5-Hydroxytryptophan (5-HTP) and tryptophan in combination with nialamide, induce a dose-dependent increase in extensor reflex activity. Similar results are obtained with 5-HT receptor-stimulating agents such as LSD and 5-methoxy-N-dimethyl-tryptamine (5-Meo-DMT). An investigation was, therefore, conducted to clarify how chlordiazepoxide, diazepam, and phenobarbital influence this reflex, and the 5-HTP-induced and 5-Meo-DMT-induced increase in reflex activity.

III. RESULTS

A. Effects on DA neurons

1. *Turnover studies.* It was found that all barbiturates, benzodiazepines, and meprobamate decreased the H44/68-induced disappearance of DA (Fig. 1). Histochemical studies revealed this effect in the DA nerve terminals of the neostriatum and the limbic forebrain (mainly tuberculum

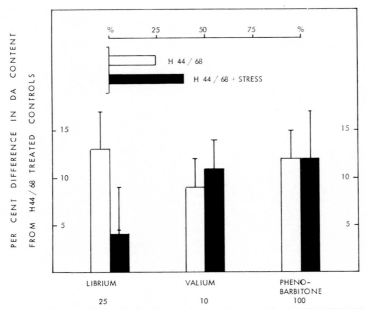

FIG. 1. Biochemical analysis of the effect of chlordiazepoxide (LIBRIUM; 25 mg/kg), diazepam (VALIUM; 10 mg/kg), and phenobarbital (phenobarbitone; 100 mg/kg) on the H 44/68 (250 mg/kg, i.p., 4 hr)-induced DA depletion in whole brain of stressed and unstressed rats. DA levels in the different groups were calculated as the percentage of normal levels (mean ± S.E.M., N: 4-8). The upper two horizontal bars represent the values of the two control groups. The difference between these percentage values and those of the corresponding drug-treated groups are given below.

olfactorium and nucleus accumbens) but were not able to ascertain it to any definite extent in the median eminence. These results suggest a decrease in DA turnover in the forebrain after treatment with minor tranquilizers and barbiturates. The doses used were observed to cause a clear-cut behavioral depression and, in the case of barbiturates, a hypnotic action. Doses causing only minor sedation had only slight if any effects on DA turnover (see also Corrodi et al., 1966, 1971).

In a recent study, it was found that immobilization stress caused a decrease in the H44/68-induced DA disappearance from the forebrain (Corrodi et al., 1971). Treatment with minor tranquilizers and barbiturates resulted in an additional decrease in the H44/68-induced DA disappearance in the forebrain (Fig. 2). Stress also caused a slight decrease in the H44/68-induced DA disappearance in the median eminence but this action was not to any certain extent influenced by minor tranquilizers or barbiturates.

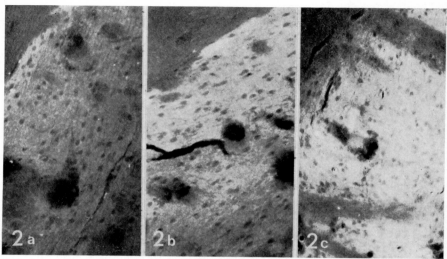

FIG. 2. *a*: Neostriatum of rat after treatment with H 44/68 (250 mg/kg, i.p., 2 hr before sacrifice). A weak DA fluorescence is observed (1+). × 140.
    b: Same area and treatment as described in text for Fig. 2*a*, except that diazepam (VALIUM) was given i.p. at a dose of 10 mg/kg, 15 min earlier. A moderate DA fluorescence (2+) is observed, due to retardation of H 44/68-induced depletion. × 140.
    c: Same area and treatment as described in text for Fig. 2*b*, except that the rats were stressed immediately after the H 44/68 injection. A fairly strong DA fluorescence (2½+) is observed, due to retardation of H 44/68-induced DA depletion. × 120.

2. *Functional studies.* None of the barbiturates or the minor tranquilizers induced rotations or any asymmetry in the rats with a unilateral lesion of the nigro-neostriatal DA pathway. Phenobarbital (50 and 100 mg/kg), chlordiazepoxide (10 and 25 mg/kg), and diazepam (1 and 10 mg/kg), on the other hand, reduced somewhat the rotations induced by both amphetamine (4 mg/kg) and apomorphine (5 mg/kg), in doses that had been found to decrease DA turnover.

    These doses were also found to reduce locomotion and amphetamine- and apomorphine-induced increases in locomotion. The same held true for the dopa-induced increases in locomotion in reserpine-treated animals, after pretreatment with a peripheral dopa-decarboxylase inhibitor, without influencing the selective DA accumulation occurring in the DA neurons.

B. Effects on NA neurons

1. *Turnover studies.* Biochemically, whole brain analysis did not reveal any changes resulting from treatment with barbiturates (Corrodi et al., 1966;

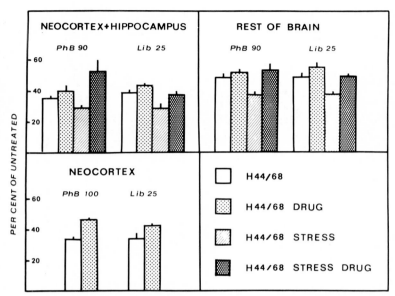

FIG. 3. Biochemical analysis of the effect of chlordiazepoxide (LIBRIUM; 25 mg/kg) and phenobarbital (90 or 100 mg/kg) on the H 44/68 (250 mg/kg, i.p., 4 hr)-induced NA depletion in stressed and unstressed rats. NA levels are expressed as the percentage of untreated groups (mean ± S.E.M., n: 4-8). PhB = phenobarbital, Lib = chlordiazepoxide (LIBRIUM).

Lidbrink et al., 1972) or the minor tranquilizers (Corrodi et al., 1971) of the H44/68-induced NA depletion (Fig. 3). However, histochemical studies selectively revealed decreases in the H44/68-induced fluorescence disappearance in the cortical NA nerve terminals (neocortex, mesocortex, archicortex, and cortex cerebelli (Figs. 4 and 6). No effects were found in the hypothalamus (Fig. 5) and in the lower brainstem. Phenobarbital had the most potent effects. The histochemical results could be confirmed by biochemical analysis of the cortex cerebri alone (Fig. 3). Histochemically, similar results on the effects of barbiturates have been obtained with FLA63. It is of interest that meprobamate, even in doses of 200 mg/kg, did not have this action on cortical NA nerve terminals (Fig. 4).

The stress-induced acceleration of H44/68-induced NA depletion found all over the brain was blocked by all the minor tranquilizers and barbiturates studied (Figs. 3, 4, 5, 6, and 8).

2. Functional studies. The minor tranquilizers and the barbiturates markedly reduced the dopa- and clonidine-induced increase in the flexor reflex activity in the same doses as used in the turnover experiments.

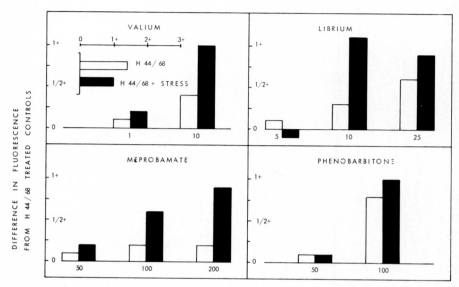

FIG. 4. Histochemical analysis of the effect of diazepam (VALIUM; 1 or 10 mg/kg), chlordiazepoxide (LIBRIUM; 5, 10, or 25 mg/kg), meprobamate (50, 100, or 200 mg/kg), and phenobarbital (phenobarbitone; 50 or 100 mg/kg) on the H 44/68 (250 mg/kg, i.p., 4 hr)-induced NA depletion in the cerebral cortex in stressed and unstressed rats. The two bars in the upper left field represent the mean fluorescence of the control groups. The effect of the drugs is shown as the difference in fluorescence between the experimental and control groups (N: 4-8) (For the representation of fluorescence-intensity estimations, see Material and Methods, and Lidbrink and Jonsson, 1971.)

C. Effects on 5-HT neurons

1. Turnover studies. Biochemical studies revealed a clear-cut reduction of the H22/54-induced 5-HT disappearance in the cortical areas after treatment with chlordiazepoxide and phenobarbital (Fig. 7). A previous study (Corrodi et al., 1967) with pentobarbital revealed also a reduction in H22/54-induced 5-HT disappearance in analysis on whole brain. These effects, however, may be restricted to the cortical areas or are at least not general with regard to chlordiazepoxide and diazepam, since no differences have been found in the H22/54-induced disappearance of 5-HT fluorescence in the nucleus suprachiasmaticus after treatment with these drugs.

2. Functional studies. Chlordiazepoxide, diazepam, and phenobarbital reduced the 5-HTP- and 5-MeO-DMT-induced increase in the extensor hindlimb reflex of the spinalized rat in doses that caused reduction of cortical 5-HT turnover.

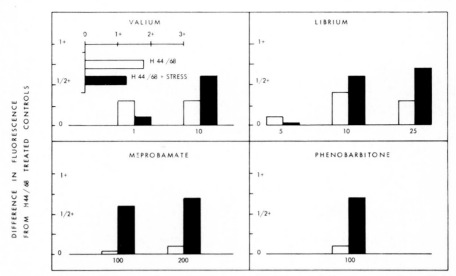

FIG. 5. Histochemical analysis of the effect of diazepam (VALIUM; 1 or 10 mg/kg), chlordiazepoxide (LIBRIUM; 5, 10, or 25 mg/kg), meprobamate (100 or 200 mg/kg), and phenobarbital (phenobarbitone; 100 mg/kg) on the H 44/68 (250 mg/kg, i.p., 4 hr)-induced depletion of NA in the hypothalamus in stressed and unstressed rats. Same procedure as in Fig. 4 (N: 4-8).

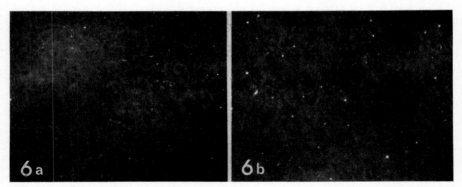

FIG. 6. *a*: Smear of the cerebellar cortex of a rat after treatment with H 44/68 (250 mg/kg, 2 hr before sacrifice). Only a few weakly fluorescent dots, representing the NA varicosities, are observed. × 300.

b: Smear of the cerebellar cortex after treatment with H 44/68 plus phenobarbital (100 mg/kg, i.p.) 15 min earlier. Moderately fluorescent dots in increased numbers are observed, compared with the controls (H 44/68 alone), due to retardation of H 44/68 NA depletion by phenobarbital. × 300.

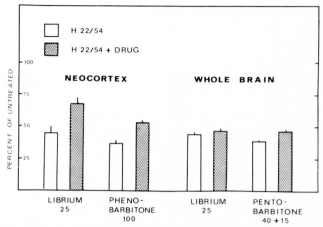

FIG. 7. Biochemical analysis of the effect of chlordiazepoxide (LIBRIUM) (25 mg/kg), phenobarbital (100 mg/kg), and pentobarbital (40 + 15 mg/kg) on the H 22/54 (500 mg/kg, i.p., 3 hr) induced depletion of 5-HT in the *cerebral cortex* and in *whole brain*. 5-HT levels are expressed as the percentage of untreated groups (mean ± S.E.M., N: 4).

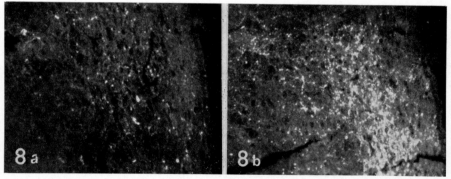

FIG. 8. *a*: The paraventricular area of the hypothalamus in a stressed rat after treatment with H 44/68 (250 mg/kg, i.p., 4 hr before sacrifice). Only a very weak fluorescence intensity is observed in a small number of NA nerve terminals. × 140.

b: The same area and treatment as in *a*, except that chlordiazepoxide (25 mg/kg, i.p.) was given 15 min before H 44/68. Moderately fluorescent NA nerve terminals are observed, due to blockade of stress-induced acceleration of H 44/68-induced NA depletion by chlordiazepoxide. × 120.

IV. DISCUSSION

The present results demonstrate that minor tranquilizers and barbiturates have a very similar action on monoamine neurons, probably causing reductions of DA turnover in the forebrain and NA and 5-HT turnover in the cortical areas. It is known that DA, NA, and 5-HT receptor-stimulating agents cause reduction of DA, NA, and 5-HT turnover, respectively (Andén et al., 1969; Fuxe et al., 1970a). The present functional studies suggest that minor tranquilizers and barbiturates do not have any monoamine receptor stimulating properties. The changes in monoamine turnover observed can, therefore, be due either to inhibition of transmitter release (Banna, 1970) by an action at the monoamine nerve terminal or to effects causing decreases in nervous impulse flow (Narahashi et al., 1971). It is of great interest that recently Pfenninger et al. (1971) have found that barbiturates decrease the number of protuberances which the presynaptic membrane makes into the boutons, an action which could contribute to the reduction of transmitter release. Furthermore, it has been reported that phenobarbital increases the absolute numbers of NA vesicles found in adrenergic nerve terminals of the vas deferens (Côté et al., 1970). MAO inhibition, which also retards monoamine turnover, cannot be involved since the monoamine levels are not increased after treatment with these drugs. Also, the work of others indicates that the minor tranquilizers do not inhibit the amine membrane pump (Chase et al., 1970; Taylor and Laverty, *this volume*). Thus, such actions are unlikely to be responsible for the observed amine-turnover changes.

It is important to note that in the normal animal selective decreases of turnover in NA and 5-HT nerve terminals were obtained in the cortical areas after treatment with chlordiazepoxide, diazepam, and barbiturates but not with meprobamate. Similar results on NA neurons have also been obtained by Taylor and Laverty (1969) using an isotope technique. The cortical areas seem to receive their NA innervation exclusively from the principal locus coeruleus, and the terminals are very fine in contrast to the fine to fairly thick terminals of the hypothalamus, the lower brainstem, and the subcortical limbic system (see Olson and Fuxe, 1971; Ungerstedt, 1971a). Thus, the possibility exists that the benzodiazepines and the barbiturates selectively decrease nervous impulse flow in the cortical NA neurons. Similarly, there may exist cortical 5-HT neurons which are selectively affected. Such effects, caused either directly or indirectly, would explain our results. However, it must also be remembered that the NA turnover is higher in the very fine NA nerve terminals of the cortical areas than in the fairly large NA nerve terminals in other parts of the brain (see Iversen and Glowinski, 1966; Corrodi et al., 1971). This may also be true for cortical 5-HT turnover. Therefore, an action on transmitter release would reveal itself first in terminals with a high turnover such as the cortical NA nerve terminals and the telencephalic DA nerve terminals.

It has not been possible to obtain any evidence that minor tranquilizers and barbiturates diminish CA or 5-HT release in functional experiments, since reductions of reflex activity, locomotion, and rotational behavior were obtained not only when the compounds were given together with monoamine precursors or monoamine releasing agents, but also when combined with monoamine receptor-stimulating agents. Thus, the minor tranquilizers and barbiturates have potent effects beyond the monoamine receptors, which are capable of reducing the reflex and behavioral functions dependent on monoamine-receptor activity. Thus, it seems clear that these drugs must act also on nonmonoamine neuron systems.

The drugs under study have well-known subcellular metabolic effects [e.g., inhibition of a reduced nicotinamide adenine dinucleotide phosphate-linked aldehyde reductase in the brain (Erwin et al., 1971], which could result in accumulations of biogenic aldehydes. Such aldehydes have been postulated to have a depressant action on CNS functions (Sabelli et al., 1969). Uncoupling of oxidative phosphorylation also occurs (see Doggett et al., 1970). It is not known if these effects can be related to decreases in transmitter release.

In stressed rats, no selectivity was observed in the responses of the central NA neurons to minor tranquilizers and barbiturates. There was a blockade of stress-induced acceleration of NA turnover all over the brain, as revealed both in the H44/68 and in the FLA63 model. In the cortical areas, the stress-induced acceleration of the H44/68-induced NA depletion was even reversed into a retardation of the H44/68-induced NA depletion. These effects are common to all barbiturates and minor tranquilizers studied and are in good agreement with the work of Taylor and Laverty (1969), who reported blockade of electroshock-induced acceleration of NA turnover with minor tranquilizers, using labeled tyrosine and DA in their turnover studies.* It must be pointed out that species differences may exist in the response of the NA neurons to barbiturates, since Persson and Waldeck (1971) have found a reduction of NA turnover after FLA63 treatment but not after H44/68 treatment in the mouse with whole brain analysis. However, it may also be that the high doses of FLA63 given intraperitoneally by these authors can induce stress in mice, explaining the retardation seen after combined barbiturate-FLA63 treatment.

---

* A stress-induced depression of hexobarbital metabolism has been found, but is probably of little importance for the present results in view of the high doses used (Wei and Wilson, 1971).

It is of great interest that the reduction of cortical NA turnover induced by benzodiazepines and barbiturates is blocked by the psychostimulant drug amphetamine (5 mg/kg), which can induce its normal increase in cortical NA turnover in spite of the presence of benzodiazepines and barbiturates (*unpublished data*). These results suggest that the latter compounds cannot block extragranular release of amines but probably only a release from the granular pool, since amphetamine is a releasor of extragranular amines (Carlsson et al., 1966). On the other hand, the increase of H44/68-induced NA depletion caused by the anxiety-producing agent yohimbine (1 mg/kg) in the cortical areas is reduced by the barbiturates and benzodiazepines (Lidbrink and Fuxe, *unpublished data*). This could suggest that yohimbine brings about a granular release via a direct or indirect action. The data from the stress experiments suggest that such a release is sensitive to the action of the drugs under study, since nervous impulse-induced amine release only occurs from the granules (see Malmfors, 1965).

On the other hand, stress induced a deceleration of the DA turnover which was potentiated by the minor tranquilizers and barbiturates. Similarly these drugs also caused a reduction of DA turnover in nonstressed rats. In agreement with this it has been found that another form of anesthesia, N_2O-O_2-penthane, probably has an inhibitory effect on the release of ^3H-DA from the neostriatum (Besson et al., 1971). From the present results, it seems clear that minor tranquilizers and barbiturates in the relatively high dose range used diminish DA, NA, and 5-HT release, particularly in the cortical areas and especially in relation to stress. It is possible that part of their pharmacological effects could be mediated via activity in the central monoamine neurons.

Functional studies indicated that these drugs also affect other neuron systems, counteracting increases in behavior and reflex activity in spinalized rats caused by monoamine-dependent processes.*

One of the effects claimed for the benzodiazepines is their taming action on monkeys, cats, and rats (Heise and Boff, 1961; Randall et al., 1961; Schallek et al., 1962). Overstimulation of NA receptors with nialamide-dopa or clonidine treatment is associated with aggressive behavior (see Randrup and Munkvad, 1970; Andén et al., 1970), and sham-rage behavior, induced by brainstem transection, is dependent on activation of central NA receptors (Reis and Fuxe, 1969). Therefore, it seems likely that the decrease of NA turnover produced by minor tranquilizers and barbiturates,

---

* These studies also show that benzodiazepines and barbiturates have depressant actions on reflex activity not only by a direct action on the spinal cord but also by an action on the reticular formation of the lower brainstem (Tseng and Wang, 1971).

especially in relation to stress, could contribute to the taming action of these drugs in the high dose range. With chronic treatment, however, an increased aggressiveness has instead been observed (see Fox et al., 1970).

Extinction or acquisition of fear-motivated behavior such as active and passive avoidance and various types of conflict behavior are highly influenced by treatment with minor tranquilizers or barbiturates. Thus, these drugs increase the rate of extinction of a conditioned avoidance response and diminish the suppressant effects of shock punishment (see Miller, 1964, 1966; Kumar, 1971; Cook and Davidsson, *this volume*). On the other hand, amphetamine, a CA-releasing agent, seems to increase the amount of fear and retards the extinction of fear-motivated behavior. There is no doubt that rats displaying fear-motivated behavior are in a highly stressful situation. Therefore, their NA turnover should be increased, which has, in fact, also been found to be true in rats performing lever pressing in negatively reinforced conditioned avoidance behavior (Fuxe and Hansson, 1967).

In view of the above, it seems likely that part of the mechanism for the fear-reducing properties of the minor tranquilizers and the barbiturates could be the reduction of NA turnover induced by these drugs. In agreement with this view it has been found that amphetamine, which counteracts the inhibitory effects of barbiturates and benzodiazepines on NA release, also antagonizes to a considerable extent the fear-reducing properties of these drugs. The mechanism by which the NA neurons influence these behavioral processes could lie in their inducing increased cortical arousal (Jones et al., 1969; see also review by Fuxe et al., 1970; Fuxe and Lidbrink, 1971). The increased degree of cortical arousal could, in turn, be responsible for a higher degree of memory of punishment, resulting in, e.g., retardation of extinction rates of negatively reinforced behavior. In agreement with this discussion, the increased and decreased degree of fear-motivated responses, found in adrenalectomized and hypophysectomized rats (de Wied, 1969; Weiss et al., 1970), can be related to an increased and decreased NA turnover, respectively (Fuxe et al., 1971). It seems possible that the antianxiety activities of these drugs in man could partly be mediated via a reduction of NA neuronal activity. Further support for this view has recently been obtained in studies on yohimbine, a drug that causes anxiety (Lidbrink and Fuxe, *unpublished data*). As mentioned above, minor tranquilizers and barbiturates diminish the increased NA turnover induced by this drug.

Benzodiazepines increase the threshold for behavioral arousal produced by electrical brain stimulation (Randall et al., 1961). Since behavioral arousal is dependent on activity in the nigro-neostriatal DA neurons (see review by Carlsson, 1966; Randrup and Munkvad, 1968; Andén, 1970;

Fuxe et al., 1970*a*; Fuxe and Ungerstedt, 1970; Ungerstedt, 1971*b*), the reduction of DA turnover in the forebrain after benzodiazepines seems to give an explanation for these effects of the minor tranquilizers. On the other hand, both amphetamine- and apomorphine-induced locomotion is reduced by these drugs and also by barbiturates in the higher dose range. In view of the fact that these compounds have hardly any DA receptor blocking properties (see above), these behavioral effects may be due to interference with nonmonoaminergic systems and may not involve DA receptors directly. In agreement with this the amphetamine and apomorphine induced stereotyped behavior persists (see also Babbini et al., 1971) probably because this type of behavior only involves the head and not the body.

The decrease of cortical 5-HT turnover caused by minor tranquilizers and barbiturates could also be important for the reduction of "fear-motivated" behavior brought about by these drugs. It has been shown that electrical stimulation of the mesencephalic raphe nuclei, which to a large extent are built up of 5-HT cell bodies and have ascending axons (Dahlström and Fuxe, 1964), results in blockade of habituation processes (Sheard and Aghajanian, 1968). This is probably related to the increased release of 5-HT. In support of this view, a 5-HT receptor-stimulating agent, such as D-LSD, induces similar effects (Bradley and Key, 1958). The mechanism of this blockade of habituation processes could be that activation of 5-HT neurons indirectly, via descending pathways, results in decreased functioning of the gating processes in the reticular formation, allowing a higher degree of the sensory input to reach the cortex cerebri. Such events could contribute to the reduction of "fear-motivated" responses, e.g., by increase of pain threshold. The ascending 5-HT neurons could also be visualized as a nonreinforcement system (Poschel and Ninteman, 1971), with the cortical projection being selectively affected by barbiturates and benzodiazepines.

Throughout the present study one is impressed by the similarities found in the action of the minor tranquilizers and barbiturates on the central monoamine neurons. This was not unexpected since these two groups of compounds have many pharmacological effects in common and both have an antianxiety action in man. Meprobamate differed from the benzodiazepines only in its effectiveness on cortical NA turnover in unstressed rats; it caused no reduction in NA turnover. This is in agreement with the fact that the benzodiazepines are more active as sedative agents than meprobamate.

It must be remembered that the effects observed on monoamine turnover were obtained only with relatively high doses. Therefore, low doses of these drugs may have other, not yet discovered, effects on central monoamines. In low doses, both barbiturates and minor tranquilizers have been found to potentiate the actions of amphetamine on behavior (Rushton and Steinberg, 1963, 1967). This behavioral action suggests that, in low doses, these compounds could have an effect on the CA neurons, promoting a release which is masked with higher doses. Such an action could also explain an increased aggressiveness found in male mice when chronically treated with chlordiazepoxide and diazepam (Fox et al., 1970; Guaitani et al., 1971).

V. SUMMARY

A combined biochemical and histochemical approach with amine-synthesis inhibitors was used to study the effects of minor tranquilizers and barbiturates on central monoamine turnover. The influence of these drugs on central monoaminergic mechanisms was evaluated in various functional models. It was found that the compounds were particularly potent in stressed rats, in which they counteracted stress-induced increases in NA turnover all over the brain. In unstressed rats, only cortical NA turnover was reduced, leaving the NA turnover in the brainstem unaffected. The DA and 5-HT turnover in the forebrain and in the cortex cerebri was also reduced respectively. Studies in functional models reveal that these compounds also act beyond the monoamine receptors, depressing brain functions such as behavior and reflex activity. It is conceivable that the reduction of fear-motivated behavior and aggressiveness after these compounds could be due to (1) a reduction of NA turnover, since NA neurons are probably involved in the processes mediating cortical arousal and (2) to the reduction of cortical 5-HT turnover since the cortical 5-HT neurons may be involved in some, e.g., nonreinforcement mechanism. The reduction of DA turnover could contribute to the behavioral depression observed after these compounds.

ACKNOWLEDGMENT

This work has been supported by grants from the Swedish Medical Research Council (B72-14X-1015-05 and B72-14X-3185-02), Ollie och Elof Ericssons Stiftelse, Magn. Bergvalls Stiftelse and Stiftelsen Therese och Johan Anderssons Minne.

REFERENCES

Andén, N.-E. (1968): Discussion of serotonin and dopamine in the extrapyramidal system. *Advances in Pharmacology,* 6A:347–349.

Andén, N.-E. (1970): Effects of amphetamine and some other drugs on central catecholamine mechanisms. In: *Amphetamines and Related Compounds,* edited by E. Costa and S. Garattini. Raven Press, New York, pp. 447–462.

Andén, N.-E., Corrodi, H., Dahlström, A., Fuxe, K., and Hökfelt, T. (1966a): Effects of tyrosine hydroxylase inhibition on the amine levels of central monoamine neurons. *Life Sciences,* 5:561–568.

Andén, N.-E., Corrodi, H., and Fuxe, K. (1969): Turnover studies using synthesis inhibition. In: *Metabolism of Amines in the Brain,* edited by G. Hooper. Macmillan, London.

Andén, N.-E., Corrodi, H., Fuxe, K., Hökfelt, B., Hökfelt, T., Rydin, C., and Svensson, T. (1970): Evidence for a central noradrenaline receptor stimulation by clonidine. *Life Sciences,* 9:513–523.

Andén, N.-E., Dahlström, A., Fuxe, K., and Larsson, K. (1966b): Functional role of the nigro-neostriatal dopamine neurons. *Acta Pharmacologica et Toxicologica,* 24: 263–274.

Andén, N.-E., Fuxe, K., and Hökfelt, T. (1966c): The importance of the nervous impulse flow for the depletion of the monoamines from central neurons by some drugs. *Journal of Pharmacy and Pharmacology,* 18:630–632.

Andén, N.-E., Fuxe, K., Hökfelt, T., and Rubensson, A. (1967): Evidence for dopamine receptor stimulation by apomorphine. *Journal of Pharmacy and Pharmacology,* 19: 335–337.

Andén, N.-E., Jukes, M. G. M., and Lundberg, A. (1964): Spinal reflexes and monoamine liberation. *Nature,* 262:1222–1223.

Andén, N.-E., Jukes, M. G. M., and Lundberg, A. (1966d): The effect of DOPA on the spinal cord. 2. A pharmacological analysis. *Acta Physiologica Scandinavica,* 67: 387–397.

Andén, N.-E., Jukes, M. G. M., Lundberg, A., and Vyklický, L. (1966e): The effect of DOPA on the spinal cord. 1. Influence on transmission from primary afferents. *Acta Physiologica Scandinavica,* 67:373–386.

Babbini, M., Montanaro, N., Strocchi, P., and Gaiardi, M. (1971): Enhancement of amphetamine-induced stereotyped behavior by benzodiazepines. *European Journal of Pharmacology,* 13:330–340.

Banna, N. R. (1970): Antagonism of barbiturate depression of spinal transmission of catechol. *Experientia,* 26:1330–1331.

Bertler, Å. (1961): Effect of reserpine on the storage of catecholamines in brain and other tissues. *Acta Physiologica Scandinavica,* 51:75–83.

Bertler, Å., Carlsson, A., and Rosengren, E. (1958): A method for the fluorimetric determination of adrenaline and noradrenaline in tissues. *Acta Physiologica Scandinavica,* 44:273–292.

Besson, M.-J., Cheramy, A., Feltz, P., and Glowinski, J. (1971): Dopamine: Spontaneous and drug-induced release from the caudate nucleus in the cat. *Brain Research,* 32:407–424.

Bolme, P., Fuxe, K., and Lidbrink, P. (1972): On the function of central catecholamine neurons—their role in cardiovascular and arousal mechanisms. *Research Commission on Chemical Pathology and Pharmacology (in press).*

Bradley, P. B., and Key, B. J. (1958): The effect of drugs on arousal responses produced by electrical stimulation of the reticular formation of the brain. *Electroencephalography and Clinical Neurophysiology,* 10:97–110.

Carlsson, A. (1966): Drugs, which block the storage of 5-hydroxytryptamine and related amines. In: *Handbook of Experimental Pharmacology,* edited by O. Eichler and A. Farah. Springer-Verlag, Berlin.

Carlsson, A., Fuxe, K., Hamberger, B., and Lindqvist, M. (1966): Biochemical and histochemical studies on the effects of imipramine-like drugs and (+)-amphetamine on central and peripheral catecholamine neurons. *Acta Physiologica Scandinavica,* 67:481–497.

Carlsson, A., and Lindqvist, M. (1962): A method for the determination of normeta-nephrine in brain. *Acta Physiologica Scandinavica,* 54:83–86.

Carlsson, A., Magnusson, T., and Rosengren, E. (1963): 5-hydroxytryptamine of the spinal cord normally and after transection. *Experientia,* 19:359–360.

Carlsson, A., and Waldeck, B. (1958): A fluorimetric method for the determination of dopamine (3-hydroxytryptamine). *Acta Physiologica Scandinavica,* 44:293–298.

Chase, T. N., Katz, R. I., and Kopin, I. J. (1970): Effect of diazepam on fate of intra-cisternally injected serotonin-C^{14}. *Neuropharmacology,* 9:103–108.

Corrodi, H., Fuxe, K., Hamberger, B., and Ljungdahl, Å. (1970a): Studies on central and peripheral noradrenaline neurons using a new dopamine-β-hydroxylase inhibitor. *European Journal of Pharmacology,* 12:145–155.

Corrodi, H., Fuxe, K., and Hökfelt, T. (1966): The effect of barbiturates on the activity of the catecholamine neurons in the rat brain. *Journal of Pharmacy and Pharmacology,* 18:556–558.

Corrodi, H., Fuxe, K., and Hökfelt, T. (1967): The effect of some psychoactive drugs on central monoamine neurons. *European Journal of Pharmacology,* 1:363–368.

Corrodi, H., Fuxe, K., and Hökfelt, T. (1968): The effect of immobilization stress on the activity of central monoamine neurons. *Life Sciences,* 7:107–112.

Corrodi, H., Fuxe, K., Lidbrink, P., and Olson, L. (1971): Minor tranquillizers, stress and central catecholamine neurons. *Brain Research,* 29:1–16.

Corrodi, H., Fuxe, K., Ljungdahl, Å., and Ögren, S.-O. (1970b): Studies on the action of some psychoactive drugs on central noradrenaline neurones after inhibition of dopamine-β-hydroxylase. *Brain Research,* 24:451–470.

Corrodi, H., and Hansson, L. (1966): Central effects of an inhibition of tyrosine hydroxylation. *Psychopharmacologia* (Berlin), 10:116-125.

Corrodi, H., and Jonsson, G. (1967): The formaldehyde fluorescence method for the histochemical demonstration of biogenic amines. *Journal of Histochemistry and Cytochemistry,* 15:65–78.

Costa, E., and Neff, N. H. (1970): Estimation of turnover rates to study the metabolic regulation of the steady-state level of neuronal monoamines. In: *Handbook of Neurochemistry* IV, edited by A. Lajtha. Plenum Press, New York, London.

Côté, M. G., Blovin, A., and Gascon, A. (1970): Influence of pretreatment with phenobarbitone on the ultrastructure of adrenergic nerve endings in guinea-pig seminal vesicles. *Journal of Pharmacy and Pharmacology,* 22:129–130.

Dahlström, A., and Fuxe, K. (1964): Evidence for the existence of monoamine containing neurons in the central nervous system. I. Demonstration of monoamines in the cell bodies of brain stem neurons. *Acta Physiologica Scandinavica,* 62, *Suppl.*:232.

De Wied, D. (1969): Effects of peptide hormones on behaviour. In: *Frontiers in Endocrinology,* edited by W. F. Ganong and L. Martini. Oxford University Press, New York, pp. 97–140.

Doggett, N. S., Spencer, P. S. J., and Waite, R. (1970): Hypnotic activity of centrally administered barbiturate and uncouplers of oxidative phosphorylation. *European Journal of Pharmacology,* 13:23–29.

Eakins, K. E., Costa, E., Katz, R. L., and Reyes, C. L. (1968): Effect of pentobarbital anaesthesia on the turnover of (^{3}H) noradrenaline in peripheral tissues of cats. *Life Sciences,* 7:71–76.

Ernst, A. M. (1967): Mode of action of apomorphine and dexamphetamine on gnawing compulsion in rats. *Psychopharmacologia* (Berlin), 10:316–323.

Erwin, V. G., Tabakoff, B., and Bronaugh, R. L. (1971): Inhibition of a reduced nicotinamide adenine dinucleotide phosphate-linked aldehyde reductase from bovine brain by barbiturates. *Molecular Pharmacology*, 7:169–176.

Falck, B., Hillarp, N.-Å., Thieme, G., and Torp, A. (1962): Fluorescence of catecholamines and related compounds condensed with formaldehyde. *Journal of Histochemistry and Cytochemistry*, 10:348–354.

Florvall, L., and Corrodi, H. (1970): Dopamine-β-hydroxylase inhibitors. The preparation and the dopamine-β-hydroxylase inhibitory activity of some compounds related to dithiocarbamic acid and thiuramdisulfide. *Acta Pharmaceutica Suecica*, 7:7–22.

Fox, K. A., Tuckosh, J. R., and Wilcox, A. H. (1970): Increased aggression among grouped male mice fed chlordiazepoxide. *European Journal of Pharmacology*, 11:119–121.

Fuxe, K., Corrodi, H., Hökfelt, T., and Jonsson, G. (1970a): Central monoamine neurons and pituitary-adrenal activity. *Progress in Brain Research*, 32:42–56.

Fuxe, K., and Hansson, L. C. F. (1967): Central catecholamine neurons and conditioned avoidance behavior. *Psychopharmacologia* (Berlin), 11:439–447.

Fuxe, K., Hökfelt, T., and Jonsson, G. (1970b): Participation of central monoaminergic neurons in the regulation of anterior pituitary secretion. In: *Neurochemical Aspects of Hypothalamic Function,* edited by L. Martini and J. Meites. Academic Press, New York-London, pp. 61–83.

Fuxe, K., Hökfelt, T., Jonsson, G., and Lidbrink, P. (1972): Brain-endocrine interaction: Are some effects of ACTH and adrenocortical hormones on neuroendocrine regulation and behavior mediated via central catecholamine neurons? *Congress of the International Society of Psychoneuroendocrinology* (in press).

Fuxe, K., and Jonsson, G. (1967): A modification of the histochemical fluorescence method for the improved localization of 5-hydroxytryptamine. *Histochemie,* 11:161–166.

Fuxe, K., and Ungerstedt, U. (1970): Histochemical, biochemical and functional studies on central monoamine neurons after acute and chronic amphetamine administration. In: *Amphetamines and Related Compounds,* edited by E. Costa and S. Garattini. Raven Press, New York, pp. 257–288.

Guaitani, A., Marcucci, F., and Garattini, S. (1971): Increased aggression and toxicity in grouped male mice treated with tranquilizing benzodiazepines. *Psychopharmacologia* (Berlin), 19:241–245.

Heise, G. A., and Boff, E. (1961): Taming action of chlordiazepoxide. *Federation Proceedings,* 20:393.

Hillarp, N. -Å., Fuxe, K., and Dahlström, A. (1966): Central monoamine neurons. In: *Mechanisms of Release of Biogenic Amines,* edited by U. S. von Euler, S. Rosell, and B. Uvnäs. Oxford Pergamon Press.

Hökfelt, T., and Ungerstedt, U. (1969): Electron and fluorescence microscopical studies on the nucleus caudatus putamen of the rat after unilateral lesions of ascending nigroneostriatal dopamine neurons. *Acta Physiologica Scandinavica,* 76:415–426.

Iversen, L. L., and Glowinski, J. (1966): Regional studies of catecholamines in the rat brain. II. Rate of turnover of catecholamines in various brain regions. *Journal of Neurochemistry,* 13:671–682.

Jones, B. E. (1969): Catecholamine-containing neurons in the brain stem of the cat and their role in waking. M. A. Thesis. Lyon, France.

Jones, B. E., Bobillier, P., and Jouvet, M. (1969): Effets de la destruction des neurones contenant des catécholamines du mésencéphale sur le cycle veille—sommeils du chat. *Comptes Rendus Société Biologie,* 163:176–180.

Jonsson, G. (1969): Microfluorimetric studies on the formaldehyde-induced fluorescence of noradrenaline in adrenergic nerves of rat iris. *Journal of Histochemistry and Cytochemistry,* 17:714–723.

Kumar, R. (1971): Extinction of fear I: Effects of amylobarbitone and dexamphetamine given separately and in combination on fear and exploratory behaviour in rats. *Psychopharmacologia* (Berlin), 19:163–187.

Lidbrink, P., Corrodi, H., Fuxe, K., and Olson, L. (1972): Barbiturates and meprobamate: Decreases in catecholamine turnover of central dopamine and noradrenaline neuronal systems and the influence of immobilization stress. *Brain Research,* 45:507–524.

Lidbrink, P., and Jonsson, G. (1971): Semiquantitative estimation of formaldehyde-induced fluorescence of noradrenaline in central noradrenaline nerve terminals. *Journal of Histochemistry and Cytochemistry,* 19:747–757.

Malmfors, T. (1965): Studies on adrenergic nerves. The use of rat and mouse iris for direct observations on their physiology and pharmacology at cellular and subcellular levels. *Acta Physiologica Scandinavica,* 64 (*Suppl.* 248): 1–93.

Meek, J., Fuxe, K., and Andén, N. -E. (1970): Effects of antidepressant drugs of the imipramine type on central 5-hydroxytryptamine neurotransmission. *European Journal of Pharmacology,* 9:325–332.

Miller, N. E. (1964): The analysis of motivational effects illustrated by experiments on amylobarbitone sodium. In: *Animal Behaviour and Drug Action,* edited by H. Steinberg, A. V. S. De Reuck, and J. Knight.

Miller, N. E. (1966): Some animal experiments pertinent to the problem of combining psychotherapy with drug therapy. *Comprehensive Psychiatry,* 7:1–12.

Narahashi, T., Frazier, D. T., Deguchi, T., Cleaves, C. A., and Ernau, M. C. (1971): The active form of pentobarbital in squid giant axons. *Journal of Pharmacology and Experimental Therapeutics,* 177:25–33.

Olson, L., and Fuxe, K. (1971): On the projections from the locus coeruleus noradrenaline neurons: The cerebellar innervation. *Brain Research,* 28:165–171.

Olson, L., Hamberger, M., Jonsson, G., and Malmfors, T. (1968): Combined fluorescence histochemistry and ^{3}H-noradrenaline measurements of adrenergic nerves. *Histochemie,* 15:38–45

Olson, L., and Ungerstedt, U. (1970a): Monoamine fluorescence in CNS smears: Sensitive and rapid visualization of nerve terminals without freeze-drying. *Brain Research,* 17:343–347.

Olson, L., and Ungerstedt, U. (1970b): A simple high capacity freeze-drier for histochemical use. *Histochemie,* 22:8–19.

Persson, T., and Waldeck, B. (1970): Some problems encountered in attempting to estimate catecholamine turnover using labelled tyrosine. *Journal of Pharmacy and Pharmacology,* 22:473–478.

Persson, T., and Waldeck, B. (1971): A reduced rate of turnover of brain noradrenaline during pentobarbitone anaesthesia. *Journal of Pharmacy and Pharmacology,* 23:377–378.

Pfenniger, K., Akert, K., Moore, H., and Sandri, C. (1972): The fine structure of presynaptic membranes. *Journal of Ultrastructure Research* (*in press*).

Poschel, B. P. H., and Ninteman, F. W., (1971): Intracranial reward and the forebrain's serotonergic mechanism. Studies employing *para*-chlorophenylalanine and *para*-chloroamphetamine. *Physiology and Behaviour,* 7:39–46.

Randall, L. O., Heise, G. A., Schallek, W., Bagdon, R. E., Banziger, R., Boris, A., Moe, R. A., and Abrams, W. B. (1961): Pharmacological and clinical studies on valium, a new psychotherapeutic agent of the benzodiazepine class. *Current Therapeutic Research. Clinical and Experimental,* 3:405–425.

Randrup, A., and Munkvad, I. (1968): Behavioural stereotypes induced by pharmacological agents. *Pharmakopsychiatrie und Neuro-Psychopharmakologie,* 1:18–26.

Randrup, A., and Munkvad, I. (1970): Biochemical, anatomical and psychological investigations of stereotyped behavior induced by amphetamines. In: *Amphetamines and Related Compounds,* edited by E. Costa and S. Garattini. Raven Press, New York, pp. 695–713.

Reis, D. J., and Fuxe, K. (1969): Brain norepinephrine: Evidence that neuronal release is essential for sham rage behaviour following brainstem transection in cat. *Proceedings of the National Academy of Sciences* (Washington), 64:108–112.

Rushton, R., and Steinberg, H. (1963): Mutual potentiation of amphetamine and amylobarbitone measured by activity in rats. *British Journal of Pharmacology and Chemotherapy,* 21:295–305.

Rushton, R., and Steinberg, H. (1967): Drug combinations and their analysis by means of exploratory activity in rats. In: *Neuropsychopharmacology,* edited by H. Brill et al. Excerpta Medica, Amsterdam.

Sabelli, H. C., Giardani, W. J., Alivisatos, S. G. A., Seth, P. K., and Ungar, F. (1969): Indolacetaldehydes: Serotonin-like effects on the central nervous system. *Nature,* 223:73.

Schallek, W., Kuehn, A., and Jew, N. (1962): Effects of chlordiazepoxide (*LIBRIUM*) and other psychotropic agents on the limbic system of the brain. *Annals of the New York Academy of Sciences,* 96:303–312.

Sheard, M. H., and Aghajanian, G. K. (1968): Stimulation of midbrain raphé: Effect on serotonin metabolism. *Journal of Pharmacology and Experimental Therapeutics,* 163:425–430.

Svensson, T. H., and Waldeck, B. (1969): On the significance of central noradrenaline for motor activity. Experiments with a new dopamine-β-hydroxylase inhibitor. *European Journal of Pharmacology,* 7:278–282.

Taylor, K. M. (1969): The effect of minor tranquillizers on stress-induced noradrenaline turnover in the rat brain. *Proceedings of the University of Otaga Medical School,* 47:33–35.

Taylor, K. M., and Laverty, R. (1969): The effect of chlordiazepoxide, diazepam and nitrazepam on catecholamine metabolism in regions of the rat brain. *European Journal of Pharmacology,* 8:296–301.

Tseng, T.-C. and Wang, S. C. (1971): Locus of central depressant action of some benzodiazepine analogues. *Proceedings of the Society for Experimental Biology and Medicine,* 137:526–531.

Ungerstedt, U. (1971a): Stereotaxic mapping of the monoamine pathways in the rat brain. *Acta Psysiologica Scandinavica, Suppl.* 367:1–48.

Ungerstedt, U. (1971b): Striatal dopamine release after amphetamine nerve degeneration revealed by rotational behavior. *Acta Physiologica Scandinavica, Suppl.* 367:49–68.

Wei, E., and Wilson, J. T. (1971): Stress-mediated decrease in liver hexobarbital metabolism: The role of corticosterone and somatotropin. *The Journal of Pharmacology and Experimental Therapeutics,* 177:227–233.

Weiss, J. M., McEwen, B. B., Silva, M. T., and Kalkut, M. (1970): Pituitary-adrenal alterations and fear responding. *American Journal of Physiology,* 218:864–868.

The Benzodiazepines, Raven Press, New York © 1973

EFFECT OF BENZODIAZEPINES ON UPTAKE AND EFFLUX OF SEROTONIN IN HUMAN BLOOD PLATELETS *IN VITRO*

O. Lingjaerde, Jr.

*Department of Neurochemistry, The University Psychiatric Clinic,
Vinderen, Oslo 3, Norway*

I. INTRODUCTION

Little is known about the biochemical effects of benzodiazepine tran-quilizers in the brain. Some earlier studies seemed to indicate that, unlike neuroleptics and antidepressants, benzodiazepines do not interfere signifi-cantly with the turnover and function of catecholamines or serotonin (5HT) in brain, although a slight increase of 5-hydroxyindoleacetic acid (5HIAA) and a slight decrease of homovanillic acid (HVA) was found in the brain-stem of rats pretreated with chlordiazepoxide or diazepam (Prada and Pletscher, 1966). Studies on blood platelets, which have often been used successfully as a model of serotoninergic neurons, were also not very re-vealing. Thus, Pletscher (1969) has reported that chlordiazepoxide and diazepam in concentrations up to 10^{-4}M do not inhibit 5HT uptake in guinea pig platelets (incubated for 1 hr with 0.5 μg/ml ^{14}C-5HT at 37°). As distinct from phenobarbital and meprobamate, however, chlordiazepoxide was found to have a certain stimulating effect on the spontaneous release of 5HT from rabbit platelets during incubation for 2 hr (Pletscher, Prada and Voglar, 1967).

Several recent studies seem to indicate that the benzodiazepines do interfere with monoamine turnover and function in the brain after all. Thus, several benzodiazepines were found to decrease the turnover of do-pamine (DA) and noradrenaline (NA) in the rat brain, and to diminish the increase of NA turnover induced by electric shock (Taylor, 1969; Taylor and Laverty, 1969). As to 5HT, Chase, Katz, and Kopin (1970) have found that diazepam increases whole brain levels of ^{14}C-5HT and ^{14}C-5HIAA in rats decapitated 3 hr after intracisternal injection of ^{14}C-5HT, whereas the drug does not seem to affect the relatively rapid uptake of

225

5HT from the cistern. The result is interpreted as being due to two different mechanisms: (1) an inhibition of the transport of 5HIAA from the brain, and (2) a decreased release or utilization of 5HT.

Because of these recent findings, which strongly indicate that benzodiazepines interfere with both catecholamine and 5HT function in the brain, the author found it of interest to reevaluate the effects of benzodiazepines on uptake and release of 5HT in blood platelets. Our studies have been performed on human platelets, with the benzodiazepines, chlordiazepoxide and medazepam (both as hydrochlorides, kindly supplied by F. Hoffmann-La Roche & Co., Basel). We have not studied diazepam, because of the very poor water solubility of this compound. It will be shown that the test drugs, and especially medazepam, do indeed interfere with uptake and release of 5HT in human blood platelets *in vitro*.

II. METHODS

A. Measurement of net uptake

Blood was taken from healthy persons and mixed with 1/10 the volume of 3% EDTA in 0.5% NaCl. Platelet-rich plasma (PRP) was obtained by centrifugation (250 $\times$ g) for 15 min at room temperature. Aliquots of 1 ml PRP were mixed with 1 ml sodium phosphate buffer, pH 6.4 (in order to increase the stability of the platelets), in thin-walled polypropylene tubes, and the platelets were isolated by centrifugation (2500 $\times$ g) for 15 min at 4°C. The supernatant was discarded, the residue drained onto filter paper for a few minutes, and the remaining supernatant carefully wiped off with filter paper. The platelet pellets were then resuspended in 2 ml incubation medium, containing (if not otherwise stated) 0.055 M phosphate buffer, pH 7.3, with Na$^+$ and K$^+$ in the molar ratio of 9 to 1, 0.075 M NaCl, and the drugs under study in the concentrations which are specified in the description of each experiment. When the effect of varying concentrations of chloride was investigated, NaCl was replaced by Na$_2$SO$_4$, since the sulfate anion does not affect 5HT uptake (Lingjærde, 1971a). Incubation with ^{14}C-5HT (5-hydroxytryptamine-3-^{14}C creatinine sulfate (from the Radiochemical Center, Amersham), specific activity adjusted to 10 μC/μmole) was carried out in a shaking water bath for 5 min after temperature equilibration for 5 min. The ^{14}C-5HT was always added in a volume of 20 μl to each sample, *i.e.,* 1% of the total incubation volume. The incubation was interrupted by transferring the test tubes to ice water, and the platelets were isolated by centrifugation (2500 $\times$ g) for 25 min at

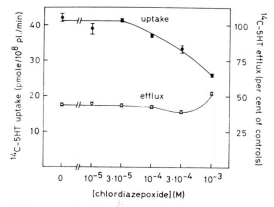

FIG. 1. Effect of chlordiazepoxide on uptake (*filled circles*) and efflux (*open circles*) of ¹⁴C-5HT in platelets. In the uptake experiment, the platelets were incubated for 5 min with 1.0 μM ¹⁴C-5HT. Each point represents the mean of three parallel samples ± standard error of mean (S.E.M.).

4°C. After draining, 1 ml of distilled water was added to the pellet, and the 5HT was completely released by freezing at −20°C, followed by thawing and centrifugation. One fifth ml of the supernatant was mixed with Bray's solution and the radioactivity measured in a Packard Tri-Carb liquid scintillation spectrometer. Blank values were obtained by adding ¹⁴C-5HT to samples kept in ice water, just prior to centrifugation. Samples to which ¹⁴C-5HT was added, after the final separation, were used as internal standards.

Platelets in PRP were counted in a phase contrast microscope. Platelet recovery after resuspension was fairly constant at 85%; this figure has been used when calculating the absolute uptake per 10^8 platelets.

B. Measurement of efflux

Aliquots of 2 ml PRP were mixed with 1 ml sodium phosphate buffer, 0.14 M, pH 6.4, and the platelets loaded with a certain amount of ¹⁴C-5HT by incubating with 0.67 μM ¹⁴C-5HT for 8 min. The platelets were then isolated by centrifugation (as above) and resuspended in 2 ml phosphate buffer, 0.11 M, pH 7.3, again with Na^+ and K^+ in the molar ratio of 9 to 1. After pooling of the resuspensions, aliquots of 1 ml were mixed with 1 ml 0.15 M NaCl with or without the test drugs added. In addition, unlabeled 5HT was added to a final concentration of 10 μM, in order to competitively block the reuptake of released ¹⁴C-5HT; without blocking of this reuptake,

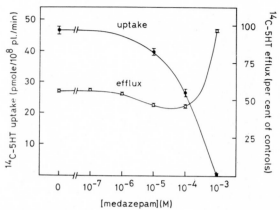

FIG. 2. Effect of medazepam on uptake (*filled circles*) and efflux (*open circles*) of ^{14}C-5HT in platelets. The experimental conditions were as described in the legend to Fig. 1. Each point represents the mean of three parallel samples ± S.E.M.

the 5HT released from the platelets would be rapidly taken up again (see Lingjærde, 1971*b*). The platelet suspensions were then reincubated at 37°C for 30 min, except for control samples kept in ice water, under which condition no efflux occurs. The percentage decrease of platelet-bound ^{14}C-5HT during 30-min incubation was taken as a measure of efflux.

III. RESULTS

Fig. 1 shows the effect of chlordiazepoxide on net uptake and on efflux (from separate experiments). The uptake is inhibited about 10% by 10^{-4} M chlordiazepoxide, whereas a concentration of 10^{-3} M gives about 50% inhibition. Efflux is not influenced by concentrations up to 10^{-4} M, while 3×10 M has a weak inhibitory effect, and 10^{-3} M a somewhat stronger, albeit rather slight, enhancing effect.

The effect of chlordiazepoxide hydrochloride on 5HT uptake and efflux in human blood platelets is thus rather weak, but not quite negligible.

As seen from Fig. 2, medazepam exerts much stronger effects both on uptake and on efflux. The uptake is significantly inhibited by 10^{-5} M of this drug, and about 50% inhibition (at this substrate concentration) is obtained with 10^{-4} M. In these concentrations, medazepam also has a slight, but significant inhibitory effect on 5HT efflux, so that the decrease of net uptake is not the result of increased efflux. When the drug concentration is increased to 10^{-3} M, however, there is no net uptake at all, which is probably due mainly to the marked increase of efflux.

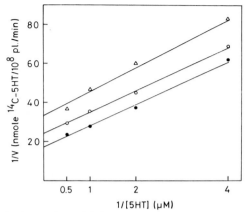

FIG. 3. Double reciprocal plots (Lineweaver-Burk plots) of ^{14}C-5HT uptake versus concentrations of ^{14}C-5HT, in the absence of drugs (*filled circles*) and in the presence of medazepam in concentrations of 10^{-5} M (*open circles*) and 10^{-4} M (*triangles*). The platelets were incubated for 5 min with ^{14}C-5HT at concentrations of 0.25, 0.50, 1.0, or 2.0 μM. Each point represents the mean of two parallel samples.

Since medazepam exerts the stronger effect on uptake, this effect has been subjected to some further experiments, using concentrations not higher than 10^{-4} M, so that the releasing effect could be avoided.

The kinetics of the inhibitory effect of medazepam are shown in the Lineweaver-Burk plot of Fig. 3. The inhibition is uncompetitive versus 5HT (or perhaps slightly uncompetitive). This type of inhibition is in striking contrast to the simple competitive inhibition exerted by such drugs as the tricyclic antidepressants and the phenothiazines, whereas it is of the same type as shown for ergotamine and dihydroergotamine (Lingjærde, 1970).

We have previously shown that 5HT uptake in human blood platelets is chloride-dependent, although chloride may be replaced by other small but unphysiological anions (Lingjærde, 1969, 1971*a*), and that the effect of chloride is in accordance with simple saturation kinetics. When platelets are incubated with or without medazepam at varying concentrations of chloride, the inhibitory effect of the drug is found to be uncompetitive also versus chloride.

The uptake of 5HT in platelets is also sodium-dependent (Lingjærde, 1969), as are many other transport processes, and is stimulated by potassium (Weissbach and Redfield, 1960; Rysanek, Svehla and Doubravova, 1967; Lingjærde, 1971*c*). We have found that the percentage inhibition by medazepam is the same in the presence and in the absence of potassium. The same effect has been observed for ergotamine and dihydroergotamine

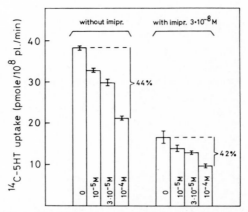

FIG. 4. Inhibitory effect of medazepam on ^{14}C-5HT uptake in the absence and presence of 3×10^{-8} M imipramine hydrochloride. The concentrations of medazepam are indicated at the bottom of each column. The platelets were incubated for 5 min with 1.0 μM ^{14}C-5HT. The mean of three parallel samples $\pm$ S.E.M. is given.

(Lingjærde, 1970), but differs from the effect of ouabain, which seems to inhibit mainly the potassium-stimulated part of the uptake (Lingjærde, *unpublished observations*).

In clinical practice, benzodiazepines are often used together with tricyclic antidepressants, and it seems to be generally recognized that this combination is very effective in agitated, anxious, depressive states. (This is also the case with medazepam; Lingjærde, 1968.) Therefore, we found it of interest to study how medazepam and imipramine interact as to the inhibitory effect on 5HT uptake in platelets. The result is shown in Fig. 4. Under the experimental conditions used, imipramine alone inhibited about 57% in the low concentration of 3×10^{-8} M. Medazepam 10^{-4} M inhibited 44% in the absence of imipramine, and this *percentage* inhibition was not significantly altered in the presence of imipramine. Thus, the inhibitory effects of imipramine and medazepam do not potentiate each other, but probably should be regarded as additive.

IV. DISCUSSION

The present results add to other recent findings which indicate that benzodiazepines have an effect on monoamine function. Medazepam was found to have an inhibitory effect on 5HT uptake in platelets in the concentration range of 10^{-5} to 10^{-4} M. Within the same concentration range, medazepam gives a small decrease of spontaneous 5HT efflux. With higher concentrations (10^{-3} M), there is a complete blocking of net 5HT uptake, and a strong depleting effect. This latter effect is probably rather unspecific, as the same effect is seen with various other drugs, such as tricyclic antidepressants, phenothiazines, and ergot alkaloids (Lingjærde, 1970).

The inhibitory effect on 5HT uptake is uncompetitive. If the 5HT uptake process is treated as analogous to an enzymatic process, then, according to the enzyme kinetic models of Cleland (1963), this would indicate that the inhibitor reacts with an enzyme form (in this case a carrier form) which is different from, and irreversibly connected with, the form which reacts with the variable substrate, in this case 5HT or chloride. (The meaning of the term "irreversibly connected with" is explained in Cleland's paper.) This model excludes the possibility that medazepam interferes with the free carrier, as does, for example, imipramine, but it does not tell much about the mechanism beyond that.

Are the demonstrated effects of medazepam characteristic of benzodiazepines with tranquilizing properties? This cannot be stated with certainty at present. Chlordiazepoxide was found to be about 10 times less potent than medazepam, and other derivatives have not been studied with this technique so far, partly because many of them have a very low water solubility. It would be especially interesting to know whether diazepam shares the potency of medazepam with regard to the effect on 5HT uptake and efflux. Since diazepam is chemically more closely related to medazepam than to chlordiazepoxide, it seems reasonable to expect that diazepam has about the same effect on 5HT uptake and efflux as has medazepam.

Now the rather crucial question is this: have the demonstrated effects on 5HT uptake and release any relevance to the clinical effects of benzodiazepines, especially medazepam? This question cannot, of course, be answered now. Thus, it is not known for certain whether any of the tested drugs, in therapeutic dosages, reach high enough concentrations in the body to exert these effects. In the case of chlordiazepoxide, this seems rather dubious, whereas medazepam may possibly reach concentrations which are high enough to give an effect *in vivo*. Lauffer and Schmid (1969) measured the blood level of medazepam in a 24-year-old man after a single dose of 30 mg, and found the peak concentration to be about 10^{-6} M. Perhaps the concentration may be higher after prolonged use, and perhaps the concentrations at the central receptors may be higher than in the blood. These are questions to be answered by future research, as are the problems of whether the uptake and efflux of 5HT in central serotoninergic neurons are affected in the same way as they are in platelets, and whether the uptake and release of catecholamines are similarly affected.

V. SUMMARY

The effects of chlordiazepoxide hydrochloride and medazepam hydrochloride on uptake and efflux of 5HT in human blood platelets have been studied. For uptake studies, platelets from healthy persons were isolated by differential centrifugation and resuspended in a simple phosphate-buffered medium, and the suspensions incubated at 37°C for 5 min with low concentrations of ^{14}C-5HT. The net uptake was subsequently assayed by liquid scintillation spectrometry. About 50% inhibition of the uptake was obtained with 10^{-3} M chlordiazepoxide and with 10^{-4} M medazepam. The inhibitory effect of medazepam was found to be uncompetitive versus 5HT and versus chloride, and unaffected by potassium. Moreover, the inhibitory effect of medazepam seems to be additive with that of imipramine.

Spontaneous efflux of ^{14}C-5HT was measured by incubating platelets preloaded with a small amount of ^{14}C-5HT in a medium with a high concentration of unlabeled 5HT (to block reuptake of released ^{14}C-5HT). Chlordiazepoxide in the concentration of 3×10^{-4} M slightly inhibited efflux, whereas a concentration of 10^{-3} M had a weak enhancing effect. Medazepam, in concentrations of 10^{-5} and 10^{-4} M, decreased the efflux somewhat, while in the concentration of 10^{-3} M, the drug exerted a strong depleting effect.

The possible relevance of these findings to the clinical effects of benzodiazepines is briefly discussed.

REFERENCES

Chase, T. N., Katz, R. I., and Kopin, I. J. (1970): Effect of diazepam on fate of intracisternally injected serotonin-C^{14}. *Neuropharmacology*, 9:103–108.

Cleland, W. W. (1963): The kinetics of enzyme-catalyzed reactions with two or more substrates or products. III. Prediction of initial velocity and inhibition patterns by inspection. *Biochimica et Biophysica Acta*, 67:188–196.

Lauffer, S., and Schmid, E. (1969): Dünnschichtchromatographisch-spektrophotofluorometrische Bestimmung von 7-Chlor-2,3-dihydro-1-methyl-5-phenyl-1H-1,4-benzodiazepin (Medazepam). *Arzneimittel-Forschung*, 19:740–741.

Lingjærde, O. (1968): Some clinical experiences with a new benzodiazepine. Paper presented at the 6th International Congress of the Collegium Internationale Neuro-Psychopharmacologicum, Tarragona, May 24-27.

Lingjærde, O. (1969): Uptake of serotonin in blood platelets: Dependence on sodium and chloride, and inhibition by choline. *FEBS Letters*, 3:103–106.

Lingjærde, O. (1970): Effects of ergotamine and dihydroergotamine on uptake of 5-hydroxytryptamine in blood platelets. *European Journal of Pharmacology*, 13:76–82.

Lingjærde, O. (1971a): Uptake of serotonin in blood platelets in vitro. I: The effects of chloride. *Acta Physiologica Scandinavica*, 81:75–83.

Lingjærde, O. (1971b): *Studies on the uptake of serotonin in human blood platelets in vitro.* Universitetsforlaget, Oslo.

Lingjærde, O. (1971c): Uptake of serotonin in blood platelets in vitro. III: Effects of acetate and other monocarboxylic acids. *Acta Physiologica Scandinavica,* 83: 309–318.

Pletscher, A. (1969): Biochemistry and psychosomatic medicine: The effects of psychotropic drugs on neurohumoral transmitters. In: *Psychotropic Drugs in Internal Medicine,* edited by A. Pletscher and A. Marino. Excerpta Medica Foundation, Amsterdam.

Pletscher, A., Prada, M. da, and Voglar, G. (1967): Differences between neuroleptics and tranquilizers regarding metabolism and biochemical effects. In: *Neuro-Psycho-Pharmacology,* edited by H. Brill. Excerpta Medica Foundation, Amsterdam.

Prada, M. da, and Pletscher, A. (1966): On the mechanism of chlorpromazine-induced changes of cerebral homovanillic acid levels. *Journal of Pharmacy and Pharmacology,* 18:628–630.

Rysanek, K., Svehla, C., and Doubravova, J. (1967): Der Einfluss von Kalium und Natrium Transport auf die Aufnahme von Serotonin durch Thrombocyten. *Activitas Nervosa Superior,* 9:414–416.

Taylor, K. M. (1969): The effect of minor tranquilizers on stress-induced noradrenaline turnover in the rat brain. *Proceedings of the University of Otago Medical School,* 47:33–35.

Taylor, K. M., and Laverty, R. (1969): The effect of chlordiazepoxide, diazepam and nitrazepam on catecholamine metabolism in regions of the rat brain. *European Journal of Pharmacology,* 8:296–301.

Weissbach, H., and Redfield, B. G. (1960): Factors affecting the uptake of 5-hydroxytryptamine by human platelets in an inorganic medium. *Journal of Biological Chemistry,* 235:3287–3291.

The Benzodiazepines, Raven Press, New York © 1973

THE EFFECT OF DIAZEPAM ON THE TURNOVER OF CEREBRAL DOPAMINE

G. Bartholini, H. Keller, L. Pieri, and A. Pletscher

Research Department, F. Hoffmann-La Roche & Co. Ltd., Basel, Switzerland

I. INTRODUCTION

In various brain regions, chlordiazepoxide (LIBRIUM) and diazepam (VALIUM) prevent the enhancement of the norepinephrine turnover caused by immobilization stress (Corrodi et al., 1971) or by foot shock (Taylor and Laverty, 1969). Diazepam also antagonizes the changes of the cerebral norepinephrine turnover induced by the administration of 6-hydroxydopamine into the brain ventricles (Nakamura and Thoenen, 1972). In addition, diazepam and chlordiazepoxide seem to reduce the turnover of dopamine in the striatum as well as to potentiate the decrease of this turnover due to immobilization stress (Corrodi et al., 1971).

It is not clear whether changes in cerebral catecholamine turnover induced by benzodiazepines are related to the hypothermic effect of the drugs. Nor has it been elucidated whether the benzodiazepines act on monoaminergic structures directly or indirectly through other neuronal pathways. Therefore, in this chapter, the alterations in body temperature induced by diazepam in rats have been correlated with the turnover of cerebral dopamine as reflected by changes in endogenous homovanillic acid (HVA), which probably originate mainly in the striatum. In addition, the effect of diazepam on the turnover of cerebral dopamine has been investigated during cortical spreading depression.

II. EFFECT OF DIAZEPAM ON BODY TEMPERATURE AND CEREBRAL HVA

As seen in Fig. 1, diazepam (10 mg/kg) administered to rats kept at a constant environmental temperature of 23°C reduces rectal temperature by about 2°C. Under these conditions, diazepam also causes a slight but significant ($p < 0.01$) decrease in the concentration of endogenous HVA in

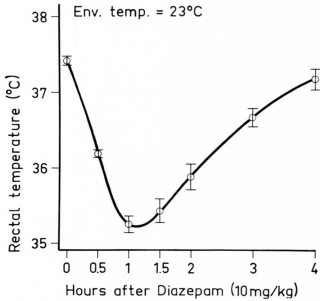

FIG. 1. Effect of diazepam on the rectal temperature of rats. The drug was injected i.p. as a suspension in water. The points indicate averages with S.E. obtained from 21 animals. The values of 0.5 to 3 hr are significantly different from the control values (0 hr; $p < 0.01$).

the brain. However, if the diazepam-treated rats are kept at an elevated environmental temperature ($+ 32°C$), neither the rectal temperature nor the cerebral HVA content change compared with untreated rats kept at 23°C.

TABLE I. *Cerebral homovanillic acid (HVA) and rectal temperature of untreated and diazepam-treated rats*

Environmental temp. (°C)	Treatment	HVA ($\mu g/g$)	Rectal temp. (°C)
23	None[a]	0.102 ± 0.002	37.2 ± 0.1
	Diazepam[b]	0.086 ± 0.004	35.4 ± 0.3
32	None[c]	0.117 ± 0.001	37.8 ± 0.2
	Diazepam[d]	0.101 ± 0.003	37.2 ± 0.2

Diazepam (10 mg/kg as suspension in water) was injected i.p. 90 min before decapitation. The rectal temperature of the animals at the beginning of the experiment was $37.3 \pm 0.2°C$; the values indicated in the table were measured before sacrifice. HVA was determined in the brain (without cerebellum and medulla oblongata) by a fluorometric procedure (Keller et al., 1972). The values of HVA indicate averages with S.E. of four experiments, each of them performed with four to six single brains.
Significance level of HVA and temperature values: *a* versus *b* and *c*, $p \leq 0.01$; *d* versus *a*, $p > 0.05$; *c* versus *d*, $p < 0.01$.

Untreated animals exposed to 32°C show a slight but significant increase in both rectal temperature and brain HVA ($p < 0.01$) (Table I).

These results indicate that the content of endogenous cerebral HVA varies with changes in body temperature and that the diazepam-induced decrease of the acid in the brain is probably due to hypothermia caused by the drug. The diminution of cerebral HVA and the hypothermia after diazepam depends on several factors, e.g., the initial body temperature, changes (even small) in environmental temperature in the course of the experiment, the number of rats per cage.

The variations in the cerebral HVA content described above probably reflect changes in the turnover of dopamine, since the brain concentration of this amine remains equal in normo-, hypo-, and hyperthermic conditions (Keller, *unpublished results*). An enhanced turnover due to hyperthermia has also been reported for brain 5-hydroxytryptamine (Weiss and Aghajanian, 1971).

III. EFFECT OF DIAZEPAM ON SPREADING DEPRESSION

Local application of KCl (1 to 25% solution) on the cortex of one cerebral hemisphere produces a depolarization which spreads from the point of application to the whole ipsilateral cortex, but not to that of the contralateral hemisphere. This phenomenon has therefore been called spreading depression (SD) (Leão, 1944; Bureš and Hartman, 1967).

TABLE II. *Effect of diazepam on the cerebral concentration of homovanillic acid (HVA) in rats with spreading depression (SD) of the right cortex*

Environmental temp. (°C)	Treatment	HVA (μg/g)		Rectal temp. (°C) before sacrifice
		L	R	
23	NaCl	0.093 ± 0.004	0.104 ± 0.007 (*a*)	37.3 ± 0.2
	KCl	0.107 ± 0.004 (*d*)	0.163 ± 0.004 (*b*)	37.2 ± 0.3
32	Diazepam + KCl	0.108 ± 0.004 (*e*)	0.189 ± 0.008 (*c*)	37.1 ± 0.4

NaCl (20%) or KCl (25%) was applied 1 hr before sacrifice on the dura of the right cerebral cortex through a cannula implanted 24 hr earlier. Diazepam was injected i.p. (10 mg/kg as a suspension in water) 30 min before the application of KCl. Hypothermia of diazepam-treated animals was prevented by keeping the animals at an environmental temperature of 32°C. The values of HVA represent averages with S.E. of four to six duplicate experiments, each of them performed with two pooled cerebral hemispheres. R = right hemisphere; L = left hemisphere.

a versus *b* and *c*, $p < 0.01$; *b* versus *c*, $p < 0.01$; *d* versus *e*, $p > 0.05$; *a* versus concentration on the left side of all groups, $p > 0.05$. The temperature values are not significantly different from each other ($p > 0.05$).

Application of a 25% KCl solution for 1 hr to the right cerebral cortex of rats (Keller et al., 1972) induces a marked increase of HVA in the ipsilateral compared with the contralateral hemisphere. An equimolar solution of NaCl does not significantly change the HVA content compared with the untreated hemisphere (Table II). The HVA rise during SD lasts about 3 hr (Keller et al., 1972) and is correlated in time with a marked decrease in the electrical activity of the ipsilateral, but not of the contralateral cortex (Bureš and Hartman, 1967; Keller et al., 1972). The HVA increase is probably due to an enhanced turnover of striatal dopamine, since the cerebral concentration of the amine is not changed (Pieri et al., 1972). An unspecific inhibition of the outflow of acidic amine metabolites from the brain by SD is unlikely, since the cerebral 5-hydroxyindoleacetic acid, the major metabolite of 5-hydroxytryptamine (present in considerable concentration in the striatum) does not change as a result of the application of KCl (Keller et al., 1972).

Diazepam (10 mg/kg, i.p.) applied to rats kept normothermic significantly enhances the HVA rise induced by KCl in the ipsilateral hemisphere. In the contralateral hemisphere, the content of the acid does not differ whether KCl is administered alone or after pretreatment with diazepam (Table II). The drug does not change the dopamine content in either of the hemispheres, indicating that the enhancement of the HVA increase in the ipsilateral hemisphere is due to an enhanced dopamine turnover, probably mainly in the striatum.

IV. DISCUSSION AND CONCLUSIONS

The present results indicate that diazepam does not affect the turnover of cerebral dopamine in normal animals provided that the development of hypothermia is prevented by elevation of the environmental temperature. The drug seems, however, to enhance the turnover of brain dopamine in the hemispheres treated with KCl, i.e., in conditions of SD. The effect of diazepam may be explained by the tentative hypothesis that SD removes an inhibitory cortical influence on the nigrostriatal dopaminergic pathway. In addition or alternatively, depression of the activity of the striatum due to the arrival of the inhibitory wave (initiated in the cortex by KCl) in this structure might lead to a disinhibition of a pathway involved in a positive striato-nigral feedback mechanism. In fact, the depolarization wave seems to reach the striatum through the pyriform cortex, the amygdala, and the claustrum, as indicated by the almost complete disappearance of the electrical activity of the caudate nucleus during cortical SD and by prevention of this disappearance after lesions of the above-mentioned structures (pyriform cortex, etc.) (Fifková and Syka, 1964). In both cases, i.e., facilitation by SD of a cortico- and/or a striato-nigral projection, cholinergic mechanisms may be involved. Thus, either intraventricular or supranigral injection of atropine enhances the turnover of dopamine in the striatum (Bartholini and Pletscher, 1971). Diazepam might enhance the firing of the dopaminergic pathway, but only if the inhibitory (possibly cholinergic) influences normally present are removed by SD. Besides the proposed effect on a dopaminergic pathway, diazepam has other facilitatory actions in the brain. Thus the drug has been reported to increase the firing of the Purkinje cells leading to an inhibition of experimental seizures in the cerebral cortex (Julien, 1972). This facilitatory action seems to be shared by other benzodiazepines. Thus, chlordiazepoxide increases the ponto-geniculo-occipital waves in the cat; an effect which is blocked by atropine (Monachon et al., *this volume*).

V. SUMMARY

Treatment of rats with diazepam induces hypothermia and a concomitant decrease of the cerebral level of HVA without affecting that of dopamine. If the animals are kept normothermic by elevation of the environmental temperature, the drug has no effect on the brain HVA. However, diazepam enhances the increase of cerebral HVA seen in SD without influencing the content of dopamine.

In conclusion, diazepam does not change the turnover of dopamine in rats kept normothermic. However, the drug enhances the increase in the turnover of this amine during spreading depression which probably removes an inhibitory influence on the nigro-striatal pathway.

REFERENCES

Bartholini, G., and Pletscher, A. (1971): Atropin-induced changes of cerebral dopamine turnover. *Experientia*, 27:1302–1303.
Bureš, J., and Hartman, G. (1967): Conduction block in capsular interna fibres caused by striatal spreading depression in rats. *Experientia*, 23:736–737.
Corrodi, H., Fuxe, K., Lidbrink, P., and Olson, L. (1971): Minor tranquillizers, stress, and central catecholamine neurons. *Brain Research*, 29:1–16.
Fifková, E., and Syka, J. (1964): Relationship between cortical and striatal spreading depression in rat. *Experimental Neurology*, 9:355–366.
Julien, R. M. (1972): Cerebellar involvement in the antiepileptic action of diazepam. *Neuropharmacology*, 11:683–691.
Keller, H. H., Bartholini, G., Pieri, L., and Pletscher, A. (1972): Europ. J. Pharmacol. (*in press*).

Leão, A. A. P. (1944): Spreading depression of activity in the cerebral cortex. *Journal of Neurophysiology*, 7:359–390.
Monachon, M. -A., Jalfre, M., and Haefely, W. (1973): A modulating effect of chlordiazepoxide on drug-induced PGO spikes in the cat. *This volume.*
Nakamura, K., and Thoenen, H. (1972): Increased irritability: A permanent behavior change induced in the rat by intraventricular administration of 6-hydroxydopamine. *Psychopharmacologia*, 24:359–372.
Pieri, L., Bartholini, G., Keller, H. H., and Pletscher, A. (1972): Effect of spreading depression on electrical activity and dopamine turnover in the striatum of rats. *Experientia, in press.*
Taylor, K. M., and Laverty, R. (1969): The effect of chlordiazepoxide, diazepam, and nitrazepam on catecholamine metabolism in regions of the rat brain. *European Journal of Pharmacology*, 8:296–301.
Weiss, B. L., and Aghajanian, G. K. (1971): Activation of brain serotonin metabolism by heat: Role of midbrain raphé neurons. *Brain Research*, 26:37–48.

The Benzodiazepines, Raven Press, New York © 1973

INCREASE IN MOUSE AND RAT BRAIN ACETYLCHOLINE LEVELS BY DIAZEPAM

H. Ladinsky, S. Consolo, G. Peri, and S. Garattini

Istituto di Ricerche Farmacologiche "Mario Negri," Milan, Italy

The anticonvulsant activity of diazepam in animals and man is well established. Marcucci et al. (1968) demonstrated that the anti-pentylenetetrazol (METRAZOLE) activity of diazepam lasts much longer in mice than in rats, because the metabolites N-desmethyldiazepam and oxazepam accumulate in mouse brain but not in rat brain.

We have recently shown that diazepam (5 to 40 mg/kg, i.p.), similarly to other anticonvulsants, increased mouse whole-brain acetylcholine levels 10 min after administration without affecting choline levels (Consolo et al., 1972).

On the basis of the above results, the duration of the effect of diazepam in increasing acetylcholine in the mouse and rat brain was compared in an attempt to determine if a relationship exists between the anti-pentylenetetrazol and acetylcholine-increasing effects of diazepam.

Female Sprague-Dawley rats (weighing 150 g) and albino Swiss mice (20 g) were used. They were sacrificed by the near-freezing method of Toru and Aprison (1966). The brains were rapidly removed, frozen in liquid nitrogen, and pulverized with a specially designed mortar and pestle (Ladinsky et al., 1972). The radiochemical method of Saelens et al. (1970) was then used to measure brain acetylcholine.

Diazepam was obtained from Ravizza, Italy, and it was dissolved in a solvent consisting of propylene glycol, glycofurol, benzyl alcohol, and water (6:6:1:87).

Table I compares the effect of a single intravenous injection of diazepam (5 mg/kg) on mouse and rat brain acetylcholine levels with time. In mice, brain acetylcholine was significantly increased up to 4 hr, whereas in rats the effect lasted only 60 min. There thus appears to be a qualitative relationship between the duration of the anti-pentylenetetrazol effect of diazepam in these two species and the duration of increased acetylcholine levels in the brain.

TABLE I. *Effect of diazepam on mouse and rat brain acetylcholine concentration*

| Time (min) | Acetylcholine | |
	Rat brain (μg/g wet wt)	Mouse brain (μg/g wet wt)
0[a]	2.13 ± 0.07 (16)	2.12 ± 0.03 (20)
1	[b]3.01 ± 0.19 (7)	[b]3.01 ± 0.12 (7)
5	[b]2.68 ± 0.11 (13)	[b]3.12 ± 0.20 (13)
30	[b]2.77 ± 0.18 (13)	[b]2.98 ± 0.14 (12)
60	[b]2.40 ± 0.10 (13)	[b]2.71 ± 0.10 (18)
120	2.14 ± 0.09 (6)	—
180	2.06 ± 0.07 (6)	—
260	—	[b]2.51 ± 0.08 (8)
360	—	2.34 ± 0.13 (6)

Figures are mean values and S.E.M.
Figures in parentheses represent the number of animals used.
Diazepam was administered intravenously at a dose of 5 mg/kg.
[a] The 0 min figure represents the combined values of the saline-treated controls at times shown after administration of the diazepam solvent. Brain acetylcholine levels were found not to change with time under these conditions.
[b] $p < 0.01$.

Preliminary experiments indicate that acetylcholine was increased in the cerebral hemispheres and diencephalon but not in the mesencephalon or cerebellum of the mouse 30 min after an intravenous injection of diazepam (5 mg/kg).

ACKNOWLEDGMENT

We acknowledge the excellent technical assistance of Miss L. Gessaroli and Miss A. Bellini.

REFERENCES

Consolo, S., Ladinsky, H., Peri, G., and Garattini, S. (1972): Effect of central stimulants and depressants on mouse brain acetylcholine and choline levels. *European Journal of Pharmacology,* 18:251–255.

Ladinsky, H., Consolo, S., and Sanvito, A. (1972): A simple apparatus for the pulverization and rapid quantitative transfer of frozen tissue. *Analytical Biochemistry,* 49: 294–297.

Marcucci, F., Guaitani, A., Kvetina, J., Mussini, E., and Garattini, S. (1968): Species difference in diazepam metabolism and anticonvulsant effect. *European Journal of Pharmacology,* 4:467–470.

Saelens, J. K., Allen, M. P., and Simke, J. P. (1970): Determination of acetylcholine and choline by an enzymatic assay. *Archives Internationales de Pharmacodynamie et de Thérapie,* 186:279–286.

Toru, M., and Aprison, M. H. (1966): Brain acetylcholine studies: A new extraction procedure. *Journal of Neurochemistry,* 13:1533–1544.

The Benzodiazepines, Raven Press, New York © 1973

EFFECT OF BENZODIAZEPINES ON CARBOHYDRATE METABOLISM IN RAT BRAIN

K. F. Gey

F. Hoffmann-La Roche & Co. Ltd., Basel, Switzerland

I. INTRODUCTION

Since the brain has been considered to use 40% of its energy production for physiologically important ion-transport mechanisms (Whittam, 1961) and since the energy metabolism of the brain depends almost completely on glucose (Quastel, 1969; Balázs, 1970), alterations of the cerebral glucose metabolism by psychotropic drugs deserve some interest. Rutishauser (1963) found in our laboratory that chlordiazepoxide hydrochloride (LIBRIUM) in a dosage of 30 mg/kg causes an extensive accumulation of intracellular glucose and a discrete decrease of intracellular pyruvate in rat brain. Recently, Hutchins and Rogers (1970) described a slight increase of glycogen in the brain of mice treated with chlordiazepoxide hydrochloride 25 mg/kg. Furthermore, Young, Albano, Charnecki, and Demcsak (1969) reported that doses of diazepam 2 to 20 mg/kg (VALIUM) induce a rapid, strong and dose-dependent increase of glucose as well as a marked decrease of lactate and malate in total brain of mice. The present chapter extends these studies mainly in two respects: (1) simultaneous assay of all parameters mentioned above together with the most important phosphorylated intermediates of glycolysis and ATP, and (2) comparison of chlordiazepoxide hydrochloride and diazepam in dosages (20 mg/kg and 4 mg/kg, respectively) which are equipotent regarding 60-70% reduction of the aggression of foot-shocked rats (Blum, 1971) but cause only slight or almost no hypothermia (Figs. 1, 3).

II. METHODS

Groups of six to 12 female albino rats (from a closed colony of Wistar origin, bred in Füllinsdorf, Switzerland) with a body weight of 110 to 140 g (fed *ad libitum*) were injected i.p. with either chlordiazepoxide hydrochlo-

244

CARBOHYDRATE METABOLISM

TABLE I. Absolute levels in blood and brain of placebo-treated rats

| Intermediate | Blood | | Brain | | |
	μmoles/ml	mg %	Total μmoles/g wet weight	Intracellular μmoles/g wet weight	mg %
Glycogen (calculated as glucose)	—		1.71 ± 0.08	—	31.3 ± 1.5
Glucose	5.8 ± 0.2	106 ± 4	1.15 ± 0.16	0.55 ± 0.16	10 ± 3
Glucose-6-phosphate	0.042 ± 0.003		0.123 ± 0.008	0.121 ± 0.007	
Fructose-6-phosphate	0.009 ± 0.001		0.040 ± 0.006	0.040 ± 0.006	
Fructose-1,6-diphosphate	0.020 ± 0.002		0.174 ± 0.007	0.173 ± 0.007	
Dihydroxyacetone phosphate	0.021 ± 0.001		0.095 ± 0.012	0.094 ± 0.012	
Pyruvate	0.083 ± 0.006	0.73 ± 0.05	0.122 ± 0.018	0.114 ± 0.018	1.00 ± 0.16
Lactate	0.97 ± 0.11	8.74 ± 0.10	2.13 ± 0.12	2.03 ± 0.12	18.3 ± 0.1
Malate	0.129 ± 0.009	1.73 ± 0.12	0.46 ± 0.04	0.45 ± 0.04	6.2 ± 0.5
α-Amino nitrogen (calculated for a mean MW of 133, e.g., aspartate)	0.94 ± 0.03	12.5 ± 0.4	4.9 ± 0.5	4.8 ± 0.5	64 ± 7
ATP	0.48 ± 0.3		2.23 ± 0.55	2.21 ± 0.03	

The figures represent mean ± S.E. of 20 to 24 values each from a pool of three to six animals. The control values of all time intervals were pooled since they showed no time-dependent variations.

ride 20 mg/kg, diazepam 4 mg/kg, or saline 10 ml/kg, and kept at room temperature of 25°C.

Blood, which was obtained from half the animals of each group by decapitation, was immediately heparinized and deproteinized by ice-cold 0.6 N $HClO_4$. The remaining animals of each animal group were killed by immersion in liquid nitrogen, their brains were chiseled out and powdered under liquid nitrogen and then (except for glycogen) homogenized in ice-cold 0.6 N $HClO_4$. After removal of $HClO_4$ (as potassium salt), glucose was measured by the hexokinase procedure; glucose-6-phosphate, fructose-6-phosphate, fructose-1,6-diphosphate, dihydroxyacetone phosphate, pyruvate, lactate, malate, and ATP were determined by conventional enzymatic methods (Bergmeyer, 1970). 2,4-Dinitrofluorobenzene was used for the assay of α-amino nitrogen according to Goodwin (1968). The cerebral glycogen was precipitated by ethanol (after placing chiseled frozen brain into 30% KOH at 95°C) and assayed by means of amyloglucosidase and glucose oxidase (Bergmeyer, 1970).

As in previous studies for the calculation of intracellular levels of phosphorylated metabolites (Gey, Rutishauser, and Pletscher, 1965), 5% of the blood values were deduced from values of total brain, according to the blood volume of brain. For the approximation of intracellular values of the remaining metabolites, 10% of blood values was deduced from that of total brain according to the extracellular space in rat brain.

The absolute levels of metabolites in control animals (Table I) showed the same order as in previous publications (Balázs, 1970). The sedation was approximated by the observation of the gross behavior of the animals according to the following scale: $+$ spontaneous motor activity as well as motility after tactile stimuli and repeated local displacements slightly diminished; and $\ddagger$ little spontaneous motor activity; markedly reduced reactions. The body temperature was measured by a thermoelement in the vagina.

III. RESULTS AND DISCUSSION

In the present experiments with chlordiazepoxide hydrochloride 20 mg/kg and diazepam 4 mg/kg, both drugs had almost identical effects on glucose and its metabolites (Figs. 1 through 4).

A. Changes in blood

Alterations were not extensive and consisted in:
1. *An immediate but very slight (20%) and transient increase in*

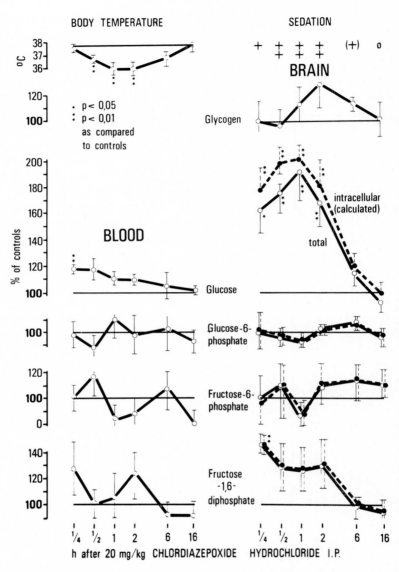

FIG. 1. Effect of chlordiazepoxide on body temperature, motor activity as well as various glucose metabolites in blood and brain of fed rats. The abscissa indicates hours after i.p. injection on a logarithmic scale. The metabolic parameters are expressed in % of placebo-treated controls (ordinate) and represent means ± S.E. of four measurements each from a pool of 4 to 6 rats. Solid lines represent total values, dotted lines intracellular levels calculated as described in METHODS.

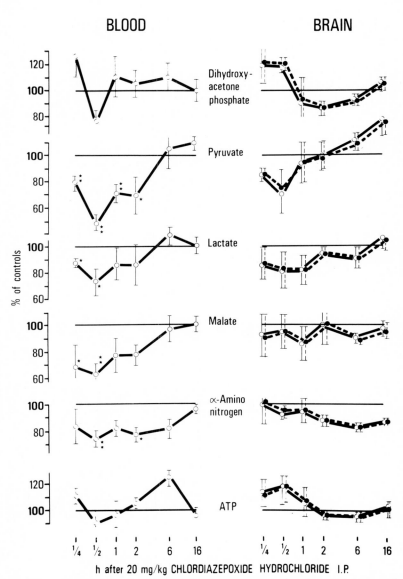

BLOOD   **BRAIN**

% of controls

h after 20 mg/kg CHLORDIAZEPOXIDE HYDROCHLORIDE I.P.

FIG. 2. Effect of chlordiazepoxide on dihydroxyacetone phosphate, various carboxylic acids, and ATP in blood and brain of fed rats. Details correspond to legend of Fig. 1.

glucose, which showed statistical significance only after 15 min and was markedly smaller than with higher doses (30 mg/kg) of chlordiazepoxide

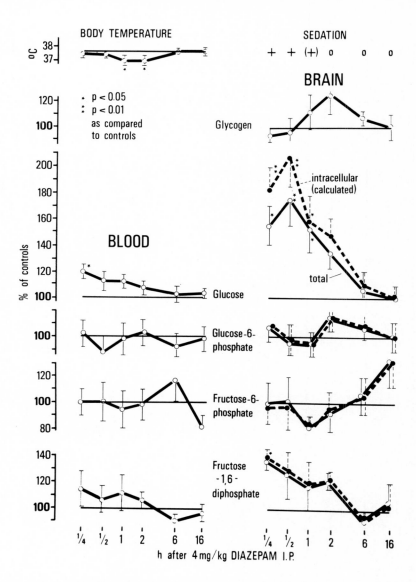

FIG. 3. Effect of diazepam on body temperature, motor activity as well as various glucose metabolites in blood and brain of fed rats. The abscissa indicates hours after i.p. injection on a log scale. The metabolic parameters are expressed in % of placebo-treated controls (ordinate) and represent means ± S.E. of four measurements each from a pool of 4 to 6 rats. Solid lines represent total values, dotted lines intracellular levels calculated as described in METHODS.

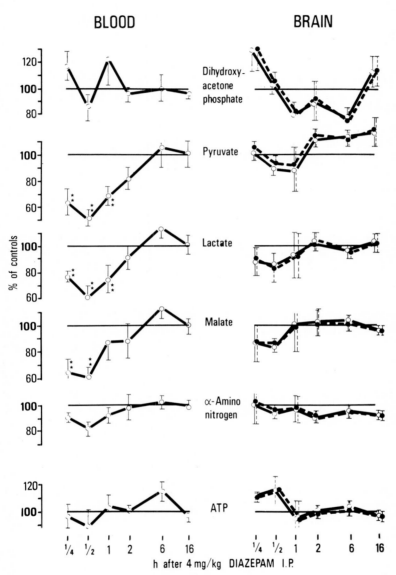

FIG. 4. Effect of diazepam on dihydroxyacetone phosphate, various carboxylic acids, and ATP in blood and brain of fed rats. Details correspond to legend of Fig. 3.

(Gey and Pletscher, 1961; Rutishauser, 1963).

    2. *A distinct decrease (by about 20 to 45%) of pyruvate, lactate, and*

malate being maximal after 30 min and losing statistical significance after 1 to 2 hr.

3. *A moderate drop (by about 20%) of free amino acids (α-amino nitrogen)* with statistical significance 30 min and 2 hr after chlordiazepoxide and with none after diazepam.

B. Changes in brain

After both benzodiazepines the following observations were made:

1. *A rapid and extensive increase (up to 100%) of intracellular glucose being maximal and statistically significant from 15 min to 2 hr after chlordiazepoxide and 15 min to 1 hr after diazepam.* Although the initial increase of glucose occurred earlier in blood than in brain, the hyperglycemia appears to be too weak (20%) and too transient to explain the glucose accumulation in brain. After both benzodiazepines, the time-course of cerebral glucose accumulation paralleled fairly well the obvious sedation. However, the degree of glucose accumulation (being equal after both treatments) was not proportional to the degree of sedation (being higher after chlordiazepoxide 20 mg/kg than after diazepam 4 mg/kg). Therefore, it remains to be elucidated by further pharmacological tests which of the various components of obvious sedation (*e.g.*, muscle relaxation, antiaggression, reduced motoric coordination, hypnotic properties) are actually involved in the cerebral glucose accumulation. With regard to this relationship, two observations deserve interest. At first, Young et al. (1969) reported that the diazepam-induced increase of glucose shows in mouse brain regional differences which are, however, actually rather small (glucose increase in brainstem and spinal cord is at least ½ to ⅔ of that in cerebral cortex 15 min after diazepam 1 to 5 mg/kg). Secondly, typical central depressants, *e.g.*, the hypnotics phenobarbital and methyprylon, or the neuroleptics chlorpromazine, reserpine, and the benzoquinolizine RO 4-1284, also increase the cerebral glucose (Mathé, Kassay, and Hunkar, 1961*a, b*; Rutishauser, 1963; Gey et al., 1965).

2. *A very slight (25%) enlargement of glycogen* which (in contrast to findings in mice by Hutchins and Rogers, 1970) was not yet statistically significant. The rise in glycogen occurred markedly later than that in glucose and might thus be the consequence of the increase in intracerebral glucose. This seems also to be true for depressants since, after phenobarbital and reserpine, the increase of cerebral glycogen (Albrecht, 1957; Svorad, 1959; Estler, 1961; Mathé et al., 1961*a, b*; Balzer and Palm, 1962; Hutchins and Rogers, 1970) occurs much later than the glucose peak and maximal central

depression, respectively (Gey et al., 1965).

3. *No significant change in the level of glucose-6-phosphate and fructose-6-phosphate.* This indicates that the benzodiazepines reduce the glucose phosphorylation neither directly nor by an accumulation of glucose-6-phosphate, *i.e.* by the physiological feedback inhibitor. Phenobarbital, chlorpromazine, and reserpine, in contrast to benzodiazepines, cause a distinct (20 to 40%) decrease in the cerebral level of the easily interconvertible glucose- and fructose-6-phosphate in fed rats (Vladimirov and Rubel, 1957; Gey et al., 1965), probably by an inhibition of hexokinase (Bernsohn, Namajuska, and Cochrane, 1956; Yanagawa, 1958; Masurat, Greenberg, Rice, Herndon, and van Loon, 1960). In conclusion, the mechanism of glucose accumulation by benzodiazepines may differ from that of hypnotics and neuroleptics.

4. *An immediate, distinct (30 to 45%) but transient rise of fructose-1,6-disphosphate,* with statistical significance after ¼ hr and with none up to 2 hr. Since the phosphorylation of fructose-6-phosphate to fructose-1, 6-diphosphate is irreversible this result indicates two facts, namely (*a*) the brain of benzodiazepines-treated rats responds initially in a rather normal way to the increase of glucose, *i.e.* by formation of fructose-1,6-diphosphate (Balázs, 1970; McGilvery, 1970). In consequence, the phosphorylation of neither glucose (by hexokinase) nor fructose-6-phosphate (by phosphofructokinase) can be inhibited by the benzodiazepines; and (*b*) the physiologically important regulation of the fructose-1,6-diphosphate level (Balázs, 1970) seems to remain intact after benzodiazepines, *i.e.* the regulation is ready to prevent a progressive accumulation of fructose-1,6-diphosphate which is typical for damaged yeast cells (McGilvery, 1970).

In the brain of benzodiazepine-treated rats, fructose-1,6-diphosphate was the third hexose phosphate which did not behave the same as after administration of neuroleptics. Thus, chlorpromazine and reserpine do not increase but rather decrease fructose-1,6-diphosphate (Gey et al., 1965).

5. *Initial tendency of dihydroxyacetone phosphate toward increase (up to 25%), later on toward decrease (by 10 to 25%)* although statistical significance is lacking. Nevertheless, the initial increase of the triose corresponded qualitatively to the initial rise in fructose-1,6-diphosphate. This excludes an inhibition of phosphofructoaldolase and rather suggests an increased glycolysis in the brain of benzodiazepine-treated rats, at least in the initial phase.

6. *No significant change of pyruvate and lactate* in spite of a weak initial tendency toward decrease (by 15 to 20% from 15 min to 1 hr). Therefore, the significant diminution of pyruvate in rat brain as seen after

higher doses of chlordiazepoxide (by 28% after 30 mg/kg; Rutishauser, 1963) seems to disappear at lower doses. Since in the present experiments the decrease of pyruvate and lactate became insignificant in spite of a persisting increase (of about 100%) in glucose, the latter cannot primarily be due to an inhibition of glycolysis. The same was already indicated by the behavior of phosphorylated hexoses mentioned above. With regard to pyruvate and lactate, respectively, benzodiazepines seem to differ—at least quantitatively—from some hypnotics and neuroleptics which show a distinct decrease in the total and intracellular pyruvate and lactate, respectively, in the brain of mice and rats (Stone, 1938; Heim and Estler, 1961; Lowry and Passonneau, 1962; Estler, Heim, and Strubelt, 1962; Rutishauser, 1963), in spite of a glucose accumulation of comparable degree (Rutishauser, 1963; Gey et al., 1965).

7. *No significant change of malate in rat brain.* This is not in accord with the finding of 30 to 50% decrease of malate in total brain of mice treated with diazepam 2 to 20 mg/kg (Young et al., 1969). The discrepancy may be due to species differences.

8. *Lack of significant alterations of free amino acids (α-amino nitrogen)*, although a tendency toward decrease was observed 6 to 12 hr after chlordiazepoxide. This is in agreement with measurements of brain glutamate in mice treated with diazepam 20 mg/kg (Young et al., 1969). Therefore, if the benzodiazepines were able to alter the quantitatively important transformation of glucose into glutamate and other amino acids (Baláazs, 1970), this would not result in a change of steady state levels of amino acids. Nevertheless, it remains to be disproven that benzodiazepines decrease the rate of this transformation, as reported for barbiturates and chlorpromazine (Bachelard and Lindsay, 1966; Yoshino and Elliot, 1970).

9. *No significant changes in total ATP* although an initial tendency toward increase (about 12 to 18% after 15 to 30 min) was obvious. If the latter were due to an increase in a particular intracellular compartment, it would deserve interest, especially with regard to the well-known inhibition of phosphofructokinase by ATP (Quastel, 1969; McGilvery, 1970). In any case, the absence of an ATP decrease indicates that the benzodiazepines do not affect the energy balance in rat brain. In this respect, benzodiazepines do not differ from phenobarbital, chlorpromazine, and reserpine (Gey et al., 1965).

C. Changes in body temperature

Although in the present experiments a relatively low dosage of chlordiazepoxide hydrochloride was used (20 mg/kg), it still caused a slight hypothermia (decrease of $37.7°$ to $36.0°C$). After diazepam 4 mg/kg, the hypothermia ($37.0°C$) seems to lack practical importance. The benzodiazepine-induced changes of cerebral glucose, fructose-1,6-diphosphate, *etc.* can hardly be attributed to hypothermia for two reasons: both benzodiazepines differed markedly with respect to their hypothermic potency but not at all regarding their effect on glucose metabolites; and prevention of the chlordiazepoxide-induced hypothermia did not abolish the alteration of glucose metabolites (Rutishauser, 1963).

IV. CONCLUSIONS

In the present experiments, an extensive accumulation of intracellular glucose was the most pronounced effect of moderate doses of chlordiazepoxide and diazepam in the brain. A corresponding effect, however, has also been demonstrated for some hypnotics and neuroleptics (Rutishauser, 1963). Theoretically, glucose accumulation may result from an absolute or relative enhancement of glucose uptake by brain cells and/or reduction of glucose metabolism. The glucose accumulation by hypnotics and neuroleptics appears to be accompanied, if not caused, at least in part, by an inhibition of glycolysis as indicated by a simultaneous drop of hexose phosphates as well as of pyruvate (Gey et al., 1965). In contrast, in the present experiments with two benzodiazepines no evidence could be found for a reduction of glycolysis since both drugs decreased neither glucose- and fructose-6-phosphate nor pyruvate and lactate significantly. The benzodiazepine-induced rise of fructose-1,6-diphosphate suggests even an augmented glycolysis, at least in the initial phase. In consequence, it should be considered that the mechanism of glucose accumulation by benzodiazepines differs from that of hypnotics and neuroleptics and consists conceivably in an enhanced glucose uptake. This hypothesis needs confirmation by experiments using labeled glucose, with the measurement of the specific radioactivity of glucose and its glycolytic metabolites.

V. SUMMARY

Fed rats received i.p. diazepam 4 mg/kg of chlordiazepoxide hydrochloride 20 mg/kg, and placebo, respectively. After various time intervals, half of each animal group was decapitated for the determination of blood levels. The remaining animals were immersed in liquid nitrogen in order to measure the total value of several metabolites in brain and to calculate their intracellular level. Both benzodiazepines caused a very slight and transient rise in blood glucose and an 80 to 100% increase of intracellular glucose in brain for 15 min to 2 hr. This pronounced accumulation of glucose occurred without a significant change of cerebral glycogen, glucose-6-phosphate, fructose-6-phosphate, dihydroxyacetone phosphate, pyruvate, lactate, α-amino nitrogen, and ATP but was initially accompanied by a slight rise of fructose-1,6-diphosphate. These results suggest that both benzodiazepines do not markedly reduce glycolysis and may thus be considered to increase the cerebral glucose by an enhanced glucose uptake. In consequence, the mechanism of the benzodiazepine-induced cerebral glucose accumulation appears to differ from that of hypnotics and neuroleptics which presumably inhibit glycolysis. The time-course of the benzodiazepine-induced glucose increase paralleled fairly with the obvious sedation of the animals.

ACKNOWLEDGMENT

The excellent technical assistance of Mr. Horst Georgi is gratefully acknowledged.

REFERENCES

Albrecht, W. (1957): Erhöhung der Glykogenkonzentration im Gehirn und das Verhalten verschiedener Fermente im Gehirn und Darm der Maus nach Reserpin. *Klinische Wochenschrift,* 35:588.
Bachelard, H. S., and Lindsay, J. R. (1966): Effects of neurotropic drugs on glucose metabolism in rat brain *in vivo. Biochemical Pharmacology,* 15:1053.
Balázs, R. (1970): Carbohydrate metabolism. In: *Handbook of Neurochemistry,* Vol. 3, edited by A. Lajtha. Plenum Press, New York, p. 1.
Balzer, H., and Palm, D. (1962): Ueber den Mechanismus der Wirkung des Reserpins auf den Glykogengehalt der Organe. *Naunyn-Schmiedeberg's Archiv für experimentelle Pathologie und Pharmakologie,* 243:65.
Bergmeyer, H. U. (1970): *Methoden der enzymatischen Analyse.* Vol. I & II. Verlag Chemie, Weinheim/Bergstr.
Bernsohn, J., Namajuska, I., and Cochrane, L. S. G. (1956): Inhibition of brain cytochrome oxidase and ATP-ase by chlorpromazine analogues. *Proceedings of the Society for Experimental Biology and Medicine,* 92:201.
Blum, J. E. (1971): Time-course of the antiaggressive effect of chlordiazepoxide and diazepam in fighting rats (foot-shock method). *Personal communication.*
Estler, C. J. (1961): Glykogengehalt des Gehirns und Körpertemperatur weisser Mäuse unter dem Einfluss einiger zentral dämpfender und erregender Pharmaka. *Medicina Experimentalis,* 4:209.
Estler, C. J., Heim, F., and Strubelt, O. (1962): Sauerstoffverbrauch, Körpertemperatur, Futteraufnahme und Metabolitgehalt des Gehirns weisser Mäuse nach Reserpin. *Medicinia Experimentalis,* 6:395.
Gey, K. F., and Pletscher, A. (1961): *Unpublished results.*

Gey, K. F., Rutishauser, M., and Pletscher, A. (1965): Suppression of glycolysis in rat brain *in vivo* by chlorpromazine, reserpine and phenobarbital. *Biochemical Pharmacology*, 14:507.

Goodwin, J. F. (1968): The colorimetric estimation of plasma amino nitrogen with DNFB. *Clinical Chemistry*, 14:1080.

Heim, F., and Estler, C. J. (1961): Der Einfluss von Phenobarbital (LUMINAL) auf einige Funktionen und Gehirns normaler und zentral erregter weisser Mäuse. *Klinische Wochenschrift*, 39:698.

Hutchins, D. A., and Rogers, K. J. (1970): Physiological and drug-induced changes in the glycogen content of mouse brain. *British Journal of Pharmacology*, 39:9.

Lowry, H. O., and Passonneau, J. V. (1962): The application of quantitative histochemistry to the pharmacology of the nervous system. *Biochemical Pharmacology*, 9:173.

Masurat, T., Greenberg, S. M., Rice, E. G., Herndon, J. F., and van Loon, E. J. (1960): The action of chlorpromazine on yeast hexokinase. *Biochemical Pharmacology*, 5:20.

Mathé, V., Kassay, G., and Hunkar, K. (1961a): Die Wirkung des Chlorpromazins auf den Kohlehydratstoffwechsel des Rattenhirns. *Psychopharmacology*, 2:334.

Mathé, V., Kassay, G., and Hunkar, K. (1961b): Die Wirkung des Reserpins auf den Gehalt des Rattenhirns an gesamtreduzierenden Stoffen und Glykogen. *Zeitschrift für die gesamte experimentelle Medizin*, 134:249.

McGilvery, R. W. (1970): *Biochemistry. A functional approach*. W. B. Saunders Company, Philadelphia, London, Toronto.

Quastel, J. H. (1969): Carbohydrate metabolism in the nervous system. In: *The Structure and Function of Nervous Tissue*, Vol. III, edited by G. H. Bourne. Academic Press, New York and London, p. 62.

Rutishauser, M. (1963): Beeinflussung des Kohlenhydratstoffwechsels des Rattenhirns durch Psychopharmaka mit sedativer Wirkung. *Naunyn-Schmiedeberg's Archiv für experimentelle Pathologie und Pharmakologie*, 245:396.

Stone, W. E. (1938): The effects of anaesthetics and of convulsants on the lactic acid content of the brain. *Biochemical Journal*, 32:1908.

Svorad, D. (1959): The relation of tranquilizers to some cerebral inhibitory states in topical distribution of brain glycogen. *Archives Internationales de Pharmacodynamie et de Thérapie*, 121:71.

Vladimirov, G. E., and Rubel, L. N. (1957): The turnover of hexosemonophosphate in the brain and the effect of stimulation, narcosis and hypothermia. In: *Metabolism of the Nervous System*, edited by D. Richter. Pergamon Press, London and New York.

Whittam, R. (1961): Active cation transport as a pace-maker of respiration. *Nature*, 191:603.

Yanagawa, M. (1958): Effect of chlorpromazine on tissue respiration and carbohydrate metabolism of rat brain. *Nippon Yakurigaku Zasshi*, 54:1141.

Yoshino, Y., and Elliott, K. A. C. (1970): Incorporation of carbon atoms from glucose into free amino acids in brain under normal and altered conditions. *Canadian Journal of Biochemistry*, 48:228.

Young, R. L., Albano, R. F., Charnecki, A. M., and Demcsak, G. (1969): Effect of diazepam on regional levels of glucose and malate in the central nervous system. *Federation Proceedings*, 28:444.

The Benzodiazepines, Raven Press, New York © 1973

RELATIONSHIPS OF CHLORDIAZEPOXIDE BLOOD LEVELS TO PSYCHOLOGICAL AND BIOCHEMICAL RESPONSES

Louis A. Gottschalk, Ernest P. Noble, Gordon E. Stolzoff, Daniel E. Bates, Claire G. Cable, Regina L. Uliana, Herman Birch, and Eugene W. Fleming

Department of Psychiatry and Human Behavior, University of California at Irvine, and the Orange County Medical Center, Irvine, California

I. INTRODUCTION

This is a report on one aspect of a research program involving the exploration of psychoactive drug blood levels and the associated clinical and biochemical responses. The special focus of this study is placed on the relationship of chlordiazepoxide blood levels (after the ingestion of 25 mg of chlordiazepoxide in 18 chronically anxious, fasting subjects) to changes in anxiety, hostility, and other psychological levels, as well as to certain biochemical substances considered responsive to alterations in the magnitude of emotions.

Very few studies are available on the relationship of chlordiazepoxide blood levels and clinical response in humans. No one has reported finding any definite relationship between the blood level of this drug and the degree of emotional response. Recent studies of chlordiazepoxide blood levels and half-life (Schwartz, Postma, and Gaut, 1971) have not examined the relationship of these to clinical response.

On the other hand, many studies have been carried out examining the effect of the oral administration of chlordiazepoxide on anxiety or hostility. In general, there is substantial evidence that chlordiazepoxide in single or repeated doses can reduce anxiety and fear. Depending on the research design and measuring instruments, some investigators report a decrease in hostility or aggression with chlordiazepoxide (Tobin and Lewis, 1960; Gleser, Gottschalk, Fox, and Lippert, 1965; Randall and Schallek, 1968), whereas others report increases in certain kinds of hostility (Tobin, Bird, and

257

Boyle, 1960; Gardos, DiMascio, Salzman, and Shader, 1968; Salzman, DiMascio, Shader, and Harmatz, 1969).

Previous studies, using the same objective content-analysis method of measuring anxiety and three kinds of hostility (outward, inward, and ambivalent) as in the present investigation, have demonstrated that chlordiazepoxide (20 mg orally) significantly decreased anxiety, overt hostility outward, and ambivalent hostility in 16- to 17-year-old juvenile delinquent boys (Gleser et al., 1965) when compared to a placebo. Also, in studies using the same content-analysis procedure for measuring psychological states, anxiety scores in individuals fasting 10 to 12 hr have been found to be significantly correlated with average free fatty acid levels over a 50-min period as well as with the increase in free fatty acid levels from 0 to 20 min after a venipuncture with an intravenous indwelling catheter (Gottschalk, Cleghorn, Gleser, and Iacono, 1965; Gottschalk, Stone, Gleser, and Iacono, 1969a; Stone, Gleser, Gottschalk, and Iacono, 1969). In these studies, positive correlations were also found between hostility scores, especially inward and ambivalent hostility, and plasma triglycerides, cholesterol, and corticosteroid levels, but not with free fatty acid levels.

The aim of the present study was not only to replicate findings from these earlier investigations, but also to extend them to explore other psychobiochemical relations and to investigate the influence of blood levels of psychoactive pharmacological agents on these psychobiochemical relationships.

II. PROCEDURE

Subjects: The participating subjects were paid volunteers, in good physical health. Eighteen anxious subjects (14 male and 4 female) were selected; they were psychiatrically screened by a preliminary interview so that they all had anxiety levels equal to or greater than 4 on the anxiety rating scale of the Overall-Gorham Brief Psychiatric Rating Scale (Overall and Gorham, 1962). In addition, they were carefully interviewed to ascertain that they were not currently taking drugs of any kind and had not done so for at least 4 weeks; they were told not to drink alcohol or coffee or to smoke tobacco for 24 hr before each testing period. Subjects who were using any hormonal medication, such as birth control pills, were rejected from participation in this study.

III. METHODS

Subjects were told, in simple terms, that the purpose of this study was to explore the relationship of various emotions, especially anxiety, to body chemistry, as well as the effect of a small oral dose of a mild tranquilizer or placebo on these psychochemical associations. Furthermore, they were advised that they would have to come to the Mental Health Outpatient Clinic, Orange County Medical Center on two mornings, at least one week apart, for about 90 min on each occasion, after eating supper the previous night between 6:00 and 8:00 p.m. and fasting thereafter until the experimental session began.

The research utilized a double-blind, crossover design, balanced for order. Nine of the subjects took a placebo capsule first and the other nine first took 25 mg of chlordiazepoxide in capsule form. Since all subjects were chronically anxious, paid volunteers, they were given two identical capsules (one containing a placebo and the other chlordiazepoxide) after being accepted for the study, and fully instructed which capsule to take before their first and second visits to the Clinic. Instructions were so designed that ingestion of the capsules took place about 75 min before the collection of psychological and biochemical data. The actual average time between ingestion and the start of data collection was 92.05 min, the standard deviation was $\pm$ 19.67, and the range was from 60 to 145 min. Actual measurement of chlordiazepoxide blood levels reassuringly revealed that subjects did take their placebo and chlordiazepoxide capsules at designated times so that the crossover design was balanced.

On arrival for the experimental procedure, the subjects were first questioned briefly about when they had had supper the previous evening, what time they had taken their capsule, and whether or not they had used other drugs of any kind. If they had taken any other drugs than those specified or if the fasting since the preceding night's supper had been violated, the experimental procedure was terminated. Such subjects were either dropped from the study and replaced by another individual or, if it could be arranged without compromising the research design, the subject was asked to return on another occasion.

Subjects were then asked to lie down on a bed, and they were permitted to relax and get comfortable for about 15 min. One arm was put through a cloth screen so that the subject could not see the next procedure although he was told what was going on. An indwelling intravenous catheter was then inserted in an antecubital vein, and normal saline solution was allowed to infuse slowly into the vein to prevent clotting of blood. The system was so arranged that blood could be withdrawn from the subject without his knowledge. Immediately after insertion of the intravenous catheter (time zero), 20 to 25 cc of blood was withdrawn and put into a heparinized tube. At 10-min intervals thereafter, for a total of 50 min, similar amounts of blood were withdrawn for biochemical determinations. At time zero and at 50 min, the subject was asked to give a 5-min sample speech into the microphone of a tape recorder in response to standardized instructions to talk about any interesting or dramatic personal life experiences. This is the data collection procedure for the objective measurement of psychological states according to the method of Gottschalk and Gleser (1969).

The heparinized blood samples were immediately placed in ice. They were then subjected to centrifugation in the cold, and plasma was separated and stored frozen for subsequent biochemical determinations. Plasma free fatty acids were determined by the method of Antonis (1965), triglycerides by the method of Kessler and Lederer (1966), glucose by the method of Frings, Ratcliff, and Dunn (1970), cortisol by the method of Butte and Noble (1969), cholesterol by the method of Huang, Chen, Wefler, and Raftery (1961), uric acid by the method of Musser and Ortigoza (1966), insulin by the method of Hales and Randle (1963), calcium by the method of Gitelman (1967), chlordiazepoxide by the method of Schwartz and Postma (1966), creatinine by the Technicon method (1965), glutamic oxalacetic transaminase (SGOT) and lactic dehydrogenase (LDH) according to Kessler, Rush, Leon, Delea, and Cupiola (1970), and alkaline phosphatase following the method of Morgenstern, Kessler, Auerbach, Flor, and Klein (1965).

Typescripts of the 5-min speech samples were scored blindly for the magnitude of various psychological states according to the Gottschalk-Gleser method (1969) by content-analysis technicians who were unfamiliar with the purpose of the research. Scores were obtained from all speech samples on anxiety, hostility outward (total, overt, and covert), hostility inward, ambivalent hostility (based on self-references to others making hostile remarks about the speaker), human relations, and achievement strivings (Gottschalk and Gleser, 1969).

All these data were punched on IBM cards, and various statistical analyses were carried out using a 360 IBM computer and an 1130 IBM computer. Dixon (1969) computer programs or other statistical analyses were used as indicated.

IV. RESULTS

A. Chlordiazepoxide blood levels

Chlordiazepoxide blood levels resulting from the ingestion of 25 mg of

TABLE I. Chlordiazepoxide blood levels at beginning of experimental period (time zero) and at termination (time 50 min)

Subjects	Sex	Weights (lb)	Chlordiazepoxide blood level at time zero (μg/ml)	Time after ingestion of drug (min)	Chlordiazepoxide blood level at time 50 min (μg/ml)	Time after ingestion of drug (min)
1	F	107	1.46	105	1.63	155
2	F	130	1.43	90	1.09	140
3	M	135	1.34	59	1.16	109
4	M	180	1.29	70	1.06	120
5	F	120	1.22	105	0.88	155
6	M	175	1.15	105	0.87	155
7	M	157	1.00	75	1.02	125
8	M	180	0.98	90	0.77	140
9	M	165	0.88	100	1.07	150
10	M	185	0.80	105	0.80	155
11	F	135	0.77	145	0.76	195
Mean*			1.12	95.36	1.01	145.36
s.d.			0.24	23.01	0.25	23.01
12	M	160	0.79	90	0.67	140
13	M	172	0.68	78	0.80	128
14	M	198	0.62	90	0.80	140
15	M	170	0.42	90	0.36	140
16	M	208	0.40	80	0.34	130
17	M	140	0.34	70	0.89	120
18	M	165	0.26	110	0.32	160
Mean*			0.50	86.86	0.59	136.86
s.d.			0.20	12.75	0.25	12.75
Mean total			0.88	92.05	0.85	142.05
s.d.			0.38	19.67	0.31	19.67

* Two sets of means are given to distinguish those subjects whose chlordiazepoxide blood levels exceeded or were less than 0.70 μg/ml.

chlordiazepoxide were obtained at 0- and 50-min points of the experimental sessions. These blood levels ranged from 0.26 to 1.63 μg/ml. Blood levels of metabolites of chlordiazepoxide were absent to negligible.

Table I shows that if the chlordiazepoxide blood level at the beginning (time zero) of the experimental period was low, it was likely to be low also 50 min later. When initial blood levels were high, the subsequent blood levels were also high. This point is emphasized in the high correlation ($r = 0.81$) between the first and second chlordiazepoxide blood levels. The chlordiazepoxide blood levels were not correlated with the time interval between ingestion of the drug and the time of drawing the blood sample for chlordiazepoxide determination ($r = 0.05$ at 0 min and -0.12 at 50 min). Also, the body weight or sex of the subject did not appear to have an appreciable influence on the drug blood level.

B. Chlordiazepoxide blood levels and anxiety scores

Taking all 18 subjects at zero time and comparing anxiety scores between when the subjects were on placebo and when they were on chlordiazepoxide, essentially no difference between the anxiety scores on these two occasions was obtained (anxiety score was 2.49 with placebo versus 2.47 with chlordiazepoxide).

Again, comparing the anxiety of all 18 patients when they were on placebo or on chlordiazepoxide, at 50 min after the beginning of the experiment, there was a trend for anxiety scores to be lower (see Table II) when the subjects were on chlordiazepoxide, but this trend was not statistically significant ($t = -1.56$, $p < 0.10$). Also, there was a trend, nonsignificant but still of interest, for higher chlordiazepoxide blood levels at time zero to be associated with greater decreases in anxiety from 0 to 50 min ($r = 0.26$).

It is reasonable to assume that there would be no noticeable effect on anxiety scores if the blood levels of chlordiazepoxide of our subjects did not reach levels which were effective pharmacologically during the experimental period. Accordingly, if we eliminated from consideration all subjects who had chlordiazepoxide blood levels at either the beginning (time zero) or the end of the experiment (time 50 min) that were less than 0.70 μg/ml, we obtained somewhat different results than if we included all 18 subjects. (A chlordiazepoxide blood level of 0.70 μg/ml was chosen as a rough cutoff level because of findings by Schwartz, Postma, and Gaut (1971) that peak blood levels ranged from 0.78 to 1.24 μg/ml over a period of 2 to 4 hr after the ingestion of 20 mg of chlordiazepoxide in six normal subjects.) Comparing the anxiety scores of the 11 subjects whose blood levels of chlordiazepoxide were consistently above 0.70 μg/ml when they were on placebo and on chlordiazepoxide 50 min after the beginning of the experiment, chlordiazepoxide resulted in a significant decrease in the anxiety scores ($t = -2.52$, $p < 0.025$). Also, there was a significantly greater decrease in anxiety scores between the beginning of the experiment and 50 min later in the chlordiazepoxide-treated subjects ($t = -2.05$, $p < 0.05$) (Table II).

TABLE II. *Average anxiety scores with anxious subjects on either chlordiazepoxide (25 mg) or placebo*

Group	Time (min)	Anxiety scores		Mean difference	Significance
		Placebo	Chlordiaze- poxide		
18	0	2.49	2.47	−0.02	$t = -0.07$, $p =$ n.s.
	50	2.62	2.24	−0.38	$t = -1.56$, $p < 0.10$
Mean difference		+0.13	−0.23	−0.36	$t = -1.61$, $p < 0.10$
11	0	2.48	2.47	−0.01	$t = 0.03$, $p =$ n.s.
	50	2.59	2.07	−0.52	$t = -2.25$, $p < 0.025$
Mean difference		+0.11	−0.40	−0.51	$t = -2.05$, $p < 0.05$

Group 18 constitutes the total group, whereas Group 11 includes only those 11 subjects whose chlordiazepoxide blood levels were > 0.70 μg/ml.

Of the 11 subjects whose chlordiazepoxide blood levels were 0.70 μg/ml or above during the experimental period, there was a slight, nonsignificant trend ($r = 0.16$) for higher chlordiazepoxide blood levels at time zero to be correlated with greater decreases in anxiety scores from the beginning of the experiment to 50 min later.

C. Chlordiazepoxide blood levels and hostility scores

There was a trend for hostility scores to decrease more with chlordiazepoxide than with placebo, especially total hostility outward ($t = -1.37$, $p < 0.10$, N = 18; $t = -1.69$, $p < 0.10$, N = 11) and ambivalent hostility ($t = -1.27$, $p < 0.15$, N = 11). However, these decreases did not reach a statistically convincing level of significance (Tables III-V).

With all 18 subjects, the greater the chlordiazepoxide level at time zero, the greater the decrease in ambivalent hostility scores ($r = -0.42$, $p < 0.05$, one-tail test) (Table VI).

No other correlations of interest approaching significance were found with other hostility scores using all 18 subjects. When only the 11 subjects

TABLE III. *Average hostility outward scores with subjects on either 25 mg chlordiazepoxide (Chl) or a placebo (Pl)*

		Hostility outward scores							
		Overt		Covert		Total		Mean difference	Significance
Group	Time (min)	Pl	Chl	Pl	Chl	Pl	Chl		
18	0	1.13	1.33	1.02	0.99	1.60	1.69	+0.09	$t = 0.45, p =$ n.s.
	50	1.23	1.15	1.04	0.81	1.72	1.45	−0.27	$t = -1.28, p < 0.15$
Mean difference		+0.10	−0.18	+0.02	−0.18	+0.12	−0.24	−0.36	$t = -1.37, p < 0.10$
11	0	1.04	1.40	1.06	1.05	1.60	1.77	+0.17	$t = 0.56, p =$ n.s.
	50	1.17	1.08	1.04	0.75	1.69	1.36	−0.33	$t = -1.32, p < 0.15$
Mean difference		+0.13	−0.32	−0.02	−0.30	+0.09	−0.41	−0.50	$t = -1.69, p < 0.10$

For explanation of groups, see Table II.

TABLE IV. *Average hostility inward scores with subjects on either chlordiazepoxide (25 mg) or a placebo*

Group	Time (min)	Placebo	Chlordiaze-poxide	Mean difference	Significance
		Hostility scores			
18	0	1.16	1.05	−0.11	$t = -0.54, p =$ n.s.
	50	1.02	1.17	+0.15	$t = 0.61, p =$ n.s.
Mean difference		−0.14	+0.12	+0.26	$t = 0.97, p =$ n.s.
11	0	1.05	1.03	−0.02	$t = -0.05, p =$ n.s.
	50	0.90	1.12	0.22	$t = 0.85, p =$ n.s.
Mean difference		−0.15	+0.09	+0.24	$t = 0.70, p =$ n.s.

For explanation of groups, see Table II.

TABLE V. *Average ambivalent hostility scores with subjects on either chlordiazepoxide (25 mg) or a placebo*

Group	Time (min)	Placebo	Chlordiaze-poxide	Mean difference	Significance
		Ambivalent hostility scores			
18	0	1.34	1.36	+0.02	$t = 0.08, p =$ n.s.
	50	1.33	1.27	−0.05	$t = -0.25, p =$ n.s.
Mean difference		−0.01	−0.09	−0.08	$t = -0.35, p =$ n.s.
11	0	1.38	1.38	0	$t = 0.02, p =$ n.s.
	50	1.42	1.08	−0.34	$t = -1.33, p < 0.15$
Mean difference		+0.04	−0.30	−0.34	$t = -1.27, p < 0.15$

For explanation of groups, see Table II.

TABLE VI. *Correlations between chlordiazepoxide blood levels (at time zero) and changes in hostility scores*

Blood level	Total hostility out $_{50-0}$	Overt hostility out $_{50-0}$	Covert hostility out $_{50-0}$	Hostility in $_{50-0}$	Ambivalent hostility $_{50-0}$
	Changes in hostility scores from zero to 50 min				
Chlordiazepoxide$_0$ (N = 11)	0.82	0.80	0.37	−0.47	−0.15
Chlordiazepoxide$_0$ (N = 18)	0.07	0.02	0.02	−0.25	−0.42

whose 0- and 50-min chlordiazepoxide levels were above 0.70 μg/ml were used, however, significant correlations occurred between chlordiazepoxide blood levels (at time zero) and total hostility outward ($r = 0.82$), due principally to the correlation with *overt* hostility outward ($r = 0.80$). The correlation between chlordiazepoxide blood levels at time zero and ambivalent hostility scores fell to -0.15 when only those subjects ($N = 11$) were considered (Table VI).

D. Chlordiazepoxide blood levels and other psychological states

There was a trend for chlordiazepoxide blood levels, using all 18 subjects, to correlate negatively (-0.31) with achievement strivings scores derived from the 5-min speech samples and obtained at time zero and positively (0.28) with difference scores from time 0 to time 50. Correlations between chlordiazepoxide blood levels and human relations scores were essentially zero.

E. Chlordiazepoxide blood levels and biochemical correlates

The biochemical correlates of chlordiazepoxide blood levels are given in Table VII. Those of possible interest are boldface. It should be noted that chlordiazepoxide blood levels among the total group ($N = 18$) and the group of 11 whose blood levels exceeded 0.70 μg/ml are included in these calculations as well as the chlordiazepoxide level at zero (chlordiazepoxide$_0$) and 50 min (chlordiazepoxide$_{50}$). There tended to be a negative correlation between chlordiazepoxide blood levels and the average plasma free fatty acids ($r = -0.26, -0.15, -0.32, -0.19$), average triglycerides ($r = -0.43, 0.01, -0.39, -0.33$), average glucose ($r = -0.73, -0.50, -0.80, -0.71$), and average insulin ($r = -0.11, -0.55, -0.17, -0.52$). The correlations between initial chlordiazepoxide blood levels (time zero) and average cortisol levels tended to be positive ($r = 0.21, 0.26$), but between chlordiazepoxide levels at 50 min and average cortisol, these correlations tended to be negative ($r = -0.30$). The correlations between chlordiazepoxide levels and SGOT ($r = -0.44, -0.34, -0.28, -0.16$), and creatinine ($r = -0.37, -0.25, -0.20, -0.25$) also tended to be negative. The correlations between chlordiazepoxide blood levels and triglycerides$_{40\text{-}20}$ tended to be positive ($r = 0.38, 0.11, 0.48, 0.42$).

One must exercise caution in drawing final conclusions about the significance of any of the correlations in Table VII without further replication, especially because of the large number run and the probability that 5 out of 100 correlations could be expected to be significant by chance alone. Nevertheless, many of these correlations provide research leads and hypotheses

TABLE VII. *Correlations between chlordiazepoxide blood levels and biochemical variables*

Blood levels	Chlordiaze-poxide$_0$ (N=18)	Chlordiaze-poxide$_0$ (N=11)	Chlordiaze-poxide$_{50}$ (N=18)	Chlordiaze-poxide$_{50}$ (N=11)
FFA av.	—0.26	—0.15	—0.32	—0.19
Triglyceride av.	**—0.43**	0.01	—0.39	—0.33
Trig$_{20-0}$	—0.36	**—0.46**	—0.38	—0.23
Trig$_{40-20}$	0.38	0.11	**0.48**	**0.42**
Glucose av.	**—0.73**	**—0.50**	**—0.80**	**—0.71**
Cortisol av.	0.21	0.26	—0.05	—0.30
Insulin	—0.11	**—0.55**	—0.17	**—0.52**
SGOT	**—0.44**	—0.34	—0.28	—0.16
Creatinine	—0.37	—0.25	—0.20	—0.25
LDH	0.17	0.44	0.12	**0.53**

(N=18) = All 18 subjects
(N = 11) = The 11 subjects whose blood levels of chlordiazepoxide exceeded 70 μg/ml.
FFA, free fatty acid; Trig$_{20-0}$, difference between Trig at 20 and 0 min; LDR, lactic dehydrogenase; SGOT, glutamic oxalacetic transaminase.

that may serve as guidelines for further studies. Let us point out some of the possible guideposts.

Any set of correlations (e.g., along the rows in Table VII) that is uniform and at least moderately negative or positive may describe a linear relationship between two variables (e.g., chlordiazepoxide and average blood glucose). Any set of correlations that uniformly changes sign when chlordiazepoxide levels at zero are compared with levels at 50 min may indicate that chlordiazepoxide has a biphasic relationship with the dependent variable (e.g., chlordiazepoxide may initially stimulate and later depress average cortisol secretion). Any set of correlations that changes sign when one biochemical variable is compared over time in its relationship with chlordiazepoxide (e.g., triglycerides$_{20-0}$ and triglycerides$_{40-20}$) may indicate a curvilinear (biphasic) association with chlordiazepoxide. Any set of correlations that changes consistently in magnitude with the blood concentration of chlordiazepoxide being considered (e.g., chlordiazepoxide and insulin) may indicate a curvilinear association (e.g., with blood insulin).

Comparing average plasma free fatty acid and triglyceride levels every 10 min when the subjects were on chlordiazepoxide or placebo (see Figs. 1 and 2 and Table VIII), there was a significant elevation, as statistically assessed by analysis of variance, of both of these lipids when the subjects were on chlordiazepoxide (free fatty acids, $F = 9.166$, $p < 0.01$; triglycerides, $F = 11.496$, $p < 0.001$ two-tailed test). Also, creatinine blood levels

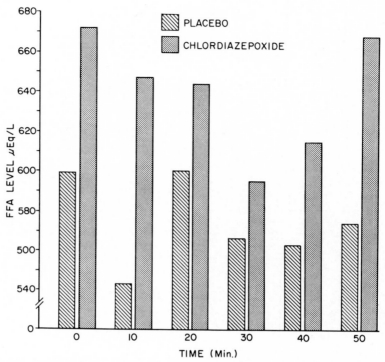

FIG. 1. Comparison of average free fatty acid levels under chlordiazepoxide and placebo conditions (N = 18).

were significantly lower under drug as compared to placebo conditions ($t = 7.22; p < 0.001$).

No significant differences were found, however, in blood glucose levels under the drug as compared to the placebo conditions.

In Table IX we have given the correlations between blood insulin and glucose, under chlordiazepoxide and placebo conditions for the 18 subjects. The expected negative correlations between insulin levels and glucose are revealed under placebo conditions, but with chlordiazepoxide this hypoglycemic effect of insulin no longer prevails.

F. Psychological states and biochemical correlates

Correlations between anxiety, hostility, other psychological states, and various biochemical substances in plasma, during the placebo and drug conditions, are presented in Table X.

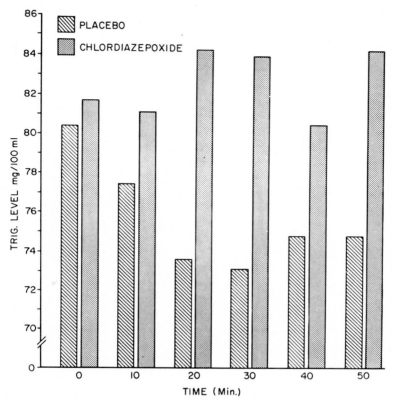

FIG. 2. Comparison of average triglyceride levels under chlordiazepoxide and placebo conditions (N = 18).

Except for correlations that have been observed in previous studies, the correlations listed in Table X should be interpreted with caution. (See also above comments regarding Table VII.) The boldface numbers do, however, suggest hypotheses that are useful for heuristic purposes and merit further exploration.

The positive correlation ($r = 0.35$), during placebo conditions, between anxiety scores and average plasma free fatty acids, although not quite significant (one-tailed test), was typically significant in our previous studies (Gottschalk et al. 1965, 1969a). Also, the positive correlation of hostility outward scores with cortisol level ($r = 0.42$) corresponded with previous findings (Gottschalk et al., 1969a), but the negative correlation of hostility outward scores with cholesterol level during the placebo condition ($r = -0.35$) was counter to a usually positive correlation in earlier studies.

TABLE VIII. *Means and standard deviations of biochemical variables under chlordiazepoxide and placebo conditions*

Biochemical variable	Conditions		Significance
	Drug	Placebo	
FFA av.	640.02 (196.02)	574.19 (223.75)	$p < 0.01$
FFA_{20-0}	−27.61 (176.25)	1.11 (194.50)	—
FFA_{40-20}	−29.44 (135.72)	−37.78 (113.71)	—
Triglyceride av.	82.66 (32.74)	75.71 (25.68)	$p < 0.001$
$Trig_{20-0}$	2.67 (20.50)	−6.67 (20.40)	—
$Trig_{40-20}$	−3.78 (13.14)	1.17 (15.90)	—
Glucose av.	85.30 (8.87)	85.34 (8.78)	—
$Gluc_{20-0}$	0.61 (6.16)	1.72 (17.14)	—
$Gluc_{40-20}$	−2.01 (7.50)	−3.11 (8.92)	—
Cortisol av.	11.69 (5.86)	11.62 (6.08)	—
$Cortisol_0$	11.84 (6.55)	12.79 (7.41)	—
$Cortisol_{50}$	11.52 (7.45)	9.76 (5.33)	—
Cholesterol	176.11 (83.09)	179.94 (26.96)	—
Uric acid	5.48 (0.83)	5.71 (1.04)	—
Calcium	8.49 (0.71)	8.11 (1.31)	—
Insulin	16.56 (5.68)	18.17 (8.79)	—
SGOT	9.16 (4.25)	6.29 (3.01)	$p < 0.10$
Alkaline phosphatase	30.22 (10.34)	32.00 (9.36)	—
LDH	114.44 (36.47)	119.22 (28.88)	—
Creatinine	0.88 (0.12)	1.04 (0.12)	$p < 0.001$

Especially to be noted is the marked change in correlation between certain scores for psychological states and biochemical variables under chlordiazepoxide conditions as compared to placebo conditions. These signify, we believe, psychoactive drug effects influencing either one or both factors in the correlate. We can be more certain of this likelihood in the case of previously observed correlations done under no-drug conditions (e.g., anxiety scores with average blood free fatty acids, or hostility outward scores with blood cortisol). With correlations that have never been previously looked at, when a reversal of sign occurs in a fairly high correlation coefficient, under drug conditions as compared to placebo conditions, the possibility is similarly increased that this change reflects a psychoactive drug effect, acting either on the central nervous system or peripherally or both.

TABLE IX. *Correlations of blood insulin and glucose (N = 18)*

Condition	$Gluc_0$	$Gluc_{10}$	$Gluc_{20}$	$Gluc_{30}$	$Gluc_{40}$	$Gluc_{50}$
Chlordiazepoxide	−0.08	0.03	0.08	0.12	0.21	0.14
Placebo	−0.69	−0.19	−0.25	−0.19	−0.59	−0.13

$Gluc_0$ = blood glucose at zero min; $Gluc_{50}$ = blood glucose at 50 min.

TABLE X. Correlations of psychological state scores with biochemical variables under chlordiazepoxide and placebo conditions (N = 18)

Biochemical variables	Psychological state scores											
	Anxiety		Hostility out total$_0$		Hostility in$_0$		Hostility ambivalent$_0$		Achievement strivings$_0$		Human relations$_0$	
	Drug	Placebo	Drug	Placebo	Drug	Placebo	Drug	Placebo	Drug	Placebo	Drug	Placebo
FFA av.	−0.18	**0.35**	−0.15	**0.48****			**−0.35**	−0.11	−0.29	**−0.55****	0.24	−0.43*
FFA$_{20-0}$												
FFA$_{40-20}$			0.11	**−0.56****	**−0.43***	0.05			**0.42***	−0.32	0.16	**0.47****
Triglyceride av.												
Trig$_{20-0}$	−0.22	**0.41***	0.14	**−0.39**	−0.32	0.22	0.07	**0.45***	−0.12	**0.41***		
Trig$_{40-20}$	0.24	**−0.46****										
Glucose av.			**0.48****	0.14			−0.32	−0.14			**−0.40***	−0.21
Glucose$_{20-0}$											0.16	**−0.36**
Glucose$_{40-20}$											−0.36	0.27
Cortisol av.			**−0.41***	0.29			−0.19	**−0.46****			0.36	**−0.43***
Cortisol$_0$			−0.29	**0.42***			−0.27	**−0.48****			**0.58****	**−0.43***
Cortisol$_{50}$			**−0.39**	0.17			−0.05	**−0.52****			0.05	−0.32
Cortisol$_{50-0}$			−0.13	**−0.47****							**−0.45***	0.21
Cholesterol			0.11	−0.35	**0.40***	0.10	**0.33**	−0.24			**−0.35**	−0.33
Uric acid							−0.12	−0.34			0.17	**−0.51****
Calcium							**−0.38**	−0.25			**0.35**	**−0.59****
Insulin							0.29	**0.53****	0.38	**−0.44****	**−0.47****	−0.01

Boldface correlations are of possible interest.

* $p < 0.10$ (two-tailed test).

** $p < 0.05$ (two-tailed test).

Trig = Triglycerides.

V. DISCUSSION

There have been very few studies examining human individual differences in drug blood levels of new psychoactive drugs after ingestion of a standardized oral dose. Even more scarce are examinations of the relationships between drug blood levels and clinical response in human subjects.

The present study illustrates that a wide range of differences in chlordiazepoxide blood levels is obtained in 18 anxious individuals receiving identical dosages (25 mg) of this pharmacological agent by mouth. The differences in blood levels cannot be attributed, in our study, to differences in time elapsed between ingestion of the drug and drawing the blood sample for chlordiazepoxide determination, nor to body weight, nor to sex. Differences in drug absorption time might account for these widely ranging chlordiazepoxide blood levels; more likely, we believe, are individual differences in drug metabolism.

Animal studies have revealed that genetic factors can account for the marked differences in drug blood levels achieved with standardized dosages. Also, in human subjects given similar parenteral doses of psychoactive drugs, marked differences in blood levels have been found using imipramine, chlorpromazine, thioridazine, and meperidine (Green, Forrest, Forrest, and Serra, 1965; Usdin, 1970; Cash and Quinn, 1970; Elliott, 1971). Moreover, the presence in the body of other psychoactive agents (e.g., sedatives, hypnotics, or tranquilizers) has been found to influence the metabolism and the half-life of certain psychoactive drugs.

The findings of the present study suggest the importance of individual differences in drug metabolism and the bearing such differences can have on drug efficacy. Our study points the way toward the importance of further investigations of the type reported here, and invites the eventual exploration of how to detect and influence psychoactive drug metabolism to maximize desired pharmacological effects.

The present investigation reveals a low correlation ($r = -0.26$) between chlordiazepoxide blood levels in 18 anxious subjects and a decrease in anxiety scores, over a 50-min observation period, and a significant correlation ($r = -0.42$, $p < 0.05$, one-tailed test) between drug blood levels and a decrease in ambivalent hostility scores. Correlations between drug blood levels and psychological reactions of 11 of the subjects whose drug blood levels exceeded 0.70 μg/ml revealed an interesting positive correlation ($r = 0.82$) between hostility outward scores and drug blood levels.

Examination of the decreases in anxiety and hostility scores over the 50-min observation period with all 18 subjects taking chlordiazepoxide (25 mg), compared to changes in these scores with a placebo, showed not quite significant ($p < 0.10$) decreases in anxiety, hostility outward, and ambivalent hostility scores. However, when changes in psychological studies were examined in those 11 anxious subjects whose chlordiazepoxide blood levels were 0.70 mg/ml or above, a significant decrease occurred in anxiety scores ($p < 0.025$) when subjects were on chlordiazepoxide rather than placebo. Comparative decreases in hostility scores (N = 11) with chlordiazepoxide still did not reach a convincing level of significance.

Previous studies using the same content-analysis method of deriving anxiety and hostility scores (Gleser et al., 1965) also showed a significant reduction in anxiety, overt hostility outward, and ambivalent hostility after 20 mg of chlordiazepoxide in 16- to 17-year-old delinquent boys. The present study strongly suggests that the blood level of an antianxiety agent such as chlordiazepoxide must be sufficient or exceed a certain level before an antianxiety effect is achieved by most individuals.

Our findings that chlordiazepoxide blood levels in 18 subjects (Table VII) tend to be negatively correlated with average blood free fatty acids ($r = -0.26$), triglycerides ($r = -0.43$), glucose ($r = -0.73$), creatinine ($r = -0.37$), and, in the 11 subjects having higher chlordiazepoxide levels, with insulin ($r = -0.55$), point to rather varied metabolic and neurogenic effects that cannot be easily understood by one explanatory mechanism.

This study contributes some information regarding the ability of chlordiazepoxide to increase hostility levels in anxious subjects. Apparently, a single oral dose of 25 mg of chlordiazepoxide does not raise hostility levels while it lowers anxiety scores in anxious subjects. In fact, the higher the blood level of chlordiazepoxide, the greater the decrease in ambivalent hostility scores. Curiously enough, however, when one uses for such a calculation the 11 subjects whose chlordiazepoxide levels were 0.70 μg/ml or above, a correlation of 0.82 is obtained between the chlordiazepoxide blood level at the beginning of the experimental period (time zero) and the increase in hostility outward scores from 0 to 50 min during the experimental session. Most of this correlation is accounted for by the overt hostility outward component ($r = 0.80$) rather than the covert hostility outward component ($r = 0.37$). These findings occur in the face of a slight but not quite significant decrease in hostility outward scores and ambivalent hostility when subjects are on chlordiazepoxide as compared to placebo. Whereas our findings on hostility are in contrast to those of DiMascio and his co-workers, our results on anxiety supplement them. In their studies (DiMascio and Barrett, 1965; Barrett and DiMascio, 1966; Gardos et al., 1968; Salzman et al., 1969), subjects received mild tranquilizers for 1 week (chlordiazepoxide or oxazepam or a placebo) in contrast to a single dose in the present study. In the most recent report by this group (Gardos et al., 1968), chlordiazepoxide (10 mg t.i.d.) for 1 week reduced anxiety (as measured by the Scheier and Cattell, 1960, self-report test) in high and medium anxious subjects (as assessed by the Taylor Manifest Anxiety Scale). Chlordiazepoxide was found to increase significantly ambivalent hostility scores in highly anxious subjects, whereas oxazepam increased hostility inward scores (Salzman et al., 1969), using the Gottschalk content-analysis hostility scale (1963). Hostility scores from the Buss-Durkee Hostility Inventory (1957) also, apparently, increased with chlordiazepoxide. Our study cannot disprove that, perhaps with more prolonged administration of minor tranquilizers, an increase in some kind of hostility may occur in some individuals.

Besides the fact that there may well be different effects on psychological states of single and multiple doses of psychoactive pharmacological agents, there are also differences in results that may be obtained depending on the psychological measuring technique used. For instance, the Buss-Durkee, the Scheier and Cattell, and Taylor Manifest Anxiety Scales are all self-report procedures for assessing psychological states, and, as such, these measurement procedures are subject to the effects of suggestion or covering up. Individuals reporting about themselves, using these methods, are aware of what is being measured and may expose only that which they want to reveal instead of how they actually feel. The Gottschalk-Gleser Content-Analysis Method (1969), on the other hand, does not reveal to the subject what psychological dimensions are really being measured, and this method is designed to give an assessment of a psychological trait. Hence, the latter procedure may not necessarily give identical results to the former measurement methods.

The elevation of plasma free fatty acid levels (and triglycerides) with a single dose of chlordiazepoxide, as compared to a placebo, in a relatively nonstressful experiment is of interest. Other investigators (Martelli and Corsico, 1969) have noted that chlordiazepoxide (40 mg/kg, i.p.) elevates plasma free fatty acids in female albino rats. There is evidence (Martelli and Corsico, 1969) that this lipomobilizing effect of chlordiazepoxide may occur through the mechanism of blocking the enzyme phospodiesterase which, in turn, like theophylline, prevents the inactivation of cyclic 3′, 5′-AMP and, hence, of adipose tissue lipase (Hynie, Krishna, and Brodie, 1966; Butcher and Baird, 1969). Oxazepam (30 mg in a single dose) has been demonstrated to keep plasma free fatty acid levels lower than a placebo (Brown, Sletten, Kleinman, and Korol, 1968) during a planned stressful experiment in which human subjects received mild electroshocks or the threat of such. Similarly, Khan, Forney, and Hughes (1964) have demonstrated that chlordiazepoxide (20 mg/kg) administered to rats will block the rise in free fatty acids more effectively than pentobarbital (5 mg/kg), or ethanol (0.55 g/kg), or after repeated electroshocks (2 sec every min for 90 min). Our study is a natural-history, relatively nonstressful study; under these circumstances, chlordiazepoxide, in a single dose, elevates blood free fatty acids and triglycerides rather than having a notable effect on blocking the elevation of these lipids through stress-induced adrenergic mechanisms. In our study, at the same time, anxiety scores were significantly reduced when chlordiazepoxide blood levels exceeded 0.70 μg/ml. In the study of Brown et al. (1968), when a stressful electroshock stimulus was applied to human subjects, oxazepam blocked free fatty acid elevation as compared to a placebo. The free fatty acid level with placebo in the Brown et al. study reached 670 μequiv/liter with electroshock, whereas in our study average free fatty acids ranged from 540 to 600 μequiv/liter when the subjects were on placebo and from 595 to 670 on chlordiazepoxide. Apparently, a single dose of chlordiazepoxide, possibly in contrast to oxazepam, may mobilize lipids as much as a stress stimulus. We propose that the lipomobilizing effect of a single dose of chlordiazepoxide involves metabolic mechanisms peripheral to the central nervous system, possibly through blocking the influence of phosphodiesterase, which permits cyclase to activate cyclic AMP, which in turn activates adipose tissue lipase. At the same time, the lowering of anxiety scores in our study demonstrates the simultaneous neurochemical inhibition of adrenergic activation through the central nervous system. Since our subjects were not purposely stressed, the capacity of chlordiazepoxide to block the lipomobilizing effect of stress was not elicited; rather, the peripheral metabolic effect of chlordiazepoxide appeared to predominate.

In contrast to other investigators (Rutishauser, 1963; Sternbach, Randall, and Gustafson, 1964; Satoh and Iwamoto, 1966) who observed that chlordiazepoxide produced short-lasting increases in blood glucose in fasted rats, we found no significant differences in blood glucose levels with chlordiazepoxide or a placebo. This hyperglycemic effect of chlordiazepoxide is believed to be an indirect action, due in part to activation of adrenergic mechanisms. In female rats, such a hyperglycemic effect does not occur when the dose level is less than 30 mg/kg, i.p., and it lasts for at least 4 hr when the dosage reaches 60 mg/kg, i.p. (Arrigoni-Martelli and Toth, 1969). The absence of such a hyperglycemic effect in our subjects, who received only 25 mg per mouth, is most likely the result of a relatively low dosage.

Our findings suggest that chlordiazepoxide (25 mg) may have a mild initial stimulating influence on cortisol blood levels and a suppressing effect on insulin secretion. More definitely, it blocks the hypoglycemic effect of insulin (Table IX). These and other biochemical effects of 25 mg of chlordiazepoxide seem to be considerable, although not heretofore noted. We have no evidence that these biochemical effects are advantageous or disadvantageous for human beings, any more than suppressing anxiety or other psychological states may or may not be deleterious. We believe strongly that these effects of this benzodiazepine should be investigated further to improve our understanding of mechanism of action and side effects.

The capacity of a small oral dose of chlordiazepoxide to change radically the typical psychobiochemical correlations of anxiety with blood free fatty acids, and hostility with blood cortisol levels, should be of interest. Our demonstration that other less well-substantiated correlations between psychological states and biochemical variables are also markedly influenced tends to corroborate the likelihood that small doses of psychoactive drugs such as chlordiazepoxide may have widespread peripheral and central nervous system effects as well as a variety of metabolic effects. Such findings underline how careful an investigator should be about attributing psychoactive drug effects to only one biological mechanism or pathway and, particularly, to the one on which he has focused attention. Biological pathways other than the intended ones may be influential in leading to the observed effects.

The apparent biochemical correlates of various psychological states that we have demonstrated here and the modification of these correlations with a psychoactive drug suggests a possible research avenue involving the detection and estimation of the magnitude of various psychological states through biochemical measures.

VI. SUMMARY

(1) We performed a double-blind, drug-placebo, crossover study examining the effects on several psychological states and on a variety of biochemical substances in the blood, including chlordiazepoxide blood levels, of a single oral dose of chlordiazepoxide (25 mg) on 18 anxious subjects.

(2) Blood levels of chlordiazepoxide ranged from 0.26 to 1.63 μg/ml. These blood levels were not correlated with the time interval between ingestion of the drug and time of drawing the blood sample for chlordiazepoxide determination (average time interval 92 min, range 60 to 145

min), nor with body weight or sex.

(3) Anxiety, hostility outward, and ambivalent hostility scores tended to decrease ($p < 0.10$) over the 50-min observation period of this study when subjects were on chlordiazepoxide, as compared to a placebo. However, a statistically significant decrease in anxiety scores ($p < 0.025$) during this time period occurred only in those 11 subjects whose chlordiazepoxide blood levels exceeded 0.70 μg/ml. In these subjects, the decrease in hostility outward and ambivalent hostility scores, when the subjects were on chlordiazepoxide, did not reach a convincing level of significance.

(4) Chlordiazepoxide blood levels of all 18 subjects correlated with scores on anxiety (-0.26), ambivalent hostility (-0.42), achievement strivings (-0.31), average blood free fatty acids (-0.26), average triglycerides (-0.43), glucose (-0.73), and creatinine (-0.37). Chlordiazepoxide blood levels of the 11 subjects whose levels exceeded 0.70 μg/ml also correlated with hostility outward scores (0.82), hostility inward scores (-0.47), and with blood insulin (-0.55).

(5) Chlordiazepoxide, at this single oral dosage (25 mg), in comparison to a placebo, significantly elevated plasma free fatty acids ($p < 0.01$) and triglycerides ($p < 0.001$) but had no effect on glucose. This lipomobilizing effect is thought to be due to blocking of the enzyme phosphodiesterase, allowing cyclase to activate cyclic AMP and, hence, adipose tissue lipase.

(6) The significant negative correlations of blood insulin levels with blood glucose, when the same subjects were on placebo, were blocked when the subjects were administered the chlordiazepoxide. There was some slight evidence that chlordiazepoxide might stimulate cortisol secretion initially and suppress it 50 min later.

(7) A series of correlations between various psychological states and biochemical variables in the blood under placebo conditions were reversed when the subjects were on chlordiazepoxide: e.g., anxiety correlated with average free fatty acids (placebo 0.35, chlordiazepoxide -0.18); hostility outward with cortisol (placebo 0.42, chlordiazepoxide -0.29); achievement strivings with average triglyceride (placebo -0.32, chlordiazepoxide 0.42); achievement strivings with insulin (placebo -0.44, chlordiazepoxide 0.38); and human relations with cortisol (placebo -0.43, chlordiazepoxide 0.58). These changes in psychobiochemical relationships, associated with ingestion of 25 mg of chlordiazepoxide, suggest the influence of a variety of intervening variables, including metabolic and neurochemical influences on the peripheral and central nervous system.

(8) Implications of these findings are discussed as well as directions for future research.

ACKNOWLEDGMENTS

This study has been supported in part by U.S. Public Health Service research grant MH-20174 from the National Institute of Mental Health, and by a grant-in-aid from Hoffmann-La Roche Inc.

REFERENCES

Antonis, A. (1965): Semiautomated method for the colorimetric determination of plasma free fatty acids. *Journal of Lipid Research,* 6:307

Arrigoni-Martelli, E., and Toth, E. (1969): Effect of chlordiazepoxide on glucose metabolism in rats. In: *European Society for the Study of Drug Toxicity. Toxicity and Side-Effects of Psychotropic Drugs; Proceedings of the Meeting Held in Paris,* Series No. 145, p. 73. Excerpta Medica Foundation, Amsterdam.

Barrett, J. E., and DiMascio, A. (1966): Comparative effects on anxiety of the "minor tranquilizers" in "high" and "low" anxious student volunteers. *Diseases of the Nervous System,* 27:483.

Brown, M. L., Sletten, I. W., Kleinman, K. M,. and Korol, B. (1968): Effect of oxazepam on physiological responses to stress in normal subjects. *Current Therapeutic Research,* 10:543.

Buss, A. H., and Durkee, A. (1957): An inventory for assessing different kinds of hostility. *Journal of Consulting Psychology,* 21:343.

Butcher, R. W., and Baird, C. E. (1969): The regulation of cyclic AMP and lipolysis in adipose tissue by hormones and other agents. In: *Drugs Affecting Lipid Metabolism,* edited by W. L. Holmes, L. A. Carlson, and R. Paoletti. Plenum Press, New York, p. 5.

Butte, J. C., and Noble, E. P. (1969): Simultaneous determination of plasma or whole blood cortisol and corticosterone. *Acta Endocrinologica,* 61:678.

Cash, W. D., and Quinn, G. P. (1970): Clinical-chemical correlations: An overview. *Psychopharmacology Bulletin,* 6:26.

Cohen, I. M., and Harris, T. H. (1969): Effects of chlordiazepoxide in psychiatric disorders. *Southern Medical Journal,* 54:1271.

Cuparencu, B., Ticsa, I., Safta, L., Rosenberg, A., Mocan, R., and Brief, G. A. (1969): Influence of some psychotropic drugs on the development of experimental atherosclerosis. *Cor et Vasa, International Journal of Cardiology,* 11:112.

DiMascio, A., and Barrett, J. E. (1965): Comparative effects of oxazepam in "high" and "low" anxious student volunteers. *Psychosomatics,* 6:298.

Elliott, H. W., Gottschalk, L. A., and Uliana, R. L. (1973): Relationship of plasma meperidine levels to changes in anxiety and hostility. *Psychopharmacologia,* (in press).

Frings, S. F., Ratcliff, C. R., and Dunn, R. T. (1970): Automated determination of glucose in serum or plasma by a direct O-toluidine procedure. *Clinical Chemistry,* 16:282.

Gardos, G., DiMascio, A., Salzman, C., and Shader, R. I. (1968): Differential actions of chlordiazepoxide and oxazepam on hostility. *Archives of General Psychiatry,* 18:757.

Gitelman, H. J. (1967): An improved automated procedure for the determination of calcium in biological specimens. *Analytical Biochemistry,* 18:521.

Gleser, G. C., Gottschalk, L. A., Fox, R., and Lippert, W. (1965): Immediate changes in affect with chlordiazepoxide in juvenile delinquent boys. *Archives of General Psychiatry,* 13:291.

Gottschalk, L. A., Cleghorn, J. M., Gleser, G. C., and Iacono, J. M. (1965): Studies of relationships of emotions to plasma lipids. *Psychosomatic Medicine,* 27:102.

Gottschalk, L. A., and Gleser, G. C. (1969): *The Measurement of Psychological States Through the Content Analysis of Verbal Behavior.* University of California Press, Berkeley.

Gottschalk, L. A., Gleser, G. C., and Springer, K. J. (1963): Three hostility scales applicable to verbal samples. *Archives of General Psychiatry,* 9:254.

Gottschalk, L. A., Stone, W. N., Gleser, G. C., and Iacono, J. M. (1969a): Anxiety and plasma free fatty acids (FFA). *Life Sciences,* 8:61.

Gottschalk, L. A., Winget, C. N., and Gleser, G. C. (1969b): *Manual of Instructions for Using the Gottschalk-Gleser Content Analysis Scales: Anxiety, Hostility, Social Alienation-Personal Disorganization.* University of California Press, Berkeley.

Green, D. E., Forrest, I. S., Forrest, F. M., and Serra, M. T. (1965): Interpatient variation in chlorpromazine metabolism. *Experimental Medicine and Surgery,* 23:278.

Hales, C. N., and Randle, P. J. (1963): Immunoassay of insulin with insulin antibody precipitate. *Lancet,* 1:200.

Hales, C. N., and Randle, P. J. (1963): Immunoassay of insulin with insulin antibody precipitate. *Biochemical Journal,* 88:137.

Huang, T. C., Chen, C. P., Wefler, V., and Raftery, A. (1961): A stable reagent for the Liebermann-Burchard reaction. *Analytical Chemistry,* 33:1405.

Hynie, S., Krishna, G., and Brodie, B. B. (1966): Theophylline as a tool in studies of the role of cyclic adenosine 3',5'-monophosphate in hormone-induced lipolysis. *Journal of Pharmacology and Experimental Therapeutics,* 153:90.

Kessler, G., and Lederer, H. (1966): Fluorometric Measurement of Triglycerides. In: *Automation in Analytical Chemistry,* edited by L. T. Skeggs, Jr., Mediad, Inc., New York, p. 341.

Kessler, G., Rush, R. L., Leon, L., Delea, A., and Cupiola, R. (1970): Automated 340-nm measurement of SGOT, SGPT, and LDH. *Clinical Chemistry,* 16:530.

Khan, A. U., Forney, R. B., and Hughes, F. W. (1964): Plasma free fatty acids in rats after shock as modified by centrally active drugs. *Archives of International Pharmacodynamics,* 151:466.

Martelli, E. A., and Corsico, N. (1969): On the mechanism of lipomobilizing effect of chlordiazepoxide. *Journal of Pharmacy and Pharmacology,* 21:59.

Morgenstern, S., Kessler, G., Auerbach, J., Flor, R. V., and Klein, B. (1965): Automated p-nitrophenylphosphate serum alkaline phosphatase procedure for the auto analyzer. *Clinical Chemistry,* 11:876.

Murray, N. (1962): Covert effect of chlordiazepoxide therapy. *Journal of Neuropsychiatry,* 3:168.

Musser, A. W., and Ortigoza, C. (1966): Automated determination of uric acid by the hydroxylamine method. *Technical Bulletin of Registry of Medical Technologists,* 36:21.

Opitz, K. (1965): Einfluss von Psychopharmaka auf die nichtveresterten Fettsäuren im Blutsplasma. *Experientia,* 21:462.

Overall, J. E., and Gorham, D. R. (1962): The brief psychiatric rating scale. *Psychological Reports,* 10:799.

Randall, L. O., and Schallek, W. (1968): Pharmacological activity of certain benzodiazepines. In: *Psychopharmacology: A Review of Progress, 1957-1967,* edited by D. H. Efron, *et al.,* p. 153. Public Health Service Publication No. 1836, U.S. Government Printing Office, Washington, D.C.

Rutishauser, M. (1963): Beeinflussung des Kohlenhydralstoffwechsels des Rattenhirns durch Psychopharmaka mit sedativer Wirkung. *Naunyn-Schmiedebergs Archiv für Pharmakologie und Experimentelle Pathologie,* 245:396.

Salzman, C., DiMascio, A., Shader, R. I., and Harmatz, J. S. (1969): Chlordiazepoxide, expectation, and hostility. *Psychopharmacologia,* 14:38.

Satoh, T., and Iwamoto, T. (1966): Neurotropic drugs, electroshock, and carbohydrate metabolism in the rat. *Biochemical Pharmacology,* 15:323.

Scheier, I. H., and Cattell, R. B. (1960): *Handbook and Test Kit for the IPAT 8-Parallel Form Anxiety Battery.* Institute for Personality and Ability Testing, Champaign, Illinois.

Schwartz, M. A., and Postma, E. (1966): Metabolic N-demethylation of chlordiazepoxide. *Journal of Pharmacological Sciences,* 55:1358.

Schwartz, M. A., Postma, E., and Gaut, Z. (1971): Biological half-life in chlordiazepoxide and its metabolite, demoxepam, in man. *Journal of Pharmaceutical Sciences,* 60:1500.

Sternbach, L. H., Randall, L. O., and Gustafson, S. R. (1964): 1,4-Benzodiazepines (chlordiazepoxide and related compounds). In: *Psychopharmacological Agents,* edited by M. Gordon., Vol. 1. Academic Press, New York, p. 199.

Stone, W. N., Gleser, G. C., Gottschalk, L. A., and Iacono, J. M. (1969): Stimulus, affect and plasma free fatty acids. *Psychosomatic Medicine,* 31:331.

Technicon Methodology N-14b. (1965): Technicon Instrument Corporation, Tarrytown, New York.

Tobin, J. M., Bird, I. F., and Boyle, D. E. (1960): Preliminary evaluation of LIBRIUM (RO-0690) in the treatment of anxiety reactions. *Diseases of the Nervous System.* 21:11.

Tobin, J. M., and Lewis, N. D. C. (1960): New psychotherapeutic agent, chlordiazepoxide. Use in treatment of anxiety states and related symptoms. *Journal of the American Medical Association,* 174:1242.

Usdin, E. (1970): Absorption, distribution, and metabolic fate of psychotropic drugs. *Psychopharmacology Bulletin,* 6:4.

DISCUSSION SUMMARY

Walter B. Essman

Department of Psychology, Queens College of the City University of New York, Flushing, New York

Several interesting features of benzodiazepine action, at both tissue and cellular levels, have been presented. It seems apparent that the metabolic concomitants or sequelae of benzodiazepines acting behaviorally can be quite distinct from any conventional biochemical profile of sedative or tranquilizing effect. This may be related to a number of factors, e.g., problems in generalizing biochemical effects of the benzodiazepines across species, particularly in view of dosage differences required to account for adequate or effective fluid or tissue levels. Additionally, the apparent absence of a direct drug-related alteration in the levels of catecholamines or indoleamines—with reduced turnover indicated for both norepinephrine and serotonin—presents problems concerning enzymes involved in activation and reuptake of such amines. Since there are indications that monoamine oxidase (MAO) activity in the rat was unaffected by some benzodiazepines (Taylor and Laverty, *this Volume*) and evidence that there were no appreciable alterations in cell body or intraterminal mitochondrial constituents as a consequence of chlordiazepoxide treatment (guinea pigs), or in detectable levels of this compound associated with these organelles (Essman, *this Volume*), it is perhaps of interest to consider membrane effects of the benzodiazepines more specifically, *i.e.*, within the context of their enzyme specificity. In this regard, the *in vitro* interaction of serotonin with diazepam and the formation of an electrostatically bound complex (Galzigna, 1969) have been suggested as bases for the *in vitro* drug-induced MAO inhibition, using serotonin as a substrate. There are, however, other molecular interactions which *in vivo* are consistent with some data presented. The binding of serotonin by nucleotides and nucleic acids (Bittman et al., 1969) has been shown to be favored by the reduction of serotonin turnover or increases in the ionic strength of Mg^{++} (Essman, 1970; Essman et al., 1971). These conditions have been noted to be consistent with benzodiazepine effects, and they appear appropriate to a general condition of reduced synaptic excitability. The issue of binding or uptake of serotonin again arises in data specific for the *in vitro* interference with this event in platelets by benzodiazepines (Lingjaerde, *this Volume*). It might be appropriate to distinguish between platelet *binding* of serotonin and platelet *uptake*; the former might be explained by the formation of a donor-acceptor complex, favored in a medium with a low dielectric constant. The platelet could meet such criteria. It is noteworthy that the interaction of serotonin with platelets appears to be a chloride-dependent process, which perhaps makes this organelle a more appropriate peripheral analog of glia rather than a model for central neurons.

The effects of benzodiazepines on carbohydrate metabolism (Gey, *this Volume*), especially with respect to data on the brain, indicate several changes that could represent a reverse profile from that commonly associated with seizures or convulsant agents; an accumulation of intracellular glucose concomitant with a decrease in lactate, and a rise in the ATP level after diazepam and chlordiazepoxide, are in the reverse direction to changes in brain-carbohydrate metabolism produced by convulsant agents or events. This aspect of these data could warrant investigation of the interaction of benzodiazepines with convulsants, on both biochemical and electrophysiological grounds.

Measures of anxiety and/or stress, as obtained from plasma indices in man, and alterations therein by benzodiazepines (Gottschalk, *this Volume*) present problems with regard to experimental design, intrasubject variability, and interpretation of results. Whereas plasma levels of a drug in man appear closely related to its therapeutic effect, even individual variations in the level of other endogenous plasma constituents can present a rather fragile baseline upon which drug effects can be further imposed and clearly interpreted.

The methodological significance of certain variables such as thermolytic properties of the drug and central effects due to thermal change rather than drug action should probably be emphasized more strongly. Bartolini et al., (*this Volume*) have not only reported a source of amine-turnover change in the thermolytic properties of diazepam, but have also raised questions about the measurement of turnover; e.g., how many time points are necessary, what substrate is measured and how, etc.?

The elevation of brain choline and acetylcholine levels by benzodiazepines (Ladinsky, *this Volume*) implicates not only cholinergic nerve endings in benzodiazepine action, but also again supports, in principle, the view that these compounds exert synaptic effects, probably of a damping nature. Such a view would certainly be consistent with previous data, particularly as these have concerned amine-turnover decrements related to a benzodiazepine effect and synaptosomal electrolyte changes induced by this group of drugs.

There are certainly other issues of probable relevance to both the peripheral and the central biochemistry of benzodiazepines. One such point concerns differences in acute versus chronic administration of these compounds; such differences, particularly as they may concern biochemical parameters, have not been sufficiently examined; they may, however, have further interactive significance in regard to the route by which such compounds are administered. It has been shown that differences in behavioral effects can be related to the method of administration, i.e., oral or parenteral, perhaps owing to differences in drug level or biochemical effect or even to some special sensory effect conferred by oral administration of the drug.

Questions of direct relevance to the neurobiochemistry of benzodiazepines may depend upon differentiation between and specificity of the regional, cellular, and subcellular effects of these compounds. Such limitations assume that biochemical changes induced by benzodiazepines, when assessed from the whole brain or gross areas thereof, provide little information regarding localization of effect or specificity of its locus. This issue is of direct relevance to the deposition, action, and biotransformation of benzodiazepines transported to the brain and the extent to which such specificity for site of drug action is related to specificity for site of correlated biochemical changes.

A final significant point regarding the biochemistry of benzodiazepine action involves the general area of protein synthesis and how this process may be altered by compounds of this class. Since the benzodiazepines are strongly implicated in some of the data presented as exerting a damping effect upon some synaptic systems, it would appear likely that, consistent with altered cerebral excitability, there are also alterations in the rate of cerebral protein synthesis. The behavioral as well as physiological role of specific proteins, both in cellular and in membrane events, would seem to warrant consideration in the study of benzodiazepine action.

It becomes apparent that an entire symposium could be based on the neuro-biochemistry of benzodiazepines, as well as overlapping and interrelating with physiological and behavioral aspects of the actions of this class of compounds. Interesting observations concerning some of the action of benzodiazepines, such as thermolytic effects, effects upon food intake, pain threshold, induced convulsions, changes in aggressive behavior, modification of sleep behavior etc., all represent phenomena with implied neurochemical corollaries or changes, which can provide a fertile field for further investigation that could add meaningful insights to the range of actions and neurobiological events associated with this significant group of psychoactive compounds.

REFERENCES

Bittman, R., Essman, W. B., and Golod, M. I. (1969): *Abstracts of the American Society,* 330.
Essman, W. B. (1970): *Transactions of the New York Academy of Sciences,* 32:948.
Essman, W. B., Bittman, R., and Heldman, E. (1971): *Proceedings of the Fourth Annual Winter Conference on Brain Research.*
Galzigna, L. (1969): *FEBS Letters,* 3:97.

The Benzodiazepines, Raven Press, New York © 1973

TRIAZOLOBENZODIAZEPINES, A NEW CLASS OF CENTRAL NERVOUS SYSTEM-DEPRESSANT COMPOUNDS

A. D. Rudzik, J. B. Hester, A. H. Tang, R. N. Straw, and W. Friis

Research Laboratories, The Upjohn Company, Kalamazoo, Michigan

I. INTRODUCTION

The discovery of the broad pharmacological and clinical spectrum of activity of chlordiazepoxide (Randall et al., 1960) and diazepam (Randall et al., 1961) has led medicinal chemists and biologists to an intense investigation of their activity. The mechanism of action of the antianxiety activity of benzodiazepines is difficult to characterize because of the complex types of behavior produced by these compounds. In animals, the following pharmacological properties of antianxiety agents can be noted: anticonvulsant, muscle-relaxant, and sedative-hypnotic activities, antiaggressiveness or taming behavior, effects on operant behavioral test systems, and increased food intake. Chlordiazepoxide and diazepam produce all these pharmacological effects, and recently a new series of compounds, the triazolobenzodiazepines (Hester et al., 1971a; Hester et al., 1971b), has been found to produce effects similar to known antianxiety compounds. This manuscript will report the pharmacological activity of these triazolobenzodiazepines and compare their activities with that of diazepam.

II. METHODS

A. Anticonvulsant activity

Antagonism of induced convulsions or lethality by nicotine (nicotine salicylate 2 mg/kg, i.v.), thiosemicarbazide (20 mg/kg, i.p.), bemegride (40 mg/kg, s.c.), electroshock (25 mA for 0.2 sec), or pentylenetetrazole (85 mg/kg, s.c.) was measured in groups of six CF-1 male albino mice (18 to 22 g). The test compound was administered 30 min before the convulsant in all cases, except for thiosemicarbazide which was administered simultaneously with the test compound. A cutoff time of 4 hr was used for the thiosemicarbazide and 15 min for the bemegride and pentylenetetrazole experiments. The number of mice protected against the convulsions or lethality was used to determine ED_{50} values.

Groups of four male albino rats, Sprague-Dawley strain, weighing 80 to 120 g were injected with pentylenetetrazole (100 mg/kg, s.c.) 30 min after the administration of the test compound. The rats were observed over the next 30 min for clonic convulsions.

Antagonism of pentylenetetrazole-induced seizures in cats was measured, as described by Straw (1968), using a four-point bioassay to estimate potency ratios (Finney, 1952).

B. Muscle-relaxant activity

Antagonism of strychnine lethality was measured in groups of six CF-1 male albino mice (18 to 22 g). Thirty min after the administration of the test compound, strychnine (3 mg/kg, i.p.) was injected and 4 hr later the number of deaths was recorded.

The traction response was also used as a test for muscle-relaxant or depressant activity. The front feet of mice were placed on a small twisted wire rigidly supported above the bench top. A normal mouse grasps the wire with his forepaws and, when allowed to hang free, places at least one hind foot onto the wire within 5 sec. Failure to put up at least one hind foot constitutes failure to traction. Failure to traction for each individual mouse was used as a quantal response to determine the traction dose 50.

The mid-collicular decerebrate cat was used as a test for muscle-relaxant activity in cats. The procedure employed was that of Ngai et al. (1966) modified to measure contralateral extension of the lower leg following stimulation of the central cut end of the sciatic nerve. Stimulation was elicited continuously with square-wave pulses of 0.2-msec duration, 10 V, at a frequency of 0.5 pps.

The test compounds were injected as a fine suspension in 0.25% aqueous methylcellulose. Repeated injections were made in cumulative doses at 0.5-log intervals about 10 min apart. Four animals were used for each compound and potency ratios were estimated using a four-point bioassay (Finney, 1952).

C. Antiaggressive behavior

Aggressive behavior was produced in male albino mice weighing 18 to

22 g using a modified method of Tedeschi et al. (1959). Pairs of mice were placed on a grid and covered by a Petri dish (2 in high and 4 in in diameter). Footshock was then administered intermittently through the grid for a 3-min period or until the mice fought spontaneously or dominance was established by one of the pair. The dominant mouse was used within 48 hr for drug studies; the other mouse was discarded. Multiple dose levels at 0.3-log intervals of the test compound, prepared in 0.25% aqueous methylcellulose, were injected i.p. or p.o. into groups of eight such dominant or spontaneously fighting mice. Thirty min later, pairs of drug-treated mice were placed on the grid and exposed to footshock for a 30-sec period. The number of pairs of mice which fought at each dose level was recorded and used to calculate the ED_{50} value.

Taming effects were measured in young male rhesus monkeys weighing between 3.6 and 6.4 kg. All animals were housed individually in wire-mesh cages (18 $\times$ 24 $\times$ 30 in) with a transparent Plexiglas front. Test agents or placebo (cornstarch) were prepared in 0.25% aqueous methylcellulose and administered by stomach tube in a randomized design to four monkeys. Observations were scored blind. The compounds were administered orally at 10:00 a.m., and behavioral observations were made 1, 3, and 5 hr later. Aggressiveness in response to pulling on the monkeys' chains was scored subjectively.

D. Depressant or sedative activity

Potentiation of ethanol narcosis was measured using a subhypnotic dose of ethanol (5 ml/kg of a 50% aqueous solution). The ethanol was administered orally to groups of six mice, 30 min after the test compound. Thirty min later, each mouse was examined for loss of the righting reflex. The number of animals in each group that exhibited loss of the righting reflex for 30 sec or more was used for calculating the ED_{50} of the test compound.

Potentiation of either pentobarbital or chlorprothixene narcosis was measured using a subhypnotic dose of pentobarbital (10 mg/kg, i.v.) or chlorprothixene (15 mg/kg, p.o.), which was administered to groups of six mice 30 min after the test compound (for pentobarbital) or immediately after the test compound (for chlorprothixene). Fifteen min after pentobarbital or 30 min after chlorprothixene, each mouse was examined for loss of the righting reflex. The number of animals in each group that exhibited loss of the righting reflex for a period of 30 sec or more was used for calculating the ED_{50} of the test compound.

E. Behavioral activity

Attenuation of conflict behavior was determined by the method of Geller and Seifter (1962) in male albino rats of the Upjohn Sprague-Dawley strain. These rats were deprived of water except for 15 min of drinking time per day, and for 24 hr prior to each experimental session.

The apparatus consisted of a Skinner box containing a lever for the rat to press, a dipper feeder for delivering a water reward, a small speaker for presentation of an auditory stimulus, and an electrifiable grid floor. Programming and recording of the experiment was automated using a Lehigh Valley solid-state programming system and Harvard cumulative recorders.

Drugs were administered orally as suspensions in 0.25% aqueous carboxymethylcellulose 1 hr before experimental sessions began. Control injections of the vehicle were administered on Tuesday and Wednesday of each week, and drug sessions were conducted on Thursday and Friday.

Experiments were conducted in groups of four rats, each animal receiving every dose of the compound being tested. Dosage schedules were randomized in a Latin-Square design.

F. Toxicity

Acute toxicity was determined in groups of 10 CF-1 male albino mice (18 to 22 g) per dose level. The mice were dosed with the test compounds and placed in plastic cages with free access to food and water. After 48 hr, the number of mice that died during the test period was counted and the LD_{50} values calculated.

ED_{50} and LD_{50} values and 95% confidence limits in mice and rats were calculated according to the method of Spearman and Karber (Finney, 1952).

III. RESULTS

The structures of the four triazolobenzodiazepine compounds and diazepam are given in Fig. 1. The triazolobenzodiazepines differ from diazepam by the addition of a 5-membered heterocyclic ring at positions 1 and 2 of the benzodiazepine moiety. The synthesis and chemistry of these compounds have been published elsewhere (Meguro and Kuwada, 1970; and Hester et al., 1971).

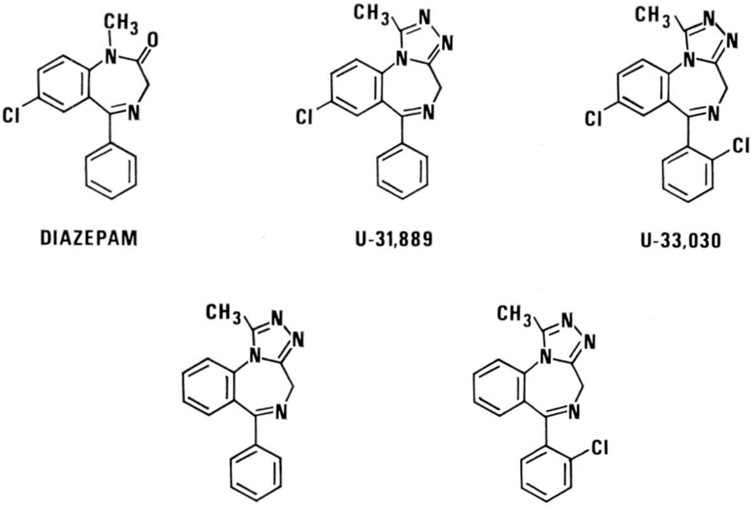

DIAZEPAM   U-31,889   U-33,030

U-31,957   U-35,005

FIG. 1. Structures of four triazolobenzodiazepines and diazepam.

A. Anticonvulsant activity

Benzodiazepines are potent antagonists of chemically induced convulsions but are somewhat less active against electrically induced seizures. They are especially active in antagonizing pentylenetetrazole-induced seizures in mice (Swinyard and Castellion, 1966), rabbits (Banziger, 1965), and cats (Straw, 1968). The antagonism of pentylenetetrazole-induced seizures has been found to correlate well with the clinical efficacy of compounds as antianxiety agents (Zbinden and Randall, 1967). Table I shows

TABLE I. *Antagonism of pentylenetetrazole-induced convulsions in various species*

	ED_{50} value (mg/kg)			Potency ratio
	Mouse			
Compound	i.p.	oral	Rat	Cat
U-31,889	0.2 (0.13–0.3)	0.2 (0.14–0.29)	0.3 (0.2–0.5)	5 (3.7–7.6)
U-33,030	0.04 (0.03–0.06)	0.07 (0.05–0.1)	0.2 (0.1–0.3)	12.3 (9–21)
U-31,957	1.0 (0.6–1.6)	1.4 (1.0–2.0)	1.7 (1.0–2.9)	0.4 (0.03–7.4)
U-35,005	0.36 (0.2–0.5)	0.28 (0.2–0.4)		1.0 (0.6–1.8)
Diazepam	0.8 (0.5–1.3)	0.7 (0.4–1.1)	1.4 (0.9–2.3)	1.

Ninety-five percent confidence limits are given in parentheses.

effects of the four triazolobenzodiazepines and diazepam on clonic seizures produced by pentylenetetrazole in mice, rats, and cats. In mice, three of the triazolo derivatives, U-31,889, U-33,030, and U-35,005, were more potent than diazepam, while U-31,957 was equipotent to diazepam. U-33,030 was the most active of the compounds shown, being 20 times more potent than diazepam administered intraperitoneally.

In rats and cats, U-31,889 and U-33,030 were more active than diazepam in antagonizing pentylenetetrazole-induced seizure. U-31,957 was equipotent to diazepam in rats but in cats it had a very flat dose-response curve. This resulted in extremely wide confidence limits for this compound. U-35,005 was equipotent to diazepam in cats.

Table I also shows a comparison in mice of the intraperitoneal and oral activity of the compounds against pentylenetetrazole-induced clonic convulsions in mice. U-33,030 appeared to be the only compound which was slightly less active orally than intraperitoneally.

These triazolobenzodiazepines were also potent antagonists of other chemically induced seizures and lethality in mice. They were active against nicotine, thiosemicarbazide, and bemegride, but were much less active against maximal electroshock seizures (Table II). U-33,030 was 14 to 28 times

TABLE II. *Antagonism of various convulsants in mice*

Compound	ED$_{50}$ value (mg/kg, i.p.) for antagonism of			
	Nicotine	Thiosemicarbazide	Bemegride	Electro-shock
U-31,889	0.02 (0.013–0.029)	0.2 (0.12–0.35)	0.06 (0.04–0.08)	25 (16–39)
U-33,030	0.01 (0.005–0.013)	0.03 (0.017–0.046)	0.04 (0.03–0.06)	23 (17–31)
U-31,957	0.3 (0.25–0.40)	0.8 (0.48–1.3)	0.8 (0.5–1.2)	>200
U-35,005	0.18 (0.12–0.26)	0.9 (0.5–1.7)	0.28 (0.19–0.41)	>200
Diazepam	0.28 (0.19–0.46)	0.7 (0.48–1.03)	0.56 (0.33–0.95)	20 (16–25)

Ninety-five percent confidence limits are given in parentheses.

more active than diazepam against chemically induced seizures and lethality, but it was only equipotent to diazepam against electroshock-induced seizures. The activity against electroshock occurred at markedly depressant doses.

B. Muscle-relaxant activity

The antagonism of strychnine lethality, the inhibition of the traction response in mice, and the depression of spinal reflexes in the decerebrate cat were used to measure the potential muscle-relaxant activity of the triazolobenzodiazepines (Table III). U-31,889 and U-33,030 were extremely

TABLE III. *Effects correlating with muscle-relaxant activity*

Compound	Antistrychnine ED_{50} (mg/kg, i.p.) mouse	Traction (mg/kg, i.p.) mouse	Spinal reflexes potency ratio (decerebrate cat, i.v.)
U-31,889	0.3 (0.17–0.5)	0.6 (0.4–1.0)	6
U-33,030	0.2 (0.16–0.29)	0.6 (0.4–1.0)	11
U-31,957	89 (61–131)	25 (18–34)	
U-35,005	>100	>200	30
Diazepam	8.0 (4.8–13.4)	7.0 (4.2–12.)	1

Ninety-five percent confidence limits are given in parentheses.

potent in antagonizing strychnine lethality and inhibiting the traction response, although U-35,005 was devoid of this activity at doses as high as 100 to 200 mg/kg.

In cats, however, both U-31,889 and U-33,030 were more potent than diazepam, and U-35,005 was the most active of the triazolobenzodiazepines tested on depressing spinal reflexes. U-35,005 was 30 times more active than diazepam in the cat. This discrepancy in the activity of U-35,005 in mice and cats is under investigation.

C. Antiaggressive behavior

One of the most striking features of the activity of benzodiazepines is their ability to decrease aggressiveness in mice and monkeys (Randall, 1961). Using a modified form of the footshock-induced aggressive behavior test in mice (Tedeschi et al., 1959), we have found that benzodiazepines and other minor tranquilizing agents antagonize aggressiveness at doses lower than those producing overt signs of depression or neurotoxicity. U-31,889 was approximately 14 times more active than diazepam in antagonizing footshock-induced aggressive behavior in mice and was one of the most active compounds that we have tested in this procedure (Table IV).

TABLE IV. *Effect on aggressive behavior in various species*

Compound	Footshock-induced aggression, mouse (ED_{50} mg/kg, i.p.)	Taming effect, monkey (Effective dose, p.o.)
U-31,889	0.13 (0.07–0.25)	0.3
U-33,030	0.3 (0.16–0.55)	0.3
U-31,957	8.4 (4.9–14.3)	
U-35,005	8.8 (4.2–18.6)	10
Diazepam	1.8 (1.1–2.9)	3

Ninety-five percent confidence limits are given in parentheses.

In monkeys, the triazolobenzodiazepines U-31,889 and U-33,030 were 10 times more active than diazepam in inhibiting aggressive behavior; U-35,005 was only one-third as active as diazepam in this test system.

D. Sedative effects

The potentiation of the hypnotic effects of certain central nervous system-depressant compounds such as ethanol or pentobarbital (Janssen et al., 1959), and, more recently, chlorprothixene (Zbinden and Randall, 1967) has been used as a method of assessing potential sedative effects of compounds. In mice, diazepam was effective in potentiating ethanol, pentobarbital, and chlorprothixene (Table V). The most active of the triazolo-derivatives in these test systems was U-31,889 which was 5 to 15 times more active than diazepam. U-31,957 and U-35,005 were weakly active in potentiating ethanol, pentobarbital, and chlorprothixene. Although U-33,030 was the most active analog in many test systems, it was much less active than U-31,889 on end points measuring sedative potential. U-33,030 was only one and one-half times more active than diazepam in potentiating ethanol narcosis in mice, but it was 20 times more active than diazepam in antagonizing pentylenetetrazole convulsions.

E. Effects on operant behavior

Meprobamate, chlordiazepoxide, diazepam, and oxazepam, all clinically active antianxiety agents, are effective in attenuating conflict behavior in rats (Geller and Seifter, 1962; Geller, 1964). The method was suggested as a preclinical evaluation of potential antianxiety agents (Geller, 1962). Diazepam produced an increase in the number of shocks taken but at higher doses (20 mg/kg) it also produced a decrease in the response rate (Fig. 2). A reduction in response rate is taken as an indication of side

TABLE V. *Potentiation of depressants in mice*

| Compound | ED_{50} (mg/kg, i.p.) to potentiate | | |
	Ethanol	Pentobarbital	Chlorprothixene
U-31,889	0.16 (0.1–0.26)	0.45 (0.28–0.73)	0.16 (0.1–0.24)
U-33,030	0.6 (0.4–1.0)	2.0 (1.4–2.9)	0.3 (0.19–0.55)
U-31,957	13 (7.7–22)	56 (36–87)	20 (14–29)
U-35,005	>100	79 (52–121)	>100
Diazepam	0.9 (0.5–1.5)	5.0 (2.7–9.2)	2.3 (1.4–3.7)

Ninety-five percent confidence limits are given in parentheses.

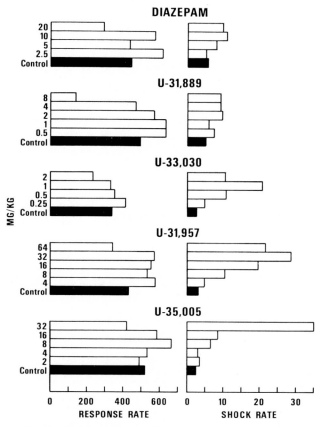

FIG. 2. Effect of oral administration of diazepam. U-31,889, U-33,030, U-31,957, and U-35,005 on the response and shock rates in a conflict-behavior test system.

effects. A dose of 10 mg/kg of diazepam was the lowest dose which produced a significant increase in the number of shocks taken.

All the triazolobenzodiazepines tested produced an increase in the shock rate. The most active of the compounds, U-33,030, increased the number of shocks taken at 0.25 mg/kg but did not depress the response rate until 2 mg/kg. U-31,889 was less effective than U-33,030 in increasing shock rate and also produced a greater suppression of the response rate. The results obtained with U-31,957 were of special interest because the compound produced effects over a much wider dose range than U-31,889, U-33,030, or diazepam. U-31,957 was more active than diazepam in increasing the shock rate. U-35,005 was effective in increasing the shock rate but a dose of 32 mg/kg also decreased the response rate.

F. Toxicity

Benzodiazepines are characterized by their low toxicity in laboratory animals (Owen et al., 1970). The acute toxicity of the triazolobenzodiazepines was determined in mice (Table VI) and was found to be of a low order. Three of the derivatives, U-31,889, U-31,957, and U-35,005, were significantly more toxic than diazepam after 48 hr. U-33,030 did not cause deaths at doses of 800 mg/kg. No loss of righting reflex was noted with either U-31,889 or U-33,030 at the highest dose (800 mg/kg) tested.

IV. DISCUSSION

Triazolobenzodiazepines represent a new series of psychotherapeutic agents. The addition of a 5-membered heterocyclic ring at positions 1 and 2 of the benzodiazepine moiety produces compounds which not only are more potent than the corresponding diazepam derivatives (Rudzik et al., 1971) but also have better therapeutic ratios. For instance, U-31,889, which is at least four times more active than diazepam on pharmacological end points in mice, has only a twofold LD_{50} value.

Minor tranquilizing agents such as chlordiazepoxide and diazepam produce a broad spectrum of pharmacological activity. Many of the triazolobenzodiazepines exhibit the same broad spectrum of activity as diazepam but there are some major differences in their activities. U-31,889 has a profile most like that of diazepam but is more active than expected on antiaggressive nicotine and strychnine end points. In fact, U-31,889 was one of the most active compounds which we have examined in the footshock-induced aggressive behavior test in mice.

Addition of a chloro group to the phenyl ring at position 6 in U-31,889

TABLE VI. *Acute toxicity in mice*

Compound	LD_{50} value (mg/kg, i.p.)
U-31,889	374 (342–415)
U-33,030	>800
U-31,957	429 (343–536)
U-35,005	528 (459–607)
Diazepam	747 (644–867)

Values in parentheses are 95% confidence limits.

not only increased the potency of the compound but also changed the pharmacological profile. U-33,030 was more active than U-31,889 on covert end points such as antagonism of nicotine, thiosemicarbazide, bemegride, and pentylenetetrazole-induced convulsions in mice, pentylenetetrazole seizures in rats and cats, and conflict behavior in rats, but was less active or only equipotent on aggressive behavior and on most muscle-relaxant and sedative-hypnotic end points. Thus, U-33,030 is more active on those end points postulated to measure antianxiety activity in animals (Geller, 1962; Zbinden and Randall, 1967). The acute toxicity of U-33,030 was less than the acute toxicity of either U-31,889 or diazepam. Therefore, the therapeutic ratio of U-33,030 is much more favorable than that of either U-31,889 or diazepam.

The most striking changes in profile of activity were observed with U-31,957 and U-35,005. U-31,957 is the 8-deschloro derivative of U-31,889 and it has potency equal to diazepam on many end points: nicotine, thiosemicarbazide, and bemegride convusions in mice, and pentylenetetrazole-induced seizures in mice and rats, and conflict behavior in rats. In the conflict behavior procedure, a depressed response rate is produced with a dose of 64 mg/kg. Therefore, U-31,957 is at least as active as diazepam in increasing the shock rate but much less depressant than diazepam. U-31,957 is much less active than diazepam on other depressant end points as well.

It is interesting that the 7-deschloro derivative of desmethyldiazepam is a weakly active compound (Sternbach et al., 1968), although the deschloro derivative in the triazolobenzodiazepine series (U-31,957) was as active as diazepam on end points postulated with antianxiety activity. Thus, it appears that the addition of the triazolo- heterocyclic ring to the benzodiazepine moiety in the four compounds reported here not only increased potency but also specificity of activity.

The marked change in specificity of these triazolobenzodiazepines is also evident in U-35,005. This compound is the 6-*o*-Cl-phenyl of U-31,957. U-35,005 possesses pronounced activity against nicotine, thiosemicarbazide, bemegride, and pentylenetetrazole-induced convulsions in mice. It is less active than diazepam on muscle-relaxant and sedative end points in mice. In cats, while the compound is still equipotent to diazepam in antagonizing pentylenetetrazole-induced seizures, it is 30 times more active than diazepam on depressing spinal reflexes in the decerebrate cat. This discrepancy between mouse and cat muscle-relaxant activity may be due to a difference in the metabolism of the compound in the two species. This effect is currently under investigation.

Because of the specificity of the activity of this group of triazolobenzodi-azepines, the profile of response of these compounds in humans may aid in elucidating the animal correlations of antianxiety activity in humans.

V. CONCLUSIONS

Triazolobenzodiazepines represent a new series of psychotherapeutic agents. Addition of the 5-membered heterocyclic ring to the benzoidiazepine moiety of the compounds reported here not only resulted in an increased potency but, more importantly, also marked differences in the spectrum of activity of these compounds.

Of the four triazolo derivatives, U-31,889 most closely resembles di-azepam in its profile of pharmacological effects. It was one of the most active compounds used for the footshock-induced aggressive behavior test in mice. U-33,030 was the most active of the triazolobenzodiazepines tested and its activity is most pronounced on end points correlating with clinical antianxiety activity.

Two of the derivatives, U-31,957 and U-35,005, differ significantly in their profile of activity from diazepam, U-31,889, and U-33,030. U-31,957 has activity equivalent to diazepam on antianxiety end points, but is weakly active on muscle-relaxant and depressant end points. U-35,005 resembles U-31,957 in activity but possesses potent muscle-relaxant activity in cats but not in mice. The species difference encountered with U-35,005 may be a result of differences in the metabolism of the compound in the two species.

Like other minor tranquilizing agents, these triazolobenzodiazepines have a low acute toxicity in laboratory animals.

The activity in man of these compounds should be invaluable in elu-cidating further the animal correlations of antianxiety effects.

REFERENCES

Banziger, R. F. (1965): Anticonvulsant properties of chlordiazepoxide, diazepam, and certain other 1,4-benzodiazepines. *Archives Internationales de Pharmacodynamie et de Thérapie,* 154:131–136.
Finney, D. J. (1952): *Statistical Method in Biological Assay.* Hafner Publishing Company, New York.
Geller, I. (1962): Use of approach avoidance behavior (conflict) for evaluating depres-sant drugs. In: *Psychosomatic Medicine,* edited by J. H. Nodine and J. H. Moyer. Lea and Febiger, Philadelphia, pp. 267–274.
Geller, I (1964): Relative potencies of benzodiazepines as measured by their effects on conflict behavior. *Archives Internationales de Pharmacodynamie et de Thérapie,* 149: 243–247.

Geller I., and Seifter, J. (1962): The effects of mono-urethans, di-urethans and barbiturates on a punishment discrimination. *Journal of Pharmacology and Experimental Therapeutics,* 136:284–288.

Hester, J. B., Duchamp, D. J., and Chidester, C. G. (1971a): A synthetic approach to new 1,4-benzodiazepine derivatives. *Tetrahedron Letters,* 1609–1612.

Hester, J. B., Rudzik, A. D., and Kamdar, B. V. (1971b): 6-Phenyl-4H-s-triazolo[4,3-a][1,4]benzodiazepines which have central nervous system depressant activity. *Journal of Medicinal Chemistry,* 14:1078–1081.

Janssen, P. A. J., can de Westeringh, C., Jageneau, A. H. M., Demoen, P J. A., Hermans, B. K. F., van Daele, G. H. P., Schellekens, K. H. L., van der Eycken, C. A. M., and Neimegeers, C. J. E. (1959): Chemistry and pharmacology related to 4-(4-hydroxy-4-phenylpiperidino) butyrophenone. *Journal of Medicinal Chemistry,* 1:281–297.

Meguro, K., and Kuwada, Y. (1970): Syntheses and structures of 7-chloro-2-hydrazino-5-phenyl-3H-1,4-benzodiazepines and some isomeric 1,4,5-benzotriazocines. *Tetrahedron Letters,* 4039–4042.

Ngai, S. H., Tseng, D. T. C., and Wang, S. C. (1966): Effects of diazepam and other central nervous system depressants on spinal reflexes in cats. *Journal of Pharmacology and Experimental Therapeutics,* 153:344–351.

Owen, G., Smith, T. H. F., and Agersborg, H. P. K. (1970): Toxicity of some benzodiazepine compounds with CNS activity. *Toxicology and Applied Pharmacology,* 16:556-570.

Randall, L. O. (1961): Pharmacology of chlordiazepoxide (LIBRIUM). *Diseases of the Nervous System,* 22:7–10.

Randall, L. O., Heise, G. A., Schallek, W., Bagdon, R. E., Banziger, R., Boris, A., Moe, R. A., and Abrams, W. B. (1961): Pharmacological and clinical studies on diazepam, a new psychotherapeutic agent of the benzodiazepine class. *Current Therapeutic Research,* 3:405–425.

Randall, L. O., Schallek, W., Heise, G. A., Keith, E. F., and Bagdon, R. E. (1960): The psychosedative properties of methaminodiazepoxide. *Journal of Pharmacology and Experimental Therapeutics,* 129:163–171.

Rudzik, A. D., Hester, J. B., and Friis, W. (1971): Pharmacologic activity of a series of 6-phenyl-4H-s-triazolo[4,3-a][1,4]benzodiazepines in mice. *Pharmacologist,* 13:205.

Steel, R. G. D., and Torrie, J. H. (1960): *Principles and Procedures of Statistics.* McGraw-Hill Book Company, Inc., New York.

Sternbach, L. H., Randall, L. O., Banziger, R., and Lehr, H. (1968): Structure-activity relationships in the 1,4-benzodiazepine series. In: *Drugs Affecting the Central Nervous System,* edited by A. Burger. Marcel Dekker, Inc., New York, pp. 237–264.

Straw, R. N. (1968): The effect of certain benzodiazepines on the threshold for pentylenetetrazol-induced seizures in cats. *Archives Internationales de Pharmacodynamie et de Thérapie,* 175:465–469.

Swinyard, E. A., and Castellion, A. W. (1966): Anticonvulsant properties of some benzodiazepines. *Journal of Pharmacology and Experimental Therapeutics,* 151:369–375.

Tedeschi, R. E., Tedeschi, D. H., Mucha, A., Cook, L., Mattis, P. A., and Fellows, E. J. (1959): Effects of various centrally acting drugs on fighting behavior in mice. *Journal of Pharmacology and Experimental Therapeutics,* 125:28–34.

Zbinden, G., and Randall, L. O. (1967): Pharmacology of benzodiazepines: Laboratory and clinical correlations. In: *Advances in Pharmacology,* edited by S. Garattini and P. A. Shore. Academic Press, New York, pp. 213–291.

ANTIANXIETY ACTION OF BENZODIAZEPINES: DECREASE IN ACTIVITY OF SEROTONIN NEURONS IN THE PUNISHMENT SYSTEM

Larry Stein, C. David Wise, and Barry D. Berger

Wyeth Laboratories, Philadelphia, Pennsylvania

I. INTRODUCTION

Benzodiazepine derivatives and other minor tranquilizers (barbiturates, meprobamate) exert opposing effects on goal-directed behavior. At high doses, these agents reduce the tendency to act by a depressant action; at lower doses, or after repeated administration of high doses (Margules and Stein, 1968), tranquilizers often facilitate behavior, especially if the tendency to respond has been suppressed by punishment or nonreward. Because facilitation is best demonstrated on baselines of inhibited behavior, we assume that benzodiazepines do not directly increase the tendency to act, but rather release behavior from suppression by a disinhibitory action (Margules and Stein, 1967). This action may correspond in man to the therapeutic or antianxiety activities of these agents, whereas the high-dose depressant action may be responsible for their sedative and behaviorally suppressant side effects.

In this chapter, we attempt to relate the behavioral actions of the tranquilizers to their recently discovered effects on monoamine turnover in the brain (Corrodi et al., 1967, 1971; Taylor and Laverty, 1969; Chase et al., 1970). Specifically, we will review evidence which suggests that tranquilizers (1) exert their disinhibitory (antianxiety) effects by reducing the activity of serotonin neurons in a behaviorally suppressant "punishment" system, and (2) exert their response-depressant effects by reducing the activity of norepinephrine neurons in a behaviorally facilitatory "reward" system.

II. NEUROCHEMISTRY OF REWARD AND PUNISHMENT

Over the past two decades, physiological (Delgado et al., 1954; Olds

299

and Milner, 1954), anatomical (Nauta, 1963), histochemical (Dahlström and Fuxe, 1965; Fuxe, 1965), and psychopharmacological (Stein, 1962; Poschel and Ninteman, 1963; Wise and Stein, 1970) research findings have resulted in the identification of monoamine systems tentatively associated with reward and punishment. While the reward system has been characterized as mainly noradrenergic (Stein, 1967, 1968; Wise and Stein, 1969; Berger et al., 1971), the punishment system appears to be at least partially serotonergic (Wise et al., 1970*a, b, in press*). In accordance with the concept of Brodie and Shore (1957) that norepinephrine and serotonin may serve generally as chemical mediators of antagonistic systems in the brain, we have assumed that goal-directed behavior is under the reciprocal control of behaviorally facilitatory norepinephrine neurons and behaviorally suppressant serotonin neurons (Wise et al., *in press*) (Fig. 1).

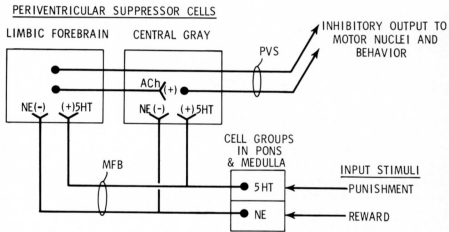

FIG. 1. Diagram representing hypothetical relationships between reward and punishment mechanisms and behavior. Signals of positive reinforcement release behavior from periventricular system (PVS) suppression by the following sequence of events. (1) Activation of norepinephrine-containing cells in lower brainstem by stimuli previously associated with reward (or the avoidance of punishment) causes release of norepinephrine (NE) into various periventricular suppressor areas via the medial forebrain bundle (MFB). (2) Inhibitory (—) action of NE decreases activity of PVS suppressor cells in diencephalic and midbrain central gray regions either directly or by reducing their excitatory (+) input from cholinergic (ACh) suppressor cells in the limbic forebrain. (3) Reduction of PVS inhibitory influence over motor nuclei of the brainstem facilitates behavior. Signals of punishment or failure increase behavioral suppression by activation of the dorsal raphe and other 5-HT cell groups. Consequent release of 5-HT excites PVS suppressor cells in the central gray area either directly or indirectly via activation of cholinergic suppressor cells in the limbic forebrain. See Dahlström and Fuxe (1965) for a detailed description of the localization of norepinephrine and serotonin cell bodies in the lower brainstem.

The cells of origin of the reward system are located in the locus coeruleus (dorsal branch) (Ritter and Stein, 1972) and reticular formation of the lower brainstem (ventral branch) (Arbuthnott et al., 1971), and the axons ascend via the medial forebrain bundle to form noradrenergic synapses in the hypothalamus, limbic system, and neocortex (Fuxe et al., 1970). The cells of the serotonergic punishment system probably originate in the raphe nuclei, and the fibers ascend in the medial forebrain bundle and distribute in the brain in a roughly similar fashion to those of the noradrenergic system. Both systems richly innervate periventricular regions of the brain; however, serotonin terminals tend to be concentrated in the central gray of the midbrain, whereas noradrenergic terminals tend to be concentrated in hypothalamic and limbic forebrain regions that surround the third and lateral ventricles (Fuxe, 1965; Aghajanian and Bloom, 1967).

Evidence that norepinephrine neurons comprise an important component of a behaviorally facilitatory reward system has been summarized elsewhere (Stein, 1968; Wise and Stein, 1970; Arbuthnott et al., 1971; Stein, 1971). Evidence that serotonin neurons form part of a behaviorally suppressant punishment system is reviewed below and in a forthcoming publication (Wise et al., *in press*).

III. EVALUATION OF ANTIANXIETY ACTIVITY IN ANIMALS

Objective tests of behavior (variously termed "passive avoidance," "punishment," or "conflict" tests) have been developed for the precise measurement of the antianxiety activity of drugs and other agents in the laboratory animal (Masserman and Yum, 1946; Miller, 1957; Geller and Seifter, 1960). In these tests, the rate of occurrence of some easily measured response is suppressed by punishment; the ability of a drug to release or disinhibit the punished behavior defines its antianxiety activity. In an elegant form of the punishment test, which permits concurrent measurement of antianxiety and depressant actions of drugs, animals are trained on a program in which punished and nonpunished reinforcement schedules are alternated (Geller and Seifter, 1960; Fig. 2). In the nonpunished schedule (15 min), a lever-press response is rewarded with sweetened milk at infrequent and variable intervals—on the average, once every 2 min. In the punished schedule (3 min), which is signaled by a tone, every response is rewarded with milk but also punished with a brief electrical-foot-shock (0.1 to 0.6 mA, 0.25 sec in duration). The rate of response in the tone period may be precisely regulated by individual adjustment of shock level, and any degree of behavioral suppression may be obtained in well-trained animals. After

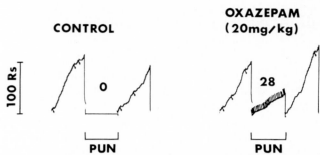

FIG. 2. Sample records of performance in the Geller and Seifter (1960) punishment test. Lever-press responses are cumulated over time: the slope of the curve at any point gives the response rate. Punishment periods (PUN) of 3-min duration, signaled by a tone, are flanked by equivalent periods of nonpunished responding. Punished responses are numbered and indicated by upward strokes of the pen; presentations of the milk reward in the unpunished schedule are indicated by downward strokes of the pen. Oxazepam causes a marked disinhibition of punishment-suppressed behavior, but does not importantly influence unpunished behavior in this drug-sophisticated rat (compare with Fig. 3) (Stein, 1968).

several weeks of training, stable low rates of response are obtained in the punished periods and stable high rates in the nonpunished periods. Drug-induced increases in the rate of punished behavior are taken as an index of antianxiety activity, while decreases in the nonpunished rate are taken as an index of depressant activity.

Geller and his associates (1960, 1962, 1963) have studied the effects of different psychotherapeutic drugs in the conflict test. In general, a substantial release of punished behavior is produced only by the minor tranquilizers. Neuroleptics or antipyschotic agents (e.g., chlorpromazine and other phenothiazine derivatives) usually do not have this effect and often cause a further suppression of punished behavior (Geller et al., 1962; Cook and Davidson, *this Volume*). Stimulant drugs of the amphetamine type also decrease the frequency of punished behavior. Even morphine, a powerful analgesic, fails to disinhibit punished behavior in this test (Geller et al., 1963; Kelleher and Morse, 1964). This observation and the fact that a well-trained animal usually will cease to respond at the onset of the warning stimulus and before any painful shocks are delivered make it clear that it is not pain itself but rather the threat or fear of pain that suppresses behavior in this test.

In order to determine if the punishment deficit induced by tranquilizers is due to a more general disinhibitory action, Margules and Stein (1967) studied the effects of oxazepam in several different situations, using different response measures and means of producing response suppression. Oxazepam caused a marked increase in the occurrence of previously suppressed behavior regardless of whether the behavior was inhibited by footshock, nonreward (extinction), punishing brain stimulation, or the bitter taste of quinine. Furthermore, oxazepam caused a substantial disinhibition of feeding in satiated rats. Margules and Stein (1967) concluded that the disinhibitory action of oxazepam and related tranquilizers is clearly a general one. They also suggested that tranquilizers may act on a final common pathway for the suppression of behavior, although the possibility that different inhibitory pathways in the brain have a common pharmacology was not excluded.

IV. CHRONIC ADMINISTRATION OF BENZODIAZEPINES

In clinical practice, it has long been recognized that the relative importance of the antianxiety and depressant components of benzodiazepine activity changes with chronic dosing. In initial administration, particularly at high dose levels, depressant side effects often predominate. With repeated doses, the depressant effects tend to disappear, although the antianxiety action persists (Warner, 1965). In confirmation of these clinical observations, Margules and Stein (1968) found that the antianxiety and depressant actions of oxazepam in the rat punishment test followed opposite courses during chronic administration (Fig. 3). The depressant action (measured by a decrease in the rate of nonpunished behavior) was observed to undergo tolerance after three to six doses, while the antianxiety action (measured by an increase in the rate of punished behavior) failed to show tolerance and even increased throughout the chronic series. The increase in the rate of punished behavior was attributed, at least in part, to a progressive unmasking of the antianxiety action as tolerance to the depressive action developed. In any case, in view of their opposite course during chronic administration, it would seem likely that the antianxiety and depressant actions of tranquilizers are mediated by different biochemical mechanisms.

V. EFFECTS OF TRANQUILIZERS IN CENTRAL MONOAMINE TURNOVER

Biochemical and histochemical studies indicate that benzodiazepines and barbiturates decrease the turnover of norepinephrine, serotonin, and other biogenic amines in the brain. Taylor and Laverty (1969) used intraventricular injections of labeled dopamine to study the effects of benzodiazepines on catecholamine metabolism. Significant decreases in norepinephrine and dopamine turnover were obtained in cortex and striate region,

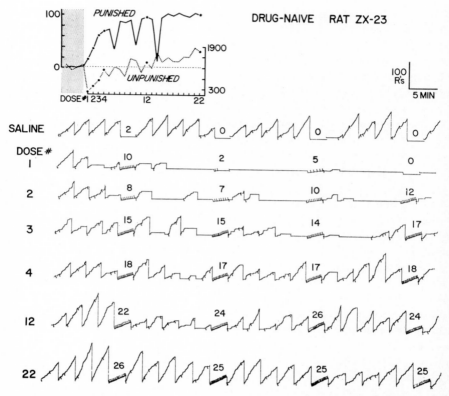

FIG. 3. Increase of antianxiety activity and tolerance of depressant activity after repeated administrations of oxazepam (20 mg/kg, i.p., immediately before each daily test). The graph (upper left) shows the daily output of punished and nonpunished responses for a representative drug-naive rat during 4 control days (stippled) and 22 days of oxazepam. Punished responses are shown on the left-hand ordinate, and nonpunished responses are shown on the right. Cumulative response records are shown for the control (saline) day immediately preceding the drug series and for various drug days as numbered. The pen resets automatically every 3 min. Note virtually complete suppression of unpunished responses after first dose of oxazepam, although punished responses increase from 2 to 17 (Margules and Stein, 1968).

respectively. Corrodi et al. (1971) showed by histochemical and biochemical means that chlordiazepoxide, diazepam, and nitrazepam blocked stress-induced increases in norepinephrine turnover and potentiated stress-induced decreases in dopamine turnover. In unstressed rats, they demonstrated by histochemical methods that norepinephrine turnover in the coeruleo-cortical (dorsal) branch of the norepinephrine system was selectively decreased by the tranquilizers, in agreement with the biochemical observations of Taylor and Laverty (1969). These effects on catecholamine turnover were attributed to a tranquilizer induced decrease in the nervous activity of norepinephrine and dopamine neurons, and were assumed to account for the sedation and EEG slowing induced by the drugs.

In addition to these effects on catecholamine neurons, benzodiazepines and barbiturates also may decrease the turnover of serotonin in the brain. Corrodi et al. (1967) found that anesthetic doses of pentobarbital partially blocked the depletion of brain serotonin normally obtained after administration of the tryptophan hydroxylase inhibitor α-propyldopacetamine, and they suggested that pentobarbital probably decreases the activity of serotonin neurons. More recently, Chase et al. (1970) demonstrated that diazepam greatly increased the retention of ^{14}C-serotonin and ^{14}C-5-hydroxyindoleacetic acid in rat brain following an intracisternal injection of labeled serotonin, although the drug had no effect on the uptake of the radioisotope. These findings were taken to suggest that diazepam may interfere with the metabolism of brain serotonin, as well as with the mechanism for the active transport of 5-hydroxyindoleacetic acid from the brain.

Thus, work from several laboratories indicates that benzodiazepines and barbiturates decrease the turnover of norepinephrine, serotonin, and other biogenic amines in the brain. Since norepinephrine and serotonin play an important role in the control of behavior, it seems reasonable to speculate that the decreased turnover of these substances may be at least partly responsible for some of the behavioral effects of tranquilizers. To test this idea, we varied norepinephrine and serotonin activities in the brain in several ways, and compared the behavioral effects of these changes with those of the benzodiazepines. On the assumption that a decrease in monoamine turnover was in fact involved, our chief aim was to determine whether the antianxiety effects of tranquilizers are principally associated with a reduction of norepinephrine activity or with a reduction of serotonin activity.

VI. ANTIANXIETY ACTIVITY AND NOREPINEPHRINE

If the antianxiety effects of benzodiazines depend on a reduction of norepinephrine activity, then other agents that reduce norepinephrine activity ought also to exert an antianxiety effect. Phenothiazine derivatives such as chlorpromazine exert strong central antiadrenergic effects, but, as already indicated, do not release punished behavior from suppression. Since phenothiazines also antagonize other neurotransmitter actions, we tested the relatively selective α-noradrenergic antagonist phentolamine and the β-noradrenergic antagonist propranolol. Both antagonists failed to release punish-

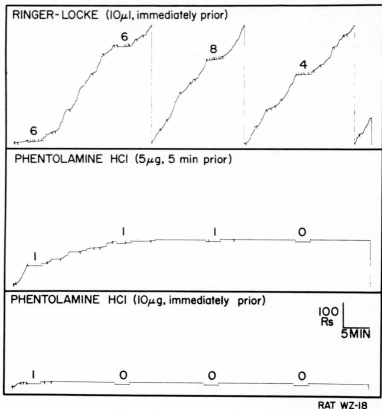

FIG. 4. Suppression of punished and nonpunished behavior by intraventricular administration of the α-noradrenergic antagonist phentolamine. In this and in subsequent figures, each record represents a complete 72-min session consisting of four 3-min punishment periods (punished responses are numbered) and five nonpunished periods of different durations, in which responses are reinforced at variable intervals (on the average, every 2 min). Pen resets automatically after about 550 responses. Unless otherwise noted, all drugs are dissolved in 10 μl of Ringer-Locke solution and injected in lateral ventricle via a permanently indwelling cannula immediately before the start of the test. Phentolamine drastically suppresses nonpunished behavior and reduces the total number of punished responses from a control level of 24 (Ringer-Locke) to 3 (5 μg) or 1 (10 μg).

ment-suppressed behavior in the rat conflict test. Intraventicular injections of propranolol (20 to 100 μg) had no important effects, but intraventricular phentolamine (5 to 10 μg) strongly suppressed both punished and nonpunished behaviors (Fig. 4; Table I). These experiments do not support the idea that antianxiety activity depends on noradrenergic blockade. Indeed,

TABLE I. *Suppression of punished and nonpunished behavior by the α-noradrenergic antagonist, phentolamine*

Dose (μg)	No. of rats	Mean % of predrug rate ± S.E.	
		Punished responses	Nonpunished responses
5	5	59.9 ± 15.8[a]	111.6 ± 26.6
10	5	9.2 ± 5.4[a]	30.9 ± 19.5[a]

Phentolamine HCl in 10 μl of Ringer-Locke solution was injected intraventricularly 10 min before the start of the Geller and Seifter (1960) test.
[a] Differs significantly from predrug control at $p < 0.01$.

they suggest instead that α-noradrenergic blockade causes suppression of behavior.

Experiments with L-norepinephrine itself supported this idea. Intraven-

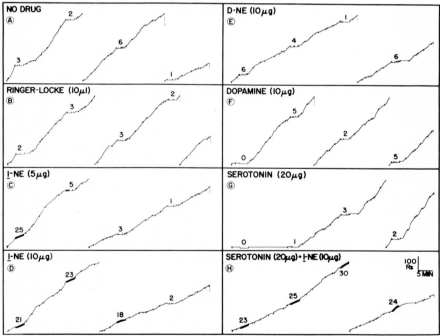

FIG. 5. Effects of intraventricular administration of various monoamine transmitters (as hydrochloride salts) on punished and nonpunished behavior in the same rat. Note especially dose-related antipunishment effect of L-norepinephrine (L-NE) (compare panels B–D), relative inactivity of D-norepinephrine (D-NE) and dopamine (panels E and F), initial suppressant action of serotonin (panel G), and paradoxical potentiation of L-norepinephrine antipunishment effect by serotonin (compare panels D and H). For details of procedure, see Fig. 4.

tricular injections of this substance caused very large increases in the rate of punished behavior, with somewhat larger effects at 10 μg than at 5μg (Fig. 5; Table II). Neurochemical specificity is suggested since d-norepinephrine and

TABLE II. *Effects of neurotransmitters in the rat conflict test*

Treatment	Dose (μg)	No. of rats	Mean responses per test $\pm$ S.E.	
			Punished	Nonpunished
No injection	—	9	7.8 $\pm$ 1.1	1,411 $\pm$ 332
Ringer-Locke	—	9	11.4 $\pm$ 3.3	1,555 $\pm$ 296
L-Norepinephrine	5	7	32.5 $\pm$ 9.0^a	1,238 $\pm$ 222
	10	8	47.6 $\pm$ 8.0^a	1,400 $\pm$ 315
D-Norepinephrine	10	9	13.2 $\pm$ 4.4	1,375 $\pm$ 370
Dopamine	10	6	7.3 $\pm$ 1.6	1,145 $\pm$ 132

Hydrochloride salts of the indicated substances were dissolved in 10 μl of Ringer-Locke solution and injected in the lateral ventricle immediately before the start of the 72-min test. See Fig. 5 for a typical experiment.
^{a} Differs from Ringer-Locke at $p < 0.01$.

dopamine had negligible effects. Most significantly, intraventricular injections of norepinephrine increased rather than decreased the antipunishment activity of systemically administered benzodiazepines (Fig. 6; Table III); this finding clearly contradicts the idea that the antianxiety activity of tranquilizers depends on a reduction of noradrenergic activity. At the same time, intraventricular norepinephrine antagonized the depressant effect of oxazepam on nonpunished behavior (Fig. 6; Table III); this observation suggests that the depressant action of tranquilizers may be mediated by reduction of norepinephrine turnover.

TABLE III. *Norepinephrine increases the antianxiety activity and decreases the depressant activity of oxazepam in the rat conflict test*

Treatment	Mean responses per test $\pm$ S.E.	
	Punished	Nonpunished (in hundreds)
No drug	3.9 $\pm$ 0.7	38.2 $\pm$ 10.1
Oxazepam	15.9 $\pm$ 7.6^a	13.6 $\pm$ 5.4^a
Norepinephrine	9.1 $\pm$ 4.7	30.7 $\pm$ 4.4^b
Oxazepam + norepinephrine	34.9 $\pm$ 7.3^{ab}	25.4 $\pm$ 7.8^{ab}

^{a} Differs from "no drug" mean ($p < 0.01$).
^{b} Differs from oxazepam mean ($p < 0.01$).
L-Norepinephrine HCl (5 μg in 10 μl of Ringer-Locke solution) was injected in the lateral ventricle, either alone, or 10 min before oxazepam (10 mg/kg, i.p., suspended in 0.5% Tween 80, immediately before conflict test). One week after the NE-oxazepam experiment, the same rats were tested with oxazepam alone. Prior to this study, rats had received oxazepam only once.

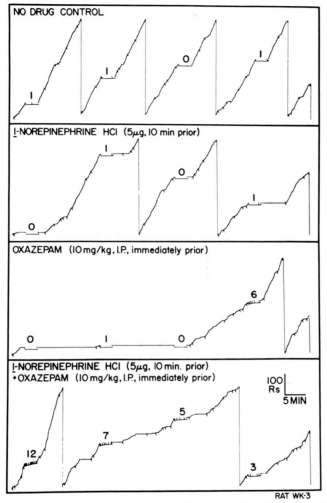

FIG. 6. Reversal of depressant action and potentiation of antipunishment action of oxazepam by intraventricular administration of norepinephrine. Because the effects of the injection had partly worn off, the 5-μg dose of norepinephrine alone had no obvious facilitating effect on behavior (second panel). For details of procedure and summary data, see Table III.

VII. ANTIANXIETY ACTIVITY AND SEROTONIN

Since the antianxiety activity of benzodiazepines did not appear to depend on a decrease in norepinephrine activity, we next considered the possibility that serotonin is involved. If so, then antagonists of serotonin or inhibitors of its synthesis should counteract the suppressive effects of punishment. Such seems to be the case. Graeff and Schoenfeld (1970) observed very large increases in the punished response rates of pigeons after intramuscular administrations of the serotonin antagonists methysergide and 2-bromo-D-lysergic acid (BOL); according to these authors, "this effect was of the same magnitude as that produced by chlordiazepoxide, diazepam, and nitrazepam" (p. 281). We also have obtained strong antipunishment effects with methysergide in the rat conflict test (Fig. 7).

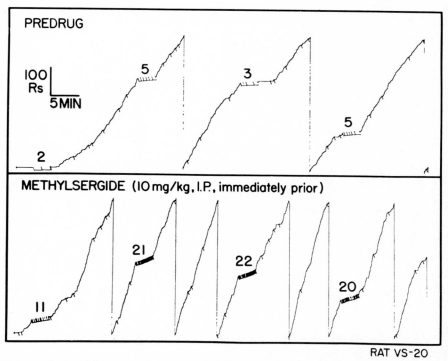

FIG. 7. Antipunishment and response-facilitatory actions of methysergide in the rat conflict test.

Similarly, several groups have reported large releases of punishment-suppressed behavior after administration of the serotonin synthesis inhibitor, *p*-chlorphenylalanine (PCPA) (Tenen, 1967; Robichaud and Sledge, 1969; Stevens and Fechter, 1969; Geller and Blum, 1970). The time courses of behavioral disinhibition and serotonin depletion after PCPA coincide closely. In confirmation of the report of Geller and Blum (1970), we observed that intraperitoneal administration of the serotonin precursor 5-hydroxytrytophan (5-HTP) reversed the disinhibitory effect of PCPA (Wise et al., *in press*). Specificity of the 5-HTP reversal was suggested by tests with the catecholamine precursor dihydroxyphenylalanine (DOPA). The rate of punished behavior in the PCPA-treated animals was increased rather than decreased by DOPA, both directly after injection and, unexpectedly, on the following day.

Consistent with these findings, elevation of brain serotonin levels by combined administration of 5-HTP and a monoamine oxidase inhibitor causes marked suppression of food-rewarded behavior in the pigeon (Aprison and Ferster, 1961). Furthermore, the long-lasting, centrally active serotonin agonist α-methyltryptamine strongly suppresses punished and nonpunished behaviors in the pigeon (Graeff and Schoenfeld, 1970) and in the rat (Fig. 8). These findings with serotonin agonists, antagonists, and the synthesis inhibitor PCPA clearly support the possibility of a behaviorally suppressive serotonin "punishment" system, whose activity may be reduced by benzodiazepines and other antianxiety agents (Wise et al., 1970*b*).

The powerful suppressant effects of intraventricular serotonin in the rat self-stimulation test generally support these conclusions (Wise and Stein, 1969; Wise et al., *in press*); however, in the conflict test, the action of the neurohormone is complex and apparently triphasic (Fig. 9). At doses of 5 to 20 μg, one observes an initial phase of intense behavioral suppression, which lasts for about 10 min (Table IV), then a period of normal response or facilitation for several hours (frequently including release of punished behavior), and finally, in some cases, a prolonged period of behavioral suppression that may persist for 1 to 2 days. Biphasic actions of serotonin on single spinal neurons have been reported which, on a reduced time scale, resemble these behavioral effects (Weight and Salmoiraghi, 1968) (Fig. 10). Also interesting are the behavioral effects of combined administrations of serotonin and norepinephrine (Fig. 5H). The facilitatory effects of norepinephrine predominated throughout the test and even seemed to be prolonged by serotonin.

As a further test of the idea that the antianxiety effects of tranquilizers are mediated by reduction of serotonin activity, we tried to antagonize the antipunishment action of oxazepam by intraventricular administration of serotonin. We focused attention on performance in the first programmed punishment period, 3 to 6 min after the injections, because the acute

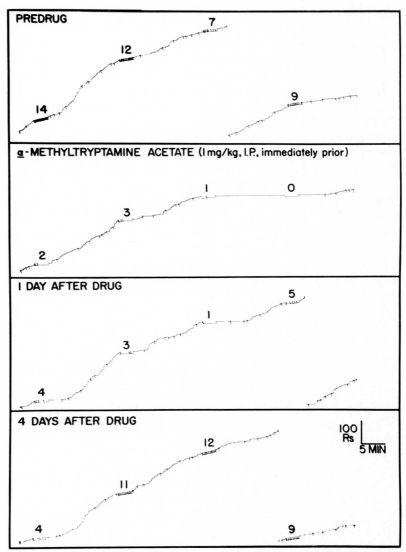

FIG. 8. Prolonged suppressant effect of the serotonin agonist α-methyltryptamine on punished behavior.

suppressant effect of serotonin persists for only about 10 min. In six out of eight cases, serotonin reduced the antianxiety action of oxazepam (Fig. 11; Table V). On the other hand, in similar tests on the same animals, intraven-

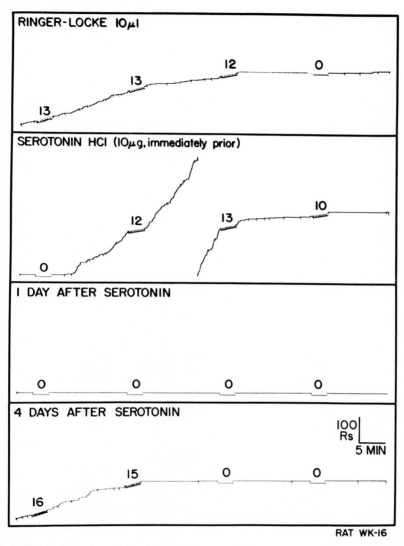

FIG. 9. Triphasic action of intraventricular serotonin in the conflict test. An initial 10-min period of intense behavioral suppression is followed by a longer period of behavioral facilitation (second panel). A third very long phase of behavioral suppression is observed on the following day. Compare with Fig. 5G.

tricular L-norepinephrine increased the antianxiety effect of the tranquilizer in seven out of eight cases (also see Table III).

TABLE IV. *Acute effects of serotonin (5-HT) and L-norepinephrine (L-NE) on punished behavior in the conflict test*

Rat no.	No drug	Ringer-Locke	5-HT (10 or 20 μg)	L-NE (10 μg)
		Experiment 1 (10 μg of 5-HT)		
WZ 16	2	0	0	
WK 13	23	19	4	
WK 16	13	13	0	
WK 18	23	21	5	
WM 42	18	18	0	
		Experiment 2 (20 μg of 5-HT)		
WZ 38	2	2	0	19
WZ 32	3	2	0	21
WZ 13	2	2	0	13
WZ 7	2	2	0	15
WZ 5	2	3	0	23
WZ 23	1	1	0	1

Hydrochloride salts of the neurotransmitters were dissolved in 10 μl of Ringer-Locke solution and injected in the lateral ventricle immediately before the test. Numbers indicate punished responses in the first tone period, 3 to 6 min after the injection.

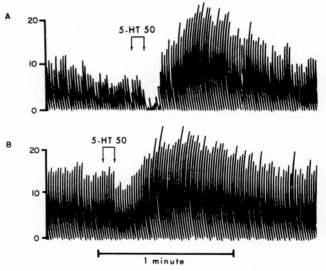

FIG. 10. Polygraph record of two spinal interneurons illustrating biphasic action of electrophoretically administered serotonin (5-HT). The height of each pen deflection indicates the number of spikes per second; calibration scale shown to the left of each record. 5-HT was administered with 50 μA of current to each neuron and produced an initial depression of firing followed by a facilitation of longer duration. (From Weight and Salmoiraghi, 1968, with permission of the publisher.)

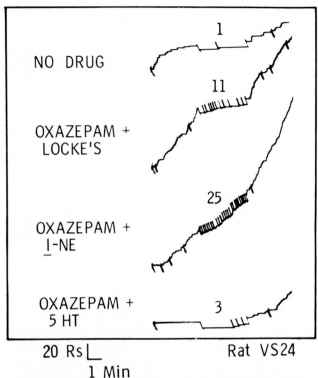

FIG. 11. Intraventricular administration of serotonin (5-HT) decreases and L-norepinephrine (L-NE) increases the antipunishment effect of oxazepam in the rat conflict test. For details of procedure and summary data, see Table V.

VIII. SUPPRESSION OF BEHAVIOR BY DORSAL RAPHE STIMULATION AND REVERSAL BY OXAZEPAM

The foregoing data are consistent with the idea that a serotonin system in the brain mediates the suppressive effects of punishment. If so, and if it were correct to assume that tranquilizers act partly by reduction of serotonin activity, then it might be possible to demonstrate (1) suppression of behavior by direct activation of serotonin systems and (2) antagonism of such directly induced behavioral suppression by administration of tranquilizers.

As a first test of these ideas, minute quantities (about 5 μg) of crystalline carbachol were applied via permanently indwelling cannulas to the dorsal raphe region of five rats with stable rates of punished and unpunished

TABLE V. *Serotonin decreases and norepinephrine increases the antianxiety effect of oxazepam in the rat-conflict test*

Rat no.	Punished responses in first tone period after oxazepam plus		
	Ringer-Locke	Serotonin	Norepinephrine
VS 27	2	0	10
WI 30	3	2	5
VS 24	11	3	25
VS 26	7	3	13
VS 28	22	4	24
VS 4	2	4	24
VS 2	3	5	10
VS 1	14	6	6
Mean	8.0	3.4[a]	14.6[a]

Eight rats received intraventricular injections of serotonin HCl (5 μg), L-norepinephrine HCl (5 μg), or Ringer-Locke solution (10 μl) immediately before oxazepam (10 mg/kg, i.p.). The behavioral test followed immediately. All rats received all drug combinations during 10 days of testing, but in different sequences. See Fig. 11 for typical experiment.
[a] Differs significantly from Ringer-Locke ($p < 0.05$).

responses in the conflict test. The dorsal raphe nucleus was selected for initial study because, as one of the two major serotonin cell groups in the midbrain, it is the site of origin of many of the ascending serotonin fibers to the diencephalon and forebrain; furthermore, this nucleus consists almost entirely of serotonin cell bodies (Fuxe et al., 1968). Crystalline carbachol was used for chemical stimulation because powerful behavioral suppressive effects are readily obtained by direct application of this substance to the medial hypothalamic region (Margules and Stein, 1967).

In three out of three cases, application of carbachol to the dorsal raphe nucleus caused marked suppression of both punished and unpunished behaviors within 1 to 2 min after treatment (Fig. 12; Table VI). In all of these cases, oxazepam reversed the suppressive effects of raphe stimulation on both types of behavior. In two rats, the tip of the cannula penetrated into the reticular formation, about 1 mm lateral or caudal to the dorsal raphe. In these cases, carbachol had a much smaller or negligible suppressive effect on behavior. Although these are only preliminary experiments, the data clearly support the idea that serotonin neurons in the dorsal raphe nucleus may be involved in the mediation of behavioral suppression.

IX. EFFECTS OF REPEATED DOSES OF OXAZEPAM ON MONOAMINE TURNOVER

As already noted, studies on animals and man clearly indicate that the

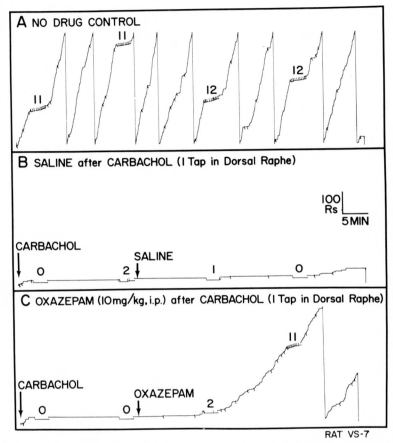

FIG. 12. Suppression of behavior in the conflict test by application of crystalline carbachol (1 tap = about 5 μg) to the ventral border of the dorsal raphe, and reversal
of carbachol suppression by systemically injected oxazepam. For summary data, see
Table VI.

antianxiety and depressant actions of benzodiazepines follow different courses
during chronic administration. The depressant action rapidly undergoes
tolerance after a few doses (Goldberg et al., 1967), while the antianxiety
action fails to show tolerance and even may increase with repeated doses
(Margules and Stein, 1968). If these behavioral effects were selectively mediated by decreases in serotonin and norepinephrine turnover as we
suggest, then it might be possible to show that benzodiazepine-induced
decreases in serotonin turnover will persist with repeated doses, while the

TABLE VI. *Behavioral suppression by carbachol stimulation of the dorsal raphe nucleus (DR), and reversal by oxazepam (10 mg/kg)*

Rat no.	Site of cannula tip	Punished responses (last two periods)			Unpunished responses (total)		
		No drug	Carbachol + saline	Carbachol + oxazepam	No drug	Carbachol + saline	Carbachol + oxazepam
7	DR	24	1	13	4,500	58	690
12	DR	32	6	31	2,956	418	1,627
22	DR	15	3	22	2,835	2,295	3,400
9	RF[a]	18	20	27	700	441	524
16	RF[b]	46	28	20	2,945	2,465	2,613

Crystalline carbamylcholine chloride (carbachol, about 5 μg) was directly applied to the brain via a permanently indwelling 27-gauge cannula immediately before the conflict test. Oxazepam or saline was injected intraperitoneally after the second tone period. See Fig. 12 for typical experiment.

[a] Reticular formation just caudal and 1 mm ventrolateral to DR.

[b] Reticular formation 0.5 to 1 mm lateral and ventral to DR.

decrease in norepinephrine turnover will undergo tolerance.

The methods used by Chase et al. (1970) to measure benzodiazepine effects on serotonin turnover were modified to permit concurrent determinations of norepinephrine turnover. Twenty-four rats, with chronically indwelling cannulas in the lateral ventricle, were injected either once or daily for six consecutive days with oxazepam (20 mg/kg, i.p.) or saline. Ten min before the single or sixth dose of oxazepam (or saline), rats in all four groups received an intraventricular injection of ^{14}C-serotonin and ^{3}H-norepinephrine. Three hr after the injection of the radioisotopes, the rats were killed by decapitation.

The brains were rapidly removed and divided into forebrain-diencephalon and midbrain-hindbrain pieces by a vertical knife cut caudal to the mammillary bodies and rostral to the superior colliculus (the cerebellum was discarded). Each brain region was homogenized in ice-cold 0.4 N hydrochloric acid. After centrifugation, 0.5 ml aliquots of supernatant fluid were taken for determination of total ^{14}C and ^{3}H by liquid scintillation spectrometry. For analysis of ^{14}C-serotonin, a 1-ml aliquot was mixed with 50 μl of concentrated hydrochloric acid and then extracted with 5 ml of amylalcohol cyclohexane (8:2). With standards, 97.8% of the ^{14}C-serotonin was recovered in the aqueous phase and virtually all of the 5-hydroxyindoleacetic acid was recovered in the organic phase. For analysis of ^{3}H-norepinephrine, a 2-ml aliquot was treated with 0.2 ml of 4 N perchloric acid and centrifuged. The supernatant was used for analysis of ^{3}H-norepinephrine and ^{3}H-metabolites as described elsewhere (Wise and Stein, 1970).

The effects of single or repeated doses of oxazepam on the fate of previously injected ^{14}C-serotonin and ^{3}H-norepinephrine are reported in Table VII. Consistent with previous work (Taylor and Laverty, 1969; Chase et al., 1971), total ^{14}C and ^{3}H in the midbrain-hindbrain region of rats treated with a single dose of oxazepam substantially exceeded that of the saline controls. This increase reflected significant elevations of ^{14}C-serotonin and ^{3}H-norepinephrine and their major metabolites ^{14}C-5-hydroxyindoleacetic acid and ^{3}H-normetanephrine. On the other hand, in rats treated with repeated doses of oxazepam, significant increases were obtained only in the levels of ^{14}C-serotonin and ^{14}C-metabolites, but not ^{3}H-norepinephrine or ^{3}H-metabolites. Thus, a decrease in norepinephrine turnover was no longer detectable after six doses of oxazepam, although serotonin turnover still was substantially reduced. As noted, in the conflict test the depressant action of this dose of oxazepam similarly disappears after six daily injections, while the antianxiety action persists. (Although the reduction of serotonin turnover

TABLE VII. *Effects of single and repeated doses of oxazepam (20 mg/kg, i.p.) on fate of intraventricularly injected* ^{14}C*-serotonin* (^{14}C*-5-HT, 1.1 µg, 0.35 µC) and* ^{3}H*-L-norepinephrine* (^{3}H*-NE, 1.1 µg, 0.26 µC)*

| | Midbrain-hindbrain | | | Diencephalon-forebrain | | |
| | Saline (N = 10) | Oxazepam | | Saline[b] (N = 10) | Oxazepam | |
		Single-dose (N = 8)	Repeated-doses (N = 6)		Single-dose (N = 8)	Repeated-doses (N = 6)
Total ^{14}C	20.7 ± 1.48	31.9 ± 1.71[a]	27.1 ± 1.08[a]	16.0 ± 1.47	17.8 ± 1.01	16.3 ± 0.49
^{14}C-5-HT	10.1 ± 0.77	15.0 ± 0.90[a]	13.4 ± 0.41[a]	7.2 ± 0.65	8.0 ± 0.43	7.8 ± 0.23
^{14}C-5-HIAA	10.6 ± 0.76	16.8 ± 0.99[a]	13.6 ± 0.71[a]	8.7 ± 0.84	9.7 ± 0.67	8.3 ± 0.41
Total ^{3}H	28.4 ± 2.11	42.6 ± 3.66[a]	33.1 ± 2.25	28.0 ± 3.26	25.9 ± 1.89	24.9 ± 0.95
^{3}H-NE	9.4 ± 1.10	16.6 ± 1.99[a]	10.9 ± 0.92	10.6 ± 1.16	10.8 ± 0.94	9.7 ± 0.48
^{3}H-NMN	10.3 ± 1.14	14.7 ± 1.51[a]	12.3 ± 1.02	10.5 ± 1.40	9.8 ± 0.69	9.7 ± 0.56

Rats were injected with radioisotopes 10 min before a single or sixth daily dose of oxazepam (or saline) and were killed 3 hr later. Control groups received single or repeated doses of saline instead of oxazepam; means of control groups did not differ significantly and were combined. For details, see text. Numbers indicate mean dpm per mg of brain; N = number of rats; ^{14}C-5-HIAA = ^{14}C-5-hydroxyindoleactic acid + neutral metabolites; ^{3}H-NMN = ^{3}H-normetanephrine.

[a] Differs from saline mean, $p < 0.05$.
[b] Includes one case for which all values substantially exceed the range of the other nine cases. If this case is excluded, resulting saline means for total ^{14}C and ^{14}C-5-HT are significantly smaller than corresponding means of both oxazepam groups.

decreases slightly after repeated doses of oxazepam, the anxiety-reducing activity actually increases. Such increase may be due partly to progressive unmasking of the anxiety-reducing action as tolerance to the depressant action develops, and partly to supersensitivity to norepinephrine, which may develop as a result of drug-induced disuse of noradrenergic synapses.)

Although oxazepam may have exerted small effects on serotonin turnover in the diencephalon-forebrain region (Table VII), its largest effects were concentrated in the midbrain-hindbrain region. This result seems consistent with the observation that serotonin administered intraventricularly is selectively accumulated in the central gray area of the midbrain (Aghajanian and Bloom, 1967), a region frequently associated with punishment and aversive behavior (de Molina and Hunsperger, 1962; Olds and Olds, 1963; Stein, 1965). Such considerations focus attention on serotonergic synapses in the central gray area as possible sites of action for the antianxiety effects of benzodiazepines and related tranquilizers.

X. GENERAL DISCUSSION AND SUMMARY

Our major findings in the rat conflict test, may be summarized as follows. (1) The antianxiety effects of minor tranquilizers were mimicked by serotonin antagonists or the synthesis inhibitor p-chlorophenylalanine, and the response-depressant effects of the tranquilizers were mimicked by norepinephrine antagonists. (2) Intraventricular injections of serotonin suppressed behavior, and also antagonized the antianxiety action of benzodiazepines. Intraventricular injections of L-norepinephrine released punished behavior from suppression, and also antagonized the depressant action of benzodiazepines. (3) Intense suppression of behavior was produced by direct application of crystalline carbachol to the dorsal raphe nucleus, a major serotonin cell group in the midbrain. The suppressive effects of raphe stimulation were antagonized by systemic administration of oxazepam.

In biochemical experiments, the decrease in norepinephrine turnover induced by oxazepam rapidly underwent tolerance, whereas the decrease induced in serotonin turnover was maintained over repeated doses. These results parallel findings in the rat conflict test which indicate that the depressant action of tranquilizers rapidly undergoes tolerance, whereas the antianxiety action is maintained over repeated doses. Oxazepam-induced decreases in serotonin turnover were concentrated in the midbrain-hindbrain region of rat brain.

All of these results are consistent with the idea that tranquilizers exert their antianxiety effects by reduction of serotonin activity in a behaviorally suppressive punishment system and their depressant effects by reduction of norepinephrine activity in a behaviorally facilitatory reward system. These experiments also suggest that serotonin nerve terminals in the diencephalic and midbrain central gray regions, which probably arise at least in part from cell bodies in the dorsal raphe nucleus, are important sites of action for the antianxiety effects of tranquilizers.

The concept that punishment effects are mediated by a central serotonin system may illuminate some hitherto puzzling actions of other drugs on anxiety-related behaviors. Amphetamine has long been known to intensify the suppressive effects of punishment in the conflict test (Geller and Seifter, 1960; Kelleher and Morse, 1964), despite the fact that the drug in the same dose range markedly facilitates self-stimulation and other rewarded behaviors (Stein, 1964; Stein and Wise, 1970). Amphetamine-induced facilitation of self-stimulation has been attributed to augmented release of norepinephrine from the reward system (Stein, 1964; Wise and Stein, 1970). Since amphetamine also releases serotonin from neurons in the dorsal raphe nucleus (Fuxe and Ungerstedt, 1970) and, presumably, from other parts of the serotoninergic punishment system as well, it would be logical to attribute the punishment-increasing effects of the drug to an increased release of serotonin. Whether amphetamine predominantly facilitates the release of norepinephrine or the release of serotonin in any given case, probably depends to a large extent on the nature of the test situation. Norepinephrine release will be predominantly facilitated by amphetamine in tests that emphasize reward (e.g., self-stimulation), whereas serotonin release will be selectively facilitated in tests that emphasize punishment (e.g., conflict).

A similar analysis may explain the findings in the pigeon that imipramine strongly facilitates rewarded behaviors but also strongly suppresses punished behaviors (Wuttke and Kelleher, 1970). Since imipramine blocks the reuptake of both norepinephrine and serotonin, the drug could selectively increase noradrenergic activity in reward situations and serotonergic activity in punishment situations.

Finally, it is tempting to speculate that the anxiety-reducing effects of morphine and other narcotic analgesics may depend at least in part on the blockade of central serotonin receptors. (Like the minor tranquilizers, morphine also exerts depressant effects on behavior which may result from a reduction of noradrenergic activity; this depressant action is particularly strong in the rat and may partly explain why morphine exhibits no antipunishment activity in the rat conflict test). According to Gaddam (1957), morphine blocks "Type-M" tryptaminergic receptors in nervous ganglia. If it is correct to assume that the serotonergic punishment system in the brain also contains Type-M receptors, then morphine could block these receptors to exert its antianxiety action. In this regard, it is interesting to note that the chronic administration of morphine causes a substantial increase in serotonin synthesis (Shen et al., 1970). The resulting increase in the availability of serotonin would increase the competition for serotonin receptors, and, together with disuse supersensitivity, might partly account for the development of tolerance to morphine. Furthermore, after morphine is withdrawn, the supernormal amounts of serotonin would strongly activate the "supersensitive" serotonin receptors in the punishment system. Such intense activity of the serotonergic punishment system could explain the intolerable psychological stress that characterizes the morphine-withdrawal syndrome.

ACKNOWLEDGMENTS

The authors thank N. S. Buonato, W. J. Carmint, H. Morris, and A. T. Shropshire for expert technical assistance.

REFERENCES

Aghajanian, G. K., and Bloom, F. E. (1967): Localization of tritiated serotonin in rat brain by electron-microscopic autoradiography. *Journal of Pharmacology and Experimental Therapeutics*, 156:23–30.

Aprison, M. H., and Ferster, C. B. (1961): Neurochemical correlates of behavior: II. Correlation of brain monoamine oxidase activity with behavioral changes after iproniazid and 5-hydroxytryptophan. *Journal of Neurochemistry*, 6:350–357.

Arbuthnott, G., Fuxe, K., and Ungerstedt, U. (1971): Central catecholamine turnover and self-stimulation behavior. *Brain Research*, 27:406–413.

Berger, B. D., Wise, C. D., and Stein, L. (1971): Norepinephrine: Reversal of anorexia in rats with lateral hypothalamic damage. *Science*, 172:281–283.

Brodie, B. B., and Shore, P. A. (1957): A concept for a role of serotonin and norepinephrine as chemical mediators in the brain. *Annals of the New York Academy of Sciences*, 66:631–642.

Chase, T. N., Katz, R. I., and Kopin, I. J. (1970): Effect of diazepam on fate of intracisternally injected serotonin-C^{14}. *Neuropharmacology*, 9:103–108.

Cook, L., and Davidson, A. B. (1972): Effects of behaviorally active drugs in a conflict-punishment procedure in rats (*This Volume*).

Corrodi, H., Fuxe, K., and Hokfelt, T. (1967): The effect of some psychoactive drugs on central monoamine neurons. *European Journal of Pharmacology*, 1:363–368.

Corrodi, H., Fuxe, K., Lidbrink, P., and Olson, L. (1971): Minor tranquilizers, stress and central catecholamine neurons. *Brain Research*, 29:1–16.

Dahlström, A., and Fuxe, K. (1965): Evidence for the existence of monoamine-containing neurons in the central nervous system. I. Demonstration of monoamines in the cell bodies of brain stem neurons. *Acta Physiologica Scandinavica*, 232:62, *Suppl.* 1–55.

Delgado, J. M. R., Roberts, W. W., and Miller, N. E. (1954): Learning motivated by electrical stimulation of the brain. *American Journal of Physiology*, 179:587–593.

324   ANTIANXIETY ACTION OF BENZODIAZEPINES

Fuxe, K. (1965): The distribution of monoamine nerve terminals in the central nervous system. *Acta Physiologica Scandinavica,* 64, *Suppl* 247:37–102.

Fuxe, K., Hökfelt, T., and Ungerstedt, U. (1968): Localization of indolealkylamines in CNS. In: *Advances in Pharmacology,* edited by S. Garattini and P. A. Shore. Academic Press, New York, pp. 235–251.

Fuxe, K., Hökfelt, T., and Ungerstedt, U. (1970): Morphological and functional aspects of central monoamine neurons. *International Review of Neurobiology,* 13:93–126.

Fuxe, K., and Ungerstedt, U. (1970): Histochemical, biochemical, and functional studies on central monoamine neurons after acute and chronic amphetamine administration. In: *Amphetamines and Related Compounds,* edited by E. Costa and S. Garattini. Raven Press, New York, pp. 257–288.

Gaddam, J. H. (1957): Serotonin–LSD interactions. *Annals of the New York Academy of Sciences,* 66:643–648.

Geller, I., Bachman, E., and Seifter, J. (1963): Effects of reserpine and morphine on behavior suppressed by punishment. *Life Sciences,* 4:226–231.

Geller, I., and Blum, K. (1970): The effects of 5-HT on *para*-chlorophenylalanine (*p*-CPA) attenuation of "conflict" behavior. *European Journal of Pharmacology,* 9:319–324.

Geller, I., Kulak, J. T., Jr., and Seifter, J. (1962): The effects of chlordiazepoxide and chlorpromazine on a punishment discrimination. *Psychopharmacologia* (Berlin), 3:374–385.

Geller, I., and Seifter, J. (1960): The effects of meprobamate, barbiturates, *d*-amphetamine and promazine on experimentally induced conflict in the rat. *Psychopharmacologia* (Berlin), 1:482–492.

Goldberg, M. E., Manian, A. A., and Efron, D. H. (1967): A comparative study of certain pharmacological responses following acute and chronic administration of chlordiazepoxide. *Life Sciences,* 6:481–491.

Graeff, F. G., and Schoenfeld, R. I. (1970): Tryptaminergic mechanisms in punished and nonpunished behavior. *Journal of Pharmacology and Experimental Therapeutics,* 173:277–283.

Kelleher, R. T., and Morse, W. H. (1964): Escape behavior and punished behavior. *Federation Proceedings,* 23:808–835

Margules, D. L., and Stein, L. (1967): Neuroleptics vs. tranquilizers: evidence from animal studies of mode and site of action. In: *Neuro-Psychopharmacology,* edited by H. Brill, J. O. Cole, P. Deniker, H. Hippius, and P. B. Bradley. Excerpta Medica Foundation, Amsterdam, pp. 108–120.

Margules, D. L., and Stein, L. (1968). Increase of "antianxiety" activity and tolerance of behavioral depression during chronic administration of oxazepam. *Psychopharmacologia* (Berlin), 13:74–80.

Masserman, J. H., and Yum, K. S. (1946): An analysis of the influence of alcohol on experimental neurosis in cats. *Psychosomatic Medicine,* 8:36–52.

Miller, N. E. (1967): Objective techniques for studying the motivational effects of drugs. In: *Psychotropic Drugs,* edited by S. Garattini and V. Ghetti. Elsevier, Amsterdam, pp. 83–102.

Nauta, W. J. H. (1963): Central nervous organization and the endocrine nervous system. In: *Advances in Neuroendocrinology,* edited by A. V. Nalbandov. University of Illinois Press, Urbana, pp. 5–21.

Olds, J., and Milner, P. (1954): Positive reinforcement produced by electrical stimulation of septal area and other regions. *Journal of Comparative Physiological Psychology,* 47:419–427.

Poschel, B. P. H., and Ninteman, F. W. (1963): Norepinephrine: A possible excitatory neurohormone of the reward system. *Life Sciences,* 3:782–788.

Ritter, S., and Stein, L. (1972): Self-stimulation of the locus coeruleus. *Federation Proceedings,* 31:820.

Robichaud, R. C., and Sledge, K. L. (1969): The effects of *p*-chlorophenylalanine on experimentally induced conflict in the rat. *Life Sciences,* 8:965–969.

Shen, F-H., Loh, H. H., and Way, E. L. (1970): Brain serotonin turnover in morphine tolerant and dependant mice. *Journal of Pharmacology and Experimental Therapeutics,* 175:427–434.

Stein, L. (1962): Effects and interactions of imipramine, chlorpromazine, reserpine, and amphetamine on self-stimulation: Possible neurophysiological basis of depression. In: *Recent Advances in Biological Psychiatry, Vol. 4,* edited by J. Wortis. Plenum Press, New York, pp. 288–308.

Stein, L. (1964): Self-stimulation of the brain and the central stimulant action of amphetamine. *Federation Proceedings,* 23:836–850.

Stein, L. (1967): Psychopharmacological substrates of mental depression. In: *Antidepressant Drugs,* edited by S. Garattini and N. M. G. Dukes. Excerpta Medica Foundation, Amsterdam, pp. 130–140.

Stein, L. (1968): Chemistry of reward and punishment. In: *Psychopharmacology, A Review of Progress: 1957-1967,* edited by D. H. Efron. U.S. Government Printing Office, Washington, D.C., pp. 105–123.

Stein, L. (1971): Neurochemistry of reward and punishment: Some implications for the etiology of schizophrenia. *Journal of Psychiatric Research,* 8:345–362.

Stein, L., and Wise, C. D. (1970): Behavioral pharmacology of central stimulants. In: *Principles of Psychopharmacology,* edited by W. G. Clark and J. del Giudice. Academic Press, New York, pp. 313–325.

Stevens, D. A., and Fechter, L. D. (1969): The effects of *p*-chlorophenylalanine, a depletor of brain serotonin, on behavior: II. Retardation of passive avoidance learning. *Life Sciences,* 8:379–385.

Taylor, K. M., and Laverty, R. (1969): The effect of chlordiazepoxide, diazepam and nitrazepam on catecholamine metabolism in regions of the rat brain. *European Journal of Pharmacology,* 8:296–301.

Tenen, S. S. (1967): The effects of *p*-chlorophenylalanine, a serotonin depletor, on avoidance acquisition, pain sensitivity, and related behavior in the rat. *Psychopharmacologia* (Berlin), 10:204–219.

Warner, R. S. (1965): Management of the office patient with anxiety and depression. *Psychosomatics,* 6:347–351.

Weight, F. F., and Salmoiraghi, G. C. (1968): Serotonin effects on central neurons. In: *Advances in Pharmacology,* edited by S. Garattini and P. A. Shore. Academic Press, New York, pp. 395–413.

Wise, C. D., Berger, B. D., and Stein, L. (1970*a*): Brain serotonin and conditioned fear. *Proceedings of the 78th Annual Meeting, American Psychological Association,* 5:821–822.

Wise, C. D., Berger, B. D., and Stein, L. (1970*b*): Serotonin: A possible mediator of behavioral suppression induced by anxiety. *Diseases of the Nervous System,* GWAN Suppl. 31:34–37.

Wise, C. D., Berger, B. D., and Stein, L. (1972): α-Noradrenergic receptors for reward and serotonergic receptors for punishment in the rat brain. *Biological Psychiatry (in press).*

Wise, C. D., and Stein, L. (1969): Facilitation of brain self-stimulation by central administration of norepinephrine. *Science,* 163:299–301.

Wise, C. D., and Stein, L. (1970): Amphetamine: Facilitation of behavior by augmented release of norepinephrine from the medial forebrain bundle. In: *Amphetamines and Related Compounds,* edited by E. Costa and S. Garattini. Raven Press, New York, pp. 463–485.

Wuttke, W., and Kelleher, R. T. (1970): Effects of some benzodiazepines on punished and unpunished behavior in the pigeon. *Journal of Pharmacology and Experimental Therapeutics,* 172:397–405.

The Benzodiazepines, Raven Press, New York © 1973

EFFECTS OF BEHAVIORALLY ACTIVE DRUGS IN A CONFLICT-PUNISHMENT PROCEDURE IN RATS

Leonard Cook* and Arnold B. Davidson

Pharmacology Department, Smith Kline & French Laboratories, Philadelphia, Pennsylvania

I. INTRODUCTION

Estes and Skinner (1941) found that presentation of a stimulus preceding an unavoidable shock resulted in the suppression of ongoing food-maintained behavior. Certain drugs attenuated this response suppression, although the results were highly variable (reviewed by Cook and Kelleher, 1963; Kelleher and Morse, 1968). Geller and Seifter (1960) developed a related procedure, based on immediate punishment, in which response-contingent shocks suppressed food-maintained behavior in one component of a multiple schedule of reinforcement. The effects of a variety of pharmacological agents on punishment-induced response suppression in several species have been reported (Geller and Seifter, 1960, 1962; Geller et al., 1962, 1963; Geller, 1964; Morse, 1964; Hanson et al., 1967), and have been generally consistent (Cook and Kelleher, 1963; Kelleher and Morse, 1968).

In Geller's procedure, drugs were tested against either a high or low degree of response suppression, depending on the anticipated direction of pharmacological effect. The different baselines resulted from delivering to different groups of rats either high or low intensities of shock punishment for each response. The difficulty of comparing quantitatively the effects of different types of drugs with such procedures was partially overcome by Hanson et al. (1967), who reported comparative dose-response data on the effects of several behaviorally active pharmacological agents on a single baseline of suppression in squirrel monkeys. They used a variation of Geller's procedure in which intermittent punishment was delivered, and which resulted in a higher baseline of punished responding than Geller's maximal suppression baseline.

---

* Present address: Department of Pharmacology, Hoffmann-LaRoche Inc., Nutley, New Jersey.

327

A similar punishment procedure, for use with rats in comparative pharmacological evaluations, has been reported by Davidson and Cook (1969). In this procedure, the intermittent schedule of response-contingent shock punishment resulted in a level of suppressed responding which could be used in any of the trained rats to demonstrate dose-related increases or decreases in punished or nonpunished responding. The similar control baselines in different drug tests permitted the use of group data and facilitated the quantitative comparison of drug effects.

This chapter presents a more complete description of the characteristics of the behavior generated in rats with this procedure and a comparison of the effects of a variety of types of psychopharmacological agents, with particular emphasis on those reported to have some clinical efficacy in psychoneurotic patients.

The effects of chlordiazepoxide in another type of punishment procedure, employing squirrel monkeys, are also presented to illustrate the generality of the effects of chlordiazepoxide in attenuating the suppressant effects of punishment contingencies across species and procedures.

It appears that punished behavior is particularly useful for studying psychopharmacological properties of minor tranquilizers. This type of behavioral situation is considered by some to resemble a conflict situation since the animal's response tendency, or the probability of a lever response, is controlled by the consequences of both positive (food) and negative (shock) reinforcement. However, it is not yet possible to specify the psychodynamic factors (*e.g.* conflict) which control this behavior.

II. METHOD

Male Sprague-Dawley rats (Charles River) were gradually reduced to 80% of their free-feeding body weight by food deprivation over a period of approximately 2 weeks; they were maintained at this level (average: approximately 250 g) by food delivery in the experimental sessions and supplementary feeding at the end of each day.

The experimental chamber measured $25.5 \times 25.5 \times 23$ cm and contained a response lever and food-delivery apparatus, an electrifiable grid floor, and a stimulus light which illuminated the chamber. The experimental chamber was enclosed by a sound-resistant, ventilated box. Lever-press responses were reinforced by delivery of 45 mg Noyes food pellets. The experiment was automatically programmed and recorded, and the intensity of foot-shock punishment, delivered by a Grason-Stadler shock generator and scrambler, was adjusted for each rat during training (range: 0.8 to 2.5 mA; 0.1-sec duration), to achieve an optimal degree of suppression.

The rats were first trained to respond on a fixed-ratio schedule of 10 lever-press responses for each food reinforcement (FR 10) in the presence of a flashing light. After stable FR 10 performance had developed, 5-min periods of a variable-interval schedule of reinforcement, in the presence of a steady light, were alternated with 2-min periods of the FR 10 reinforcement in the presence of the flashing light. In the variable-interval reinforcement component, responses were reinforced by food delivery at varying intervals of time; the mean interval was 30 sec (VI 30 sec). When performance on this multiple schedule of reinforcement stabilized, a punishment contingency was added to the FR 10 component so that every 10th response resulted in the simultaneous delivery of a food pellet and an electric foot shock. Once the rats were trained, there were very few instances in which the shock level established for each rat had to be changed to maintain the desired level of suppression. The rats were used for drug testing for an average of 12 months; some rats were used for as long as 2 years.

Daily experimental sessions consisted of 6 periods of the FR 10 (punishment) component and 7 periods of the VI 30-sec (nonpunishment) component and were conducted at the same time, Monday through Friday, for each rat.

Physiological saline, administered by gavage in amounts equivalent to those used in drug test sessions, did not appreciably change responding in either component. All drugs were administered by gavage except for morphine, which was administered intraperitoneally. Water-soluble drugs were prepared in aqueous solution; others were suspended in a 0.5% tragacanth suspension. At least 3 rats were tested at each dose, and at least 1 week elapsed between treatments. Doses of drugs administered as salts are expressed in terms of their base weights, except as noted. The drugs tested (and, where applicable, the salt form administered) were amobarbital (Na), D-amphetamine (SO$_4$), benactyzine (HCl), chlordiazepoxide (HCl), chlorpromazine (HCl), diazepam, diphenhydramine (HCl), iproniazide (H$_3$PO$_4$), meprobamate, methaqualone, morphine (SO$_4$), oxazepam, phencyclidine (HCl), phenobarbital (Na), trifluoperazine (di-HCl), and tybamate.

For statistical analyses, each rat was used as its own control, and the average number of responses on the 3 days immediately preceding treatment, in each of the two schedule components, was compared to the number of responses in these components following treatment on the 4th day. Response values for the 6 FR 10 periods and 7 VI 30-sec periods were analyzed by analysis-of-variance techniques in most of the drug studies reported. In a few of the initial studies, total response over all VI 30-sec components in a session were analyzed using the *t* test for paired comparisons, due to limitations in statistical facilities. Results are reported in terms of drug-to-control ratios, rather than absolute values, to facilitate comparison of drug effects.

III. PHARMACOLOGICAL STUDIES

Representative segments of the cumulative-response records of rats in these studies are presented (Fig. 1) to illustrate typical patterns of behavior. The rats generally responded on the lever at a moderately high rate during the VI 30-sec components, and at a somewhat higher rate during the FR components, prior to the introduction of foot shocks (left panel). After the shock contingency was put into effect and delivered concurrently with the food pellet at the end of each FR 10, response rates during the VI 30-sec components were essentially unchanged; however, response rates in the FR

RAT CONFLICT-PUNISHMENT

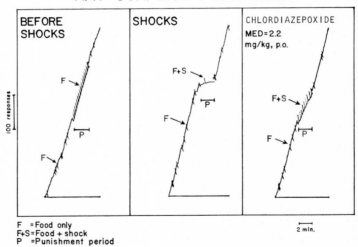

F   = Food only
F+S = Food + shock
P   = Punishment period

2 min.

FIG. 1. Segments of cumulative response records illustrating response patterns in punished FR 10 component (P) and nonpunished VI 30-sec component. *Left panel*: high response rate in P before introduction of shock; *center panel*: suppressed responding in P after introduction of concurrent food and punishing shock (F + S) in absence of drug; *right panel*: increased responding in P after treatment with chlordiazepoxide. Pen deflections indicate delivery of food (F) in VI 30-sec component and of concurrent F + S in FR 10 component.

10 punishment periods were markedly reduced. Under nontreatment control conditions, the rats used in these studies exhibited behavior qualitatively similar to that illustrated in the center panel of Fig. 1.

Drugs which attenuated the suppressant effects of punishment produced effects qualitatively similar to those illustrated in the right panel of Fig. 1. As shown, chlordiazepoxide increased responding during the punishment period, resulting in increased delivery of both food pellets and shocks.

Qualitatively different effects were observed on punished and nonpunished behavior after treatment with different classes of pharmacological agents. Chlorpromazine typically decreased responding in both the nonpunished VI 30-sec component and the punished FR 10 component, concurrently. D-Amphetamine significantly suppressed response rates in the punishment condition at doses which had no effect on nonpunished VI 30-sec response rates or, in some animals, even when VI 30-sec responses were concurrently increased. Similar effects of D-amphetamine were reported by Geller and Seifter (1960). At higher doses, D-amphetamine significantly suppressed all responding in the punishment condition, while little change occurred in the number of responses in the nonpunishment components. D-Amphetamine is the only compound tested which produced this type of effect.

Figures 2–10 present dose-effect curves of various pharmacological agents on punished and nonpunished response rates. The results are expressed as the ratio of rates on the day of drug treatment (D) to control rates (C), plotted on a log-log scale.

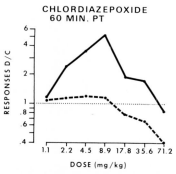

FIG. 2. Dose-effect curve: chlordiazepoxide. Ratio of response rate after drug treatment to pre-drug control response rate (D/C) shown on ordinate; dose, expressed in terms of milligrams per kilogram of free base, shown on abscissa. Time (PT: pretreatment time) shown on figure indicates time between oral treatment and start of test session. Each point represents the mean D/C for all rats at a particular dose. Solid line indicates effects in punished FR 10 component, broken line indicates effects in nonpunished VI 30-sec component; horizontal dotted line indicates control level.

Chlordiazepoxide (Fig. 2) alleviates the suppressant effects of punishment over a relatively wide range of doses. The effects shown are based on a total of 173 different rats. Significantly increased responding in the conflict-punishment condition was produced by doses ranging from 2.2 to 35.6 mg/kg, orally. Figure 2 also indicates that antisuppression effects occur at dose levels of chlordiazepoxide which do not have significant effects on nonpunished behavior and even at some higher dose levels which significantly decrease nonpunished response rates. The effects of oxazepam and diazepam are qualitatively similar to those of chlordiazepoxide in this procedure.

Meprobamate (Fig. 3) significantly enhanced response rates under

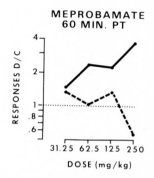

FIG. 3. Dose-effect curve; meprobamate. See Fig. 2 for legend.

punishment conditions at doses which had minimal effects on nonpunished response rates. However, it differed from chlordiazepoxide in that it produced maximum alleviation of the suppressant effects of punishment at the dose (250 mg/kg, oral) which significantly depressed nonpunished behavior. Amobarbital (Fig. 4) alleviated suppression at doses ranging from those with no effect on nonpunished responding to those which reduced nonpunished responding. Qualitatively similar effects were produced by phenobarbital.

The effects of chlorpromazine, tested over a wide range of doses, are shown in Fig. 5. At no dose tested did chlorpromazine alleviate suppressed responding. Overall, it decreased both punished and nonpunished response rates. Measurement of the effects of low doses of chlorpromazine was particularly emphasized in order to determine whether it produced an effect similar to that seen after trifluoperazine (Davidson and Cook, 1969). As shown in Fig. 6, trifluoperazine differed from chlorpromazine in that low

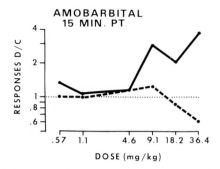

FIG. 4. Dose-effect curve: amobarbital. See Fig. 2 for legend.

doses (0.17 and 0.34 mg/kg) produced a statistically significant alleviation of the effects of punishment, with no effect on nonpunished responding. Higher doses decreased both punished and nonpunished responding, similar to chlorpromazine. Therefore, it appears that among the phenothiazines, qualitatively different effects may be produced in the same procedure, particularly at different dose levels.

Ethanol (Fig. 7) also significantly alleviated the effect of punishment at doses of 1000 and 2000 mg/kg orally. At the higher dose, the nonpunished (VI 30-sec) response rate was depressed, and animals showed some degree of ataxia. As indicated previously, D-amphetamine (Fig. 8) further

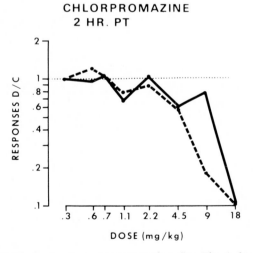

FIG. 5. Dose-effect curve: chlorpromazine. See Fig. 2 for legend.

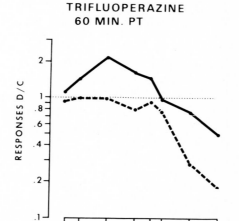

FIG. 6. Dose-effect curve: trifluoperazine. See Fig. 2 for legend.

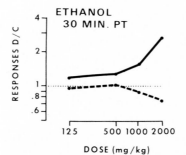

FIG. 7. Dose-effect curve: ethanol. See Fig. 2 for legend.

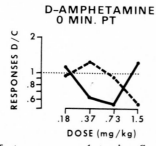

FIG. 8. Dose-effect curve: D-amphetamine. See Fig. 2 for legend.

reduced the already low response rate under the punishment conditions; that is, it resulted in greater suppression. This reduction was specific to the punished behavior, since these effects were produced at doses which had no effect on the nonpunished behavior. At doses of 1.5 mg/kg, absolute response rates of punished and nonpunished behavior tend to become equivalent; that is, punished response rates are increased and nonpunished (VI 30-sec) response rates are decreased. Rats at this higher dose showed observable signs of gross motor stimulation.

Morphine sulfate was studied in order to determine whether an analgetic pharmacological action could cause and possibly account for the increased responding under punishment conditions produced by various pharmacological agents. Morphine (Fig. 9) had only a small (significant) effect at one

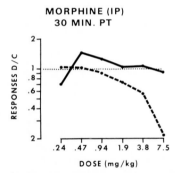

FIG. 9. Dose-effect curve: morphine (intraperitoneal administration). See Fig. 2 for legend.

dose level (0.47 mg/kg, i.p.) and, therefore, could not be considered an effective compound in this procedure. Morphine exhibits analgetic activity over a wide range of dose levels in many pain-threshold procedures, but it did not consistently alleviate the effects of punishment in this conflict-punishment procedure. These results suggest that it is unlikely that the antisuppression effect measured in this procedure can be completely accounted for by any analgetic or pain-threshold elevating activity.

The effects of a number of standard pharmacological agents tested in this procedure are summarized in Table I. Effective compounds are those which produced significant alleviation of the punishment-induced suppression at more than one dose. Ineffective compounds include those which failed to show any anticonflict effects in the range of doses tested, or those which produced effects at only one dose level. In all instances, those effects

TABLE I. *Activity of behaviorally active compounds*

	Compound	Punishment, M.E.D. (mg/kg, p.o.)	Relative potency
Effective compounds	Chlordiazepoxide	2.2	1
	Oxazepam	1.25	1.8
	Diazepam	0.625	3.5
	Tybamate	80	0.03
	Meprobamate	62.5	0.04
	Amobarbital	5	0.4
	Phenobarbital	4.5	0.5
	Methaqualone	10	0.2
	Ethanol	1000	—
	Phencyclidine	4.4	0.5
	Benactyzine	4.5	0.5
	Trifluoperazine	0.165	13.0

	Compound	Range of doses tested (mg/kg, p.o.)
Ineffective compounds	Iproniazide	20–120
	Imipramine	0.55–17.7
	Morphine	0.24– 7.5 (i.p.)
	Haloperidol	0.03– 0.48
	Diphenhydramine	0.55–17.5
	D-Amphetamine	0.18– 1.5
	Chlorpromazine	0.27–17.9

M.E.D. indicates lowest dose tested which significantly ($p < 0.05$) attenuated effects of punishment. Potency was calculated relative to chlordiazepoxide. Within the range of doses shown for ineffective compounds, no consistent attenuation of the effects of punishment was measured. Doses are expressed as free base.

which occurred at only one dose level were inconsistent upon repetition of the test.

It should be noted that the order of potency of diazepam > oxazepam > chlordiazepoxide was the same as that described by Geller (1964), although his method of comparison was more qualitative than quantitative.

IV. EXPERIMENTAL PARAMETRIC STUDIES

Several alternative mechanisms were considered to explain the increase of punished responding produced by the compounds classified as effective (Table I). These possible mechanisms were a weakening of the behavioral control exerted by the stimulus associated with punishment components; changes in the motivational states produced by food deprivation or by foot shock; and a rate-dependency effect of the type described by Dews (1958). Control studies were conducted to determine the influence of some of these factors on behavior in the punishment component.

A. Effect of removal of discriminative stimuli

The stimulus light, which continuously illuminated the experimental chamber during nonpunished components and which flashed on and off during the punishment components, was disconnected, thereby removing the stimuli used to discriminate between the two components.

The record shown in Fig. 10 illustrates the behavior generally seen

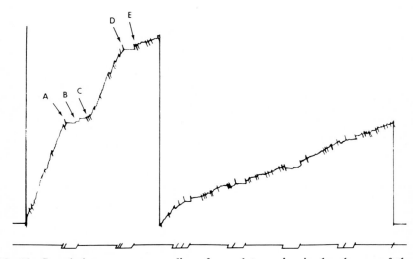

FIG. 10. Cumulative-response recording of complete session in the absence of the discriminative stimuli used to differentiate punishment and nonpunishment components. On the upper (response) tracing, and on the bottom (event) tracing, punishment components are shown by the pen offset downward and concurrent deliveries of food and shock by brief upward deflections. Delivery of food only (during nonpunished components) is shown by brief downward deflections on response tracing; response tracing automatically resets after cumulation of 500 responses. See text for further explanation.

under these conditions. After the onset of the first punishment (FR 10) component, the subject continued to respond at its VI rate until the completion of 10 responses and delivery of the first shock (A). Responding ceased abruptly for a period of time; and a brief probe (emission of a few responses) occurred at B, followed by another long pause which continued beyond termination of the punishment component. The first response of the next probe occurred in the nonpunished component. The resultant food delivery (C), unaccompanied by shock, produced a progressive return of the VI response rate, which continued until the rat received shocks in the next punishment component (D).

Similar behavior was exhibited throughout the remainder of the session. Response probes were emitted to determine the prevailing component; shocks were followed by long pauses, and nonpunished food reinforcements were followed by continued responding. After the second punishment component, the indeterminancy of the situation produced a marked decrease in VI response rate (beginning at E).

Elimination of discrete stimulus control produced effects on behavior different from those produced by chlordiazepoxide treatment. Chlordiazepoxide produced responding throughout each punishment component, even though shocks were received, and did not significantly change VI response rate over a wide range of doses.

B. Effects of increased food deprivation

Since increased food deprivation generally augments food-maintained responding, rats were tested during several days of increased food deprivation to determine the extent to which punished responding (maintained by concurrent food reinforcement) would be increased under these conditions. The only food they received was the 4.5 g obtained in the experimental situation each day. This experiment was terminated when each rat essentially stopped responding for food due to severe debilitation. Under these conditions of severe food deprivation, responding under the punishment conditions was only slightly increased over the previous control rates, and the magnitude of this increase was small compared to that produced by chlordiazepoxide, at an oral dose of 8.9 mg/kg, in these same rats (Fig. 11). In these rats, it was not possible to achieve the magnitude of anticonflict effect produced by chlordiazepoxide, even under severe food-deprivation conditions.

C. Effect of removal of shocks

To determine whether the increases of punished responding produced by various drugs might be attributable to changes in foot-shock threshold, or to an analgetic phenomenon different from that discussed with reference to morphine, the maximum effect of decreased sensitivity to foot shock was produced experimentally by disconnecting the shock generator. The results are shown in Fig. 12. Responding in the FR component (curves on right-side panels), in the presence of the stimulus which had been associated with punishment, gradually increased over a period of approximately 6 days. On the first day of shock removal, the increase in FR responding was only about

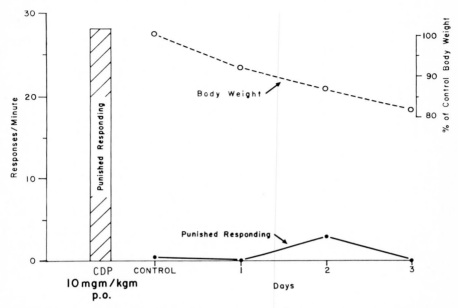

FIG. 11. Representative effects of increased food deprivation in one rat. Decreased weight under severe food deprivation (broken line, right ordinate) shown as percentage of normal running weight under limited feeding condition. Rate of punished responding (on left ordinate) shown during this 3-day period (solid line) and on the day of treatment with chlordiazepoxide (CDP) at a dose of 10 mg/kg (8.9 mg/kg expressed as base weight).

half that produced by the administration of chlordiazepoxide (8.9 mg/kg) in the presence of shocks. Another day of experience with the absence of shocks was required to increase FR responding to a level comparable to that produced by chlordiazepoxide.

Some interaction of lever-press behavior in the two components can also be seen here, in the initial effect of the reintroduction of shock punishment on responding in the nonpunished VI component (Fig. 12). The reintroduction of shocks, in the 20th session, produced an immediate return of punished FR responding to the control level seen before experimental manipulation, and also produced a marked decrease in nonpunished VI responding. Approximately 5 additional days elapsed before nonpunished VI responding returned to control level. This interaction may account for the slight elevations of nonpunished response rates often produced by drugs which significantly increase rates of punished responding.

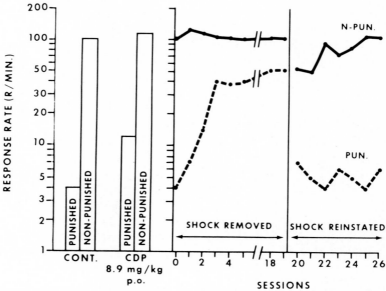

FIG. 12. Representative effect of discontinuation of shock punishment. Response rate shown on ordinate; control (cont.) rates and rates after chlordiazepoxide (CDP) treatment shown in histograms on left. Effects of shock removal for 19 successive sessions and of reinstatement of shock before the 20th session shown in right panels (broken line = punished responding, solid line = nonpunished responding).

D. Effect of pre-drug response rate on drug effect

The predominant effect of the agents studied on punished responding does not appear to be solely a function of pre-drug response rates. For example, D-amphetamine did not increase the low suppressed rate of responding, as it does in other schedules of reinforcement (discussed by Kelleher and Morse, 1968). In addition, Cook and Catania (1964) reported on experiments in squirrel monkeys in which punished response rates were made experimentally equivalent to concurrent nonpunished response rates. It was shown that meprobamate increased punished responding to a significantly greater degree than nonpunished responding.

However, comparison of the effect of chlordiazepoxide (8.9 mg/kg) on punished responding, with pre-drug punished response rates of 47 rats, showed that there was a significant negative correlation ($r = -0.604$, $p < 0.001$) between the magnitude of effect of chlordiazepoxide and pre-drug response rate. The lower the pre-drug rate of suppressed responding, the greater the attenuation of suppression produced by chlordiazepoxide.

V. PREDICTABILITY OF ANTICONFLICT EFFECTS

An attempt was made to determine the relevance of pharmacological activity, in this procedure, to the clinical application of effective compounds. The literature was surveyed (see Table II) to determine the average daily

TABLE II. *Comparison of clinical potency with rat test results*

	Rat conflict			Clinical psychoneurotics		
	M.E.D. (mg/kg, p.o.)	Relative potency	Rank order	Avg. daily dose (mg) Oral	Compar. studies, relative potency	Rank order
Diazepam	0.63	3.5	1	20	2	1
Chlordiazepoxide	2.2	1.0	3	40	1	2
Oxazepam	1.25	1.8	2	49	0.8	3
Phenobarbital	4.5	0.5	4	115	0.3	4
Amobarbital	5.0	0.4	5	175	0.17	5
Meprobamate	62.5	0.04	6	1410	0.03	6

M.E.D. indicates lowest dose tested which significantly ($p < 0.05$) attenuated effects of punishment. Average daily dose (clinical) obtained from 74 published clinical reports. Complete list of references can be obtained from the authors. Relative potency (clinical) based on studies in which drugs were compared directly to chlordiazepoxide.

dose employed in treating psychoneurotic patients. In addition, several studies were available in which the clinical protocol was designed to compare directly the relative potencies of chlordiazepoxide and other agents (Gore and McComiskey, 1961; Lorr et al., 1961; Lader and Wing, 1965, 1966; Jenner and Kerry, 1967; Wheatley, 1968; DeSilverio et al., 1969; Kingstone et al., 1969). These findings were analyzed and the compounds ranked according to relative dose potency (Table II). The rank order of relative clinical potency was found to be similar to the rank order of potency in the rat conflict-punishment procedure. It is acknowledged that it is difficult to make a precise comparison of compounds possessing different pharmacological profiles from clinical studies carried out under different conditions, and that other surveys may indicate different rank orders of relative potency and even clinical effectiveness. Indeed, no attempt was made here to compare degrees of clinical effectiveness or limiting side effects, nor to determine if all of these compounds produced the same magnitude or quality of clinical effect. The clinically effective doses shown were based only on the conclusions of the various authors. However, the question of relevance of pharmacological test procedures is a constant concern to those dealing with animal test models. Attempts at correlation and evaluation of predictability of test methodology are important, but the significant limitations on accurate quantification of clinical information must be recognized. Despite these limitations, it appears that the rat conflict-punishment procedure measures psychopharmacological properties of compounds which are related to their clinical application in psychoneurosis.

VI. EFFECT OF CHLORDIAZEPOXIDE ON PUNISHED BEHAVIOR IN SQUIRREL MONKEYS

Pharmacological attenuation of punishment-induced response suppression can also be measured in species other than rats (Cook and Catania, 1964; Morse, 1964; Hanson et al., 1967). The effects of chlordiazepoxide on punished behavior in squirrel monkeys are illustrated in Fig. 13. The characteristics of behavioral responses in this procedure are different from those used in the previous rat studies in that the food-deprived monkeys were responding on two levers, and a different schedule of reinforcement was in effect on each, concurrently. On one lever a VI 1.5-min schedule (40 food pellets/hr) for food was in effect, and on the other lever a VI 6-min schedule (10 food pellet/hr) for food was in effect. As shown in the upper left panel of Fig. 13, the lever with the higher probability of reinforcement (VI 1.5-min) produced a higher rate of responding. Response-contingent shocks were introduced, on an intermittent basis, in the VI 1.5-min schedule. Therefore, responses on the lever producing the greatest density of food reinforcements were punished intermittently. As shown in the middle left-side panel of Fig. 13, the rate of responding on this lever was suppressed, and little change in response rate occurred on the nonpunished lever. As shown in the lower left panel, chlordiazepoxide (CDP) selectively increased the response rates of punished responding (B) with no effect on the nonpunished behavioral rate (A). The dose-response effect of chlordiazepoxide is also shown in Fig. 13, and it is seen that a significant degree of alleviation of the suppressant effects of punishment occurs over a relatively wide range of doses.

VII. CONCLUSIONS AND SUMMARY

Procedures in which behavioral responses for food are punished intermittently appear to be useful in analyzing the psychopharmacological effects of a variety of types of compounds. In the rat conflict procedure described, benzodiazepines, carbamate derivatives, barbiturates, ethanol, and benactyzine are among those agents effective in alleviating behavior suppressed by

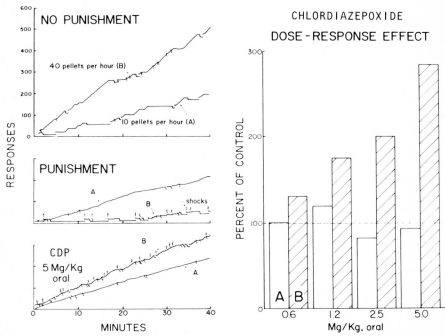

FIG. 13. Effects of chlordiazepoxide in squirrel monkeys. Left panels show traced segments of representative cumulative response records before concurrent shock punishment (top), after concurrent shock was begun (center), and after treatment with chlordiazepoxide (CDP); (bottom). Complete session typically averaged 3 hr. Responses on one lever were reinforced on a VI 1.5-min schedule (tracing B) and on another lever on a VI 6-min schedule (tracing A). Dose-related effects of chlordiazepoxide are shown in histogram on right; response rates (ordinate) are shown as percent of control, doses (abscissa) are expressed as weights of salt. Effects on nonpunished responding are shown by open bars (A), effects on punished responding by cross-hatched bars (B). See text for further explanation.

punishment contingencies. Trifluoperazine at low doses was also effective, in contrast to chlorpromazine or haloperidol, which were ineffective. Therefore, this type of behavioral procedure may be useful in differentiating effects even among compounds classified as neuroleptics. Antidepressant agents, such as imipramine and iproniazide or D-amphetamine, were ineffective, as was the antihistamine, diphenhydramine. The effects of morphine indicated that a pain-threshold elevating effect probably does not account for the increased rate of responding seen with the effective agents studied. Geller et al. (1963) also found morphine to be ineffective on punished behavior. In addition, it does not appear that loss of stimulus control or changes in motivational states, such as hunger or foot-shock threshold, play a primary role in the anticonflict effects of agents such as chlordiazepoxide. It appears that a rate-dependency effect does occur, as illustrated by the effects of chlordiazepoxide, specifically in regard to the rate of responding under the punishment conditions. The lower the control punished-response rate, the greater the effect of chlordiazepoxide in attenuating the suppressant effects of punishment. However, the differential effects of chlordiazepoxide on punished behavior, compared to nonpunished behavior, do not appear to be entirely accounted for by the rate-dependency principle. In addition, rate dependency may not entirely account for the differential effects of chlordiazepoxide over a wide range of doses, the entirely depressant effects of chlorpromazine on both low punished-response rates and higher nonpunished rates, and the effects of D-amphetamine in further decreasing the low rates of punished responding at doses which did not decrease higher nonpunished rates of responding.

The effects of chlordiazepoxide on punished behavior appear to have generality across species and specific techniques, as shown by qualitatively similar effects in the rat and squirrel monkey experiments.

Agents which have been reported to have various degrees of efficacy in psychoneurosis are effective in attenuating the suppressant effects of punishment.

The rat conflict-punishment procedure described appears to be a useful laboratory technique for identifying psychopharmacological agents such as of chlordiazepoxide and other mild tranquilizers, and the potency of certain compounds in the rat conflict procedure appears to be correlated with their clinical potency.

ACKNOWLEDGMENT

The authors gratefully acknowledge the research contributions of Charles Blum, Dixon J. Davis, John Deegan, and Charles Gill to these studies.

REFERENCES

Cook, L., and Catania, A. C. (1964): Effects of drugs on avoidance and escape behavior. *Federation Proceedings*, 23:818–835.

Cook, L., and Kelleher, R. T. (1963): Effects of drugs on behavior. *Annual Review of Pharmacology*, 3:205–222.

Davidson, A. B., and Cook, L. (1969): Effects of combined treatment with trifluoperazine-HCl and amobarbital on punished behavior in rats. *Psychopharmacologia (Berlin)*, 15:159–168.

De Silverio, R. V., Rickels, K., Raab, E., and Jameson, J. (1969): Oxazepam and meprobamate in anxious neurotic outpatients. *Journal of Clinical Pharmacology*, 9:259–263.

Dews, P. B. (1958): Studies on behavior. IV. Stimulant action of methamphetamine. *Journal of Pharmacology and Experimental Therapeutics*, 122:137–147.

Estes, W. K., and Skinner, B. F. (1941): Some quantitative properties of anxiety. *Journal of Experimental Psychology*, 29:390–400.

Geller, I. (1964): Relative potencies of benzodiazepines as measured by their effects on conflict behavior. *Archives Internationale de Pharmacodynamie*, 149:243–247.

Geller, I., Bachman, E., and Seifter, J. (1963): Effects of reserpine and morphine on behavior suppressed by punishment. *Life Sciences*, 4:226–231.

Geller, J., Kulak, J. T., Jr., and Seifter, J. (1962): The effects of chlordiazepoxide and chlorpromazine on a punishment discrimination. *Psychopharmacologia (Berlin)*, 3:374–385.

Geller, J., and Seifter, J. (1960): The effects of meprobamate, barbiturates, *d*-amphetamine and promazine on experimentally induced conflict in the rat. *Psychopharmacologia (Berlin)*, 1:482–492.

Geller, I., and Seifter, J. (1962): Effects of mono-urethans, di-urethans and barbiturates on a punishment discrimination. *Journal of Pharmacology and Experimental Therapeutics*, 136:284–288.

Gore, C. P., and McComiskey, J. G. (1961): A study of the comparative effectiveness of Librium, amylobarbitone, and a placebo in the treatment of tension and anxiety states. *Proceedings of the Third World Congress of Psychiatry (Montreal)*, Vol. 2, pp. 979–982.

Hanson, H. M., Witoslawski, J. J., and Campbell, E. H. (1967): Drug effects in squirrel monkeys trained with a multiple schedule with a punishment contingency. *Journal of the Experimental Analysis of Behavior*, 10:565–569.

Jenner, F. A., and Kerry, R. J. (1967): Comparison of diazepam, chlordiazepoxide and amylobarbitone. *Diseases of the Nervous System*, 28:245–249.

Kelleher, R. T., and Morse, W. H. (1968): Determinants of the specificity of behavioral effects of drugs. *Ergebnisse der Physiologie*, 60:1–56.

Kingstone, E., Velleneuve, A., and Kossatz, I. (1969): Double-blind evaluation of diazepam, chlordiazepoxide and placebo in non-psychotic patients with anxiety and tension: Some methodological considerations. *Current Therapeutic Research*, 11:106–114.

Lader, M. H., and Wing, L. (1965): Comparative bioassay of chlordiazepoxide and amylobarbitone sodium therapies in patients with anxiety states using physiological and clinical measures. *Journal of Neurological and Neurosurgical Psychiatry*, 28:414–425.

Lader, M. H., and Wing, L. (1966): *Physiological Measures, Sedative Drugs, and Morbid Anxiety, Maudsley Monograph No. 14*, pp. 4–27; 115–133. Oxford University Press, London.

Lorr, M., McNair, D. M., Weinstein, G. J., Michaux, W. W., and Raskin, A. (1961): Meprobamate and chlorpromazine in psychotherapy. Some effects on anxiety and hostility of outpatients. *Archives of General Psychiatry*, 4:381–389.

Morse, W. H. (1964): Effect of amobarbital and CPZ on punishment behavior in the pigeon. *Psychopharmacologia (Berlin)*, 6:286–294.

Wheatley, D. (1968): Chlordiazepoxide in the treatment of the domiciliary case of anxiety neurosis. In: *Fourth World Congress of Psychiatry, Proceedings (Madrid, 1966)*, edited by J. J. Lopez Ibor. pp. 2034–2037. International Congress Series No. 50, Excerpta Medica Foundation, Amsterdam.

The Benzodiazepines, Raven Press, New York © 1973

STUDY OF ANTIANXIETY EFFECTS OF DRUGS IN THE RAT, WITH A MULTIPLE PUNISHMENT AND REWARD SCHEDULE

A. Longoni, V. Mandelli, and I. Pessotti*

Istituto Carlo Erba per Ricerche Terapeutiche, Milan, Italy

I. INTRODUCTION

A previous study on the activity of temazepam (or methyloxazepam), carried out in our laboratory using Skinner boxes, showed that temazepam, 10 mg/kg, p.o., produces recovery of the response in rats previously trained to conditioned suppression, using a technique similar to the classic one of Estes and Skinner (1941). This finding was interpreted as an antianxiety effect of the drug at this dose. Furthermore, in an appetitive situation, using a single discrimination task, we found that the dose of 10 mg/kg, p.o., produced no significant changes in the number of correct responses, and reduced the time between the presentation of SD (discriminative stimulus) and the emission of the correct discriminative response.

On the basis of these previous results, we studied the effects of temazepam (on a stable baseline behavior) in a multiple schedule involving both punishment and reward.

II. METHODS

An electrical shock of 0.3 mA and 5-sec duration, delivered in a Sidman avoidance situation (S S 5 sec and R S 20 sec) represented the punishment. As positive reinforcement, we used water, made available at variable intervals. During the variable interval (VI) schedule, water availability was signaled by 3 sec of light from the lamp over the lever (which we call SD with limited hold 3 sec). After the preliminary training, male rats were submitted to the multiple schedule.

We alternated 15-min periods of VI with SD (box light on) in which the rat had to press the right lever with 15-min periods of the Sidman

---

* Present address: Dpto. Genetica, Fac. Medicina, Ribeirao Preto, Brasil

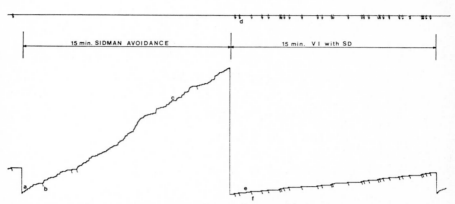

FIG. 1. Operating cumulative curve obtained in a multiple schedule with a rat working in two different schedules. The number of times the rat presses the lever is entered
on the ordinates and the time on the abscissae. a = responses made by the rat, b = shock
received, c = IRT response, d = reinforcement presented during the VI with SD
schedule, e = response during VI with SD schedule, f = reinforcement received.

avoidance schedule (signaled by darkness in the Skinner box) in which the
rat had to press the left lever. Figure 1 gives an example of a cumulative curve.
On the left, the work done by the rat during the Sidman avoidance
schedule is shown; on the right—the work during the VI with SD schedule.

Every session lasted 90 min, so that the rat had to work three times in
the Sidman avoidance schedule and three times in an appetitive situation.
Experimental sessions were conducted every day in order to reach a steady
state of performance by the animals before administration of the drug.

This multiple schedule allowed us to detect any sedative or stimulant
effects that the drugs produced and to distinguish these from the antianxiety
action, since the use of the Sidman avoidance schedule is not a sufficiently specific test for antianxiety drugs.

In the present experiment the parameters used to measure the anxiety
index were (1) rate of avoidance responses, (2) number of escape responses
(i.e., the rat's reaction speed to the shock), (3) number of responses of
avoidance and escape failures, and (4) number of intertrial responses (IRT,
i.e., the number of responses with an interval of more than 15 sec).

The parameters used to measure the motility, ability, and quickness of
the animal in reacting to the light stimulus in order to obtain water were (1)
rate of response during variable intervals with SD and (2) number of
reinforcements received in comparison with the number of SD presentations.

An antianxiety drug should produce a decrease in the rate of avoidance
response, an increase of both the number of escape responses and
number of shocks received, and an increase of IRT responses. That is, the

animal should not care about the coming punishment and should not bother to try to escape it. In addition, during the appetitive situation, the animal baseline should show no significant changes.

In order to measure the specificity of our test, we tested temazepam (a minor tranquilizer), chlorpromazine (a major tranquilizer), and phenobarbital (a barbiturate) at doses which do not induce sedation. We also tried one stimulant drug, amphetamine.

The drugs were administered only when the animal gave a stable performance for each parameter during at least 2 consecutive days, with the animals being tested every day. Therefore, the experiment is of a crossover type, since both the values were given by the same animal.

We considered a baseline stable when the variation of each parameter was not more than 10%.

III. RESULTS

The data in Table I represent the means (plus or minus standard errors) of the differences between the values found after treatment and the values before treatment. The significance given is referred to the hypothesis $\Delta = 0$ (i.e., after minus before equal 0). From the ratio between the mean and the standard error of this mean, we obtained a "T" value that gives the significance reported in this table.

In regard to the effects of temazepam, we found a reduction of activity during the Sidman avoidance schedule; in fact, a progressive increase of the number of escapes, shock, and IRT is concomitant with a reduction in the rate. This reduction is directly correlated with the increase in the doses, as is the significance. On the other hand, in a schedule of positive reinforcement, the data concerning the rate of VI at the doses 5, 10, and 20 mg/kg, p.o., show no sedative effect. Therefore, since the reduction of activity during avoidance is not due to a possible concomitant sedative effect, it may be interpreted as a specific action on the Sidman avoidance test, that is, an antianxiety effect.

The chlorpromazine effects at the doses of 1, 2, and 4 mg/kg, p.o., show no significant variation from the controls in any of the parameters used to measure the antianxiety effect.

From a comparison of the data regarding the two highest doses of temazepam and chlorpromazine, it is evident that, for the same degree of sedation, temazepam shows a decrease of activity during the Sidman avoidance test which is stronger than that of chlorpromazine.

We tried phenobarbital because we wanted to verify if its effects are similar to those of the benzodiazepines. From the results obtained, we may

TABLE 1. *Means and standard errors obtained in different parameters of multiple schedule in rats treated p.o. with different drugs at different dose levels*

Treatment	Dose mg/kg/os	No. animals	No. escapes Mean ± S.E.	No. shocks Mean ± S.E.	Av. rate Mean ± S.E.	No. tot. IRT Mean ± S.E.	No. reinf. Mean ± S.E.	VI rate Mean ± S.E.
Temazepam	5	9	3.2 ± 4.18	5.2 ± 4.0	0.3 ± 0.68	-0.6 ± 4.62	6.2 ± 3.64	1.0 ± 0.44*
"	10	8	18.4 ± 6.57*	33.1 ± 11.13*	-1.5 ± 0.74	22.2 ± 7.25*	3.1 ± 2.08	0.7 ± 0.26
"	20	8	36.0 ± 5.13**	84.1 ± 15.45**	-2.5 ± 0.45**	39.0 ± 8.22**	-5.1 ± 3.82	0.4 ± 0.38
"	40	5	52.0 ± 12.06**	106.6 ± 22.73**	-4.6 ± 0.98**	49.2 ± 5.67**	-7.2 ± 5.95	-0.7 ± 0.59
Chlorpromazine	1	5	4.4 ± 1.81	8.6 ± 3.09	-0.9 ± 0.39	16.0 ± 7.53	4.2 ± 5.99	-0.1 ± 0.20
"	2	4	5.5 ± 2.60	6.2 ± 2.17	-1.1 ± 0.24	7.7 ± 4.63	-1.2 ± 4.23	-0.3 ± 0.15
"	4	6	11.7 ± 6.68	16.7 ± 8.83	-0.8 ± 0.71	14.5 ± 5.17*	3.8 ± 2.17	-0.04 ± 0.22
"	8	4	27.0 ± 13.96*	33.2 ± 13.74*	-1.6 ± 0.37	29.7 ± 5.76*	-5.2 ± 0.75	-0.6 ± 0.10
Phenobarbital	20	5	6.2 ± 2.97	17.0 ± 4.61	-0.6 ± 0.10	9.2 ± 3.92	1.0 ± 1.52	0.5 ± 0.40
"	40	4	21.2 ± 11.93	52.7 ± 19.99*	-2.2 ± 1.24	27.2 ± 9.85*	-0.5 ± 3.93	-0.4 ± 0.38
Amphetamine	1	5	-4.4 ± 1.75	-10.0 ± 2.39*	0.3 ± 0.11	-11.0 ± 1.58*	2.4 ± 2.93	1.0 ± 0.56
"	2	4	-16.0 ± 0.41**	-19.7 ± 2.14*	2.1 ± 0.87*	-32.0 ± 5.73**	3.5 ± 2.10	0.2 ± 0.09

The above statistics express the differences between after and before the treatment values.

* Significant difference $p \leq 0.05$

** Highly significant difference $p \leq 0.01$

say that the action of this drug is similar to that of temazepam.

As regards the amphetamine effect, we found an increase of the avoidance rate. This increase is concomitant with a decrease in the number of escapes, shocks, and IRT responses. These variations are dose dependent. Concerning the variations in the parameters used to measure the change of activity, the lack of parallelism with the data relating to the antianxiety effect is not due to a sedative action, but to a concomitant stereotyped behavior that is typical of amphetamines.

In order to summarize the information given by the various parameters measured in a single value, we used an activity index for both the parameters measuring the effects in avoidance and those measuring the effects in the VI schedule. This index results from the mean of "deviations" given by each parameter. We call "deviation" (C) the difference between the mean value found and the theoretical value 0, expressed as a ratio with the standard deviation.

The equation used is the following: $\overline{C} = \dfrac{1}{p} \sum_{1}^{p} {}_i \dfrac{\overline{x}_i}{s_i}$. The variance of $\overline{C}$ is

$$V(\overline{C}) = \frac{1}{p^2} \left\{ p + \sum_{1}^{p} \frac{\overline{x}_i{}^2}{s_i{}^2} \frac{1}{2(n-1)} + \sum_{1}^{p} {}_i \sum_{1}^{p} {}_j \frac{1}{s_i s_j} s_{ij} \right\} \text{ for } j \neq i$$

where p = number of parameters, $\overline{x}_i$ = mean of ith parameter, s_i = standard deviation of ith parameter calculated with analysis of variance involving all the data obtained from the doses used, and n = number of replications.

Using this index and its standard deviation, we obtained Figs. 2 through 5 by plotting the doses on the abscissae and the mean deviations on the ordinates. Figure 2 shows the results obtained with temazepam. The function dose/antianxiety effect is given on the left of that regarding sedation. That means that the temazepam is more active as an antianxiety drug than as a sedative.

For chlorpromazine (Fig. 3), the two effects seem to be more superimposed. In this case, chlorpromazine seems to act to the same degree as a tranquilizer as it does as a sedative.

Regarding phenobarbital (Fig. 4), the situation is very similar to that of temazepam, but with a shorter distance between the two functions. This indicates a lower prevalence of the tranquilizer effect over the sedative one. We can also observe that phenobarbital was less active as a tranquilizer and as a sedative.

Figure 5 shows that the dose-effect function of amphetamine has a negative trend under the 0 point for the antianxiety effect, whereas the

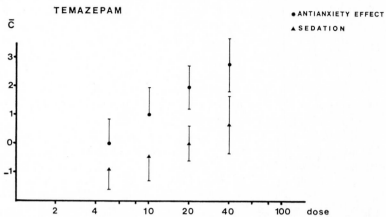

FIG. 2. Dose-effect relationship of *temazepam* administered p.o. in rats. The index values ($\overline{C}$, see text) are shown on the ordinates and the doses on the abscissae. The vertical bars represent the standard deviation.

dose-effect function relative to the activity does not seem to be dose dependent.

IV. CONCLUSIONS

By combining the Sidman avoidance test with a positive reinforcement schedule involving both activity and attentiveness, we have obtained a

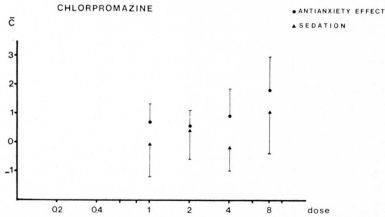

FIG. 3. Dose-effect relationship of *chlorpromazine* administered p.o. in rats. The index values ($\overline{C}$, see text) are shown on the ordinates and the doses on the abscissae. The vertical bars represent the standard deviation.

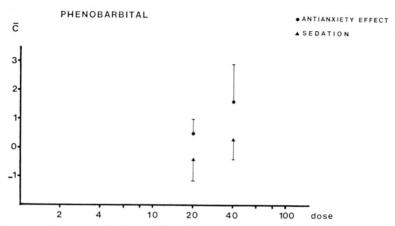

FIG. 4. Dose-effect relationship of *phenobarbital* administered p.o. in rats. The index values ($\overline{C}$, see text) are shown on the ordinates and the doses on the abscissae. The vertical bars represent the standard deviation.

specific multiple schedule which may be used for the study of antianxiety drugs. This multiple schedule offers the advantage of being able to use the Sidman avoidance test, which is very useful for provoking anxiety but which, alone, is not able to differentiate between the removal of the anxiety state and a possible sedative effect. The fact that we were able to differentiate between the four classes of drugs tested shows how specific the test is.

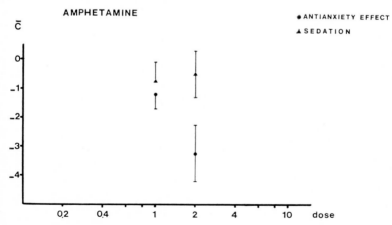

FIG. 5. Dose-effect relationship of *amphetamine* administered p.o. in rats. The index values ($\overline{C}$, see text) are shown on the ordinates and the doses on the abscissae. The vertical bars represent the standard deviation.

The index we suggest seemed to fulfil our requirements sufficiently well in that, according to the trend shown, it can indicate the type of effect exercised by the drug, at the same time condensing the various information obtained in a single figure. The standard deviation, as calculated by us, also proved to be an efficient estimate, and conformed to the deviation calculated for the individual values.

REFERENCES

Estes, W. K., and Skinner, B. F. (1941): Some quantitative properties of anxiety. *Journal of Experimental Psychology,* 29:390–400.

The Benzodiazepines, Raven Press, New York © 1973

EFFECTS OF CHLORDIAZEPOXIDE ALONE AND IN COMBINATION WITH AMPHETAMINE ON ANIMAL AND HUMAN BEHAVIOR

R. Rushton, H. Steinberg, and M. Tomkiewicz

Department of Pharmacology, University College, London, United Kingdom

I. INTRODUCTION

Chlordiazepoxide and other benzodiazepines have been widely used therapeutically and in experimental psychopharmacology for purposes similar to the use of some barbiturates. Despite a vast literature, it is still difficult to make exact comparisons between the two kinds of drugs, largely because too few investigators seem to have studied a sufficiently wide range of doses and of aspects of behavior in strictly comparable conditions. Perhaps the clearest difference to emerge so far is that benzodiazepines are behaviorally active over a much wider range of doses than barbiturates, both in animals and in man (Heise and Boff, 1962; Rushton and Steinberg, 1967; Reggiani et al., 1968; Steinberg and Tomkiewicz, 1970).

We have extensively studied chlordiazepoxide, amobarbital, cyclobarbital, and amphetamines on their own, and many combinations of amphetamines with barbiturates and with chlordiazepoxide in mice, rats, and human volunteers. All experiments were carried out under uniform laboratory conditions, with homogenous populations of both animal and human subjects, so that it should be possible to make meaningful comparisons between the effects of the drugs and between species.

Combinations of drugs were studied particularly because we have found that they can induce patterns of effects on behavior which cannot be obtained with any doses of the ingredient drugs given on their own. Combinations therefore provide a unique means of altering behavior in both subtle and gross ways.

By judiciously "titrating" doses and dose ratios of two drugs, different but orderly patterns of behavior can be consistently elicited. These effects cannot, in most instances, be predicted from the effects of the ingredient drugs even when these are known in great detail. Such patterns of effects have been found in mice (Bradley et al., 1968; Dorr et al., 1971), rats (Rushton and Steinberg, 1963, 1966, 1967; Steinberg and Tomkiewicz, 1970), human volunteers (Legge and Steinberg, 1962; Dickins et al., 1965; Besser and Steinberg, 1967), and patients (Matlin, 1963; Settel, 1963).

II. EXPERIMENTS WITH RATS

Female hooded rats were weaned at 21 days, and were then housed in groups of 16 to a cage under standard laboratory conditions and without handling. When they were approximately 120 days old, they were injected subcutaneously with various doses of amobarbital, chlordiazepoxide, dexamphetamine alone, or various combinations of dexamphetamine with amobarbital or with chlordiazepoxide. The range of doses extended from those which produced minimal effects on behavior to doses which incapacitated the animals.

All drugs were dissolved in saline, and controls received saline only. Thirty-five minutes after injection, the rats were placed singly in a symmetrical Y-shaped runway (Rushton et al., 1961), and their behavior was recorded by an observer for 5 min. Several measures of behavior were taken. One of them, the number of times the rat entered the arms of the maze with all four feet (an "entry"), was found to be one of the most robust; it is essentially a measure of the amount of coordinated walking in the maze (Rushton and Steinberg, 1963), and most of what follows will give the results obtained with "entries." Results for other measures have been described elsewhere (Rushton et al., 1961); Rushton and Steinberg, 1963, 1964; and Rushton et al., to be published).

Moderate doses of chlordiazepoxide slightly increased the number of entries, but, even with the peak stimulant dose (12.5 mg/kg), the effect was not statistically significant (Fig. 1). These slight increases were due mainly to augmented activity in the first minute or two of the trial; after this time the rats usually became inert and, sometimes, immobile (*cf.* Rushton and Steinberg, 1966). With higher doses, the number of entries was similar to that obtained with saline. But even with 400 mg/kg (not shown here), the rats were still capable of some movement, although they were obviously sedated, sluggish, flaccid, and extremely ataxic. This is different from the effects of amobarbital which at high doses completely immobilizes the animals (Fig. 2). Dexamphetamine alone had more variable effects, but the stimulation of walking was not appreciably greater than that obtained with chlordiazepoxide alone; the amphetamine-treated rats did, however, look different, and appeared "jittery" and excited, while the chlordiazepoxide and

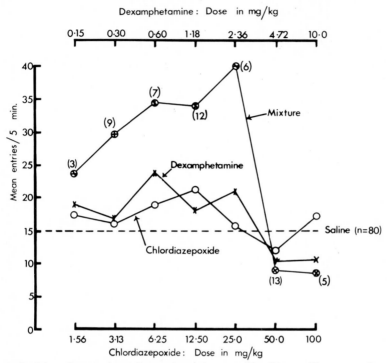

FIG. 1. Activity of rats given dexamphetamine or chlordiazepoxide separately and in combination, in a ratio of approximately 1:10 by weight. The ordinate shows the number of entries into the arms of a Y-shaped runway during a 5-min trial. Each point represents mean results for a different group of rats; the number of animals in each group is given in parentheses. The peak mixture effect was much greater than that of any dose of the separate drugs (Steinberg and Tomkiewicz, 1970).

barbiturate animals, on the whole, moved smoothly and in a way that seemed deliberate. Thus, with the doses shown in Fig. 1, neither drug given alone had any appreciable effect on the amount of walking, as measured by "entries" into the maze. But when the two drugs were given as a mixture, the amount of walking was dramatically increased. With the peak stimulant dose (25.0 mg/kg of chlordiazepoxide combined with 2.36 mg/kg of dexamphetamine), the rats' scores for entries were nearly three times as great as those of the saline controls, and also much greater than the maximum scores obtained with the ingredient doses given separately (Student's t test, $p < 0.001$ for all three comparisons).

Mixture-treated rats usually also continued to be active in this way for long periods of time; when observations were extended for up to 1 hr or

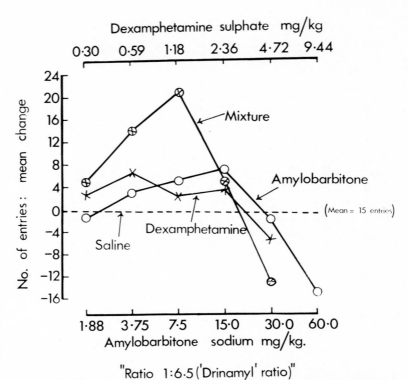

FIG. 2. Exploratory activity of rats given dexamphetamine or amobarbital (amylobarbitone) separately and in combination, in a ratio of 1:6.5 by weight, which is similar to that in a commercial preparation marketed as DRINAMYL or DEXAMYL. The ordinate shows the number of entries into the arms of a Y-shaped runway during 5 min, expressed as mean differences from the activity of the saline control group. Each point represents the mean result for a different group of eight rats. Peak mixture activity was much greater than that obtainable with either of the ingredient drugs given separately (Rushton and Steinberg, 1963).

more, no decreases were observed. With high doses of the mixtures, however, activity within the trial quickly declined (*e.g.*, Fig. 3), probably partly because the amount of amphetamine was sufficiently high to induce "stereotyped behavior," such as compulsive sniffing, head shaking, and so forth, which interfered to some extent with walking, and which chlordiazepoxide was unable to overcome (*c.f.* Rushton and Steinberg, 1963; Randrup and Munkvad, 1968; Babbini et al., 1971). Nevertheless, even with the highest doses of the chlordiazepoxide-dexamphetamine mixture shown in Fig. 1, the animals made some entries spontaneously, although they were sluggish and ataxic and moved with difficulty.

Experiments with several doses of amobarbital and of dexamphetamine, administered alone or in combination (the ratio of dexamphetamine to amobarbital was approximately 1:6.5 by weight, and was similar to a commercial preparation marked as DRINAMYL), gave results which were similar in some respects but different in others (Fig. 2). Again, moderate doses of the separate drugs had a slight stimulant effect on walking in the Y-maze. Higher doses of amobarbital depressed it. With 60 mg/kg, the rats were unable to move at all. It is interesting that with a comparable, or even higher, dose of chlordiazepoxide alone, the rats were as active as the saline controls, although the distribution of activity within the 5-min period was very different (Fig. 3). However, the dose-response curves for the drug mixtures are remarkably similar to the curves obtained with chlordiazepoxide-dexamphetamine combinations. The amounts of walking elicited by the peak dose of the mixture (1.18 mg/kg of dexamphetamine + 7.5 mg/kg of amobarbital) were much greater than those obtainable with any dose of the separate ingredients and only marginally lower than the amounts obtained with the peak dose of the chlordiazepoxide-dexamphetamine mixture shown in Fig. 1.

Thus, both amobarbital and chlordiazepoxide appeared to stimulate activity in moderate doses, and they were both capable of interacting with dexamphetamine and, in particular, of producing extremely high levels of activity; but with chlordiazepoxide a much wider range of doses was needed than with amobarbital (up to 400 mg/kg compared with 60 mg/kg of amobarbital) (Figs. 1 and 2) before activity was virtually abolished.

Figure 3 shows the results obtained with mixtures containing a constant dose of dexamphetamine (1.18 mg/kg) and four different doses of chlordiazepoxide (6.25 to 200 mg/kg), and with the constituents given separately and broken down into successive minutes of the 5-min trial. Saline-treated rats placed in the unfamiliar runway initially explored the novel environment, but this activity gradually declined, although even at the end of the 5-min trial they were still doing a little walking about in the maze. A similar trend with time occurred with dexamphetamine, but the animals were, if anything, slightly more active at the end of the trial than saline-treated animals.

In contrast, a range of doses of chlordiazepoxide induced a marked initial hyperactivity, which after a minute or two was followed by an abrupt switch to a "trance-like" immobility and a characteristic hunched posture. The animals did not appear to be asleep. They could resume walking spontaneously, and they could usually be aroused from this state by a radical change in the test environment, for example, when the enclosing Y-maze

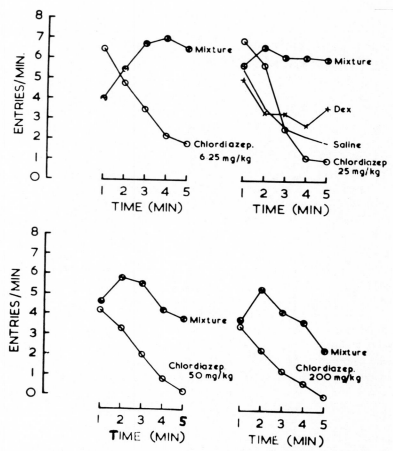

FIG. 3. Exploratory activity of rats given chlordiazepoxide alone or in combination with a constant dose of dexamphetamine (1.18 mg/kg) during successive minutes of a 5-min trial in a Y-shaped runway. The results for dexamphetamine alone and saline are shown in the top right-hand quadrant. The sharp decline in activity with chlordiazepoxide alone reflects an abrupt transition from hyperactivity to relative immobility after the first few minutes. If chlordiazepoxide is given in combination with dexamphetamine, this rapid decline is counteracted and, with several doses, activity continues at a uniformly high level (after Rushton and Steinberg, 1966).

was removed or when the animals were handled. These effects were reliable and repeatable, and they can occur in nearly all animals, although there are considerable individual differences in sensitivity to a given dose.

The immobility could be prevented and the initial hyperactivity maintained for long periods of time by administering appropriate doses of dexamphetamine together with the chlordiazepoxide (Fig. 3). It seems, therefore, that the very high levels of activity observed with drug mixtures are obtained by preventing the decline in activity which normally occurs during a trial and in a particularly exaggerated form with chlordiazepoxide. Amobarbital alone induced a much milder form of initial hyperactivity, and this may explain why amobarbital-dexamphetamine mixtures were somewhat less effective in inducing peak activity than mixtures containing chlordiazepoxide.

III. EXPERIMENTS WITH MICE

The effects described are not limited to rats tested in this particular apparatus. Figure 4 shows results which have been obtained with mice tested on a hole board (Dorr et al., 1971), The test consisted of placing naive, adult female mice singly on a wooden board with 16 evenly spaced holes and (Boissier and Simon, 1962) counting the number of times they dipped their heads into the holes during 3-min trials. Again, as was the case

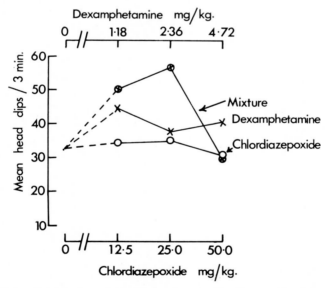

FIG. 4. Activity of mice given dexamphetamine or chlordiazepoxide alone and in combination in a ratio of approximately 1:10 by weight. The ordinate shows the number of times the mice dipped their heads into the holes on a hole board during a 3-min trial. Each point represents the mean for different groups of 8 mice given drugs and for 24 controls injected with saline. The peak mixture effect is greater than the effects of either of the constituents given separately and occurs with similar doses as in rats (Fig. 1) (Dorr et al., 1971).

with rats tested in Y-mazes, neither dexamphetamine nor chlordiazepoxide produced much effect when used alone, but the mixture resulted in a nearly twofold increase in the number of head dips. Similarly, analysis of within-trial activity showed that the mixture effect was due to prevention of the normal decline in activity. Dose-response curves, similar to our earlier amphetamine-barbiturate ones (Rushton and Steinberg, 1963) but obtained by using mice tested on hole boards instead of rats tested in Y-mazes, have been obtained by Bradley et al. (1968). Amphetamine-barbiturate mixtures have also been found to induce comparable effects in various other species and test situations as well as in man, for example, pigeons and dogs (Weiss and Laties, 1964; Rutledge and Kelleher, 1965; Joyce and Summerfield, 1967; Cooper et al., 1969; Porsolt et al., 1969, 1970; Sethy et al., 1970).

IV. EXPERIMENTS IN MAN

In normal volunteers, chlordiazepoxide alone seemed to impair performance in simple laboratory tasks. Dexamphetamine with 20 mg of chlordiazepoxide significantly improved performance in a "digit symbol substitution" test, as compared with subjects receiving placebo, and this improvement was maintained for at least 3 hr (Fig. 5). The results of other tests, such as "number copying," have been described by Besser and Steinberg (1967), and were similar although less dramatic.

Subjective effects which were recorded at the same time as test performance indicated that subjects who received the drug mixture reported feeling "happy," "efficient," "sociable," and "clear-headed" more often than controls. The incidence of undesirable subjective effects such as "anxious," "restless," and so forth was greatly reduced (Fig. 6). Findings substantially in agreement with these have previously been obtained with some barbiturate-amphetamine mixtures (Legge and Steinberg, 1962; Dickins et al., 1965). It seems possible, therefore, that the improvement in performance may have been mediated through improvement in subjective effects or mood.

These experiments seem to illustrate particular kinds of similarity between the effects of chlordiazepoxide and amobarbital, and they also show particular kinds of important differences between the two drugs. The most conspicuous difference is that chlordiazepoxide is behaviorally active over a much wider range of doses, but also, on closer inspection, there are qualitative differences which can be detected by direct observation of the animals.

The characteristic rapid transition from an initial hyperactivity to a state of relative immobility accompanied by "trance-like" postures has not, for

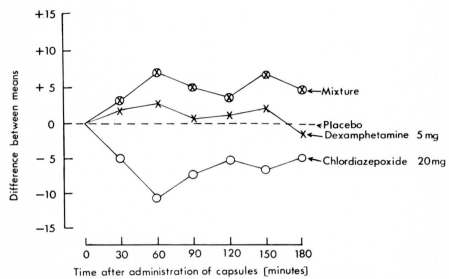

FIG. 5. Results obtained with a "digit symbol substitution" test by three groups ($n =$ 8) of male and female student volunteer subjects, who received capsules containing 5 mg of dexamphetamine, 20 mg of chlordiazepoxide, or a mixture of the two, or lactose in a double-blind crossover design, so that each subject acted as his own control. The interval between drug and placebo tests was 7 days. The number of correct substitutions during 90 sec, at different times after the administration of capsules, were counted and the results are expressed as differences between drug and placebo means. Dexamphetamine alone did not affect the level of performance, while chlordiazepoxide depressed it. Subjects given the drug mixture made a larger number of correct substitutions than those given the separate drugs (Besser and Steinberg, 1967).

example, been clearly observed with amobarbital. Further work on this is in progress (Davies and Steinberg, to be published).

As could have been expected, increasing doses of both drugs were associated with progressively greater degrees of locomotor ataxia. However, differences between the two drugs emerged when they were administered in combinations with dexamphetamine. Dexamphetamine-amobarbital mixtures tended to make the animals somewhat more ataxic than comparable doses of amylobarbitone given alone (Rushton and Steinberg, 1963). In dexamphetamine-chlordiazepoxide mixtures, the degree of ataxia was smaller than with chlordiazepoxide alone. This discrepancy was particularly noticeable with the higher doses since ataxia increased proportionately to the dose with chlordiazepoxide whereas it tended to level off with the mixtures.

The relations between ataxia and other measures of behavior tended to be rather complex. Very large numbers of entries could be obtained either

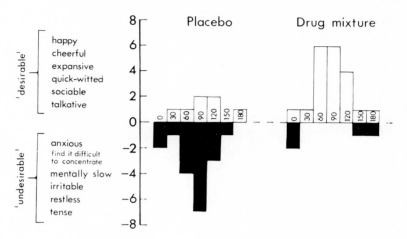

FIG. 6. Subjective effects selected from an adjective check list after administration of a placebo or a mixture of 5 mg of dexamphetamine and 20 mg of chlordiazepoxide to the same group of eight volunteers as in Fig. 5. Subjects given the placebo reported a large number of "undesirable" and few "desirable" effects, presumably because of fatigue and boredom. The drug mixture reduced the incidence of "undesirable" and greatly increased the "desirable" effects, especially at 60 and 90 min after administration (Besser and Steinberg, 1967).

with or without considerable ataxia, whereas rearing appeared to be closely correlated with the degree of ataxia; that is, if animals were ataxic, they either fell over when they attempted to rear, and/or did not even attempt it. It would seem, therefore, that walking, as measured by the number of entries, is a very robust form of behavior, which cannot readily be disrupted even by high degrees of motor incoordination. These results will be more fully described by Rushton et al. (*to be published*).

Butler, Besser and Steinberg (1968) reported that 5 mg of dexamphetamine prevented a normal decline in plasma cortisol, 20 mg of chlordiazepoxide accentuated it, and combined administration of the drugs produced a mutual cancellation of the effects in normal volunteers. However, since only one dose of the mixture and of its constitutents was used in this preliminary study, it is difficult to draw conclusions about patterns of the effects of mixtures on endocrine functions.

Our experiments have established that combinations of drugs provide a useful tool for studying the effects of drugs on behavior (*cf.* Hollister, 1969), since they are capable of producing constellations of effects, which are much more varied than those that can be obtained with the separate drugs.

Furthermore, the effects obtainable with the drug mixtures in animals may be valuable by demonstrating and possibly allowing one to predict at least some "beneficial" effects of these and similar mixtures in man.

REFERENCES

Babbini, M., Montanaro, N., Strocchi, P., and Gaiardi, M. (1971): Enhancement of amphetamine-induced stereotyped behaviour by benzodiazepines. *European Journal of Pharmacology*, 13:330–340.

Besser, G. M., and Steinberg, H. (1967): L'interaction du chlordiazépoxide et du dexamphétamine chez l'homme. *Thérapie*, 22:977–990.

Boissier, J. R., and Simon, P. (1962): La réaction d'exploration chez la souris. *Thérapie*, 17:1225–1232.

Bradley, D. W. M., Joyce, D., Murphy, E. H., Nash, B. M., Porsolt, R. D., Summerfield, A., and Twyman, W. A. (1968): Amphetamine-barbiturate mixture: Effects on the behaviour of mice. *Nature*, 220:187–188.

Butler, P. W. P., Besser, G. M., and Steinberg, H. (1968): Changes in plasma cortisol induced by dexamphetamine and chlordiazepoxide, given alone and in combination in man. *Journal of Endocrinology*, 40:391–392.

Cooper, S. J., Joyce, D., Porsolt, R. D., and Summerfield, A. (1969): Self-stimulation of the brain after administration of an amphetamine-barbiturate mixture. *British Journal of Pharmacology*, 36:192.

Davies, C., and Steinberg, H.: Trance-like states in rats treated with chlordiazepoxide. (*To be published.*)

Dickins, D., Lader, M. H., and Steinberg, H. (1965): Differential effects of two amphetamine-barbiturate mixtures in man. *British Journal of Pharmacology*, 24:14–23.

Dorr, M., Joyce, D., Porsolt, R. D., Steinberg, H., Summerfield, A., and Tomkiewicz, M. (1971): Persistence of dose-related behaviour in mice. *Nature*, 231:121–123.

Heise, G. A., and Boff, E. (1962): Continuous avoidance as a baseline for measuring behavioural effects of drugs. *Psychopharmacologia (Berlin)*, 3:264–282.

Joyce, D., and Summerfield, A. (1967): Interaction between drugs and reinforcers: Amphetamine-barbiturate mixture. *Neuropsychopharmacology*, 5:1008.

Legge, D., and Steinberg, H. (1962): Actions of a mixture of a barbiturate and amphetamine in man. *British Journal of Pharmacology*, 18:490–500.

Matlin, E. (1963): Use of chlordiazepoxide in obese patients. *Clinical Medicine*, 70:780–784.

Porsolt, R. D., Joyce, D., and Summerfield, A. (1969): Lack of tolerance to an amphetamine-barbiturate mixture and its components. *Nature*, 223:1277–1278.

Porsolt, R. D., Joyce, D., and Summerfield, A. (1970): Changes in behaviour with repeated testing under drugs: Drug-experience interactions. *Nature*, 227:286–287.

Reggiani, A., Hürlimann, A., and Theiss, E. (1968): Some aspects of the experimental and clinical toxicology of chlordiazepoxide. *Proceedings of the European Society for the Study of Drug Toxicity*, 9:79–97.

Rushton, R., and Steinberg, H. (1963): Mutual potentiation of amphetamine and amylobarbitone measured by activity in rats. *British Journal of Pharmacology*, 20:145–147.

Rushton, R., and Steinberg, H. (1964): Modification of behavioural effects of drugs by past experience. In: *Animal Behaviour and Drug Action*, edited by H. Steinberg, A. V. S. de Reuck, and J. Knight. Churchill, London, pp. 207–219.

Rushton, R., and Steinberg, H. (1966): Combined effects of chlordiazepoxide and dexamphetamine on activity of rats in an unfamiliar environment. *Nature*, 211:1312–1313.

Rushton, R., and Steinberg, H. (1967): Drug combinations and their analysis by means of exploratory activity in rats. *Neuropsychopharmacology. Proceedings of the 5th International Congress of the Collegium Internationale Neuro-Psychopharmacologicum, Washington, D. C., March 28–31, 1966,* edited by H. Brill et al. Excerpta Medica *(I.C.S. No. 129),* Amsterdam, pp. 464–470.

Rushton, R., Steinberg, H., and Tinson, C. (1961): Modification of an amphetamine-barbiturate mixture by past experience of rats. *Nature,* 192:533–535.

Rushton, R., Steinberg, H., and Tomkiewicz, M.: Patterns of spontaneous activity in a Y-maze induced by chlordiazepoxide and dexamphetamine separately and as mixtures: I. Separate drugs, II. Mixtures. *(To be published.)*

Rutledge, C. O., and Kelleher, R. T. (1965): Interaction between the effects of methamphetamine and pentobarbital on operant behaviour in the pigeon. *Psychopharmacologia (Berlin),* 7:400–408.

Sethy, V. H., Naik, P. Y., and Sheth. U. K. (1970): Effects of *d*-amphetamine sulphate in combination with CNS depressants on spontaneous motor activity of mice. *Psychopharmacologia (Berlin),* 18:19–25.

Settel, E. (1963): Combined *d*-amphetamine and chlordiazepoxide therapy in the emotionally disturbed obese patient. *Clinical Medicine,* 70:1077–1088.

Steinberg, H., and Tomkiewicz, M. (1970): Animal behaviour models in psychopharmacology. *In: Chemical Influences on Behaviour,* edited by R. Porter and J. Birch CIBA Foundation Study Group, No. 35. pp. 199–206.

Weiss, B., and Laties, V. G. (1964): Effects of amphetamine, chlorpromazine, pentobarbital, and ethanol on operant response duration. *Journal of Pharmacology and Experimental Therapeutics,* 144:17–23.

The Benzodiazepines, Raven Press, New York © 1973

ANTIANXIETY DRUGS IN CLINICAL PRACTICE

Leo E. Hollister

Veterans Administration Hospital, Palo Alto, California, and Stanford University School of Medicine, Stanford, California

I. INTRODUCTION

Antianxiety drugs are widely used, perhaps even overused, in the clinical practice of medicine. In the United States, the use of such drugs continues to increase with each passing year, the total use having doubled in the last several years. It is uncertain whether the increase is the result of the generally turbulent times which have prevailed in the past decade, or of the introduction of new drugs and their widespread promotion, or of sloppy prescribing practices of physicians.

Despite their wide popularity, antianxiety drugs continue to be controversial. Many aver that such great dependence on drug therapy for symptoms that seem to be rooted in life experience may deny patients the benefits of other treatments, particularly "psychotherapy," a more specific and permanent treatment. Although the efficacy of these drugs in relieving symptoms of anxiety is generally accepted, the difficulties in showing differences between active drugs and placebos make many doubt that the benefits of drug therapy are very great. Finally, the rapid increase of nonmedical use of mind-altering drugs has led to concern that overenthusiastic prescribing of these drugs may contribute to this problem.

II. CHEMICAL CLASSES AND PHARMACOLOGICAL PROPERTIES OF ANTIANXIETY DRUGS

A wide array of drugs claim efficacy for treating anxiety. On the one hand are those drugs which possess varying degrees of the pharmacological actions of sedative-hypnotics: sedation proceeding to hypnosis; muscle relaxation; anticonvulsant action; development of tolerance; and a potential for habituation and physical dependence. These include the barbiturates, mainly phenobarbital or amobarbital sodium; meprobamate and its cogeners; and the benzodiazepine derivatives, chlordiazepoxide, diazepam, and oxazepam.

367

On the other hand are drugs which might be termed "sedative-autonomic" which, unlike the others, affect the peripheral autonomic nervous system. They also differ in increasing muscle tone, in lowering convulsive thresholds, and in lacking the potential for habituation and physical dependence. Drugs in this class include sedative antihistaminics, such as hydroxyzine or diphen-hydramine; phenothiazine antipsychotics; and the more sedative tricyclic antidepressants, such as doxepin or amitriptyline.

Most clinicians, as well as patients, prefer drugs of the sedative-hypnotic type. Their sedation is more familiar to us, resembling that of alcohol. The sedative-autonomic drugs often create feelings of inner restlessness and mental fuzziness, as well as having bothersome side effects, such as dry mouth or blurred vision. These same effects make such drugs unlikely choices for drugs of abuse, but less acceptable as therapeutic agents. Still, some patients tolerate the side effects and gain more therapeutic benefit from these drugs than the other group (Goldstein and Brauzer, 1971).

Animal pharmacologists may complain, not without reason, about such a simple classification of these drugs. Special exception may be taken to grouping together such drugs as phenobarbital, meprobamate, and the benzo-diazepines, without taking into account the many differences that have been found in various pharmacological tests (Berger, 1968; Randall and Schallek, 1968; Irwin, 1968). The key phrase "varying degrees" is an adequate quali-fier, at least in describing those pharmacological differences of clinical im-portance. One can get into many arguments about the various similarities and differences among the pharmacological properties of sedative-hypnotic-anti-anxiety drugs, but to the clinician the similarities outweigh the differences, with exceptions to be noted below.

III. SOME PROBLEMS IN THE CLINICAL EVALUATION OF ANTIANXIETY DRUGS

A. Measurement of anxiety

Anxiety should be clearly distinguished from tension. Although the two symptoms may parallel each other, one is essentially subjective, the other objective. Thus, if we are to measure anxiety, we are generally at the mercy of what patients tell us, either by reporting their symptoms and feelings without our intervention (self-reports) or with it (interviews). In the latter situation, it is not always possible, nor necessarily desirable, to separate anxiety from tension. The quavering voice, the wet armpits, the tremulous hands, the nervous tics of our patients should not escape our notice and are certainly relevant data. Most clinical scales, however, rely on subjective data (McReynolds, 1968).

For purposes of a clinical study evaluating the efficacy of antianxiety drugs, other types of anxiety measurements are far less often used. Although projective tests may be employed to measure anxiety, their administration and interpretation is far more complex than the simpler clinical rating scales. Specialized techniques, such as verbal analysis, also have stringent clinical limitations. Physiological measurements of anxiety often leave much to be desired in that they require special equipment and often elaborate procedures. One of the more impressive attempts combined clinical, psychological, and physiological measurements of anxiety to compare chlordiazepoxide with placebo after a week of treatment (Kelly et al., 1969). Chlordiazepoxide was significantly superior to placebo in observer and self-ratings of global improvement and depression and self-ratings of anxiety (clinical measurements), decreased dizzy factor score on the Clyde Mood Scale (psychological measurement), and decreased resting and basal forearm blood flow (physiological measurement). Comparing an active drug with placebo is much simpler than comparing two active drugs. Multiple measurements in the latter case frequently show differences in either direction.

B. Selection of patients and sample size

One would think that anxious patients would be so abundant that there would be no problem in finding enough who were symptomatic that adequate drug studies could be carried out. Just because patients are abundant, however, does not necessarily mean that they are most suitable. Many anxious patients seen in physicians' offices or clinics have chronic anxiety that has been reinforced by serious job, marital, and alcohol problems, or concurrent medical illnesses. Most have already been exposed to a large variety of drugs, and the more well-to-do to some sort of psychotherapy. These patients are still the ones most often used for comparing antianxiety drugs but, when one delves closely into their social and medical problems, it seems almost unrealistic to expect them to show appreciable responses to drugs. Quite likely, many of the failures to distinguish antianxiety drugs from placebo stem from this source of negative bias.

Another reason for lack of sensitivity in distinguishing between drugs and placebo, or possibly between active drugs, is an insufficient sample size. As a general rule, the smaller the difference between treatments, the larger the sample size required for such differences to be statistically significant. Practical experience suggests that a sample size of at least 30 in each treatment group is reasonable, and that one should not miss differences of large degree given a reasonably homogeneous sample. If one has to treat 100 patients in each group to find a statistically significant difference, the actual difference involved may not be of clinical importance.

C. Experimental designs

The great variations in the course of anxiety dictate the need for controlled studies wherever possible. Most studies of antianxiety drugs employ a chronic treatment regimen, which may vary from a few weeks to a few months, with fixed doses of drugs, the latter being randomly assigned. Personally, I think this type of experimental design is not appropriate for antianxiety agents. An intriguing experimental design, to me, would be one in which drug treatments being tested would be employed for very limited periods on an *ad hoc* basis during separate episodes of anxiety. The assignment by episodes would be random, with an evaluation of relief being obtained for each episode. One might think in terms of treatment lasting no more than 5 to 7 days, and of dosage schedules which would emphasize using the major daily dose at bedtime and additional small doses during the day as needed. The treatment of successive episodes would have to be separated by no-treatment intervals of about equal length, but this procedure should not impose too great a hardship. Such a design might more closely reflect the use of these drugs in practice, and could conceivably lead to a clearer separation of efficacious drugs.

D. Other methodological approaches

Comparisons of antianxiety drugs in normal subjects are always open to the charge that the effect of major interest may be missed and that one only measures different degrees of impairment which are most likely artefacts of dose. This criticism seems valid in the light of two studies using film-provoked anxiety. College students given single doses of placebo, secobarbital, or chlordiazepoxide before viewing an anxiety-inducing film showed only sedative, but not antianxiety effects from the drugs (Pillard and Fisher, 1967). A comparison of diazepam (20 mg daily) with placebo in moderate to severe anxious patients indicated that the drug diminished several autonomic responses of anxiety induced by the film (Clemens and Selesnick, 1967). Patients were more sensitive indicators of drug action than normal subjects, at least in this experimental situation.

In general, present clinical techniques for evaluating antianxiety drugs leave much to be desired, just as do the techniques of animal pharmacological screening.

IV. EFFICACY OF ANTIANXIETY DRUGS

A. In anxiety

Considering the widespread and continuing use of antianxiety drugs, it may seem impertinent to question their efficacy, and yet a number of well-designed controlled studies have failed to show consistent differences between drug and placebo therapy of anxiety. The negative biases provided by a poor sample of patients or an insufficient sample size have already been mentioned. Fixed doses of drugs and fixed durations of treatment may also contribute a negative bias. One of the better studies of doxepin as an antianxiety drug compared it with diazepam and placebo in a large group of anxious post-alcoholic patients (Gallant et al., 1969). At the end of 4 weeks of treatment, no significant difference was found between the treatments. One would expect that anxiety would be even more transitory in this group of patients than in most, so that a briefer course of treatment might have been more sensitive to specific effects of drug as contrasted with the natural course of the symptom. Such was the case; comparisons made at 1 week showed both drugs to be better than placebo, a finding given no emphasis by the investigators. Probably most important in making it difficult to distinguish drug effects in anxious patients are the large number of metapharmacological factors, involving the patient, the physician, and the setting of treatment, which may interact with the pharmacological effects of drugs and placebo (Rickels, 1968).

Frequently, tabulations summarize the results of blind, controlled comparisons between various antianxiety drugs: in so many studies, drug *A* was better than placebo, and in so many it was not; in so many studies, drug *A* was better than drug *B*, in so many studies it was equal to drug *B*, and in so many it was inferior to drug *B*. Such tabulations make the assumption that all blind controlled trials are equally valid, which is often clearly not the case. When one considers the clinical evidence to date, one is inclined to rank the benzodiazepines ahead of meprobamate and phenobarbital, and the latter, in turn, ahead of placebo. Nonetheless, some might question if the rank order can be defined that clearly; one review of the comparative studies of chlordiazepoxide and amobarbital sodium concluded that, except for differences in potency, the drugs were approximately equally efficacious (Lader and Wing, 1966). On the other hand, another group found diazepam (15 mg per day) superior to amobarbital sodium (180 mg per day) in alleviating anxiety (McDonall et al., 1966).

As yet it is not clear if the benzodiazepines may differ among themselves. Over a series of studies involving neurotic outpatients, it was found that diazepam (15 mg per day) was preferred over chlordiazepoxide (30 mg per day) which, in turn, was preferred to amobarbital sodium (150 mg per day) (Jenner and Kerry, 1967). A comparison of oxazepam and chlordiazepoxide in various types of anxious outpatients did not demonstrate any special advantage of oxazepam in terms of rapidity of action or overall effectiveness (General Practitioner Research Group, 1967). Difficulties in detecting differences between similar types of psychotherapeutic drugs are not limited to antianxiety drugs by any means, but the difficulties are somewhat greater.

Under such conditions, it is always hazardous to assume that because no significant difference was found between two drugs, they must necessarily be equally effective. A provocative comparison of chlordiazepoxide (30 mg per day) and propranolol (90 mg per day) concluded that they were equally effective in treating anxiety, but that chlordiazepoxide was better for treating symptoms of depression and sleep disturbance. Sedative side effects were greater with chlordiazepoxide (Wheatley, 1969). Although some evidence is accumulating that propranolol has an antianxiety action, it may be premature to conclude that it is the equal of all the current drugs.

B. In depression

Quite early, meprobamate was reported to be helpful in some depressed patients (Hollister et al., 1957). A controlled comparison of meprobamate, protriptyline, a combination of the two drugs, and placebo indicated that patients improved significantly more after receiving the three drug treatments than after receiving the placebo (Rickels et al., 1967). More recent use of benzodiazepines, such as diazepam for depression, also suggested therapeutic benefits (Swenson and Gordon, 1965). The most recent study of consequence compared acetophenazine with diazepam in patients diagnosed as suffering from "anxious depressions" (Hollister et al., 1971). Previous studies by this group had established that phenothiazines were preferable to tricyclics in this most frequent type of depressive reaction. Since both groups were equally improved in the latest study, it was concluded that benzodiazepines were also effective and were probably preferred over the more hazardous phenothiazines.

V. CHOICE OF ANTIANXIETY DRUG

The extensive popularity of the benzodiazepines may stem from a few important pharmacological differences.

First, and most important, the benzodiazepines are virtually suicide-proof. Massive overdoses have resulted in very little difficulty in managing patients and no fatalities in the absence of other drugs (Davis et al., 1968). The safety of benzodiazepines is probably related to less depression of the neurogenic respiratory drive than with conventional sedative-hypnotic drugs. This same advantage is also found with the phenothiazines.

Second, metabolic tolerance to the benzodiazepines is a lesser problem than it is with meprobamate, or with phenobarbital, both of which rapidly induce their own drug-metabolizing enzymes (Conney, 1967). Although pharmacodynamic or behavioral tolerance does not appear to be a great clinical problem with the other two drugs (consider the sustained anticon-vulsant effects of phenobarbital taken chronically), it is probably minimal with the benzodiazepines. Thus, the latter drugs are less likely to lose their clinical effects upon chronic dosage.

Third, the duration of action of the benzodiazepines is somewhat longer than that of meprobamate. Most benzodiazepines have plasma half-lives in man of 24 to 48 hr, while that of meprobamate is around 12 hr (Hollister and Levy, 1964; Randall and Schallek, 1968). A prolonged duration of ac-tion is desirable in that doses need be less frequent and clinical benefits are more sustained. In this regard, the benzodiazepines resemble phenobarbital, which also persists for a long while.

Fourth, the relative lack of tolerance to the drug and long duration of action makes it a poor candidate for production of physical dependence. Although I was able to show many years ago that physical dependence to chlordiazepoxide could occur, it has generally been overlooked that it took extremes of dose and duration of treatment to produce these signs (Hollister et al., 1961). Because of its prolonged sojourn in the body, chlordiazepoxide does not produce the immediate, severe type of withdrawal reaction which follows abrupt discontinuation of meprobamate. Presumably, this reaction is related in part to the rapidity of declines of plasma and tissue concentrations of the drug. Over the years, it has been difficult to find well-documented cases of withdrawal reactions associated with clinical use of benzodiazepines. To some extent, the same is true for phenobarbital which, while as available in the black market as secobarbital sodium, is far less favored as a drug of abuse.

Finally, a number of studies from various sleep laboratories indicate that benzodiazepines produce remarkably little change in normal sleep patterns as compared with most other sedative-hypnotic drugs. Since it may be highly desirable to exploit the hypnotic effects of antianxiety drugs, one would like to minimize alterations in normal sleep patterns. Nonetheless, benzodiazepines consistently decrease the amount of time spent in slow-wave sleep, so that until the consequences of this change in sleep pattern are elucidated, their chronic use, as with other hypnotics, should be limited.

Although these factors may explain the huge preference of clinicians for the benzodiazepines, it is worth emphasizing that the patient's preference is of greater importance. I would be loathe to change a patient from a drug which he has found to be acceptable and beneficial, even though it might not be the one that I should have prescribed for him.

VI. SOME IDEAS ABOUT THE CLINICAL USE OF ANTIANXIETY DRUGS

Mies van der Rohe's famous dictum about architecture, "Less is more," applies to antianxiety drugs. Using them less may use them better. Unless physicians learn to employ these drugs with restraint, political pressures stemming from the growing problem of drug abuse may lead to unwise constraints.

First, anxiety is clearly related to life experiences. Psychotherapy and altering the environment may be more to the point. Use of drugs, at least in our present understanding, should be considered no more than symptomatic, adjunctive treatment.

Second, anxiety is often episodic, waxing and waning with changes in the patient's life. In such cases, treatment might follow the course, drugs being used only when symptoms are discomforting or disabling and not indefinitely. Such episodes of treatment might be limited to a week. If anxiety is relieved, it might remain so without drugs. The knowledge that relief is available may sustain the patient over subsequent episodes. By limiting treatment to short courses, problems of tolerance with loss of efficacy, or increased doses with the risk of physical dependence, are avoided.

Third, doses must be titrated against the patient's need for relief from symptoms and his ability to function without mental impairment. Patients vary widely in their requirements for these drugs. We have found a threefold range of plasma concentrations of meprobamate in patients receiving the same dose. We prefer to test the patient's tolerance to the drug by giving the initial doses when he is at home in the evening hours; if he gets sleepy then, little harm is done.

Fourth, traditional divided-dosage schedules of the t.i.d or q.i.d. type make little sense. Most of these drugs have rather long plasma half-lives, ranging from 12 to 30 hr, so that frequent doses are not required. Any patient so troubled by anxiety that drug treatment is required should also have trouble sleeping. By giving the major portion of the daytime dose of drug at night, one capitalizes on its hypnotic effect. The long half-life assures some daytime carryover of effects, which is precisely the type of mild sedation desired during daytime activities. *Ad hoc* doses of much smaller size may be used by the patient as his symptoms during the day may require. Many patients discover this dosage schedule for themselves and prefer it over the fixed, traditional schedules.

Fifth, these drugs must now be seriously considered as possible treatments for those common depressive syndromes in which the major attendant symptoms are anxiety, tension, and bodily complaints. Evidence so far obtained suggests that not only are the latter symptoms relieved, but concomitantly, depression is relieved as well.

No one has yet tried to exploit fully the interaction of metapharmacological influences and antianxiety drugs. From what we know already, it would seem reasonable to try to maximize the patient's belief in the treatment. On the other hand, some patients, who react to anxiety by extrovertive behavior and physical activity, tolerate sedative drugs poorly. Such patients often complain of being made more anxious by sedatives, and should be treated with smaller than usual doses, if at all.

VII. SUMMARY

Antianxiety drugs are used widely but not necessarily well. Those with more traditional sedative-hypnotic properties are usually preferred over other sedatives which also affect the peripheral autonomic nervous system. Among the sedative-hypnotic drugs, there are varying degrees of the various pharmacological effects shared by the group as a whole. The clinical evaluation of antianxiety drugs is difficult, but substantial evidence supports belief in their efficacy. In general, benzodiazepines are considered to be both the most efficacious and the safest of the various agents available. The clinical use of antianxiety drugs should be predicated on their being adjunctive treatment, on the fluctuating natural course of anxiety, on the individual dose needs of patients, on a dosage schedule designed to maximize symptomatic relief and to minimize daytime oversedation, and on their possible use in the common depressive syndromes accompanied primarily by anxiety.

REFERENCES

Berger, F. (1968): The relation between the pharmacological properties of meproba-mate and the clinical usefulness of the drug. *In: Psychopharmacology. A Review of Progress. 1957–1967*, edited by D. H. Efron, pp. 139–152. Public Health Service Publication No. 1836, U.S. Government Printing Office, Washington, D.C., pp. 139–152.

Clemens, T. L., and Selesnick, S. T. (1967): Psychological method for evaluating medi-cation by repeated exposure to a stressor film. *Diseases of the Nervous System*, 28: 98–104.

Conney, A. H. (1967): Pharmacological implications of microsomal enzyme induction. *Pharmacological Reviews*, 19:317–366.

Davis, J. M., Bartlett, E., and Termini, B. A. (1968): Overdosage of psychotropic drugs: A Review. *Diseases of the Nervous System*, 29:157–164; 246–256.

Gallant, D. D., Bishop, M. P., Guerrero-Figuero, R., Selby, M., and Phillips, R. (1969): Doxepin versus diazepam. A controlled evaluation in 100 chronic alcoholic patients. *Journal of Clinical Pharmacology*, 9:57–65.

General Practitioner Research Group (1967): Oxazepam in anxiety. *Practitioner (Lon-don)*, 199:356–359.

Goldstein, B. J., and Brauzer, B. (1971): Pharmacological considerations in the treat-ment of anxiety and depression in medical practice. *Medical Clinics of North Amer-ica*, 55:485–494.

Hollister, L. E., Elkins, H., Hiler, E. G., and St. Pierre, R. (1957): Meprobamate in chronic psychiatric patients. *Annals of the New York Academy of Sciences*, 67:789–798.

Hollister, L. E., Motzenbecker, F. P., and Degan, R. O. (1961): Withdrawal reactions from chlordiazepoxide (LIBRIUM). *Psychopharmacologia*, 2:63–68.

Hollister, L. E., and Levy, G. (1964): Kinetics of meprobamate elimination in humans. *Chemotherapia*, 9:20–24.

Hollister, L. E., Overall, J. E., Pokorny, A. D., and Shelton, J. (1971): Acetophenazine and diazepam in anxious depressions. *Archives of General Psychiatry*, 24:273–278.

Irwin, S. (1968): Antineurotics: Practical pharmacology of the sedative-hypnotics and minor tranquilizers. *In: Psychopharmacology. A Review of Progress. 1957–1967*, edited by D. H. Efron. Public Health Service Publication No. 1836, U.S. Govern-ment Printing Office, Washington, D.C., pp. 185–204.

Jenner, F. A., and Kerry, R. J. (1967): Comparison of diazepam, chlordiazepoxide and amylobarbitane (a multidose double-blind cross-over study). *Diseases of the Nervous System*, 28:245–249.

Kelly, D., Brown, C. C., and Shafter, J. W. (1969): A controlled physiological, clinical and psychological evaluation of chlordiazepoxide. *British Journal of Psychiatry*, 115: 1387–1392.

Lader, M. H., and Wing, L. (1966): *Physiological Measures, Sedative Drugs, and Mor-bid Anxiety. Maudsley Monograph No. 14*, Oxford University Press, London, pp. 1–167.

McDonall, A., Owen, S., and Robin, A. A. (1966): A controlled comparison of diaze-pam and amylobarbitane in anxiety states. *British Journal of Psychiatry*, 112:629–631.

McReynolds, P., Editor (1968): The assessment of anxiety. A survey of available techniques. *In: Advances in Psychological Assessment,* Science and Behavior Books, Palo Alto, Calif., Chapter XIII.

Pillard, R. C., and Fisher, S. (1967): Effects of chlordiazepoxide and secobarbital on film-induced anxiety. *Psychopharmacologia (Berlin),* 12:18–23.

Randall, L. O., and Schallek, W. (1968): Pharmacological activity of certain benzodiazepines. *In: Psychopharmacology. A Review of Progress. 1957–1967,* edited by D. H. Efron. Public Health Service Publication No. 1836, U. S. Government Printing Office, Washington, D.C., pp. 153–184.

Rickels, K., Editor (1968): Non-specific factors in drug therapy of neurotic patients. *In: Non-specific Factors in Drug Therapy.* Charles C. Thomas, Springfield, Ill., pp. 3–26.

Swenson, S. E., and Gordon, L. E. (1965): Diazepam: A progress report. *Current Therapeutic Research,* 7:367–391.

Wheatley, D. (1969): Comparative effects of propranolol and chlordiazepoxide in anxiety states. *British Journal of Psychiatry,* 115:1411–1412.

The Benzodiazepines, Raven Press, New York © 1973

DRUG RESPONSES IN DIFFERENT ANXIETY STATES UNDER BENZODIAZEPINE TREATMENT. SOME MULTIVARIATE ANALYSES FOR THE EVALUATION OF "RATING SCALE FOR DEPRESSION" SCORES

G. B. Cassano, P. Castrogiovanni, and L. Conti

Psychiatric Clinic, University of Pisa, Pisa, Italy

I. INTRODUCTION

Benzodiazepines alone can be given to anxious outpatients, but this is rarely possible in the treatment of inpatients. Anxiety states in psychotic or neurotic patients are usually treated with benzodiazepines in combination with major tranquilizers or antidepressants. Although there is clinical evidence that benzodiazepines have a broad therapeutic effect on anxiety symptoms, a clear-cut profile of their antianxiety effects has not yet been established as a separate entity.

In the Psychiatric Clinic of Pisa, various compounds of the benzodiazepine series were compared in several trials, in order to study the effects of these drugs on anxiety states occurring in endogenous depressive and neurotic patients, in the absence of other kinds of psychoactive treatment.

This chapter reviews our relevant experiments to date and attempts to outline the main characteristics of the response to benzodiazepine treatment, using comprehensive statistical multivariate analyses of a large sample of inpatients.

II. OBJECTIVES

The aims of the present study were (1) to compare the effects of various benzodiazepine compounds on different clinical features of anxiety; (2) to study the characteristic aspects of benzodiazepine treatment in anxious inpatients, as compared with those of major tranquillizers and antidepressants in the same population; (3) to determine if there is any correlation

379

between responses to benzodiazepines, diagnostic categories, and age; (4) to outline patterns of symptoms, which could be used to predict clinical responses to benzodiazepines; and (5) to determine what kind of treatment would offer the best combination of antianxiety effects and high drug tolerance.

III. METHODS

Some multivariate analyses, such as discriminant analysis (Veldman, 1967), factor analysis (Dixon, 1968), and multiple regression analysis (Veldman, 1967), were performed on the scores of Hamilton's "Rating Scale for Depression" (R. S. D.; 1970) administered in several drug trials. In these experiments, antidepressant drugs (Cassano et al., 1968*b*; Conti et al., 1969; Castrogiovanni et al., 1971), antipsychotic compounds (Sarteschi et al., 1972), and antianxiety compounds (Sarteschi et al., 1968; Cassano et al., 1968*a*; Riccioni et al., 1968; Castrogiovanni et al., 1968; Sarteschi et al., 1972) were investigated in anxious patients, endogenous depressives, and neurotics; no cases involving possible schizophrenic symptoms, alcohol abuse, or organic brain disease were included.

Data on patients for this study were retrieved from a computer-based data bank in which standardized clinical psychiatric records are stored, including all the data obtained from drug trials performed in the Psychiatric Clinic of Pisa.

The IBM 2741 terminal connects the Clinic with an IBM 360/67 system operating on a time-sharing basis.

It was possible to collect and store data on patients who underwent psychopharmacological experiments. Thus, data on 536 depressed or neurotic patients treated with antianxiety, antidepressant, or antipsychotic agents and scored by Hamilton's R. S. D. were retrieved and computer processed, with statistical analysis.

The types of therapy and sample sizes are reported in Table I. The scores of the changes in the psychopathological picture registered before and after treatment were used for multivariate analysis.

Discriminant analysis between groups of patients before and after treatment with various benzodiazepines, such as nitrazepam (Cassano et al., 1968*a*), oxazepam (Sarteschi et al., 1968), chlordiazepoxide (Cassano et al., 1968*b*; Castrogiovanni et al., 1968*a*), temazepam (Riccioni et al., 1968; Sarteschi et al., 1972), and medazepam (Castrogiovanni et al., 1968*b*), made it possible to carry out a comparative study of the antianxiety effects of these compounds when given to a similar population of anxious patients.

TABLE I. *Experimental sample of neurotics and endogenous depressives taken from different drug trials performed in the Psychiatric Clinic of Pisa*

Treatment	Drug	No. of patients
Benzodiazepines	nitrazepam oxazepam temazepam chlordiazepoxide medazepam	200
Antipsychotics	chlorpromazine flupenthixol droperidol	45
Antidepressants	amitriptyline imipramine doxepin Ciba 34276 BAY 1521 AF 1161 mefexamide	220
Placebo		71
	Total	536

IV. RESULTS AND COMMENTS

In Fig. 1*a*, the five groups treated with different benzodiazepine compounds are listed according to the first and second discriminant functions identified.

The experimental groups were separated mainly on the basis of pre-existing psychopathological characteristics, which showed the highest weight in the first discriminant function, such as "hypochondriasis."

The differences obtained at the beginning of the trials may be the result of some divergencies in the assessment criteria, rather than of real differences between groups of patients; the raters were not the same for all the trials, and, moreover, the trials were performed at different times.

The second function distinguished between the types of treatment on the basis of their therapeutic effect. This function reflected the spectrum of activity of these compounds, identifying the items which contribute most to a discrimination between before and after treatment, such as "insomnia," "psychic and somatic anxiety," "agitation," and "gastrointestinal symptoms." From this analysis, it appears that if a difference exists between these benzodiazepine compounds, it is not qualitative, but mainly depends on the differences in the potency of the antianxiety effect.

On the basis of these results, the benzodiazepine-treated patients were

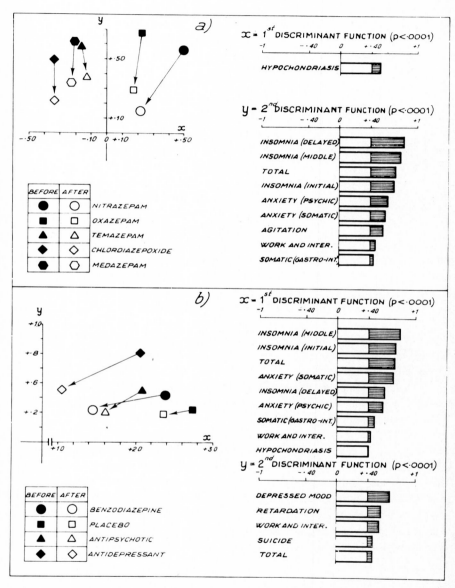

FIG. 1. Discriminant analysis between groups of neurotic and endogenous depressed patients before and after: *a*) different benzodiazepine treatments, and *b*) different treatments.

considered as a group and studied with respect to patients treated with placebo, antipsychotic, or antidepressant agents.

Before treatment, the groups investigated were discriminated mainly on the basis of the items "depressed mood" and "retardation" with positive values, and on the basis of "somatic anxiety" and "hypochondriasis" with negative values. This showed that the groups could be differentiated initially through the presence of retarded depression in those treated with antidepressants, and through the presence of neurotic symptoms of somatization in the groups treated with other drugs. In fact, the antianxiety and antipsychotic compounds and placebo were evaluated in anxiety states of neurotic and endogenous patients, whereas antidepressants were also given to subjects with retarded endogenous depressions.

The four groups before and after treatment (Fig. 1*b*) were separated mainly on the basis of improvement in "insomnia," "anxiety," "somatic symptoms," "hypochondriasis" (1st discriminant function), representing the anxiety symptoms, and by a second set of items, "depressed mood," "retardation," "work and interests," "suicide" (2nd function), representing the depressive symptomatology. The first function, measuring the changes in the anxious symptomatology, was primarily representative of therapeutic efficacy. Different treatments were distinguished at the end of the trials on the basis of the items 'insomnia," "somatic anxiety," and "hypochondriasis," and on the basis of total scores, making it possible to separate benzodiazepines, in particular, from placebo and treatment with major tranquilizers.

The items "retardation," "depressed mood," and "delayed insomnia," and the total scores appeared to separate the group treated with antidepressants from the others, particularly those treated with placebo and benzodiazepines.

The most significant difference between the effects of different kinds of treatment are given in Fig. 2: the benzodiazepines showed their typical spectrum of activity limited to "insomnia," "work and interests," "agitation," "psychic and somatic anxiety," "gastrointestinal symptoms," and "libido."

Note: in *a* on the right side, the second discriminant function which distinguishes groups before and after treatment measures the improvement and shows the spectrum of activity of benzodiazepines; the first function distinguishes between groups independently of the treatment; on the left side, the action of different benzodiazepine compounds shows only quantitative differences.

Note: in *b* on the right side, the two discriminant functions. The first corresponds to anxiety symptoms and the second to retarded depression. On the left side, benzodiazepines appear to be differentiated from antidepressant and antipsychotic compounds mainly in the second discriminant function.

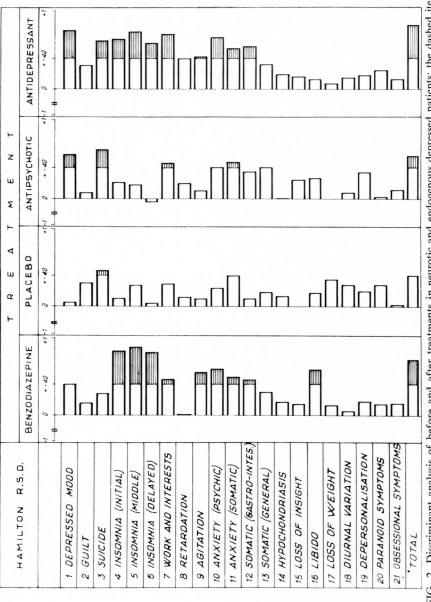

FIG. 2. Discriminant analysis of before and after treatments in neurotic and endogenous depressed patients; the dashed items are those with high loadings of the discriminant function before and after the treatments. The figure gives a clear-cut comparative feature of the different spectra of therapeutic activity of different treatments.

Antidepressant action appeared to be differentiated by the presence of depressive mood, suicidal ideas, and motor retardation, and by a general wide activity from symptoms of inhibited or agitated depression. A very limited effect on insomnia and anxiety was shown by antipsychotic treatment.

The most relevant difference between benzodiazepines and major tranquilizers in the treatment of anxious neurotics and depressive patients is that of the difference in drug tolerance, due to the greater number and severity of side effects (Table II) in patients when treated with antipsychotic agents (Sarteschi et al., 1972*b*).

TABLE II. *Number and type of side effects registered in anxious neurotic and endogenous depressive patients under benzodiazepine treatment (temazepam) or antipsychotic treatment (flupenthixol, chlorpromazine, droperidol)*

Drug	Temazepam	Flupenthixol	Chlorpromazine	Droperidol
mg/day	90.0	2.1	50.0	1.3
No. of treated patients	23	14	15	13
Side effects	No. of patients with side effects			
Asthenia	1	10	4	1
Constipation	3	1	1	3
Dry mouth	3	8	5	8
Headache		9	1	
Insomnia		8	2	5
Tremors		7	11	11
Hyperidrosis		3	4	6
Dizziness		1	1	
Accomodation disorders		1		
Tachycardia		1	1	
Salivation		1		
Nausea			2	3
Hypotension			1	

Discriminant analysis (Fig. 3*a*) made it possible to separate neurotics from endogenous depressives on the basis of their different initial psychopathological patterns and different drug responses. The first symptoms of manifest and hidden depression indicated the main improvements in anxious symptomatology under benzodiazepine treatment; in the second function, the items "depression" and "work and interests" had positive weight whereas "somatic anxiety" was negative. Therefore, endogenous depression was distinguished from neurosis mainly on the basis of depressive mood and "somatic anxiety."

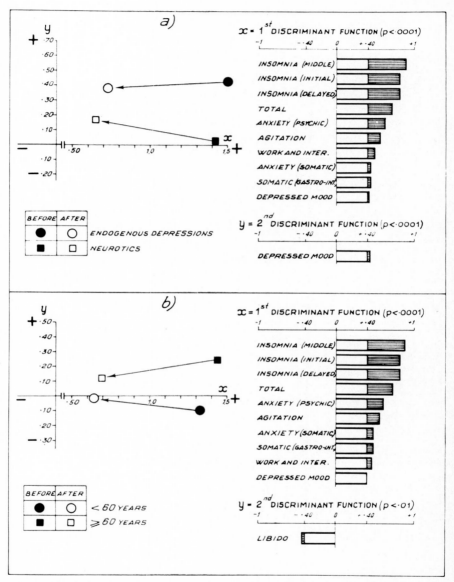

FIG. 3. Discriminant analysis between groups of patients distinguished according to: *a*) the diagnosis and *b*) the age.

Note: the first discriminant function, which measures the improvement, is practically identical independently of diagnosis and age; this shows that the benzodiazepine spectrum of activity was very similar in these four categories of patients.

Nevertheless, the effect of benzodiazepines was very similar in neurotics and endogenous depressive patients, demonstrating that the spectrum of activity of the benzodiazepines may be fairly constant in different diagnostic categories since therapeutic action concentrates selectively on "insomnia," "agitation," and "psychic anxiety" in endogenous depressives, and on "insomnia," "agitation," "psychic and somatic anxiety," and "gastrointestinal symptoms" in neurotics.

It is interesting to note that antianxiety drugs had very little effect on "depressed mood."

Comparative analysis between anxious patients over and under 60 (Fig. 3b) did not show striking differences between their responses to antianxiety treatment. "Insomnia" and "somatic and psychic anxiety" improved significantly; the older patients could be isolated on the basis of the greater improvement in depressive symptomatology and somatic symptoms of anxiety.

The patients were divided into four groups according to benzodiazepine-treatment response: stationary or worsened, improvement slight, fair, and marked. It was, therefore, possible to establish that the best response to benzodiazepine treatment was given by anxious patients, who had a high rating in their global score, guilt feelings and suicidal ideas, loss of interests and decrease in working capacity, and most severe middle insomnia. These findings were confirmed by the results of the multiple regression analysis which showed that the improvement in the total score of Hamilton's R. S. D., under benzodiazepine treatment, was closely correlated with middle insomnia and guilt ideas; it is interesting to note that, though not statistically significant, the items "motor retardation" and "paranoid symptoms" had negative weight.

Factor analysis was performed on the item scores of Hamilton's R. S. D. before and after benzodiazepine treatment. As in a previous factor analysis, it was possible to isolate three main factors at the beginning of treatment (Sarteschi et al., 1972b): the first consisted of inhibited or anxious endogenous depression; the second, "hypochondriasis" and a somatic manifestation of anxiety; and the third, "insomnia," "anxiety," and "agitation."

After antianxiety treatment, the same analysis in the same population yielded six different factors. The first was indicative of the depressive retarded and anxious symptomatology and may correspond to endogenous depressive patients who are stationary or worsened after benzodiazepine treatment.

The remaining five factors did not appear to have relevant or clear clinical meaning.

TABLE III. *Discriminant analysis between patients before and after benzodiazepine treatment (Hamilton R.S.D.)*

Item[a]	Discriminant score weight	Item[b]	Discriminant score weight
Insomnia (middle)	.88	Depressed mood	.38
Insomnia (initial)	.82	Somatic (general)	.30
Insomnia (delayed)	.80	Suicide ideas	.28
Total score	.72	Depersonalization	.18
Anxiety (psychic)	.60	Hypochondriasis	.16
Agitation	.55	Obsessional symptoms	.16
Anxiety (somatic)	.49	Guilt	.15
Work and interests	.45	Loss of insight	.14
Somatic (gastro-intest.)	.45	Paranoid symptoms	.14
		Loss of weight	.12
		Libido	.05
		Diurnal variation	.05
		Retardation	.00

[a] Items significantly influenced by antianxiety compounds.
[b] Items less influenced by antianxiety compounds.

When the effect of benzodiazepine was evaluated comparatively, by means of discriminant analysis in 200 anxious patients considered independently of age and diagnostic categories, there emerges a clear outline of the spectrum of activity which is represented by the items that have greater weight in the discriminant functions (Table III). This analysis also shows which items in the endogenous depressive and neurotic symptomatology are not influenced by benzodiazepine treatment.

V. CONCLUSIONS

(1) The benzodiazepine spectrum of activity involves "insomnia," "psychic and somatic anxiety," "agitation," and "somatic gastrointestinal symptoms" as registered by Hamilton's R. S. D.

(2) This therapeutic action is very similar; therefore it is very difficult to differentiate in the case of all these benzodiazepine compounds on the basis of qualitative items.

(3) The effect of benzodiazepines on neurotics and depressive patients may be differentiated from that of antipsychotic drugs on the basis of their effect on "insomnia" and of a stronger antianxiety effect. The benzodiazepines are more valuable than major tranquilizers for the relief of anxiety in neurotic and depressive anxious patients because of their better tolerance; in these patients, the treatment with benzodiazepines is preferable to that with "major tranquilizers" both for greater antianxiety properties and for lower frequency and severity of side effects.

(4) Benzodiazepines can be differentiated from antidepressants because the former have no effect on motor retardation and induce a very moderate improvement in depressive symptomatology.

(5) Insofar as the effect of benzodiazepines in various age groups and diagnostic categories is concerned, no striking differences were found. This supports the conclusion that antianxiety drugs exert their therapeutic effect constantly and independently of differences in age both in depressive and neurotic subjects. At the end of antianxiety treatment, the endogenous patients could be differentiated from neurotics only because of significantly high residual depression with retardation; this confirms the fact that antianxiety drugs have little or no action on the depressive-inhibited picture.

(6) The attempt to identify symptoms or sets of symptoms which would make it possible to predict different patterns of benzodiazepine response in the population of anxious patients studied showed that a better response may be expected when guilt feelings, middle insomnia, suicide ideas, and loss of work capacity are present.

VI. SUMMARY

The R. S. D. scores from 536 depressed or neurotic patients were stored in a computer data bank of the Psychiatric Clinic of Pisa and used for multivariate statistical analysis. Different treatments, such as antipsychotics and antidepressants, were compared to the effects of benzodiazepines in this sample of inpatients.

The spectrum of activity typical of benzodiazepines corresponded to "insomnia," "psychic and somatic anxiety," "agitation," and "somatic gastrointestinal symptoms," and the therapeutic action was very difficult to differentiate for various benzodiazepine compounds.

On neurotics and depressive patients, the benzodiazepine effects could be differentiated from that of antipsychotic drugs since the former appeared more valuable for the relief of anxiety and because of their better tolerance. In contrast to the antidepressant drugs, benzodiazepines did not show any significant effect on motor retardation and induced a very moderate improvement in depressive symptomatology.

Antianxiety drugs seemed to exert their therapeutic effect constantly and independently of differences in age, both in depressive and neurotic subjects. According to the multiple regression analysis in this population of anxious patients, a better response may be expected when guilt feelings, loss of work capacity, and suicide ideas are present.

REFERENCES

Cassano, G. B., Castrogiovanni, P., and Gallevi, M. (1968a): Nitrazepam (RO 4-5360) in the treatment of anxiety states, double blind trial. *Pharmakopsychiatrie und Neuro-Psychopharmakologie,* 3:195–201.
Cassano, G. B., Castrogiovanni, P., Gallevi, M., and Riccioni, R. (1968b): Sperimentazione clinica controllata di una associazione farmacologica (amitriptilina-clordiazepossido) nella terapia dell'ansia delle psicosi dipressive. In: *Atti Convegni Farmitalia "Le Associazioni di Psicofarmaci."* Edizioni Minerva Medica, Torino, pp. 93–120.
Castrogiovanni, P., Biagini, G. L., Cannella, M., Gallevi, M., and Cassano, G. B. (1968): Confronto doppio cieco tra RO 5-4556 (NOBRIUM), Clordiazepossido e placebo in pazienti ansiosi. *Neopsichiatria,* 36:1–14.
Castrogiovanni, P., Placidi, G. F., Maggini, C., Ghetti, B., and Cassano, G. B. (1971): Clinical investigation of doxepin in depressed patents. Pilot open study, controlled double blind trial versus imipramine, an all-night polygraphic study. *Pharmakopsychiatrie und Neuro-Psychopharmakologie,* 4:170–181.
Conti, L., Placidi, G. F., Biagini, G. L., Maggini, C., and Castrogiovanni, P. (1969): Studio clinico e poligrafico di un derivato del triazolo-4-3-alfa-piridina (AF 1161). *Bollettino Societa' Medico Chirurgica di Pisa,* 37:226–239.
Dixon, W. J. (1968): *BMD: Biomedical Computer Programs.* University of California Press, Berkeley and Los Angeles.
Hamilton, M. (1960): A "rating scale for depression." *Journal of Neurology, Neurosurgery and Psychiatry,* 23:56–72.
Riccioni, R., Gallevi, M., Sacchetti, G., Castrogiovanni, P., and Cassano, G. B. (1968): II temazepam: nuovo ansiolitico della serie 1-4 benzodiazepinica. Studio pilota nell'uomo. *Neopsichiatria,* 39:1–22.
Sarteschi, P., Cassano, G. B., Castrogiovanni, P., and Conti, L. (1971): Studio degli aspetti psicopatologici delle psiconevrosi e delle sindromi depressive endogene dell'età senile mediante analisi multivariate. In: *Symposio di Gerontologia,* Milano (*in press*).
Sarteschi, P., Cassano, G. B., Castrogiovanni, P., and Gallevi, M. (1968): Valutazione comparativa dell'oxazepam (Wy 3498) e del placebo su nevrotici e depressi endogeni. *Medicina Psicosomatica,* 13:3–27.
Sarteschi, P., Cassano, G. B., Castrogiovanni, P., Placidi, G. F., and Sacchetti, G. (1972): Major and minor tranquilizers in the treatment of anxiety states. *Arzneimittel-Forschung,* 22:93–97.
Veldman, D. J. (1967): *Fortran Programming for the Behavioral Sciences.* Holt, Rinehart and Winston, New York.

The Benzodiazepines, Raven Press, New York © 1973

PREDICTORS OF RESPONSE TO BENZODIAZEPINES IN ANXIOUS OUTPATIENTS

Karl Rickels

Department of Psychiatry, University of Pennsylvania, and Philadelphia General Hospital, Philadelphia, Pennsylvania

I. INTRODUCTION

Chlordiazepoxide, diazepam, and oxazepam, three benzodiazepine derivatives widely prescribed for the relief of anxious symptomatology, have been demonstrated to be significantly more effective than placebo. A number of publications report on these findings, including several studies conducted by this research group (Lorr et al., 1962; Lipman et al., 1966; Rickels and Clyde, 1967; Rickels, 1968; Klein and Davis, 1969; Hesbacher et al., 1970; Kellner, 1970). It was therefore somewhat unexpected that we were unable, in a recent study, to demonstrate significant differences in the 4-week improvement levels of 169 clinic outpatients treated with meprobamate, chlordiazepoxide, or placebo (Rickels et al., 1971b). One of the major explanations for the observed lack of active drug effects proved to be the marked influence on treatment response of the nonspecific factor "doctor warmth." In addition, patients in this study were more pretreated than patients participating in earlier studies.

The extent to which nonspecific factors affect patient response even to drugs of proven efficacy was further demonstrated to us in a study comparing diazepam, phenobarbital sodium, and placebo in 472 anxious outpatients (Hesbacher et al., 1970). Patients were treated in three settings, medical clinic, general practice, and private psychiatric practice; and while both active agents did produce significantly more improvement than placebo, the nonspecific factor "treatment setting" was found to be at least as important for treatment outcome as the active medications. This effect of treatment setting on medication response is presented graphically in Fig. 1.

Recognizing the marked influence which nonspecific factors may have on the response to benzodiazepines, we recently conducted a number of statistical analyses to determine those factors which predict improvement

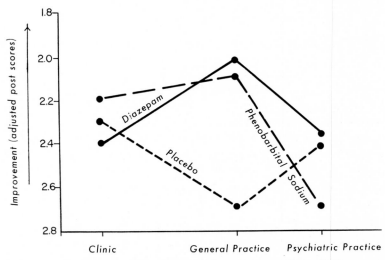

FIG. 1. Treatment setting and drug response (N = 322).

with benzodiazepines as compared to placebo and to the more sedating agents, such as phenobarbital sodium. The results of these analyses may well prove of interest both to the researcher who evaluates benzodiazepines and to the physician who prescribes them. In the present discussion, we thus plan to focus primarily on the relationship between improvement with benzodiazepines and such factors as patient demographic background, patient and physician attitudes and expectations, prognosis, illness history, treatment setting, and presenting symptomatology.

II. RESULTS AND DISCUSSION

A. Benzodiazepines—Placebo

Table I presents the results of two separate multiple-regression analyses directed toward determining significant predictors of improvement for anxious outpatients receiving chlordiazepoxide (40 mg per day) and placebo in one study (Rickels et al., 1971a), and diazepam (10 mg per day) and placebo in the other study (Downing and Rickels, *submitted*). The statistical technique employed was the nonorthogonal multiple covariance analysis (MCOV) of Uhlenhuth et al. (1969). This procedure selects from a pool of potentially relevant variables those factors which contribute most significantly to the prediction of improvement and also assesses their combined predictive effect.

As a first step in each analysis, the entire pool of variables, including their interactions with treatment, was forced into a regression equation, with a 4-week rating of global improvement as the dependent variable. The multiple correlations obtained in both analyses indicated that the total set of variables was significantly related to improvement and justified proceeding with step-search analyses.

In this search procedure, the variable contributing most to predicting the dependent variable is first selected from the pool of available predictors. Second, the variable which most improves prediction is added to the equation. The program continues to add variables in this manner until no further additions significantly increase prediction. Also, at each step, the program evaluates those predictors already selected and drops any which no longer contribute significantly to the predicted variance.

As seen in Table I, which lists those factors significantly predicting improvement,* and which also gives the direction of their predictive relationship, the chlordiazepoxide and diazepam analyses yielded rather similar results. Thus, in both analyses (see Table I-A), patients receiving the *active drug* were predicted to improve more ($p < 0.01$) than patients on placebo, and patients given a favorable *prognosis* improved more ($p < 0.01$), irrespective of treatment agent, than patients with a poor prognosis.

The variable *most suitable treatment* also emerged as a significant predictor in both analyses, although its implications for treatment response differed slightly. In the chlordiazepoxide analysis, patients for whom the drug was considered the most suitable treatment were predicted to improve more on drug and less on placebo; while in the diazepam analysis, such patients were predicted to improve more on both drug and placebo. *Time spent* with the patient, a variable which failed to enter the diazepam equation, affected both chlordiazepoxide and placebo response, but patients receiving the drug had a greater tendency to improve more when their initial interview was longer than did patients on placebo.

In regard to patient-related characteristics (Table I-B), while the two analyses did not select identical predictors, higher *education* in the chlordiaze-

---

* Additional variables used as potential predictors but not selected by our search analyses were age, marital and employment status, weight, response to previous psychiatric drugs, treatment expectations, physician comfort with and liking for the patient, and the Physician Questionnaire psychopathology items anxiety, irritability, phobia-obsession, somatic concern (hypochondriasis), appetite disturbance, and headaches.

TABLE I. *Predictors of 4-week global improvement with two benzodiazepines and placebo in anxious neurotic outpatients*

A. Treatment and physician-related variables

Selected predictors	Chlordiazepoxide $(N = 111)$	Placebo $(N = 201)$	Diazepam $(N = 105)$	Placebo $(N = 93)$
Treatment	more**	less**	more**	less**
Physician-related				
More time spent in initial interview	more*	more†		
Most suitable treatment (drug alone versus psychotherapy or drug-psychotherapy combined)	more	less	more**	
More favorable prognosis	more**		more**	

B. Patient-related and treatment setting variables

Selected predictors	Chlordiazepoxide $(N = 111)$	Placebo $(N = 201)$	Diazepam $(N = 105)$	Placebo $(N = 93)$
Patient-related				
Sex (male)	more			
Race (Caucasian)	more**	less**		
Higher education	more			
Higher income			more*	
Patient realizes problems are emotional	[more]††	[less]	more**	
Treatment desired (drug alone versus psychotherapy or drug-psychotherapy combined)			more	less
Treatment setting (general practice versus clinic or psychiatric practice)	[more]	[less]	more*	same*

C. Illness-related variables

Selected predictors	Chlordiazepoxide $(N = 111)$	Placebo $(N = 201)$	Diazepam $(N = 105)$	Placebo $(N = 93)$
Illness-related				
Duration of illness (> 6 mo.)	more*	less*		
Fewer previous drugs			more**	
Higher initial psychopathology:				
Depression	more*	less*		
Hostility			less*	
Somatization			more*	
Insomnia	more**		less*	same*

Multiple $R = 0.53$, $p < 0.001$ for the chlordiazepoxide, and 0.52, $p < 0.001$ for the diazepam analysis.

* $p < 0.05$.

** $p < 0.01$.

† $p < 0.05$; this effect more marked with drug than with placebo.

†† Brackets indicate relationships derived from zero-order correlations rather than regression analyses.

poxide group and higher *income* in the diazepam group, two highly inter-correlated social class variables, both predicted more improvement, irre-spective of treatment agent. *Race* in the chlordiazepoxide analysis, and *treatment desired* and *treatment setting* in the diazepam analysis, variables also highly intercorrelated and related to social class, emerged as significant predictors with similar implications for treatment response. White patients, patients desiring drug treatment, and general practice patients were pre-dicted to improve more on active drug and less on placebo.

The treatment setting effect observed for diazepam was supported by zero-order correlations from the chlordiazepoxide analysis. Similarly, an-other social class-related finding, namely, that patients who *realize their prob-lems are emotional* improved more on diazepam and placebo, also received zero-order correlational support from the chlordiazepoxide analysis.

These social class-related findings are illustrative of a phenomenon frequently observed in step-search procedures, namely, that variables emerg-ing as predictors in one analysis may be replaced by predictors with similar correlates in another analysis.

With regard to *sex,* males improved more than females on both chlor-diazepoxide and placebo; sex did not significantly affect response in the diazepam analysis. It should be mentioned here that findings on sex fluctuate widely and may in some cases be explained simply in terms of body weight alone. For example, in a recent study in which we were unable to differ-entiate clinically meprobamate from placebo (Rickels et al., 1971*b*), females were predicted to improve more than males (Uhlenhuth et al., 1972). One may speculate that males, being heavier than females, needed a higher drug dosage; however, it is also possible that females responded more mark-edly to the treatment situation (i.e., responded more to placebo). Similarly, in a study comparing phenobarbital to a relatively low dosage of diazepam (10 mg per day) (Downing and Rickels, *submitted for publication*), females were again observed to improve more than males. In this study, diazepam may well have been given in too low a dosage for males, while in the case of phenobarbital, given in the relatively high sedating dosage of 120 to 150 mg per day, females probably responded better than males because they were less likely than working males to be disturbed by sedative side effects.

Examining illness-related variables (Table I-C), one finds *duration of illness* entering the chlordiazepoxide equation, and *number of previous drugs,* a variable which would also seem to involve length of illness, entering the diazepam equation. The two variables have rather different implications for treatment response: with longer duration of illness, more improvement on chlordiazepoxide and less on placebo is predicted, while when fewer drugs were used previously, a better response to both diazepam and placebo is predicted. It thus seems that the more chronically ill patient, who has nevertheless received fewer psychotropic drugs, may be expected to respond rather well to benzodiazepines.

Finally, in regard to initial psychopathology (Table I-C), *insomnia* was the only area of symptomatology to emerge as a significant predictor in both analyses. While high insomnia patients were predicted to improve more in the chlordiazepoxide analysis, in the diazepam analysis, high insomnia affected drug response negatively and had no effect on placebo response. High *somatization*, on the other hand, predicted more improvement in the diazepam but not the chlordiazepoxide analysis.

The finding in the chlordiazepoxide analysis that patients higher in initial *depression* improved more with drug and less with placebo may well be considered to be representative of the relationship which generally prevails between improvement and initial level for a number of neurotic symptoms. Thus, zero-order correlations derived from the chlordiazepoxide analysis revealed that higher levels of initial anxiety, irritability, somatic concern (hypochondriasis), and headaches also predicted significantly more improvement on drug but not on placebo.

A higher level of *hostility* predicted less improvement in the diazepam analysis, but did not emerge as a predictor in the chlordiazepoxide analysis. While higher initial hostility may be regarded as a factor generally predictive of a negative response, the question arises as to whether its negative implications are, possibly, more marked for patients treated with diazepam than for patients treated with chlordiazepoxide.

In any event, our finding that higher hostility did not negatively affect the patient's global response to chlordiazepoxide supports earlier results obtained by us in several clinical trials, namely, that chlordiazepoxide, compared to placebo, reduced more markedly the patient's clinical hostility and irritability, as measured by the physician (Fig. 2) as well as by the patient. These data are, however, not consistent with the results obtained by Di Mascio and his group (Gardos et al., 1968), who reported an increase in Buss-Durkee measured hostility in highly anxious normal subjects receiving chlordiazepoxide. Differences in population, degree and type of symptomatology, and type of measuring instruments employed may well account for these inconsistencies. Some light may also be shed by a review of Clyde Mood Scale data collected by us earlier (Rickels and Clyde, 1967). We found that on chlordiazepoxide, as compared to placebo, patients became

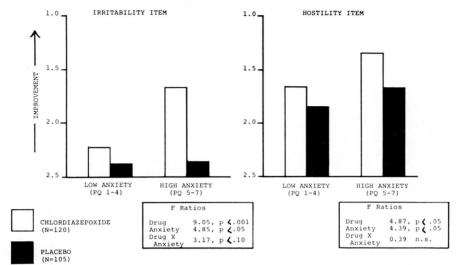

FIG. 2. Four-week improvement in physician questionnaire items as a function of initial anxiety.

more "friendly" as well as more "aggressive." It would seem that greater friendliness combined with greater aggressiveness may well denote increased, and quite possibly healthy, assertiveness rather than increased neurotic hostility or irritability.

B. Benzodiazepines—Sedatives

We should now like to consider if differential improvement can be predicted for different types of psychotropic agents. Accordingly, Table II

TABLE II. *Predictors of 4-week global improvement with chlordiazepoxide (40 mg per day) and hydroxyzine (400 mg per day)*

Selected predictors	Chlordiazepoxide ($N = 72$)		Hydroxyzine ($N = 53$)	$p <$
Treatment	more		less	0.10
Population (general practice versus clinic)		more		0.01
More favorable prognosis		more		0.05
Higher initial anxiety		more		0.01
Duration of illness ($\geq$ 1 year)	more		less	0.10
More liking of patient	more		less	0.10
Employment status (employed)	more		less	0.05

Multiple $R = 0.58$; percent predicted variance $= 34$.

presents the results of a multiple regression analysis performed for 72 patients treated with chlordiazepoxide and 53 patients treated with hydroxyzine (Rickels et al., *in press*).

As is indicated in Table II, chlordiazepoxide (40 mg per day) produced more improvement ($p < 0.10$) than hydroxyzine (400 mg per day), and patients from *general practice*, patients given more favorable *prognoses*, and patients with higher levels of initial *anxiety* were predicted to improve more on both drugs. Of greater clinical interest, however, are the three variables which predicted improvement differentially for the two agents. Patients with a longer *duration of illness*, patients *more liked* by their physician, and *employed* versus *unemployed* patients were observed to improve more on chlordiazepoxide and less on hydroxyzine. The finding on employment status, the only differential predictor to reach significance at the 5% level, is probably best explained in terms of the large number of sedative side effects produced by hydroxyzine at 400 mg per day.

The fact that *employed* patients are likely to respond poorly to a more sedating agent is seen again in Table III, which presents the results of a

TABLE III. *Predictors of global improvement with diazepam (10 mg per day) and phenobarbital (120 mg per day)*

Selected predictors	Diazepam ($N = 107$)	Phenobarbital ($N = 92$)	$p <$
Treatment setting (gen. practice versus clinic and psychiatric practice)		more	0.01
Sex (female)		more	0.05
Employment status (employed)	more	less	0.05
Duration of illness ($\leqq$ 6 mo.)	more	less	0.10
Realize problems are emotional (yes)		more	0.05
Prognosis (favorable)		more	0.01
Initial psychopathology (high)			
Hostility		less	0.01
Insomnia		less	0.01
Somatization	more	less	0.10

Multiple $R = 0.57$; percent predicted variance $= 32$.

multiple regression analysis conducted for 199 outpatients receiving either diazepam (10 mg per day) or phenobarbital (120 to 150 mg per day) (Downing and Rickels, *submitted for publication*). As mentioned earlier, the sedative side effects of phenobarbital, as well as the relatively low dosage of diazepam, also help to account for the fact that *females* in this analysis were predicted to improve more than males.

As in the chlordiazepoxide-hydroxyzine analysis, *treatment setting, prognosis,* and *duration of illness* emerged as significant predictors. Higher initial levels of *insomnia* and *hostility* were seen to have a negative effect on treatment response, while *somatization* emerged as a differential predictor. The fact that patients with higher initial somatization responded more to diazepam is probably related to the muscle-relaxant properties of this agent.

C. Chlordiazepoxide—Diazepam

The preceding discussion has concerned: (1) general predictors of improvement, irrespective of treatment agent (e.g., prognosis, realization that problems are emotional); (2) predictors of differential response to benzodiazepines and placebo (e.g., social class-related variables, treatment setting, duration of illness, initial psychopathology); and (3) predictors of differential response to the benzodiazepines, on the one hand, and to such more sedating agents as hydroxyzine (400 mg per day) and phenobarbital (120 to 150 mg per day), on the other hand (e.g., employment status, duration of illness). We would now like to focus on attempts to locate predictors which might differentiate among the different benzodiazepines.

From our data bank, we created an *artificial* study population, and using the step-search multiple regression approach discussed above (Uhlenhuth et al., 1969), compared chlordiazepoxide patients from our chlordiazepoxide prediction study (Rickels et al., 1971*a*) with diazepam patients from our diazepam prediction study (Downing and Rickels, *submitted*).

We fully realize that such a comparison can provide only very tentative results, since data on the two drugs were collected several years apart in separate placebo-controlled, double-blind studies. Yet, since the same research group conducted both studies, using identical research methodology, criterion measures, and patient-selection criteria, we felt justified in comparing both drugs in this rather preliminary way, always keeping in mind that any findings obtained would have to be confirmed in a large-scale, double-blind study comparing chlordiazepoxide with diazepam. Such a large-scale study, conducted with several hundred patients, may be planned within a factorial design employing two drugs, two diagnostic categories (anxious versus mixed anxious-depressed), and two dosage levels as factors. In addition to factorial covariance analyses, techniques such as step-search multiple regression or discriminant function should provide data of considerable clinical relevance.

Table IV presents the results of the multiple regression analysis ap-

TABLE IV. *Predictors of 4-week global improvement in anxious outpatients*

Selected predictors	Chlordiazepoxide (30 to 40 mg per day)	Diazepam (8 to 10 mg per day)	$p <$
Treatment	more	less	0.01
Patient-related:			
Race (Caucasian)	more		0.05
Employment status (employed)	less	more	0.10
Illness-related:			
Duration of illness ($>$ 6 mo.)	same	more	0.10
Prognosis (good)	more		0.001
Fewer previous drugs (0 to 1)	more		0.05
Higher initial psychopathology:			
Anxiety	more		0.10
Depression	more	less	0.05
Hostility	more	less	0.05
Hypochondriasis	more	less	0.05
Somatization	less	more	0.001

Multiple $R = 0.53$, $p < 0.001$.

plied to our *artificial* study population, and performed with global improvement after 4 weeks of treatment [scores ranging from "no improvement" (1) to "marked improvement" (5)] as an outcome criterion. Interestingly, a main *treatment* effect emerged, with chlordiazepoxide producing significantly more improvement than diazepam. This main treatment effect may well be related to the relatively low daily dosage of diazepam (8 to 10 mg per day) used, while the daily dosage of chlordiazepoxide (30 to 40 mg per day) was probably clinically a more appropriate one.

Interestingly, the present analysis confirmed our previously discussed analyses in identifying *prognosis* and *fewer previous drugs* as general predictors of improvement. *Race,* frequently a differential predictor when placebo is involved, emerged in the present analysis as a general predictor of improvement, with whites improving more than blacks.

The only indications for drug differences other than those observed in initial psychopathology, were in *employment status* and *duration of illness*; both of these findings were significant, however, only at the 10% level and thus may well represent chance occurrences. Diazepam, possibly because of lower daily dosage and consequently fewer sedative side effects, was observed to produce more improvement than chlordiazepoxide in employed patients.

Of greatest interest in the present analysis are the data on initial psy-

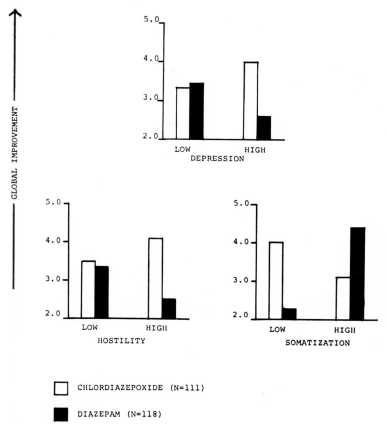

FIG. 3. Differential drug response (global improvement) as a function of initial psycho-pathology.

chopathology (see Fig. 3). Three items, *depression, hostility,* and *hypo-chondriasis,* predicted greater global improvement for chlordiazepoxide than for diazepam ($p < 0.05$). These findings may well be related to the differ-ent daily dosages in which the drugs were taken, but they also reflect the fact that a decrease in one area of symptomatology is generally accom-panied by decreases in other areas, resulting in a general reduction of neu-rotic symptomatology, similar to a "halo" effect. The finding concerning depression is of particular interest to us, since we have found in mixed anxious-depressed outpatients that chlordiazepoxide (40 mg per day) repre-sented a rather effective antidepressant (Rickels et al., 1970), while diaze-pam in dosages of 12 mg per day in one study (Rickels et al., 1969) and

20 mg per day in another (Rickels et al., *in preparation*) seemed to be slightly less effective in similar outpatients. In fact, in the latter study, phenobarbital was found to be at least as effective as diazepam.

The observation that, in contrast to the other psychopathology items, high *somatization* predicted more improvement for diazepam than for chlordiazepoxide is of particular interest. It supports findings from our diazepam-phenobarbital analysis and, as mentioned earlier, may well be related to the muscle-relaxant properties of diazepam.

The findings of the present analysis, while admittedly tentative, do seem to indicate that a comparison of two benzodiazepine derivatives may indeed allow us to find some differential predictors. Whether these drug differences are only dosage related or are, indeed, drug specific cannot be said at this time. Yet, at least in the dosages used in our research, chlordiazepoxide seemed to be more effective than diazepam in the initially sicker, anxious neurotic patients and may also, irrespective of dosage level, possess stronger antidepressant properties. Finally, drug differences, in terms of initial level of somatization, may indeed reflect valid findings related to the muscle-relaxant properties of diazepam. Whether or not these differential predictors of the clinical response of anxious neurotic outpatients to chlordiazepoxide and diazepam will hold up when more comparable drug dosages are employed has to be tested by comparing both agents in a large-scale controlled clinical trial.

TABLE V. *Summary of more consistent predictors of response to benzodiazepine treatment*

General predictors	
(Irrespective of treatment agent)	Prognosis
	Income—education
	Most suitable treatment
	Realization that problems are emotional
	Number of previous drugs
Differential predictors	
Benzodiazepines versus placebo	Time spent with patient
	Race
	Treatment setting
	Duration of illness
	Initial depression
Benzodiazepines versus sedating agents	Employment status
Chlordiazepoxide versus diazepam	Initial depression
	Initial hostility
	Initial somatization
	Initial hypochondriasis

III. SUMMARY

Step-search multiple regression analyses have enabled us to locate some nonspecific factors which appear with some consistency to predict improvement with benzodiazepine treatment (see Table V). With regard to general predictors of improvement, irrespective of the treatment agent, good response has been predicted with more favorable *prognosis,* higher *income* or *education,* the drug being considered as the *most suitable treatment,* the patient's realizing that his *problems are emotional,* and having received fewer *previous drugs.*

Greater improvement with chlordiazepoxide or diazepam than with placebo has been observed for more *time spent* with the patient, being of the Caucasian *race,* general practice as the *treatment setting,* longer *duration of illness,* and higher initial *depression.*

Better response to chlordiazepoxide or diazepam than to such sedating agents as hydroxyzine, 400 mg per day, or phenobarbital, 120 to 150 mg per day, has been predicted for being *employed.*

Finally, based on tentative results obtained with a population created from our data bank, greater improvement has been predicted with chlordiazepoxide for higher levels of initial *depression, hostility,* and *hypochondriasis,* and with diazepam for higher levels of *somatization.*

ACKNOWLEDGMENTS

This work was supported by United States Public Health Service Grants MH-08957-8. The author acknowledges with gratitude the editorial assistance provided by Miss Ellen Fisher.

REFERENCES

Downing, R. W., and Rickels, K.: Predictors of response to diazepam and placebo in anxious neurotic outpatients (*submitted for publication*).

Downing, R. W., Rickels, K., and Dreesmann, H.: Orthogonal factors *vs.* interdependent variables as predictors of drug treatment response in anxious outpatients (*submitted for publication*).

Gardos, G., DiMascio, A., Salzman, C., and Shader, R. I. (1968): Differential actions of chlordiazepoxide and oxazepam on hostility. *Archives of General Psychiatry,* 18: 757–760.

Hesbacher, P. T., Rickels, K., Hutchison, J., Raab, E., Sablosky, L., Whalen, E. M., and Phillips, F. J. (1970): Setting, patient, and doctor effects on drug response in neurotic patients. II. Differential improvement. *Psychopharmacologia (Berlin),* 18:209–226.

Kellner, R. (1970): Drugs, diagnoses, and outcome of drug trials with neurotic patients. *Journal of Nervous and Mental Disease,* 151:85–96.

Klein, D. F., and Davis, J. M. (1969): *Diagnosis and Drug Treatment of Psychiatric Disorders*. Williams and Wilkins, Baltimore.

Lipman, R. S., Park, L. C., and Rickels, K. (1966): Paradoxical influence of a therapeutic side-effect interpretation. *Archives of General Psychiatry*, 15:462–474.

Lorr, M., McNair, D. M., and Weinstein, G. J. (1962): Early effects of chlordiazepoxide (LIBRIUM) used with psychotherapy. *Journal of Psychiatric Research*, 1:257–270.

Rickels, K., and Clyde, D. J. (1967): Clyde mood scale changes in anxious outpatients produced by chlordiazepoxide therapy. *Journal of Nervous and Mental Disease*, 145: 154–157.

Rickels, K. (1968): Drug use in outpatient treatment. *American Journal of Psychiatry*, 124 (*Suppl. February*):20–31.

Rickels, K., Perloff, M., Stepansky, W., Dion, H. S., Case, W. G., and Sapra, R. K. (1969): Doxepin and diazepam in general practice and hospital clinic neurotic patients: A collaborative controlled study. *Psychopharmacologia* (*Berlin*), 15:265–279.

Rickels, K., Hesbacher, P., and Downing, R. W. (1970): Differential drug effects in neurotic depression. *Diseases of the Nervous System*, 31:468–475.

Rickels, K., Downing, R. W., and Howard, K. (1971a): Predictors of chlordiazepoxide response in anxiety. *Clinical Pharmacology and Therapeutics*, 12:263–273.

Rickels, K., Lipman, R. S., Park, L. C., Covi, L., Uhlenhuth, E. H., and Mock, J. E. (1971b): Drug, doctor warmth, and clinic setting in the symptomatic response to minor tranquilizers. *Psychopharmacologia* (*Berlin*), 20:128–152.

Rickels, K., Csanalosi, I., Downing, R. W., and Hesbacher, P. T.: Prognostic indicators of 4 week clinical improvement with chlordiazepoxide and hydroxyzine in anxious neurotic outpatients. *International Pharmacopsychiatry* (*in press*).

Rickels, K., Chung, H. R., Feldman, H. S., Gordon, P. E., Kelly, E. A., and Weise, C. C.: Amitriptyline, diazepam, and phenobarbital sodium in depressed outpatients (*in preparation*).

Uhlenhuth, E. H., Duncan, D. B., and Park, L. C. (1969): Some non-pharmacologic modifiers of the response to imipramine in depressed psychoneurotic outpatients: A confirmatory study. In: *Psychotropic Drug Response: Advances in Prediction*, edited by P. R. A. May and J. R. Wittenborn. Charles C. Thomas, Springfield, Ill., pp. 155–197.

Uhlenhuth, E. H., Covi, L., Rickels, K., Lipman, R. S., and Park, L. C. (1972): Predicting the relief of anxiety with meprobamate: An attempt at replication. *Archives of General Psychiatry*, 26:85–91.

The Benzodiazepines, Raven Press, New York © 1973

ACTIVITY OF BENZODIAZEPINES ON AGGRESSIVE BEHAVIOR IN RATS AND MICE

Luigi Valzelli

Istituto di Ricerche Farmacologiche "Mario Negri," Milan, Italy

I. INTRODUCTION

Chordiazepoxide was the first of a new class of psychotropic drugs showing a primarily antianxiety effect in human subjects (Harris, 1960; Tobin and Lewis, 1960). Even though no specific animal analogs of human anxiety exist, the pharmacological profile of antianxiety agents in animals is said to include taming, central muscle-relaxant activity, electroshock and chemical anticonvulsant activity, attenuation of fear in avoidance and conflict behavior, and increased food intake (Randall and Schallek, 1968).

As early as 1960, chlordiazepoxide was shown to have a taming effect on vicious monkeys and various wild zoo animals (Heuschele, 1961; Randall, 1960, 1961), and Heise and Boff (1961) devised a checklist for behavior to differentiate between aggression and motor activity. Chlordiazepoxide was found to have a safety margin of 20 between the antiaggressive dose and the ataxic dose, and further observations by Randall et al. (1961) showed a similar potency and characteristics for diazepam.

In general the relative taming effect in monkeys parallels the muscle-relaxant and anticonvulsant effects in mice. However, diazepam was stronger than chlordiazepoxide in these mice tests but only equal in the monkey-taming activity, whereas oxazepam was much weaker in mice (Randall et al., 1965). Diazepam is similar in potency to chlordiazepoxide in taming rhesus monkeys (Klupp and Kähling, 1965).

In fact, it is still difficult to solve the problem of defining a psychoactive drug as being antiaggressive because of the complexity of the mechanisms of spontaneous or induced aggressive behavior in animals (Valzelli, 1967; Moyer, 1968). However, a number of recent papers (Valzelli et al., 1967; Malick et al., 1969; Sofia, 1969; Christmas and Maxwell, 1970; Langfeldt and Ursin, 1971; Valzelli and Bernasconi, 1971) deal with the antiaggressive activity of benzodiazepine derivatives, with certain conflicting results.

The present study represents an attempt to characterize the activity of some benzodiazepines on aggressive behavior induced in male mice and rats by prolonged isolation, as described elsewhere (Yen et al., 1959; Valzelli, 1969a; Valzelli and Bernasconi, 1971).

II. MATERIALS AND METHODS

Male Swiss albino mice (approximately 20 ± 2 g) were isolated in single Makrolon cages (internal size $= 21 \times 15 \times 13$ cm) for 4 weeks, at a constant room temperature of 22°C and 60% relative humidity, and fed *ad libitum*. To test aggressive behavior, the animals were grouped in threes, and their aggressiveness was evaluated at different times after intraperitoneal administration of saline or drug. The degree of aggressiveness was scored as reported elsewhere (Valzelli, 1967; Valzelli et al., 1967).

The reproducibility of the results was checked on a blind basis. At the end of the isolation period, the animals were constantly aggressive in 100% of the cases, with a score of 100.

The rotarod test was performed in both normal and aggressive mice, training the animals to maintain themselves on a slowly revolving rod (5 rpm) for at least 3 min.

Wistar male rats (220 ± 10 g) were isolated in single Makrolon cages (internal size $= 14 \times 24 \times 41$ cm) for a period of 6 weeks, at a constant room temperature of 22°C and 60% relative humidity, and fed *ad libitum*. The development of the different behavioral profiles was followed during the entire period of isolation, by putting a naive normal mouse in contact with the isolated rat and evaluating the "muricide," "indifferent," or "friendly" behavior according to the score shown in Table I.

The "muricide" reaction, by which the rat kills the mouse by breaking its neck, was obtained in our experimental conditions within 10 ($\pm$ 3) min. The term "indifferent" was adopted to indicate the rat's complete lack of interest in the presence of a mouse in the cage. This behavior is quite different from the normal for rats not submitted to isolation. The so-called "indifferent" behavior might be considered as an expression of deep emotional blockade, preventing the development of any oriented behavior toward the intruder animal placed in the cage.

The behavior of those rats defined as "friendly," for lack of a more adequate definition, is more complex; they exhibit an active behavior toward the mouse, reminiscent of animals playing among themselves, with some inclination to nurse, e.g., the preparation of a nest in which they put the mouse,

TABLE I. *Different types of behavior shown by rats, submitted to 6 weeks of isolation, toward a mouse placed in their cage*

Behavioral patterns	% of rats showing pattern observed			
	N	M	F	I
Approaching	100	100	100	4
Sniffing	100	20	100	4
Licking	56	0	100	0
Grooming	15	0	64	0
Nest preparing	0	0	22	0
Picking up	0	0	91	0
Carrying around	0	0	87	0
Joking	18	0	65	0
Jumping	0	10	12	0
Incoordinated hyperactivity	0	0	34	0
Tremor	0	0	12	0
Piloerection	2	100	0	0
Digging	0	85	0	0
Compulsive eating and drinking	0	16	2	67
Attacking	0	100	0	0
Killing	0	100	0	0

Observations made by two observers in 800 rats.
N = normal, M = muricide, F = friendly, and I = indifferent rats.

Isolation produces approximately 30% "muricide," 25% "friendly," and 45% "indifferent" animals (Valzelli and Garattini, 1972).

All the drugs employed were administered intraperitoneally both in isolated mice and rats, and the effect was followed for a period of 48 hr; as reported elsewhere for aggressive mice (Valzelli et al., 1967), saline or other solvents employed did not affect the different behavioral patterns (Table IV).

Extraction of brain serotonin (5-HT) and 5-hydroxyindolacetic acid (5-HIAA) was performed simultaneously in the same brain-tissue sample, according to Giacalone and Valzelli (1969), and determined spectrofluorometrically by means of an Aminco-Bowman apparatus. Brain 5-HT turnover was calculated according to Tozer et al. (1966).

III. RESULTS AND DISCUSSION

Considering the results obtained in mice on the rotarod performance and aggressive behavior following the administration of different benzodiazepine derivatives, it seems evident that the antiaggressive activity shown by some of these compounds is not always related to their possible muscle-

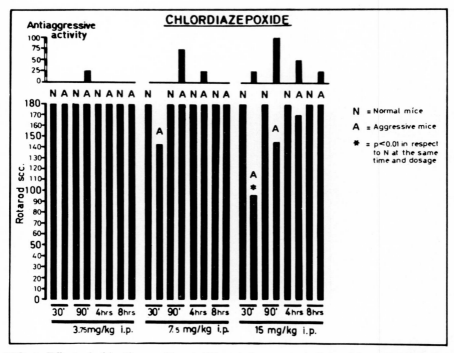

FIG. 1. Effect of chlordiazepoxide at different dosages and time of i.p. administration on rotarod performance in normal (N) and aggressive (A) mice, and on aggressive behavior of mice.

relaxant effect (Figs. 1–7). In fact, for instance, chlordiazepoxide (Fig. 1) shows a good antiaggressive activity at a dose (7.5 mg/kg, i.p.) at which no muscular impairment is present; moreover, when this last aspect becomes evident (15 mg/kg, i.p.), it is without effect on the antiaggressive properties of this compound, which instead shows its maximal antiaggressive efficacy when the muscular impairment becomes no longer significant. Even more evident is the lack of a relationship between muscular impairment and antiaggressive activity in the case of oxazepam (Fig. 2) and of medazepam (Fig. 3). Medazepam is strongly antiaggressive and at the same time absolutely inactive on the muscular strength of animals at the dosages employed.

Instead, the antiaggressive activities of diazepam (Fig. 4), nitrazepam (Fig. 5), N-methyloxazepam (Fig. 6), and N-demethyl-diazepam (Fig. 7) seem to be related to their muscle-relaxant properties. A similar picture is typical for bromazepam, not reported here, whereas *o*-chloroxazepam, a very potent anticonvulsant agent (Bell et al., 1968), does not show any anti-aggressive activity.

Another interesting observation is that the benzodiazepines employed in these experiments show a different muscle-relaxing activity in aggressive versus normal mice. Most typical in this context are the effects of diazepam (Fig. 4), N-demethyl-diazepam (Fig. 7), N-methyloxazepam (Fig. 6), and nitrazepam (Fig. 5); these results are consistent with those previously obtained on the different activities of certain psychoactive drugs in normal and aggressive mice (Consolo et al., 1965*a, b*; Valzelli, 1969*b*, 1971; Valzelli and Bernasconi, 1971).

From the neurochemical point of view, neither chlordiazepoxide nor medazepam, administered in dosages which exert the maximal antiaggressive effect (15 mg/kg, i.p.; Figs. 1 and 3), is able to modify significantly the brain content of 5-HT and 5-HIAA (Tables II and III).

Chlordiazepoxide and diazepam were described as inhibiting the behav-

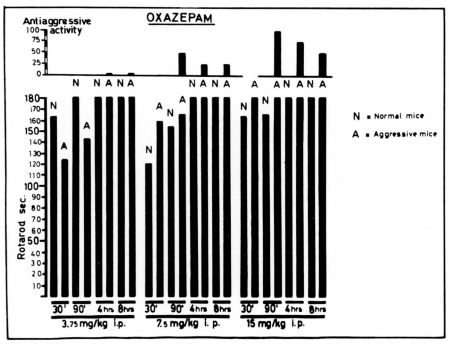

FIG. 2. Effect of oxazepam at different dosages and time of i.p. administration on rotarod performance in normal (N) and aggressive (A) mice and on aggressive behavior of mice.

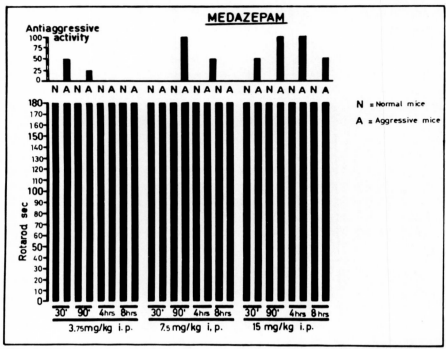

FIG. 3. Effect of medazepam at different dosages and time of i.p. administration on rotarod performance in normal (N) and aggressive (A) mice and on aggressive behavior of mice.

ior of rats made aggressive by surgical brain lesions (Malick et al., 1969; Sofia, 1969) or by painful electric shock (Christmas and Maxwell, 1970), as well as to block the "muricide" reactions of rats genetically selected for this behavior (Horovitz et al., 1966).

TABLE II. *Activity of chlordiazepoxide (15 mg/kg, i.p.) on brain 5-HT and 5-HIAA levels of normal (N) and aggressive (A) mice*

Time after administration	5-HT (γ/g $\pm$ S.E.)		5-HIAA (γ/g $\pm$ S.E.)	
	N	A	N	A
0	0.61 ± 0.02	0.58 ± 0.03	0.35 ± 0.01	0.28 ± 0.02
30 min	0.68 ± 0.05	0.63 ± 0.02	0.34 ± 0.02	0.29 ± 0.03
1 hr	0.63 ± 0.03	0.65 ± 0.03	0.35 ± 0.01	0.31 ± 0.02
2 hr	0.64 ± 0.05	0.60 ± 0.02	0.33 ± 0.03	0.30 ± 0.01
4 hr	0.61 ± 0.03	0.59 ± 0.03	0.31 ± 0.02	0.31 ± 0.02

Each figure corresponds to 10 animals.

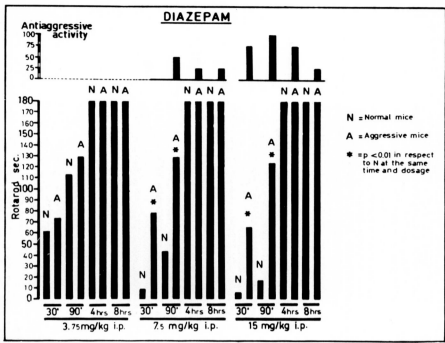

FIG. 4. Effect of diazepam at different dosages and time of i.p. administration on rotarod performance in normal (N) and aggressive (A) mice and on aggressive behavior of mice.

By contrast, the benzodiazepines seem to have no effect on Wistar rats which become "muricide" rats after prolonged isolation, (Table IV): only Sofia (1969) observed, after prolonged isolation of another strain of rats, an antimuricidal activity of chlordiazepoxide and diazepam at a dose level

TABLE III. *Activity of medazepam (15 mg/kg, i.p.) on brain 5-HT and 5-HIAA levels of normal (N) and aggressive (A) mice*

Time after administration	5-HT (γ/g $\pm$ S.E.)		5-HIAA (γ/g $\pm$ S.E.)	
	N	A	N	A
0	0.59 ± 0.02	0.58 ± 0.03	0.33 ± 0.01	0.30 ± 0.02
30 min	0.58 ± 0.03	0.61 ± 0.02	0.36 ± 0.01	0.32 ± 0.03
1 hr	0.53 ± 0.04	0.69 ± 0.04	0.32 ± 0.01	0.34 ± 0.03
2 hr	0.59 ± 0.03	0.55 ± 0.03	0.31 ± 0.01	0.28 ± 0.01
4 hr	0.59 ± 0.03	0.64 ± 0.04	0.38 ± 0.03	0.33 ± 0.02

Each figure corresponds to 10 animals.

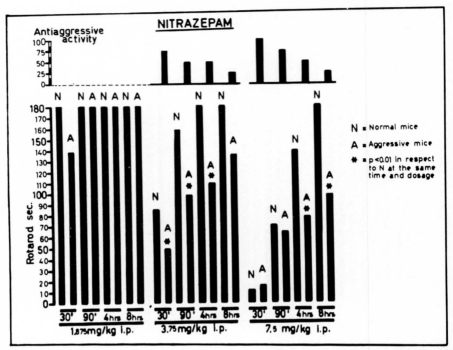

FIG. 5. Effect of nitrazepam at different dosages and time of i.p. administration on rotarod performance in normal (N) and aggressive (A) mice and on aggressive behavior of mice.

which definitely impairs the rotarod performance of the animals.

A certain percentage of the "indifferent" rats (Tables I and IV) respond to chlordiazepoxide administration with a "muricide" reaction. This observation is not easily explained but it could be related in some way to the paradoxical effect which benzodiazepines exert in animals living together rather than in isolation. In fact, it has recently been described by several authors (Fox and Snyder, 1969; Fox et al., 1970, 1971; Guaitani et al., 1971) that benzodiazepines increase spontaneous aggression and become toxic in grouped male mice. Other recent studies report hostile-aggressive tendencies among human subjects chronically treated with some benzodiazepine derivatives (Gardos et al., 1968; Di Mascio et al., 1969; Salzman et al., 1969; Di Mascio, 1971).

It is important to emphasize that in mice the hostile reaction caused by benzodiazepines takes place only in grouped male mice and only in the

TABLE IV. *Effect of some benzodiazepine derivatives on aggressive mice (A) and and "muricide" rats (M)*

Treatment	Dose (ml or mg/kg i.p.)	% inhibition or aggressive behavior	
		A	M
Saline	1.0	0	0
Tween 80 (2% sol.)	1.0	0	0
Carboxymethylcellulose (0.5% sol.)	1.0	0	0
Chlordiazepoxide	5.0	75 (2)	0
	7.5	100 (2)	0
	15.0	100 (3)	0*
Diazepam	7.5	50 (1½)	0
	10.0	100 (2)	0
	15.0	100 (3)	0
Oxazepam	7.5	50 (2)	0
	15.0	100 (4)	0
Medazepam	7.5	100 (1½)	0
	15.0	100 (5)	0
	25.0	100 (7)	0
	40.0	—	0

* 20% of indifferent rats become "muricide."
Each figure corresponds to 30 animals. Duration (hr) of the observed effect given in parentheses.

case of oral administration; chronic injection of the derivates is without effect (A. Guaitani, *to be published*). It is a curious phenomenon that this situation is almost the opposite to that of induced aggressiveness, in which the antiaggressive activity of benzodiazepines takes place in isolated mice and shows the best results by parenteral administration. On the other hand, a common trait is represented by the fact that the two opposite effects occur only in male mice and, in this connection, it could be of interest to determine if the benzodiazepine-induced hostility takes place indiscriminately in all strains of mice or only in those strains which were shown to react to isolation with aggressive behavior (Valzelli and Garattini, 1968).

In the absence of more precise information about biochemical or metabolic data responsible for these opposite effects of benzodiazepines, we can only conjecture that, at least in laboratory animals, the effect depends on the sex and the route of administration, and appears to be modulated by the social setting. This last point seems to support further the hypothesis of the role played by the social environment in reactions to drugs (Balazs et al., 1962; Consolo et al., 1965a, b; Rushton and Steinberg, 1966; Baumel et al., 1969; Katz and Steinberg, 1970; Porsolt et al., 1970).

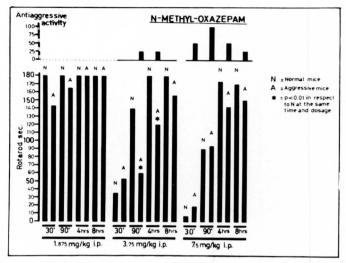

FIG. 6. Effect of N-methyl-oxazepam at different dosages and time of i.p. administration on rotarod performance in normal (N) and aggressive (A) mice and on aggressive behavior of mice.

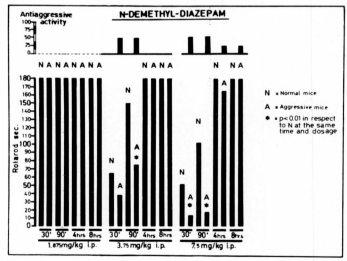

FIG. 7. Effect of N-demethyl-diazepam at different dosages and time of i.p. administration on rotarod performance in normal (N) and aggressive (A) mice and on aggressive behavior of mice.

VI. SUMMARY

The antiaggressive and muscular relaxing activity of benzodiazepines has been studied in both normal and aggressive mice and rats. The antiaggressive efficacy shown by some of these compounds appears clearly unrelated to their possible muscle-relaxing effect.

The different response to benzodiazepines, as to other psychoactive drugs, of isolated aggressive or normal grouped animals seems to be due to the changes of the biochemical and emotional characteristics of the experimental animals, induced by an alteration of their normal social environment, as reported also for human beings.

REFERENCES

Balazs, T., Murphy, J. B., and Grice, H. C. (1962): The influence of environmental changes on the cardiotoxicity of isoprenaline in rats. *Journal of Pharmacy and Pharmacology*, 14:750–755.

Baumel, I., De Feo, J. J., and Lal, H. (1969): Decreased potency of CNS depressants after prolonged isolation in mice. *Psychopharmacologia* (Berlin), 15:153–158.

Bell, S. C., McCaully, R. J., Gochman, C., Childress, S. J., and Gluckman, M. I. (1968): 3-Substituted 1,4-benzodiazepin-2-ones. *Arzneimittel-Forschung,* 11:457–461.

Christmas, A. J., and Maxwell, D. R. (1970): A comparison of the effects of some benzodiazepines and other drugs on aggressive and exploratory behavior in mice and rats. *Neuropharmacology*, 9:17–29.

Consolo, S., Garattini, S., and Valzelli, L. (1965a): Amphetamine toxicity in aggressive mice. *Journal of Pharmacy and Pharmacology*, 17:53–54.

Consolo, S., Garattini, S., and Valzelli, L. (1965b): Sensitivity of aggressive mice to centrally acting drugs. *Journal of Pharmacy and Pharmacology*, 17:594–595.

Di Mascio, A. (1971): The effects of benzodiazepines on aggression: Reduced or increased? (This Volume)

Di Mascio A., Shader, R. I., and Harmatz, J. (1969): Psychotropic drugs and induced hostility. *Psychosomatics*, 10:47–50.

Fox, K. A., and Snyder, R. L. (1969): Effect of sustained low doses of diazepam on aggression and mortality in grouped male mice. *Journal of Comparative and Physiological Psychology*, 69:663–666.

Fox, K. A., Tuckosh, J. R., and Wilcox, A. H. (1970): Increased aggression among grouped male mice fed chlordiazepoxide. *European Journal of Pharmacology*, 11:119–121.

Fox, K. A., Webster, J. C., and Guerriero, F. J. (1971): Increased aggression among grouped male mice fed nitrazepam and flurazepam. *European Journal of Pharmacology* (in press).

Gardos, G., Di Mascio, A., Salzman, C., and Shader, R. I. (1968): Differential actions of chlordiazepoxide and oxazepam on hostility. *Archives of General Psychiatry*, 18:757–760.

Giacalone, E., and Valzelli, L. (1969): A spectrofluorometric method for the simultaneous determination of 2-(5-hydroxyindol-3-yl) ethylamine (serotonin) and 5-hydroxyindol-3-yl-acetic acid in the brain. *Pharmacology*, 2:171–175.

Guaitani, A., Marcucci, F., and Garattini, S. (1971): Increased aggression and toxicity in grouped male mice treated with tranquilizing benzodiazepines. *Psychopharmacologia* (Berlin), 19:241–245.

Harris, T. H. (1960): Methaminodiazepoxide (preliminary communications). *Journal of the American Medical Association,* 172:1162–1163.

Heise, G. A., and Boff, E. (1961): Taming action of chlordiazepoxide. *Federation Proceedings,* 20:393.

Heuschele, W. P. (1961): Chlordiazepoxide for calming zoo animals. *Journal of the American Veterinary Medical Association,* 139:996–998.

Horovitz, Z. P., Piala, J. J., High, J. P., Burke, J. C., and Leaf, R. C. (1966): Effects of drugs on the mouse-killing (muricide) test and its relationship to amygdaloid function. *International Journal of Neuropharmacology,* 5:405–411.

Katz, D. M., and Steinberg, H. (1970): Long-term isolation in rats reduces morphine response. *Nature,* 228:469–471.

Klupp, H., and Kähling, J. (1965): Pharmakologische Wirkung von 7-Chlor-1,3-dihydro-3-hydroxy-5-phenyl-2H-1,4-benzodiazepin-2-on. *Arzneimittel-Forschung,* 15:359–365.

Langfeldt, T., and Ursin, H. (1971): Differential action of diazepam on flight and defense behavior in the cat. *Psychopharmacologia* (Berlin), 19:61–66.

Malick, J. B., Sofia, R. D., and Goldberg, M. E. (1969): A comparative study of the effects of selected psychoactive agents upon three lesion-induced models of aggression in the rats. *Archives Internationales de Pharmacodynamie et de Thérapie,* 181:459–465.

Moyer, K. E. (1968): Kinds of aggression and their physiological basis. *Communications in Behavioral Biology, A,* 2:65–87.

Porsolt, R. D., Joyce, D., and Summerfield, A. (1970): Changes in behaviour with repeated testing under the influence of drugs: drug-experience interactions. *Nature,* 227:286–287.

Randall, L. O. (1960): Pharmacology of methaminodiazepoxide. *Diseases of the Nervous System,* 21 *Suppl.:*7–10.

Randall, L. O. (1961): Pharmacology of chlordiazepoxide (LIBRIUM). *Diseases of the Nervous System,* 22 *Suppl.:*7–15.

Randall, L. O., Heise, G. A., Schallek, W., Bagdon, R. E., Banziger, R. E., Boris, A., Moe, R. A., and Abrams, W. B. (1961): Pharmacological and clinical studies on VALIUM (TM), a new psychotherapeutic agent of the benzodiazepine class. *Current Therapeutic Research,* 3:405–425.

Randall, L. O., Scheckel, C. L., and Banziger, R. F. (1965): Pharmacology of the metabolites of chlordiazepoxide and diazepam. *Current Therapeutic Research,* 7:590–606.

Randall, L. O., and Schallek, W. (1968): Pharmacological activity of certain benzodiazepines. In: *Psychopharmacology: a review of progress, 1957–1967,* edited by D. H. Efron. Public Health Service Publication No. 1836, pp. 153–184.

Rushton, R., and Steinberg, H. (1966): Combined effects of chlordiazepoxide and dexamphetamine on activity of rats in an unfamiliar environment. *Nature,* 211:1312–1313.

Salzman, C., Di Mascio, A., Shader, R. I., and Harmatz, J. S. (1969): Chlordiazepoxide, expectation and hostility. *Psychopharmacologia* (Berlin), 14:38–45.

Sofia, R. D. (1969): Effects of centrally-active drugs on four models of experimentally-induced aggression in rodents. *Life Sciences,* 8, pt. I:705–716.

Tobin, J. M., and Lewis, N. D. C. (1960): New psychotherapeutic agent, chlordiazepoxide. Use in treatment of anxiety states and related symptoms. *Journal of the American Medical Association,* 174:1242–1249.

Tozer, T. N., Neff, N. H., and Brodie, B. B. (1966): Application of steady state kinetics to the synthesis rate and turnover time of serotonin in brain of normal and reserpine-treated rats. *Journal of Pharmacology and Experimental Therapeutics,* 153:177–182.

Valzelli, L. (1967): Drugs and aggressiveness. In: *Advances in Pharmacology,* Vol. 5, edited by S. Garattini and P. A. Shore. Academic Press, New York, pp. 79–108.

Valzelli, L. (1969*a*): Aggressive behavior induced by isolation. In: *Aggressive Behaviour,* edited by S. Garattini and E. B. Sigg. Excerpta Medica Foundation, Amsterdam, pp. 70–76.

Valzelli, L. (1969*b*): The exploratory behavior in normal and aggressive mice. *Psychopharmacologia* (Berlin), 15:232–235.

Valzelli, L. (1971): Further aspects of the exploratory behaviour in aggressive mice. *Psychopharmacologia* (Berlin), 19:91–94.

Valzelli, L., and Bernasconi, S. (1971): Differential activity of some psychotropic drugs as a function of emotional level in animals. *Psychopharmacologia* (Berlin), 20:91–96.

Valzelli, L., and Garattini, S. (1968): Behavioral changes and 5-hydroxytryptamine turnover in animals. In: *Advances in Pharmacology, Vol. 6B,* edited by S. Garattini and P. A. Shore. Academic Press, New York, pp. 249–260.

Valzelli, L., Giacalone, E., and Garattini, S. (1967): Pharmacological control of aggressive behavior in mice. *European Journal of Pharmacology,* 2:144–146.

Valzelli, L., and Garattini, S. (1971): Biochemical and behavioral changes induced by isolation in rat. *Neuropharmacology,* 11:17–22.

Yen, C. Y., Stanger, L., and Millman, N. (1959): Ataractic suppression of isolation-induced aggressive behavior. *Archives Internationales de Pharmacodynamie et de Thérapie,* 123:179–185.

The Benzodiazepines, Raven Press, New York © 1973

ANTIAGGRESSIVE EFFECTS
OF CHLORDIAZEPOXIDE

José M. R. Delgado

Department of Psychiatry, Yale University School of Medicine, New Haven, Connecticut

I. INTRODUCTION

Studies involving antiaggressive drugs have usually been oriented toward observation of external behavioral effects rather than investigation of the intracerebral mechanisms involved in drug action. Chlordiazepoxide (LIBRIUM) is known to be clinically effective in reducing anxiety in human subjects (Harris, 1960; Tobin and Lewis, 1960), and, although it has been suggested that limbic structures could play a role in this action, little experimental evidence has been uncovered as yet. The clinical reduction in anxiety has been explained in terms of: (1) taming effects, (2) central muscle-relaxant activity, (3) anticonvulsant properties, (4) attenuation of fear in avoidance and conflict behavior, and (5) increase in food intake.

Experimental investigation of the mode of action of chlordiazepoxide requires the use of animal subjects, and some reports have been published of its effectiveness in monkeys. Taming was observed in vicious cynomologus monkeys (Randall, 1960, 1961; Heise and Boff, 1961), in rhesus monkeys (Snyder, 1969), squirrel monkeys (Scheckel and Boff, 1966), and in various wild zoo animals (Heuschele, 1961). Some of these studies attempted to differentiate antiaggressive from ataxic effects, showing a reduction in locomotor activity. Delayed responses, requiring the monkeys to match a colored light with a color-coded lever in order to obtain food reward, have been proposed as "anxiety situations" because "these monkeys are under considerable emotional stress which is reflected in a very tense appearance and seemingly anxious behavior" (Zbinden and Randall, 1967). It is doubtful, however, that simple observation of the animal could provide suitable information about its emotionality, and anthropomorphic interpretations of monkey faces are rather controversial.

Testing of the antiaggressive properties of drugs would be facilitated if we could provide an experimental situation which induced reliable hostility.

419

In monkeys, food competition, reduction in available space, or the presence of an estrous female may increase threats, but not in a predictable way. In previous studies (Delgado, 1962; Plotnik et al., 1971), we have demonstrated that electrical stimulation of the thalamus, central gray, and other cerebral structures results in well-directed and organized attacks. In the present research, the aggressive response reliably evoked by cerebral radio stimulation was used in the evaluation of antiaggressive properties of chlordiazepoxide. In addition, continuous day and night monitoring of individual and group motor activity could provide useful data in psychopharmacological studies, and in this chapter I present a method for mobility recording based on activation of a small inertia-sensitive FM instrument carried by each member of a group.

II. METHODS

Five male and three female adolescent rhesus monkeys (*Macaca mulatta*), weighing from 3.2 to 5.6 kg, were stereotaxically implanted with up to 28 intracerebral electrodes according to procedures described elsewhere (Delgado, 1955, 1964).

Two weeks after surgery, each animal was restrained in a Brady chair. Spontaneous electrical activity was recorded from all implanted points by means of an eight-channel Grass EEG machine. Electrical stimulation was delivered to each point and the observed effects were noted. Stimulations were cathodal, constant-current, square waves of 0.5-msec pulse duration at 100 Hz, applied for 3 to 5 sec. Voltage and milliamperage were monitored by a two-channel Tektronix oscilloscope.

In some experiments, free animals were stimulated by radio control according to procedures described elsewhere (Delgado, 1964, 1970). They were housed in a glass-front cage measuring $2.2 \times 1.1 \times 1.4$ m. Observations were made according to the methods of Snyder (1970) and Delgado et al. (1971). In this case, the pair of monkeys was placed in the cage and tested as follows: (1) 15 min of mildly competitive feeding, during which lever pressing delivered banana pellets on a FR 4 schedule of reinforcement; (2) 30 min of no-demand spontaneous behavior; and (3) 15 min of competitive feeding at an elevated hopper into which one pellet was automatically dispensed every 5 sec.

After completion of the experiments, the animals were anesthetized and perfused with saline and formaline. Then the brain was stereotaxically sectioned in 10-mm blocks which were placed in a freezing microtome to be sectioned at 50 μ. The location of electrodes was verified histologically (Delgado, 1961).

III. RESULTS

A. Taming effect

Rhesus monkeys are known to be spontaneously aggressive. They usually retreat when an observer approaches and launch an attack against the intruder. This typical behavior was expressed by all four pairs of rhesus that we studied: they grabbed and bit an approaching glove and escaped to avoid being touched.

In the dominant animal of each pair, 5 mg/kg, i.m., of chlordiazepoxide was administered with similar results. After injection, the monkey was placed alone in a cage where it walked around without any noticeable motor deficit, took food, and jumped to the overhead perch as usual. Its reaction toward observers, however, was drastically modified: it sat peacefully in the cage when the door was opened, and allowed the investigator to reach in and touch its head or body. The animal would watch an approaching hand without reacting except by grimacing momentarily when first touched. None of the four dominant animals treated with chlordiazepoxide attempted to bite or escape. When a rubber doll was introduced into the cage, the monkey's spontaneous reaction was to threaten, attack, and bite it, but under chlordiazepoxide the monkey paid little attention to the doll and did not show signs of hostility. This pharmacological pacification was similar to the inhibition of aggressiveness induced in the same animal by radio stimulation of the caudate nucleus (Fig. 1).

To evaluate the social consequences of this taming effect, studies of animal pairs were performed as explained under point II, Methods. On different days, each of the four dominant monkeys was injected with 5 mg/kg, i.m., of chlordiazepoxide and placed in a Brady chair where for 1 hr EEG recordings were taken and after-discharge (AD) thresholds were tested. Then the animal was released in the colony cage with its partner for a second hr during which their spontaneous motor and social behaviors were observed and quantified. The neurophysiological data obtained from EEG recordings and threshold changes indicated that the drug dose was fully effective when the hour of social behavior began. During this period, there was a total absence of aggressive acts between the paired monkeys, and behavioral data demonstrated that motor, alimentary, and social be-

FIG. 1. *A:* Introduction of a doll in the cage elicits aggressive behavior of the monkey. *B:* Administration of chlordiazepoxide (8 mg/kg) has an effect similar to radio stimulation of the caudate nucleus, inhibiting the aggressive response of the monkey without, however, having additive effects.

havior, as well as dominance, remained unchanged. We therefore concluded that this taming effect was independent of other modifications in spontaneous individual or social behavior.

Experiments were continued, injecting 20 mg/kg in each dominant animal. This amount produced marked somnolence, and the animals remained motionless for hours with eyes closed, uninterested in food or social exchange. When they did move around, they were ataxic, poorly coordinated, and fell easily. This high dose and its observed effects were considered beyond therapeutic interest.

B. Differential effects on evoked behavior

In one of the monkeys restrained in a Brady chair (female, M No. 1), electrical stimulation with 1.5 mA (100 cps, 0.5 msec) of Forel's Field (H_2), immediately dorsal to the supraoptic nucleus, elicited a complex response consisting of: (1) dilatation of the ipsilateral pupil, (2) arousal followed by restlessness and body jerking, and (3) well organized offensive-defensive behavior, including staring, frowning, ear flattening, pilo-erection, mouth opening, barking, and attempting to grab and bite. Stimulation thresholds and effects were determined every 5 min for 1 hr on different days, and they remained stable.

To test the effect of chlordiazepoxide on this complex response, 8 mg/kg, i.m., was injected into M No. 1. Stimulation thresholds for the different components of the response were monitored initially every few minutes and later at longer intervals for a total period of 8 hr. Results are shown in Fig. 2. Thresholds for ipsilateral pupil dilatation remained constant at the control level of 1.5 mA, without fatiguing, even if stimulation was applied continuously for several minutes.

After drug administration, the control threshold of 0.8 mA for arousal and body jerking rose to 1.4 mA. After 2 hr, it reached a peak of 2.8 mA, remained high for 7 hr, and then diminished, without returning to the control value before the end of the experiment. The following day, threshold was normal.

The most notable effect was total blocking of elicited offensive-defensive responses 15 min after injection. At that time, stimulation of more than 3.0 mA elicited considerable dilatation of the ipsilateral pupil accompanied by restlessness and jerking movements, without the appearance of staring, barking, or any other aggressive manifestation. This complete arrest of induced aggressiveness lasted for about 6 hr. Thresholds for staring and barking were still high (2.75 mA) 7.5 hr after drug injection.

C. Blocking of elicited aggression in the free situation

In the above experiment with animals under restraint, classification of behavioral responses as "offensive-defensive" may be open to criticism

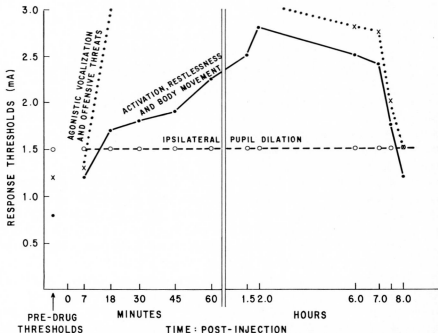

FIG. 2. Administration of chlordiazepoxide has a differential effect on the complex response elicited by electrical stimulation of Forel's Field: dilatation of the pupil was not modified, although threshold for restlessness increased considerably and agonistic responses were completely blocked.

because of the artificiality of the situation, lack of a suitable target, and absence of a socially reactive partner. The recently developed techniques for radio stimulation of the brain in free animals (Delgado, 1963a, 1964) permit cerebral exploration in monkeys free while forming part of a group, and the study of drug effects on elicited responses.

Stimulation of Forel's Field in M No. 1 was repeated with the animal free and paired with M. No. 2, also female, whose dominance had been clearly established during 6 hr of control observations on three different days. The stimulation experiment began with 30 min of control during which the animals were engaged in reciprocal grooming and locomotion without any sign of hostility for approximately 24 min. Then 1.5 mA of radio stimulation was applied to M No. 1 for 5 sec. The animal immediately became agitated, looked around, walked rapidly, jumped in front of her dominant partner, stared at her, flattened her ears, opened her mouth, and barked. This behavioral attitude resulted in a brief but violent fight between the monkeys in which M No. 2's superiority was demonstrated.

The experiment was repeated several times over two days with similar results: radio stimulation of M No. 1 elicited aggressive manifestations followed by physical confrontation and punishment of M No. 1 by her dominant partner. In the absence of brain stimulation, social relations were friendly and included long periods of reciprocal grooming. On the third day, submissive M No. 1 responded to radio stimulations by trying to avoid confrontation with dominant M No. 2; she moved away rapidly, was careful not to look at M No. 2, and stayed behind her as much as possible. In spite of this strategy, she was repeatedly attacked by M No. 2.

Administration of chlordiazepoxide (8 mg/kg, i.m.) to M No. 1 completely changed her elicited behavior. During stimulations, she remained attentive and responsive to the attitude of the dominant partner, but there was no staring or social conflict between the animals. Without modifying the animal's spontaneous behavior, the drug inhibited the aggressive effect of Forel's Field stimulation.

Similar results were obtained in M No. 3, a male dominant over M No. 2. M No. 3 had electrodes in the olfactory area, and radio stimulation elicited violent fights between the two animals. Thirty min after administration of 8 mg/kg of chlordiazepoxide to M No. 3, radio stimulation failed to elicit aggressive behavior, and the only observed responses were locomotion and restlessness, without any social conflict. In this experiment, blocking of aggressive behavior lasted a long time. About 5 hr after drug administration, the animal responded to radio stimulation with increasing speed of locomotion but without aggression. His threatening reaction reappeared 16 hr after chlordiazepoxide injection, and the first elicited attack was observed 18 hr after the drug.

D. Radio-induced caudate inhibition

In monkeys No. 2, No. 3, and also No. 4 (a male, weighing 3.4 kg), inhibitory effects induced by stimulation of the head of the caudate nucleus were explored in the free situation under chlordiazepoxide. Three responses were chosen as criteria to identify a caudate point as inhibitory: (1) interruption of spontaneous walking (the animal would sit down and remain quiet during stimulation); (2) loss of interest in food (a piece of orange, usually eaten quickly, was ignored during stimulation, and the animal looked at the fruit without trying to take it); and (3) absence of the usual offensive-

426   ANTIAGGRESSIVE EFFECTS OF CHLORDIAZEPOXIDE

defensive reaction of running away, threatening, and trying to bite when the investigator opened the cage door and tried to touch the animal (under stimulation, it was possible to reach in and touch the animal with bare hands). The monkey seemed aware of its surroundings and looked at the investigator without signs of fear or hostility.

The above-mentioned inhibitory effects were reliably obtained in all three monkeys with caudate stimulation intensities of 1.0 to 1.4 mA. One and 2 hr after administration of 8 mg/kg, i.m., of chlordiazepoxide, the effects of caudate excitation were tested again. Walking was arrested and food intake inhibited at the same electrical thresholds as before drug administration and with similar characteristics. Because the offensive-defensive reaction was blocked by the drug, its caudate inhibition could not be explored, and no additive effects of the drug plus brain stimulation could be demonstrated.

E. Radio-induced motor responses

In monkeys No. 2, No. 3, and No. 4, radio stimulation of the internal capsule in the free situation with intensities of 0.4 to 0.9 mA evoked flexion of the contralateral hind limbs. In M No. 3 and No. 4, the contralateral forelimb was also flexed. These motor effects interfered with normal walking but did not prevent general motility; although the animals limped, their handicap did not appear emotionally or socially disturbing. Administration of 8 mg/kg, i.m., of chlordiazepoxide did not modify thresholds or characteristics of the motor responses elicited by radio stimulation of the internal capsule in any of the three monkeys.

F. A method for continuous telemetric recording of mobility

Spontaneous mobility is one of the main indicators of behavioral reactivity. Anxiety is usually accompanied by motor restlessness, whereas one of the symptoms of sedation and depression is diminished motility. The techniques proposed in the literature for the study of spontaneous movement have been based on direct observation or on instrumental recording by activation of photocells, triggering of microswitches placed on the cage floor, or changes in capacitance induced by the animal's proximity (Dews, 1953; Cole and Glees, 1957; Davis, 1957). Although these methods are useful for experiments with single animals, they cannot distinguish between the activity of individuals in a group and are, therefore, unsuitable for studies of social behavior.

Social mobility is of obvious interest in psychopharmacological studies, and during the last 10 years we have developed and tested several instruments for mobility recording (Delgado, 1963*a*). The final product, described briefly here, has been used in monkeys, gibbons, chimpanzees, and humans. Its circuitry is available on request.

1. *The mobility-sensing transmitter.* The movement sensor in this instrument is a small steel ball running freely in a minute plastic chamber located within the emitting coil of a 100 MHz, FM transmitter. The movements of the ball change the inductance of the coil, resulting in 100 KHz shifts on either side of the transmitting frequency. Electronically the unit is very simple. It has only one transistor (2N 5485); it uses coil radiation and does not require an antenna; total current drain at 2.7 V is 0.8 mA/hr; and the frequency stability is 1.5 KHz/degree C. The instrument is housed in a plastic cylinder 6.7 $\times$ 2.2 cm, which also contains a 1.000 mA/hr battery to provide long life (about one month). The effective transmitting range is approximately 30 m inside buildings, and more than 300 m outside.

The transmitter is attached by means of a light-weight, nonchafing nylon rope harness on the animal's back. (In human patients we use a wrist-watch band.) Each transmitter is adjusted to an individual wavelength within the 100 MHz band, but at least 1 MHz equidistant. In some experiments, we have used up to 12 different subjects simultaneously.

2. *The recording unit.* The mobility recorder consists of a 75-ohm dipole antenna connected with high-gain FM tuners. The signal is demodulated in the tuner, and the received frequency changes, produced by the steel ball in the transmitter, are transformed into amplitude changes which are fed into a Schmitt trigger and monostable multivibrator which functions as a pulse shaper, providing a square-wave pulse, when the amplitude exceeds a preset value, in order to eliminate noise. These pulses are stored in a four-decade electronic counter and are transferred at chosen intervals (for example, every 15 min) to a buffer register and then to a print counter. The print on each channel represents the cumulative number of pulses generated by the movements of the steel ball in each animal. Our standard recording unit has six channels.

3. *Experimental setup.* A colony consisting of three adolescent gibbons (*Hylobates lar*) named *Coti* (3.3 kg, male), *Maria* (3.5 kg, female), and *José* (2.2 kg, male) was housed in a cage measuring 1.82 $\times$ 1.24 $\times$ 0.96 m with a glass front for observation and photography. The cage was located in a soundproof, temperature-controlled room with automatic 12-hr cycles of light and darkness, and it could be observed through a one-way window. The

experiments lasted for 6 weeks during which continuous day and night recordings of mobility were obtained. In addition, spontaneous behavior of the colony was recorded by time-lapse photography (Delgado, 1964) and by direct observation, scoring behavioral categories during 30-min periods before, during, and after feeding the group.

4. *Effect of chlordiazepoxide on mobility.* Scoring animal mobility by direct observation was comparable to recordings obtained by telemetry. The main advantages of radio telemetry were that the collection of data was automatic and it was therefore (1) economical because observers were not needed, (2) objective and accurate because it did not depend on personal judgment, (3) continuous, permitting day and night studies without inconvenience for the experiments, and (4) simultaneous in several subjects, allowing comparison of individual mobility and establishment of social correlations. Control studies showed the reliability of the method in different animals throughout the duration of our study.

On different control days, during 2 hr of testing, gibbon *José* moved 300 to 600 counts per 15 min, while *Coti* and *Maria* had a range of 900 to 1,500 counts for the same period. Chlordiazepoxide was less effective in gibbons than in monkeys. A 30 mg/kg dose given to *Coti,* the boss of the group, did not produce ataxia although it decreased his movements to 20 to 30 counts per 15 min. The activity of the other two nondrugged gibbons was reduced to about half of the control value. These results were repeated in three different experiments. Direct observation and time-elapse photography showed that, under chlordiazepoxide, aggressive acts were significantly lower, although playful activity, grooming, and food intake were higher than during controls. These studies have been presented in detail elsewhere (Gardner, 1971). The fact that mobility was reduced whereas play and grooming were increased by chlordiazepoxide indicated a differential effect on specific behavioral manifestations and emphasized the well-known need for multifactorial analysis in the study of drug effects.

IV. DISCUSSION

There is increasing interest in the investigation of human aggression at both social and clinical levels. Psychiatrists consider possible traumas during childhood, frustrations, repressed desires, and similar psychological events. Attention is also paid to economic, social, ideological, and political factors and to their undesirable consequences expressed as antisocial behavior. In view of this growing interest, it is surprising that the most essential link of the aggressive process, which is in the central nervous system, is usually ignored. We should understand that the environment is only the provider of sensory inputs which must then be coded, decoded, and evaluated by specific areas of the brain: any kind of behavior is also the result of intracerebral activity. Experimental investigation of the cerebral mechanisms responsible for aggressive behavior should be an essential counterpart of social and psychiatric studies, and this fact must be recognized by all scientists interested in understanding aggression.

Primates with electrodes permanently implanted in the brain permit the study of neurological mechanisms of aggression and provide a reliable model of induced hostility which may be useful for the evaluation of antiaggressive properties of drugs. In our experiments, aggression was induced in a dominant (M No. 3) against a subordinate monkey (M No. 2), and also in a submissive (M No. 1) against a dominant animal (M No. 2), proving that radio-induced aggression may be directed downward and upward in the hierarchical scale, in agreement with the previous studies by Delgado (1963*b*) and Robinson et al. (1969). In this experimental situation, we no longer rely on unpredictable spontaneous threats and we can control the timing and frequency of aggression. The fact that chlordiazepoxide blocked agonistic behavior supports the hypothesis of limbic system psychodepression (Randall and Schallek, 1968) or, alternately, a possible depression of nociceptive mechanisms in the thalamus and central gray which seem to be related to electrically induced aggression (Plotnik et al., 1971).

In any event, our experiments demonstrated that chlordiazepoxide has a selective action on cerebral mechanisms related to aggressiveness. In the same animal and structure (Fig. 2), thresholds to evoke agonistic behavior were considerably increased while thresholds for restlessness were less affected, and autonomic excitability for pupillary dilatation was not modified at all. These differential effects of the drug could be of great interest as a tool for pharmacological dissection of functional representation in determined areas of the brain.

Specificity of action was also indicated by the lack of change in thresholds for motor responses (internal capsule) and for inhibiting effects (head of the caudate nucleus). This last finding was against our provisional working hypothesis that antiaggressive compounds could facilitate the activity of known inhibiting structures such as the caudate system. In our experiments, the taming effect of caudate stimulation was very similar to the taming effect of chlordiazepoxide injection, but apparently there is no mutual potentiation.

The method that we proposed for the continuous recording of mobility by radio telemetry is the final product of a variety of progressively improved models. Its main advantage, not shared by other available mobility recorders, is that each subject is identified by its own narrow frequency band so that it is possible to record simultaneously the activity of several subjects free in a social group. Cross-correlation of data may be thus established in order to analyze social effects of experimental variables administered to an individual. For example, in previous experiments (Delgado, 1963a) chlorpromazine given to a monkey decreased the activity of the whole colony, and, similarly in the present study, injection of chlordiazepoxide to one gibbon resulted in a reduction of general mobility.

The fact that the instrument is monkey proof, very reliable, and fully automatic makes it practical for pharmacological studies of spontaneous and induced behavior, including the analysis of cycles such as sleep-wakefulness. The mobility recorder has proved its usefulness in studies performed on chimpanzees (Delgado et al., 1970) and also in clinical research (Kupfer et al., 1972).

V. SUMMARY

Rhesus monkeys with electrodes permanently implanted in the brain were tested to evaluate the effects of administration of chlordiazepoxide with the following results.

1. Using pairs of monkeys free in a cage, the dominant animal was injected with 5 mg/kg, i.m., of chlordiazepoxide. No changes were observed in motor, alimentary, or social behavior. The injected animals, however, lost their usual fear and aggressive behavior toward observers and dolls introduced in the cage. These results indicate that the taming effect of chlordiazepoxide is independent of other behavioral modifications.

2. Under restraint, the aggressive reaction elicited by electrical stimulation of Forel's Field was blocked by administration of 8 mg/kg, i.m., of chlordiazepoxide, although the concomitant evoked effect of restlessness was less affected and the threshold for pupillary dilatation was not modified. This pharmacological dissection of evoked effects may be useful for the analysis of mechanisms of drug action.

3. In pairs of monkeys free in a cage, aggressive responses were reliably elicited by radio stimulation of specific brain structures (Forel's Field). This response was blocked by administration of chlordiazepoxide. Electrically-evoked aggression is proposed as a convenient model to evaluate antiaggressive drugs.

4. Inhibition induced by radio stimulation of the caudate nucleus and motor responses elicited by radio stimulation of the internal capsule were not modified by administration of chlordiazepoxide.

5. Individual and social mobility are valuable indexes to test psychopharmacological effects. A technique is presented for the continuous recording of mobility, based on the changes in tuning of the emitting coil of a small FM transmitter caused by the displacements of a free-moving stainless steel ball. This transmitter is attached to the experimental subject. Up to 12 simultaneous recordings are possible, permitting cross-correlations of individual and social mobility. The method has proved its usefulness in monkeys, gibbons, chimpanzees, and human patients.

ACKNOWLEDGMENTS

Research described in this paper was supported by the U. S. Public Health Service grant 1-R01-MH-17408 and by the International Psychiatric Neuroresearch Foundation. Part of it was presented in the paper "Limbic excitability and elicited aggression modified by chlordiazepoxide in monkeys" by D. Snyder, H. Bracchitta, and J. M. R. Delgado (*in press*). A film entitled "Psychoactive Drugs and Radio-Controlled Behavior" was presented at the meeting of the American Psychiatric Association May 2 to 6, 1971, in Washington, D. C., and is available on request.

REFERENCES

Cole, J., and Glees, P. (1957): Some effects of methyl-phenidate (RITALIN) and amphetamine on normal and leucotomized monkeys. *Journal of Mental Science,* 103: 406–417.

Davis, G. D. (1957): Effects of central excitant and depressant drugs on locomotor activity in the monkey. *American Journal of Physiology,* 188:619–623.

Delgado, J. M. R. (1953): Evaluation of permanent implantation of electrodes within the brain. *Electroencephalography and Clinical Neurophysiology,* 7:637–644.

Delgado, J. M. R. (1961): Chronic implantation of intracerebral electrodes in animals. In: *Electrical Stimulation of the Brain,* edited by D. E. Sheer. University of Texas Press, Austin, pp. 25–26.

Delgado, J. M. R. (1962): Pharmacological modifications of social behavior. In: *Pharmacological Analysis of Central Nervous Action,* edited by W. D. M. Paton. Pergamon Press, Oxford, pp. 265-292.

Delgado, J. M. R. (1963a): Telemetry and telestimulation of the brain. In: *Bio-Telemetry,* edited by L. Slater. Pergamon Press, New York, pp. 231–249.

Delgado, J. M. R. (1963b): Effect of brain stimulation on task-free situations. In: *The Physiological Basis of Mental Activity,* edited by R. H. Peón. *Electroencephalography and Clinical Neurophysiology, Suppl. 24,* pp. 260-280.

Delgado, J. M. R. (1964): Free behavior and brain stimulation. In: *International Review of Neurobiology, Vol. VI,* edited by C. C. Pfeiffer and J. R. Smythies. Academic Press, New York, pp. 349–449.

Delgado, J. M. R. (1970): Multichannel Transdermal Stimulation of the Brain. *Technical Documentary Report No. ARL-TR-70-1*. Holloman Air Force Base, New Mexico.

Delgado, J. M. R., Johnston, V. S., Wallace, J. D., and Bradley, R. J. (1970): Operant conditioning of amygdala spindling in the free chimpanzee. *Brain Research*, 22:347–362.

Delgado, J. M. R., Lico, M., Bracchitta, H., and Snyder, D. (1971): Brain excitability and behavioral reactivity in monkeys under meprobamate. *Archives Internationales de Pharmacodynamie et de Thérapie*, 194:5–17.

Dews, P. B. (1953): The measurement of the influence of drugs on voluntary activity in mice. *British Journal of Pharmacology*, 8:46–48.

Gardner, J. J. (1971): The mobility telemetric system in psychotropic drug comparisons and behavioral monitoring in the gibbon (*Hylobates lar*). Thesis, Yale University, New Haven, Conn.

Harris, T. H. (1960): Methamino diazepoxide (preliminary communications). *Journal of the American Medical Association*, 172:1162–1163.

Heise, G. A., and Boff, E. (1961): Taming action of chlordiazepoxide. *Federation Proceedings*, 20:393.

Heuschele, W. P. (1961): Internal parasitism of monkeys with the pentastomid, *Armillifer armillatus*. *Journal of the American Veterinary Medical Association*, 139:996–998.

Kupfer, D. J., Detre, T. P., Foster, G., Tucker, G. J., and Delgado, J. M. R. (1972): The application of Delgado's telemetric mobility recorder for human studies. *Behavioral Biology*, 7:585–590.

Plotnik, R., Mir, D., and Delgado, J. M. R. (1972): Map of reinforcing sites in the rhesus monkey brain. *International Journal of Psychobiology* (*in press*).

Randall, L. O. (1960): Pharmacology of methaminodiazepoxide. *Diseases of the Nervous System*, 21:Suppl. 7.

Randall, L. O. (1961): Pharmacology of chlordiazepoxide (LIBRIUM). *Diseases of the Nervous System*, 22:Suppl. 7.

Randall, L. O., and Schallek, W. (1968): Pharmacological activity of certain benzodiazepines. In: *Psychopharmacology: A Review of Progress 1957-1967*, edited by D. H. Efron. Public Health Service Publication No. 1836, Washington, D.C., pp. 153–184.

Robinson, B. W., Alexander, M., and Bowne, G. (1969): Dominance reversal resulting from aggressive responses evoked by brain telestimulation. *Physiology and Behavior*, 4:749–752.

Scheckel, C. L., and Boff, E. (1966): Effects of drugs on aggressive behavior in monkeys. In: *Proceedings of the Vth International Congress of the Collegium Internationale Neuropsychopharmacologium*. Excerpta Medica, International Congress Series No. 129, pp. 789–795.

Snyder, D. R. (1970): Fall from social dominance following orbital frontal ablation in monkeys. *Proceedings of the American Psychological Association*, 235–236.

Snyder, D. R. (1971): Social and emotional behavior in monkeys following orbital frontal ablations. *Ph. D. Dissertation*. The University of Michigan, Ann Arbor. *Dissertation Abstracts International*, B31:5679–5680.

Tobin, J. M., and Lewis, N. D. C. (1960): New psychotherapeutic agent, chlordiazepoxide. Use in treatment of anxiety states and related symptoms. *Journal of the American Medical Association*, 174:1242–1429.

Zbinden, G., and Randall, L. O. (1967): Pharmacology of benzodiazepines: laboratory and clinical correlations. In: *Advances in Pharmacology, Vol. 5*, edited by S. Garattini and P. A. Shore. Academic Press, New York, pp. 213–291.

The Benzodiazepines, Raven Press, New York © 1973

THE EFFECTS OF BENZODIAZEPINES ON AGGRESSION: REDUCED OR INCREASED?

Alberto DiMascio

Department of Mental Health, Commonwealth of Massachusetts, and Tufts University School of Medicine, Boston State Hospital, Boston, Massachusetts

I. INTRODUCTION

The introduction of chlordiazepoxide and diazepam for use in psychiatry over 10 years ago was accompanied by considerable publicity concerning their ability to quiet, calm, and tame wild or vicious animals. Based on these data, the implicit assumption was made that chlordiazepoxide and diazepam would also reduce aggressiveness in humans.

The initial observations and their clinical derivative were neither questioned nor subjected to extensive exploration during the subsequent 6 to 8 years. A number of recent studies have been carried out in both animals and humans that involve this aspect of the benzodiazepines. Some of the findings seem to be at variance with the earlier observations and clinical expectations and therefore require a full exposition.

II. ANIMAL STUDIES

Early in the pharmacologic examination of chlordiazepoxide, it was noted that vicious cynomolgus monkeys (Randall, 1960, 1961) could be "tamed" by an injection of the drug to the degree that the animal could be handled, petted, or even poked without exhibiting the typical aggressive reaction of lunging and biting. Heuschele (1961) reported that this drug was similarly effective when given to a variety of zoo animals. Heise et al. (1961) claimed that this effect occurred at doses which did not significantly affect observable locomotor activity.

Chlordiazepoxide and other benzodiazepines—principally diazepam, oxazepam, and nitrazepam—have since been investigated using a variety of tests designed to elicit aggressive behavior and which allow for the quantitative (or semiquantitative) assessment of drug effects thereon.

The models for inducing or examining aggressive behavior include the following:

1. Electro-footshock (Tedeschi et al., 1959).
2. Isolation rearing (Yen et al., 1958).
3. Bilateral septal area lesions (Brady et al., 1953).
4. Chemical induction by apomorphine (Senault, 1970), racemic DOPA, (Yen et al., 1970), mesorgydine (Podvalovà et al., 1969), and 2-2,6-dichlorophenylamino 2-imidazoline HCl (Morpuvgo, 1968).
5. Observation of spontaneously aggressive reactions (unprovoked or provoked) in species such as ants (Kostowski, 1966), scorpions (Mercier et al., 1970), minks (Bauen et al., 1970), muricidal "killer" rats (Sofia, 1969), mice (Cole et al., 1970), cats (Langfeldt et al., 1971), and squirrel monkeys (Scheckel et al., 1967).

In addition, in order to gauge more accurately the relationship between a drug's effect on aggressive behavior and its sedative motor-inhibiting properties, sensitive tests of neuromuscular ability, inclined-screen (Hoffmeister et al., 1969), and of rotorod (Horowitz et al., 1963) and locomotor activity (Christmas et al., 1970) have often been included in the testing battery.

From a literature search on the effects of drugs on aggressive behavior, it became clear that there were two nondrug factors (the test-model employed and the species examined) and two drug factors (the dose of drug administered and whether it was given acutely or chronically) that play significant roles in determining the results and that must be taken into consideration when attempting to interpret the findings.

An obvious conclusion drawn by numerous investigators is that the various test models induce different types of aggressive behavior. Whenever more than one test model was employed in examining specific drugs (Horowitz, et al., 1963; Sofia, 1969; Hoffmeister, et al., 1969; Senault, 1970), it was found that no drug reduced aggressive behavior in all test models at doses below their respective neuromuscular toxic levels. The test models were differentially sensitive to the effects of benzodiazepines and productive of results in opposing directions.

Septal-lesion induced irritability was found to be the most resistant to alteration by benzodiazepines (Sofia, 1969; Christmas et al., 1970). *Spontaneous aggression upon provocation* (e.g., introduction of human hand into cage) was most readily reduced, at nonneurotoxic doses, by benzodiazepines (Hoffmeister et al., 1969; Bauen et al., 1970; Langfeldt et al., 1971).

Shock- or isolation-induced aggressive behavior usually was reduced after treatment with benzodiazepines, but generally only at dose levels determined by the rotorod or inclined screen tests, to be near or above the neurotoxic level (Randall, 1960; Horowitz et al., 1963; Valzelli et al., 1967; Sofia, 1969; Hoffmeister et al., 1969). An important exception to this generalization is a study by Cole and Wolf (1970), in which they examined the effects of chlordiazepoxide when given to a particularly aggresive species of mice (*Omychomys leucogaster*). At a dose half that determined to be the minimum neurotoxic level in 50% of the mice on the rotorod test, an *increase* in shock-induced aggressive behavior was documented. This increase had not been noted when the same study was conducted with considerably less aggressive albino white mice.

The ability of the *chemically induced aggression tests* to detect benzodiazepine effects is dependent on the specific chemical used: atropine-, racemic DOPA-, or mesorgydine-induced aggression could be reduced by benzodiazepines but only at or above neurotoxic doses.

Spontaneous, naturally occurring aggression (e.g., inter- or intraspecies hostility) was either not effected by benzodiazepines (Kostowski, 1966; Boissier et al., 1968; Mercier et al., 1970) or was *increased* by them (Fox and Snyder, 1969; Fox et al., 1970, *in press;* Guaitani et al., 1971). It must be pointed out that the investigators who noted the increase in aggression after benzodiazepines when using this model (Valzelli et al., 1967; Fox et al., 1970; *in press;* Guaitani et al, 1971) administered the drugs for a week prior to placing the animals (grouped four per cage) in the situation in which spontaneous fighting occurred. These investigators were the only ones to examine the effects of chronic drug administration. That they noted an increase in aggressive behavior while others observed a reduction may be due in part to their use of naturally aggressive animals and in part to some acute versus chronic drug administration differential in action. It must be remembered, however, that Cole and Wolf (1970), who also used spontaneous hyperaggressive organisms, were able to demonstrate an increase in aggressiveness to benzodiazepines when administered acutely.

As previously mentioned, many investigators had concluded that the various test models induced different aspects of aggressive behavior. Hoffmeister and Wuttke (1969) explored this further by differentiating between pure *attack* behavior and exaggerated *defensive* behavior (that is, defensive-aggressive behavior) that occurred in the shock-induction model. Langfeldt and Ursin (1971) differentiated between *defense* behavior and *flight* behavior in cats in the shock-induction model. In each instance, it was the defense behavior (i.e., an aggressive reaction to provoking stimuli) that was reduced by chlordiazepoxide or diazepam. Attack behavior was not reduced at doses less than neurotoxic levels. These data are consistent with those concerning the effects of chlordiazepoxide and diazepam in the naturally hyperaggressive animals. In addition, they lend support to those investigators who claim that the earliest preclinical trials of chlordiazepoxide and diazepam were done on fear-induced aggression—an anxiety reaction that is responsive to benzodiazepine therapy. Similarly, the positive response of many caged zoo animals may be attributed to the antianxiety effects of these drugs.

Thus, depending on the test model employed, the species examined, the dosage, and the length of time for which a benzodiazepine was given, significant differences were or could be obtained.

With regard to the aggression-altering properties of the various benzodiazepines, there seems to be merely a quantitative one among chlordiazepoxide, diazepam, and nitrazepam. Oxazepam, however, has been found to be more than merely quantitatively different. Christmas and Maxwell (1970) found that chlordiazepoxide or diazepam, but not oxazepam—even at doses up to 80 mg/kg—reduced aggression in rats with septal lesions. Bauen and Possanza (1970) found that, after an acute administration of chlordiazepoxide or diazepam, aggressive minks could be handled and petted, but not so after oxazepam. Fox (1970), by contrast, found that aggressiveness was increased in hyperaggressive mice with the chronic administration of chlordiazepoxide and diazepam, but with oxazepam the animals were even somewhat less aggressive than control mice.

III. HUMAN STUDIES

The examinations of the effects of benzodiazepines on aggressive behavior in humans are few and based more on clinical anecdotes than on systematic evaluations. The various reports, however, also contain contrasting findings.

In 1961, Boyle and Tobin used chlordiazepoxide to treat 25 patients from a variety of diagnostic categories whose symptomatology included aggressiveness, hyperactivity, and hostility. From this nonblind, uncontrolled study, they concluded that there was a 50% chance of substantial reduction in symptomatology.

Kalina (1964) administered diazepam in an open trial to 52 prisoners confined to maximum security and obtained "complete control" of violent, destructive, and assaultive behavior. The first double-blind experiment assessing the actions of chlordiazepoxide on verbal hostility was carried out in 46 male juvenile delinquents by Gleser et al. (1965). After one day of drug administration, the verbal sample analyses showed a reduction of "ambivalent hostility" (statements blaming or criticizing others) and of "outward hostility" (statements of destructive or aggressive intent).

In a recently published clinical double-blind study, Podobnikar (1971) examined the effects of chlordiazepoxide on aggressive symptoms accompanying neurotic behavior. He found a reduction after 2 weeks of drug treatment. This finding is substantiated by Rickels (1971) who also reported that chlordiazepoxide reduced anger-hostility in anxious neurotic outpatients.

These observations stand in contrast to those of other investigators. For example, Kelley and Gisvold (1960) studied chlordiazepoxide in 66 patients, using the MMPI as a measure of change. They found the drug produced a tendency toward externalization of conflict and stress. Feldman (1962) reported that diazepam had no favorable effect on 87 patients in regard to the target symptoms of combativeness and hostility. In fact, diazepam even produced in "many" patients dislikes or hates that became progressively more intense and gradually extended toward more individuals.

More recently, there have been several articles (DiMascio et al., 1967, 1970; Gardos et al., 1968; Salzman et al., 1969) from our laboratory in which double-blind, placebo-controlled comparisons were made of chlordiazepoxide and oxazepam on feelings of hostility. These studies tended to show that 1-week administration of chlordiazepoxide increased scores on a test measuring assaultiveness, direct hostility, irritability, and verbal hostility in high and medium anxious males. Oxazepam, in contrast, did not do so.

As in the animal studies, both a reduction and an increase in hostile-aggressive behavior have been found in humans given benzodiazepines. The finding of increased aggressive behavior in our studies with humans came as a surprise since, based on the initial claims, decreased aggressiveness had been expected. Subsequently, we suggested a number of reasons why the phenomenon of increased aggression had not been more universally known and recognized.

1. The amount of drug-released hostility is often not marked.

2. Clinical observations are insensitive compared to the hostility scales utilized in our studies.

3. For many patients, the release of aggressiveness may be socially or therapeutically beneficial, when channelled appropriately, and thus may be considered part of overall therapeutic movement rather than as drug-induced behavior.

We further hypothesized that it may be only in those patients with a history of poor impulse control or of aggressive, destructive behavior that chlordiazepoxide can release sufficient hostility to result in "rage reactions." Our findings and hypotheses are opposed, however, by the previously cited findings of Kalina (1964) and Boyle and Tobin (1961). No explanation for the observed differences (and it must be realized that evidence for both contentions is rather meager) could be derived from the few systematic studies available.

IV. DISCUSSION AND SUMMARY

Pharmacologists are asked to select or develop test models which will predict behavior in humans. There are a number of test models available for inducing and assessing aggressive behavior in animals, but the use of different test models, different species, or different drug-administration schedules produces differing results. The question is then: Which of the test models or which combination of these factors should be utilized for the most accurate prediction of drug efficacy in humans? Before we can answer this question, however, we must decide what results in humans to predict (i.e., aggression reduction or increase).

In addition, clinicians are asked to select the most appropriate drug for a given patient. To do this, he must know in which patient which drug will produce which effect. His ability to make such accurate predictions of drug actions requires that, with regard to the effects of benzodiazepines on aggressive behavior, much more extensive and sophisticated research in humans be carried out. The clinician's prescription of a given benzodiazepine that may alter a patient's level of aggression, as well as his level of anxiety, will have an impact not only on the internal feelings of the patient but also on his interpersonal relationships with others. Also, the social significance of the behavior of individuals who receive these medications—and there have been millions of them in all walks of life and in various decision-making positions—should not be underestimated.

REFERENCES

Bauen, A., and Possanza, G. J. (1970): The mink as a psychopharmacological model. *Archives Internationales de Pharmacodynamie et de Thérapie*, 186:133–136.

Boissier, J. R., Simon, R., and Aaron, C. (1969): A new method for rapid screening of minor tranquilizers in mice. *European Journal of Pharmacology*, 4:145–151.

Boyle, D., and Tobin, J. (1961): Pharmaceutical management of behavior disorders. *Journal of the Medical Society of New Jersey*, 58:427–429.

Brady, J. V., and Nauta, W. J. H. (1953): Subcortical mechanism in emotional behavior: Affective changes following spetal forebrain lesion in the albino rat. *Journal of Comparative Physiological Psychology,* 46:339–346.

Christmas, A. J., and Maxwell, D. R. (1970): A comparison of the effects of some benzodiazepines and other drugs on aggressive and exploratory behaviour in mice and rats. *Neuropharmacology,* 9:17–29.

Cole, H. F., and Wolf, H. H. (1970): Laboratory evaluation of aggressive behavior of the grasshopper mouse (*Onychomys*). *Journal of Pharmaceutical Sciences,* 59:969–971.

DiMascio, A., Shader, R. I., and Giller, D. R. (1970): Behavioral toxicity Part III: Perceptual-cognitive functions and Part IV: Emotional (mood) states. In: *Psychotropic Drug Side Effects,* edited by R. I. Shader and A. DiMascio. Williams & Wilkins, Baltimore, pp. 132–141.

DiMascio, A., Shader, R. I., and Harmatz, J. (1967): Psychotropic drugs and induced hostility. *Psychosomatics,* 10:46–47.

Feldman, P. E. (1962): An analysis of the efficacy of diazepam. *Journal of Neuropsychiatry* 3 (Suppl.): S62–S67.

Fox, K. A. (1971): *Personal communication.*

Fox, K. A., and Snyder, R. L. (1969): Effect of sustained low doses of diazepam on aggression and mortality in grouped male mice. *Journal of Comparative and Physiological Psychology,* 69:663–666.

Fox, K. A., Tockosh, J. R., and Wilcox, A. H. (1970): Increased aggression among grouped male mice fed chlordiazepoxide. *European Journal of Pharmacology,* 11:119–121.

Fox, K. A., Webster, J. C., and Guerriero, F. J. (1972): Increased aggression among grouped male mice fed nitrazepam and flurazepam. *European Journal of Pharmacology (in press).*

Gardos, G., DiMascio, A., Salzman, C., and Shader, R. I. (1968): Differential actions of chlordiazepoxide and oxazepam on hostility. *Archives of General Psychiatry,* 18:757–760.

Gleser, G. C., Gottschalk, L. A., Fox, R., and Lippert, W. (1965): Immediate changes in affect with chlordiazepoxide. *Archives of General Psychiatry,* 13:291–295.

Guaitani, A., Marcucci, F., and Garattini, S. (1971): Increased aggression and toxicity in grouped male mice treated with tranquilizing benzodiazepines. *Psychopharmacologia* (Berlin), 19:241–245.

Heise, G. A., and Boff, E. (1961): Taming action of chlordiazepoxide. *Federal Proceedings,* 20:393.

Heuschele, W. P. (1961): Chlordiazepoxide for calming zoo animals. *Journal of the American Veterinary Medical Association,* 139:996.

Hoffmeister, F., and Wuttke, W. (1969): On the actions of psychotropic drugs on the attack- and aggressive-defensive behaviour of mice and cats. In: *Aggressive Behavior,* edited by S. Garattini and B. Sigg. Wiley & Sons, New York, pp. 273–280.

Horowitz, Z. P., Furgiuele, A. R., Brannick, L. J., Burke, J. C., and Craver, B. N. (1963): A new chemical structure with specific depressant effects on the amygdala and on the hyperirritability of the "septal rat." *Nature,* 200:369.

Kalina, R. K. (1964): Diazepam: Its role in a prison setting. *Diseases of the Nervous System,* 25:101.

Kelley, J. W., and Gisvold, D. I. (1960): The use of M.M.P.I. in the evaluation of Librium. *Colorado GP,* 2:3–8.

Kostowski, W. (1966): A note on the effects of some psychotropic drugs on the aggressive behaviour in the ant, *Formica rufa. Journal of Pharmacy and Pharmacology,* 18:747–749.

Langfeldt, T., and Ursin, H. (1971): Differential action of diazepam on flight and defense behavior in the cat. *Psychopharmacologia* (Berlin), 19:61–66.

Mercier, J., and Dessaigne, S. (1970): Influence exercée par quelques drogues psycholeptiques sur le comportement du scorpion (*Androctonus australis Hector*). *Comptes Rendus des Séances de la Société de Biologie et de Ses Filiales* (Paris), 164:341–344.

Morpurgo, C. (1968): Aggressive behavior induced by large doses of 2-(2,6-dichlorophenylamino)-2-imindozoline hydrochloride (ST 155) in mice. *European Journal of Pharmacology*, 3:373–377.

Podobnikar, I. G. (1971): Implementation of psychotherapy by Librium in a pioneering rural-industrial psychiatric practice. *Psychosomatics*, 12:205–209.

Podvalova, I., Dlabac, A., and Votava, Z. (1969): Mesorgydine-induced aggressive behavior in rats. International Congress of Pharmacology, 4th, Basle, July 14–18, pp. 297–298 (pamphlet).

Randall, L. O. (1960): Pharmacology of methaminodiazepoxide. *Diseases of the Nervous System*, 21 (Suppl.): 7–10.

Randall, L. O. (1961): Pharmacology of chlordiazepoxide. *Diseases of the Nervous System*, 20 (Suppl.):7–15.

Rickels, K. (1971): *Personal communication*.

Salzman, C., DiMascio, A., Shader, R. I., and Harmatz, J. S. (1969): Chlordiazepoxide, expectation and hostility. *Psychopharmacologia* (Berlin), 14:38–45.

Scheckel, C. L., and Boff, E. (1967): Effects of drugs on aggressive behavior in monkeys. In: *Neuro-Psychopharmacology, V. Proceedings of the Fifth Meeting of the Collegium Internationale Neuro-Psychopharmacologicum*, edited by H. Brill, J. O. Cole, P. Deniker, et al. Excerpta Medica Series #129, Amsterdam, pp. 789–795.

Senault, B. (1970): Comportement d'aggressivité intraspécifique induit par l'amporphine chez le rat. *Psychopharmacologia* (Berlin), 18:271–287.

Sofia, R. D. (1969): Effects of centrally active drugs on four models of experimentally-induced aggression in rodents. *Life Sciences*, 8:705–716.

Tedeschi, R. E., Tedeschi, D. H., Mucha, A., Cook, L., Mattis, P. A., and Fellows, E. J. (1959): Effects of various centrally acting drugs on fighting behavior of mice. *Journal of Pharmacology and Experimental Therapeutics*, 125:28–34.

Valzelli, L., Giacalone, E., and Garattini, S. (1967): Pharmacological control of aggressive behavior in mice. *European Journal of Pharmacology*, 2:144–146.

Yen, H. C. Y., Katz, M. H., and Krop, S. (1970): Effects of various drugs on 3,4-dihydroxyphenylalanine (DL-DOPA)-induced excitation (aggressive behavior) in mice. *Toxicology and Applied Pharmacology*, 17:597–604.

Yen, H. C. Y., Stanger, R. L., and Millman, N. (1958): Isolation-induced aggressive behavior in ataratic tests. *Journal of Pharmacology and Experimental Therapeutics*, 122:85A.

DISCUSSION SUMMARY

Leo E. Hollister

Veterans Administration Hospital, Palo Alto, California, and Stanford University School of Medicine, Stanford, California

The paradox of possible release of aggressive tendencies in man by benzodiazepines and the opposite effect in animals has not been completely resolved. The curbing of aggressive displays following radiostimulation of various parts of the brain of monkeys, as shown in Delgado's film, was most impressive. These findings elaborated on the early observation of the "taming" effects of benzodiazepines on spontaneous aggression in primates. Part of the difficulty in extrapolating from various species to man may be due to the fact that metabolism of the drug may vary from one species to another. Valzelli's demonstration of an antiaggressive action in mice but not in rats suggests such an explanation.

On the whole, rhesus monkeys metabolize drugs in a fashion similar to man. It was emphasized that serious aggressive displays in man ("rage reactions") have been extremely rare phenomena, probably representing release of suppressed rage by a disinhibiting drug. Similar release has been seen following other such drugs, principally alcohol. On the other hand, mild release of aggression, as shown by increased ratings of hostile feelings in patients under therapy, may actually be therapeutically beneficial. No one could offer any explanation as to the differing effects, in this regard, between pharmacologically similar benzodiazepines.

In summary, it was felt that any possible release of aggression caused in man by benzodiazepines was a problem of little clinical importance and that the effects of suppression of aggression in man would be exceedingly difficult to evaluate.

The Benzodiazepines, Raven Press, New York © 1973

EFFECTS OF CHRONIC ADMINISTRATION OF BENZODIAZEPINES ON EPILEPTIC SEIZURES AND BRAIN ELECTRICAL ACTIVITY IN *PAPIO PAPIO*

E. K. Killam, M. Matsuzaki, and K. F. Killam

Department of Pharmacology, School of Medicine, and National Center for Primate Biology, University of California at Davis, Davis, California

I. INTRODUCTION

Naquet and the present authors have, in a series of papers, described the photically induced, epileptoid response of the Senegalese baboon *Papio papio* (Killam et al., 1966a, b, 1967a). The similarity between the elicited electroencephalographic patterns and behavioral concomitants of the baboon and those described in man (Naquet et al., 1960) led the authors to explore the usefulness of the baboon as a model for evaluating classical and new anticonvulsants. Early studies demonstrated that diazepam effectively blocked clinical seizures and abnormal EEG responses elicited by flashing light at a dosage range of 0.5 to 2.0 mg/kg (Killam et al., 1967a). These studies were extended to include the evaluation of two other benzodiazepines and their assessment in baboons with a full spectrum of stabilized responsiveness (Stark et al., 1970). From these studies, it was apparent that *Papio papio* represented an adequate model for assessing drugs within the benzodiazepine class.

In view of the scattered clinical reports that some patients controlled successfully with diazepam "escape from control" despite continued treatment, it appeared possible that the model provided by *Papio papio* might prove useful in studying this phenomenon. At the same time, the effects of chronically administered anticonvulsants could be characterized.

Over the last several years, a colony of 40 baboons has been studied, using standardized stimuli and control conditions to establish long-term patterns of response. Whenever photic stimulation elicited a marked seizure-like response, an animal was not subjected to the test conditions again for 1 week. Furthermore, the triggering stimulus pattern was never continued more than 10 to 15 sec after the onset of continuous, abnormal motor activity involving the musculature of the body. The attempt has been to stabilize the level of responsiveness in each animal over a period of several years, even though most animals do not then display as severe self-sustained responses as have been observed and described previously. This stable colony appeared ideal for the long-term study of chronic drug therapy, which, to the authors' knowledge, has not previously been accomplished in such a close animal model of human epilepsy. In this population, the effects of diazepam and 7-nitro-5-(2-chlorophenyl)-1,3-dihydro-2H-1,4-benzodiazepine-2-one (RO5-4023), administered both chronically and acutely, have been compared.

II. METHODS

The general methodology has been described previously. The animals were fitted with belts to facilitate capture for injections or transfer to the recording area. For all recording and test sessions, the animals were seated in chairs restraining movement at the hips and neck. Except for these 3 to 4 hr sessions, the animals were housed singly or in pairs. Electroencephalograms (EEG's) from most animals were monitored using needle scalp electrodes. For definitive evoked response and EEG studies, two animals had chronically implanted electrodes. In all cases, a classical EEG recording was taken. EEG's from the implanted animals were recorded in parallel on magnetic tape to allow subsequent processing using digital computational techniques.

After recording of spontaneous EEG, implanted animals were subjected to a series of flashes from a Grass photostimulator placed 30 cm from the eyes. At least 30 flashes at each of the following frequencies were used: 1, 2, 5, and 10 Hz, with an interval of recovery between sequences. All potentials in a sequence were recorded either with eyes open or with eyes closed for the entire period, since amplitudes of evoked potentials were different under the two conditions.

Moving average responses were calculated for groups of eight responses using a LINC II computer. In each response, 400 points over a 256-msec analysis period after each flash were digitized and used in calculation of the average.

Spectral analysis of the spontaneous EEG was used to characterize the distribution of power at various frequencies. With a Time/Data 100 spectral analyzer, the ensemble average of spectral densities was calculated using seven 8-sec continuous samples over 56 sec of elapsed time. Since baboons under daily drug administration were very active in the chairs, it was not possible in a half-hour period, 60 to 90 min after drug administration, to obtain long samples uninterrupted by artefacts of movement in any lead. Therefore, as an estimate of reliable power density distribution, five different 56-sec periods were analyzed, plotted, and then superimposed to show the variability of sampling.

Seizure sensitivity was tested at the end of the recording period in both implanted and nonimplanted animals, using the same light source and a precise sequence of 25-Hz light, followed by interrupted bursts of 25 Hz and then by a rapid glissando over the frequency spectrum from 2 to 40 to 2 Hz. Responses were graded for convenience as follows:

+1: EEG abnormalities with uncontrolled clonus of obicularis oculi and eyelid muscles.

+2: EEG spikes, spike and wave complexes of high voltage, and/or irregular paroxysmal activity, with muscle clonus plus twitching of face, neck, and head, and minor jerks.

+3: generalized twitching and/or jerking of eye, face, neck, and body musculature, with or without major convulsive movements and rapid contractures of the whole body.

Drug preparations were provided by the manufacturer (Roche) in injectable form, with ample supplies of the appropriate diluent to allow assessment of vehicle effects and constant volumes of injection. During the chronic administration regimens, intramuscular injections were made in the home cage or in the recording area at approximately the same time every day. Dose regimens were of two types. In one series, a dose was selected which prevented the seizure response to flashing light; this dose was given daily, with weekly testing of seizure threshold and pattern 60 to 90 min after the daily dose. In other animals, the dose selected was the minimal effective dose on single administration. The dose was given daily until evidence was obtained that it did not prevent the seizure response. The dose was then doubled, and the new dose was injected daily until the animal was no longer controlled. The dose level was again doubled in some animals to obtain complete blockade of seizures.

III. RESULTS

A. Seizure control following single doses of diazepam and RO5-4023

In the group of 22 selected animals, the individual dose-response curves to administration of single doses of diazepam and RO5-4023 were established. Irrespective of the absolute dose range, in all animals RO5-4023 was observed to be approximately 10 times more effective than diazepam (e.g., 5 μ versus 50 μ for seizure control).

B. Seizure control under chronic dosage with diazepam and RO5-4023

In a series of seven baboons maintained for more than 2 years with weekly or biweekly testing of seizure sensitivity, crossover studies were then undertaken which indicated that, at low doses of both benzodiazepines (close to threshold for control of seizures in the individual animal), there is evidence of escape from therapeutic control with daily administration.

As can be seen from Figs. 1–3, each animal in the series responds somewhat differently to chronic administration of the benzodiazepines, even

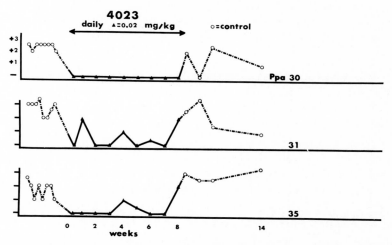

FIG. 1. Seizure responses evoked by intermittent light stimulation (ILS) at weekly intervals before (open circles), during (closed symbols), and after (open circles) daily administration of RO5-4023 at 0.02 mg/kg. "Before" drug control studies have been compressed merely to reduce illustration space. See Section II for interpretation of ordinate values.

though the initial dose given was selected because it completely prevented light-induced EEG and motor abnormalities.

Ppa 30, an animal with stable, self-limited +3 responsiveness over a long period, appeared never to have seizure responses during an 8-week period of drug administration (Fig. 1). However, instability of seizure control was evidenced in weekly testing of Ppa 31 and 35 when the drug-dose schedule remained constant at 0.02 mg/kg.

The other four animals in this series demonstrated the same phenomenon even more strikingly. It should be noted that, in this set of baboons (Fig. 2), the dose was doubled when loss of seizure control was evidenced.

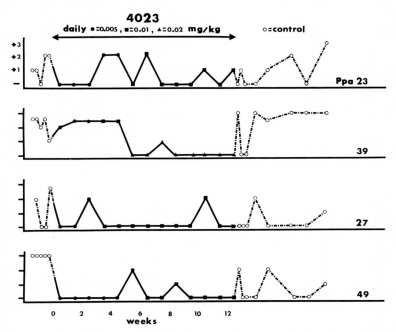

FIG. 2. Seizure responses evoked by ILS at weekly intervals before, during, and after daily administration of RO5-4023 at increasing doses. Symbols indicate response after seven daily injections at the dose shown. Rest of figure as in Fig. 1.

In Ppa 23, increasing the dose from 0.005 to 0.01 mg/kg at first blocked seizures again, but control was unstable thereafter for up to 3 months of chronic administration. In Ppa 39, the dose which in a single experiment several weeks earlier had blocked seizure activity, only slightly modified the seizure pattern after seven daily injections of the same dose. The dose was first doubled and then doubled again to 0.02 mg/kg before complete control was reestablished. Similar instability of the effectiveness of the RO5-4023 was evidenced in the other two animals. In each case, doubling the initial dose temporarily restored the anticonvulsant effect of the compound.

Further evidence of some tolerance build-up with RO5-4023 was found in the pattern of seizure response after cessation of the drug. Several animals appeared to have somewhat more marked seizure responses for 1 or more weeks. This was especially notable in Ppa 39, in which limited +3 responses were converted for about 5 weeks to full seizures outlasting the light stimulation. This is shown in Fig. 2 on the first recovery day after drug cessation in which such a severe response occurred. The animal was then unresponsive for several days. At weekly intervals thereafter, the prolonged seizure response was again elicited. A somewhat similar effect occurred in

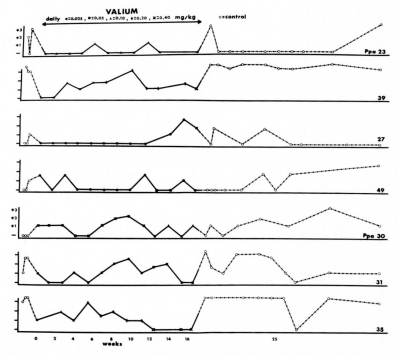

FIG. 3. Seizure responses evoked by ILS at weekly intervals before, during, and after daily administration of diazepam (VALIUM) at increasing doses. Symbols and interpretation as in Figs. 1 and 2.

Ppa 35. Abnormally severe seizure responses followed removal of the drug in this animal, whose usual response had been +1 or +2 for some months previous to this study (Fig. 1).

After several months of weekly control studies of seizure patterns, the same seven animals were tested with comparably low doses of diazepam (VALIUM) given daily. All but one of the animals showed rapid adaptation to the drug, and even that baboon (Ppa 27) eventually began to show more or less severe epileptoid responses to the flashing light (Fig. 3). Doubling the dose once or several times was necessary to restore seizure control. Again, in animals whose normal responses were not maximal, it was possible to demonstrate unusually severe or long-lasting abnormalities of EEG and motor seizures immediately after chronic drug administration was terminated. Daily drug administration, using higher initial doses, was studied in three of the animals mentioned above and in one other animal (Fig. 4). Diazepam at 0.5 mg/kg was injected intramuscularly daily for 1 month and then twice daily for an additional month. No behavioral side effects were noted at this dose, and there was no evidence of escape from seizure control. For 4 weeks after discontinuance of drug treatment, the least epileptic animal, Ppa 23, which, under control conditions before any chronic administration of drug, had never had more than a minimal motor response along with EEG paroxysms, gave evidence of a definitely increased seizure responsiveness. A self-sustained major seizure was elicited 2 days after discontinuance of diazepam. Thereafter, that animal was unresponsive for a long period of time. Several other animals also showed a disturbance of normally stable control patterns after the long chronic series of drug injections.

C. Evoked cerebral activity

In two animals with implanted electrodes, evoked responses to flashes presented at slow frequencies were analyzed over each of the experimental periods. Moving averages of eight responses to intermittent light stimulation (ILS) at 1, 2, 5, and 10 Hz were calculated for series of 30 potentials. Recordings from epidural electrodes over frontal and parietal area, with eyes open or closed, showed disappointingly small amplitudes and no changes over the course of the chronic experiments other than a slight generalized loss of amplitude.

However, occipital derivations showed a surprising change in the form of potentials over the course of the 3 months of the experiment. In Ppa 39, who received 0.005 mg/kg daily for 3 weeks and then 0.01 mg/kg for 2

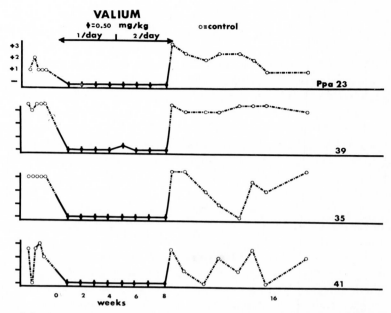

FIG. 4. Seizure responses evoked by ILS at weekly intervals before, during, and after diazepam, given in doses of 0.5 mg/kg daily or twice daily, as indicated. See Fig. 1 for details.

weeks, a slight and variable increase in amplitude of the evoked average potential was noted each week. However, after 5 weeks, when the dose was increased to 0.02 mg/kg to reinstitute control of seizures, an interesting alteration in the form of the potential was noted in *both* occipital leads. Multiple small peaks appeared approximately 100 to 125 msec after the flash of light, which will be described as a "ringing" effect for convenience. Examples from typical weekly recordings are shown in Fig. 5. The mechanism of the effect is still unknown. This altered potential persisted to the end of the experiment (12 weeks) but was no longer present 24 hr after the last dose of the drug (Fig. 5). The effect was not an expression of an altered variability of latency or amplitude in the potentials averaged. Standard deviations of each of the 400 digitized and averaged points making up each average potential were calculated and plotted. No increase in variability at the crucial time after flash can be seen when the calculated standard deviations of each point are plotted against a zero baseline (Fig. 6). Lines 2, 4, and 6 in the figure show this calculation write-out. This characteristic alter-

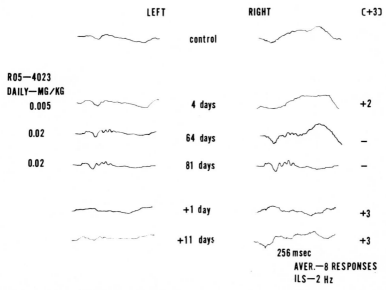

FIG. 5. Averaged evoked potentials recorded over occipital cortex following eight successive flashes of light at 2 Hz. Examples from typical weekly recordings from Ppa 39 during experiment shown in Fig. 2. Seizure response to 25-Hz ILS shown at right.

ation in occipital cortex did not seem to result from the lower doses given to the other implanted animal. At 0.005 and 0.01 mg/kg, the occipital potentials from Ppa 23 appeared to be only slightly increased in definition and amplitude.

The interesting finding just described was therefore checked in a naive animal, in which 20 cortical derivations were recorded in a manner similar to the recordings in the long-term chronic experiment. Figure 7 shows the location of the electrodes. Derivations 1 and 2 were implanted in more or less frontal areas, 10 was temporal, and 5, 6, and 9 were placed in the occipital area in an attempt to localize the phenomenon. Little if any effect other than slight depression was noted in most areas, just as in the chronically treated baboons (Fig. 8). In occipital areas, however, both 0.01 and 0.05 mg/kg in single doses induced the ringing effect which had been noted in the chronic animal. Both medial and lateral derivations showed the change in potential form. Figure 9 shows left and right medial occipital recordings. Averaged potentials selected from a series, calculated with the moving average technique, are shown here to demonstrate the extraordinary consistency

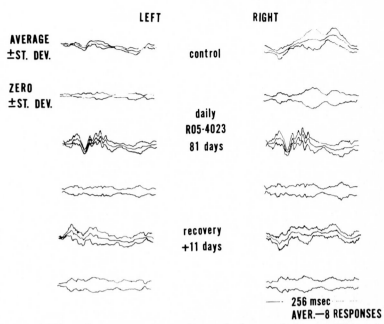

FIG. 6. Variability of responses averaged to calculate evoked potentials on occipital cortex shown in Fig. 5, lines 1, 4, and 6. For each of the 400 digitized points in an average potential, the standard deviation was calculated and plotted first against the average and immediately below against a zero baseline. Note that at the time of the "ringing" effect (see text) after 81 days of drug administration (line 3), there was no increase in variability of the potentials. This is best demonstrated in line 4, in which standard deviations at all points form almost a straight line when plotted directly against a zero baseline.

of the effect from the first eight to the last eight potentials of the series of 30 flashes. A more lateral derivation on both left and right sides also showed the effect, although to a lesser degree (Fig. 10).

D. Influence of chronic drug administration on spontaneous EEG

Spontaneous EEG tracings recorded weekly over the period of study indicated some striking shifts of frequency distribution of ongoing cortical activity under the influence of the anticonvulsants. In frontal cortex, in the control period preceding an evoked epileptoid response, consistent spectral peaks at 1 to 2, 3 to 5, and 14 to 17 Hz were seen. These are shown in Fig. 11 for one animal. During the chronic regimen of RO5-4023, as seizure control was instituted these spectral characteristics decreased. Most striking

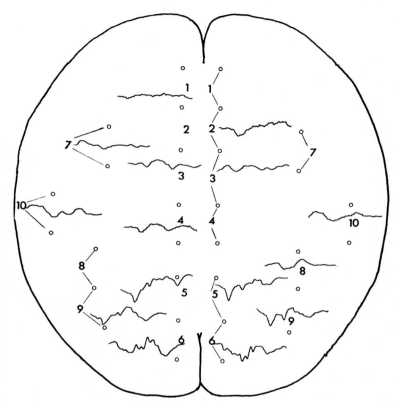

FIG. 7. Location of sites of cortical recordings shown in Figs. 8–10, with typical average responses to the first eight successive flashes at 1 Hz.

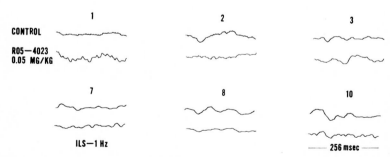

FIG. 8. Average evoked responses to eight flashes of light at 1 Hz before (control) and 90 min after a single dose of RO5-4023 at 0.05 mg/kg. Recordings from sites indicated in Fig. 7. Left and right sides were similar in form of potential and relatively small depressant effect of the drug.

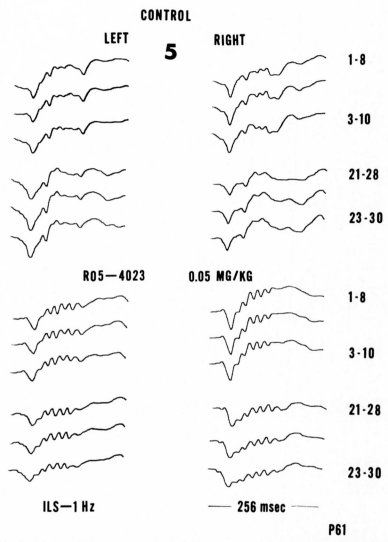

CONTROL

LEFT **5** **RIGHT**

1-8

3-10

21-28

23-30

RO5—4023 **0.05 MG/KG**

1-8

3-10

21-28

23-30

ILS—1 Hz — **256 msec** —

P61

FIG. 9. Average evoked responses to flashing light at 1 Hz from left and right medial parieto-occipital cortex before and 90 min after a single dose of RO5-4023 at 0.05 mg/ kg. Data from same experiment as that shown in Figs. 7 and 8. Sites of recordings shown in Fig. 7. Moving averages of groups of eight potentials were successively calculated and those of the first 8, 2nd through 9th, 3rd through 10th, 21st through 28th, 22nd through 29th, and 23rd through 30th are shown to demonstrate the stability of the form over time. Average of first eight was recorded simultaneously with those shown in Fig. 8.

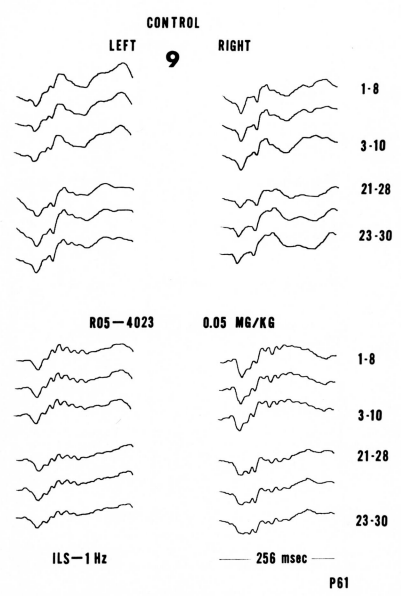

FIG. 10. Average evoked responses to flashing light at 1 Hz from left and right lateral occipital cortex before and 90 min after a single dose of RO5-4023 at 0.05 mg/kg. Data was recorded simultaneously with that shown in Fig. 9, and sites are shown in Fig. 7. See Fig. 9 for further explanation.

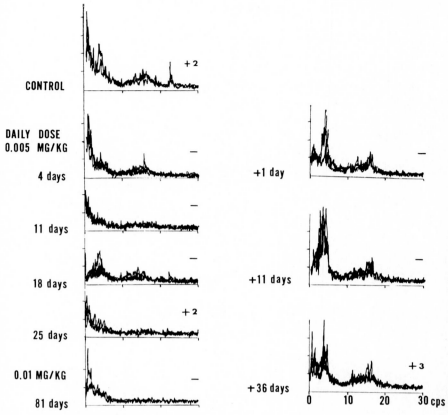

FIG. 11. Power spectral analysis of 56-sec samples of frontal cortical EEG on selected days during the experiment shown in Fig. 2, Ppa 23. Four or five individual spectra were superimposed for each trace to indicate variability. During drug administration, EEG samples were taken 60 to 90 min after the daily dose. Seizure response is given to the right of each set of traces. The right column shows traces during the recovery period.

was the loss at 14 to 17 Hz, while variable amplitudes of spectral lines were observed at lower frequencies. Following termination of the drug, there was a partial return to the control spectrum, but a single dominant peak of spectral lines at 3 to 5 Hz was evident for more than 11 days post-drug, with gradual reduction toward control levels by 36 days and a restoration of a peak at the slower frequency. In this more or less medial frontal lead, there was no evidence of any increase in power at faster frequencies. Analyses up to 60 Hz were made on all samples, but there was virtually no recordable power beyond the 30 Hz shown here.

In parietal cortex, much less effect of the drug was seen in both animals, except for loss of 1 to 2 Hz activity which failed to return within the 2-month post-drug period over which the animals were studied. Examples are shown in Fig. 12. Some increase in total power in the 3 to 5 Hz range indicated a shift of the normal 1 to 2 Hz activity.

In occipital cortex, there was no evidence of accentuation of power in any frequency band during chronic drug administration. Except in recordings taken on the 18th day of drug dosage in one animal (Fig. 13), all occipital recordings showed a gradual loss of spectral characteristics which only partially recovered in the 2 months following cessation of drug. There was no correlation observed between seizure responsiveness and power spectra, as can be seen by the notation to the right of each spectrum.

IV. CONCLUSIONS

From these studies, diazepam and the 7-nitro derivative RO5-4023 appear to share many properties. RO5-4023 was approximately 10 times more

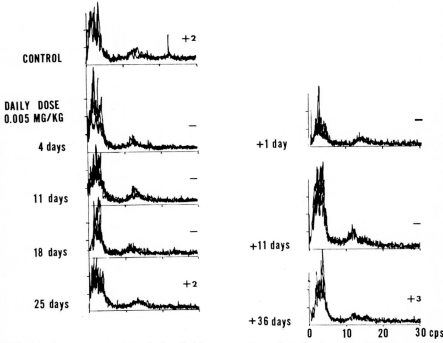

FIG. 12. Power spectral analysis of 56-sec samples of parietal cortical EEG recorded from Ppa 23 simultaneously with those shown in Figs. 11 and 13.

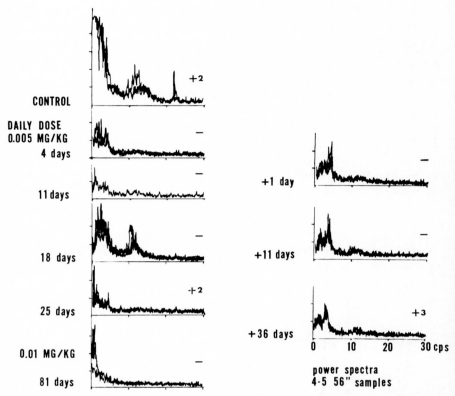

FIG. 13. Power spectral analysis of 56-sec samples of occipital cortical EEG recorded from Ppa 23 simultaneously with those shown in Figs. 11 and 12.

effective than diazepam as an anticonvulsant, under either an acute or chronic dosage schedule. Both compounds exhibited a decrease in effective control over time—the so-called escape phenomenon. However, loss of therapeutic control appeared to be related to the initial stabilization dosage range. With both larger and threshold doses, abrupt termination of chronic drug treatment lead to more striking epileptoid responses at least for a few days, often for several weeks. The data thus suggests a build-up of tolerance which should be investigated further. It should be possible to determine whether a shift in distribution or metabolism of the compounds is taking place, or whether a change in activity of brain structures involved in seizure initiation or spread are responsible for the changes in control of seizures.

From these studies, some clinical implications emerge. It is the general practice to institute anticonvulsant therapy by employing accepted "threshold" dosage regimens and adjusting the schedule upwards as the patient "escapes from control." With this class of drug, it may be more efficacious to use higher initial regimens despite possible mild neurotoxic signs. The latter would attenuate with time, but the chance of escape from therapeutic control would be lessened.

Papio papio suggests itself as a useful "clinical" model for controlled studies of this type. No other studies on the influence of chronic drug treatment on seizures in animal models of inborn epilepsies have, to our knowledge, thus far appeared.

Evoked potential studies identified changes in evoked responses in the occipital regions induced by RO5-4023, which paralleled seizure control. Early studies of such potentials had indicated that, in the epileptic baboon, the distribution of potentials induced by ILS was much more widespread than in the nonepileptic baboon or in normal man (Killam et al., 1967*a, b*). Phenobarbital in the epileptic baboon seemed to enhance potentials in visual areas and to reduce them in parietal and frontal cortex (Killam et al., 1967*a, b*, 1968). This effect, which paralleled anticonvulsant activity, tended to alter evoked potentials toward the normal condition. Similarly, single doses of diazepam were shown to reduce the amplitudes of potentials in frontal and parietal regions, while regularizing those of posterior regions (Killam et al., 1967*b*). The findings with chronically administered RO5-4023 indicate some shift toward increased occipital responsiveness to ILS, with a concomitant reduction in other (nonvisual) areas.

When one examines the effects of chronically administered RO 5-4023 on the frequency distributions of EEG activity using power spectral analysis, there appears to be a generalized reduction of spectral lines above 2 Hz in the cortex areas sampled. The data will be presented in detail elsewhere.

V. SUMMARY

Problems of maintaining seizure control with benzodiazepines administered chronically at stable dose levels have been investigated using *Papio papio* as an animal model of photomyoclonic epilepsy in man. Individual threshold doses for the control of epileptic responses to flashing light were determined in seven baboons for both diazepam and 7-nitro-5 (2 chlorophenyl) 1,3,dihydro-2H-1,4,benzodiazepine-2-one (RO5-4023). Spontaneous EEG, evoked activity to low-frequency flashing light, and seizure responses to 25 Hz were recorded weekly or biweekly for a period of 2 years, during which a 2-month period of daily dosage with diazepam, a 3-month period of daily dosage with RO5-4023, and a 3-month period of daily dosage with diazepam were studied, as were long interspersed control periods without medication. At threshold doses, both compounds failed to maintain seizure control when administered daily, but seizures were again suppressed on raising the daily dose level. In some animals, "escape" was evidenced and doses had to be increased several times. If initial doses were slightly higher, the animals failed to show the adaptation to drug. A brief rebound phenomenon occurred in some animals during the immediate post-drug periods. Spontaneous and evoked electrical activity (indicated by averaged potentials and spectral power density analysis) changed only slightly at these low dose levels. There were no behavioral changes with the drugs.

ACKNOWLEDGMENTS

Supported in part by USPHS Grants NS 08935 and MH 17471.

The authors gratefully acknowledge a grant of equipment from the Roche Foundation and express their appreciation to Drs. Harrison Langrall and Lowell O. Randall of Hoffman-LaRoche Inc., Nutley, N. J., for supplies of diazepam and RO5-4023.

It is a pleasure to acknowledge the able technical assistance of Douglas Lytle in carrying out the experiments and of Edwin Budd in computing the data.

REFERENCES

Killam, K. F., Killam, E. K., and Naquet, R. (1966a): Mise en évidence chez certains singes d'un syndrome photomyoclonique. *Comptes Rendus Hebdomadaires des Séances de l'Académie des Sciences, D. (Paris)*, 262:1010–1012.
Killam, K. F., Killman, E. K., and Naquet, R. (1966b): Etudes pharmacologiques réalisées chez des singes présentant une activité E.E.G. paroxystique particulière à la stimulation lumineuse intermittente. *Journal de Physiologie*, 58:543–544.
Killam, K. F., Killam, E. K., and Naquet, R. (1967a): An animal model of light sensitive epilepsy. *Electroencephalography and Clinical Neurophysiology*, 22:497–513.
Killam, K. F., Killam, E. K., and Naquet, R. (1967b): Evoked potential studies in response to light in the baboon *(Papio papio)*. *Electroencephalography and Clinical Neurophysiology (Suppl.)* 26:108–113.
Killam, K. F., Killam, E. K., and Stark, L. G. (1968): Studies of the effect of phenobarbital on the photic responses in the baboon *(Papio papio)*. In: *A Symposium: Use of Non-human Primates in Drug Evaluation*, edited by H. Vagtborn. University of Texas Press, Austin, pp. 425–441.
Naquet, R., Fegersten, L., and Bert, J. (1960): Seizure discharges localized to the posterior cerebral regions in man provoked by intermittent photic stimulation. *Electroencephalography and Clinical Neurophysiology*, 12:305–316.
Stark, L. G., Killam, K. F., and Killam, E. K. (1970): The anticonvulsant effects of phenobarbital, diphenylhydantoin and two benzodiazepines in the baboon, *Papio papio*. *Journal of Pharmacology and Experimental Therapeutics*, 173:125–132.

EXPERIMENTAL AND CLINICAL STUDIES OF THE ANTICONVULSANT PROPERTIES OF A BENZODIAZEPINE DERIVATIVE, CLONAZEPAM (RO 5-4023)

**Gian Franco Rossi, Concezio Di Rocco,
Giulio Maira, and Mario Meglio**

Institute of Neurosurgery, Catholic University, Rome, Italy

I. INTRODUCTION

This chapter describes the studies conducted by the authors, in the last few years, on the antiepileptic properties of a benzodiazepine derivative, 7-nitro-5-(2-chlorophenyl)-3H-1,4 benzodiazepine-2(1H) or clonazepam (RO 5-4023). Only a few papers have been published on this particular benzodiazepine derivative (Swinyard and Castellion, 1966; Gastaut, 1969; Gastaut et al., 1969; Giunta et al., 1969, 1970; Guerrero-Figueroa et al., 1969*a, b*; Poiré and Royer, 1969; Turner et al., 1969*a, b*; Fariello and Mutani, 1970; Lison and Fassoni, 1970; Bergamini et al., 1971). The data reported by the different authors are not always in agreement. However, as will be shown below, we think that, on the whole, the results so far obtained in our own as well as in other laboratories are sufficient to indicate the main characteristics of the antiepileptic property of the drug in question.

II. MATERIALS AND METHODS

Our research has been performed on epileptic patients (49 subjects of both sexes from 8 to 57 years of age) and in animals (13 adult cats and 20 rabbits). Most of our findings have been obtained by injecting the benzodiazepine derivative intravenously and searching for its immediate effect on both the behavioral and electroencephalographic symptoms of epilepsy. In 27 patients, the drug was given orally, and the treatment was protracted for some weeks. In these cases, the effects on seizure frequency and the electroencephalogram were analyzed.

Patients suffering from different types of epilepsy were examined, as described below. In some of them, in addition to standard electroencephalograms (EEG's), stereoelectroencephalographic recordings (SEEG's) from different brain structures—mainly temporal and frontal—were also performed.

For the research on animals, a simple model of experimental epilepsy was utilized: 1 strychnine or penicillin was acutely applied on the exposed cortex of the frontal lobe (sigmoid gyrus). Cortical as well as subcortical electrodes were used for recording the local epileptic discharges and their propagation. The animals were anesthetized with urethane. Both in humans and in animals, intravenously injected bemegride (MEGIMIDE) was sometimes used to enhance epileptic activity.

Some of the results obtained have been already published (Giunta et al., 1969, 1970; Di Rocco et al., 1971).

III. RESULTS

To facilitate discussion, the results of our study can be schematically subdivided into two main groups: those related to the effect of clonazepam on the generalized signs of epilepsy, and those related to the effect of the drug on the partial or focal epileptic signs.

A. Effect of clonazepam on the generalized epileptic activity

1. Generalized epilepsy produced by intravenous injection of bemegride

It is well known that intravenous bemegride is capable of provoking a generalized seizure both in cats and in man. When slowly injected in animals or in patients suffering from partial epilepsy, the convulsant drug can initially produce an enhancement of the partial epileptic activity; later on, however, generalization frequently occurs. We shall deal here only with the generalized epileptic signs.

Figure 1 shows the effect of clonazepam on the generalized epileptic EEG activity produced by bemegride in the cat. High-amplitude epileptic potentials occur apparently synchronously in all the available recordings. An immediate suppression of any epileptic activity follows the intravenous injection of 1 mg of the benzodiazepine derivative. This type of effect was always observed in all our experiments.

Figure 2A reproduces the EEG record obtained during a generalized epileptic seizure induced by bemegride; 100 mg of the drug is the lowest

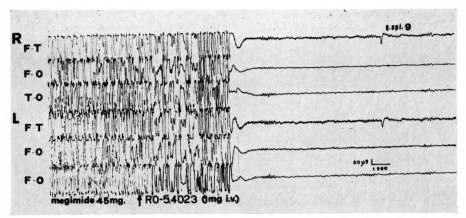

FIG. 1. Cat under urethane anesthesia. Blocking of generalized epilepsy by clonaze-pam i.v. injection. For detailed explanation, see text. In this and in the following figures: R =right, L = left, F = frontal, T = temporal, O = occipital, C = central, a = anterior, m = middle, p = posterior EEG recordings.

dose capable of producing generalized epilepsy (i.e., threshold dose). As soon as the effect of the convulsant drug has completely disappeared, 0.5 mg of clonazepam is injected intravenously. As shown in Fig. 2B, the subsequent injection of 100 mg of bemegride is no longer capable of provok-ing epileptic signs. As shown in Fig. 2C, 350 mg of bemegride is necessary at this point to precipitate a generalized epileptic seizure.

The capacity of the benzodiazepine derivative to increase markedly the threshold of resistance to the epileptogenic action of bemegride has also been proven in epileptic patients. Figure 3 illustrates one of these cases. The EEG record was obtained from a 21-year-old man suffering from generalized convulsive seizures. Twenty-five mg of bemegride constituted the thresh-old dose required to precipitate the seizure, as illustrated in Fig. 3A. In a second examination, performed one day after the first, it is shown that, if a very low amount of clonazepam (0.02 mg/kg) is injected intravenously prior to bemegride administration, the convulsive threshold increases up to 240 mg (see Fig. 3B). The phenomenon has been confirmed in all 13 patients subjected to this test (see Section IV for the diagnostic purposes of the test). The convulsive threshold was found to increase about fourfold.

2. Generalized epileptic signs due to brain disease

Our patients showing generalized epileptic signs (20 cases) formed a rather heterogeneous group, as far as type of epilepsy was concerned.

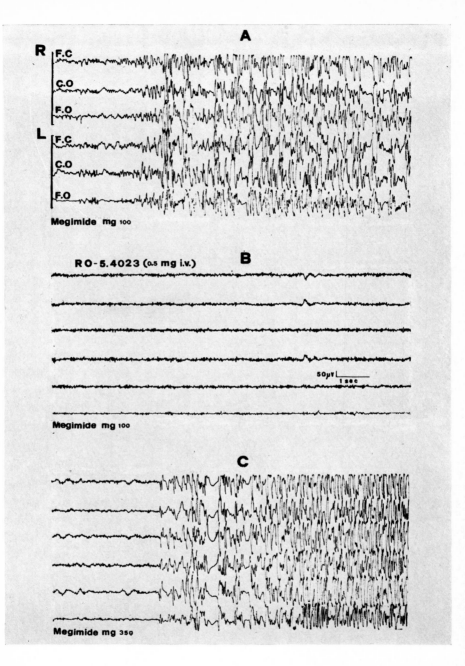

FIG. 2. Cat under urethane anesthesia. Increase of the convulsive threshold of bemegride (MEGIMIDE) following clonazepam i.v. injection. For detailed explanation, see text.

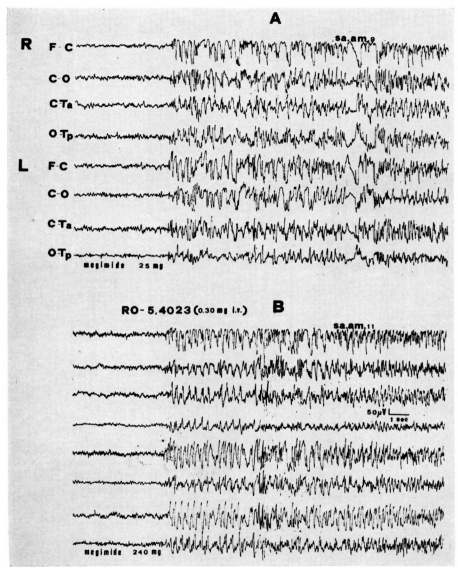

FIG. 3. EEG record of epileptic patient. Increase of the convulsive threshold of bemegride (MEGIMIDE) following clonazepam i.v. injection. For detailed explanation, see text.

However, they had in common the occurrence of generalized convulsive or nonconvulsive seizures and the presence of generalized bilateral and syn-

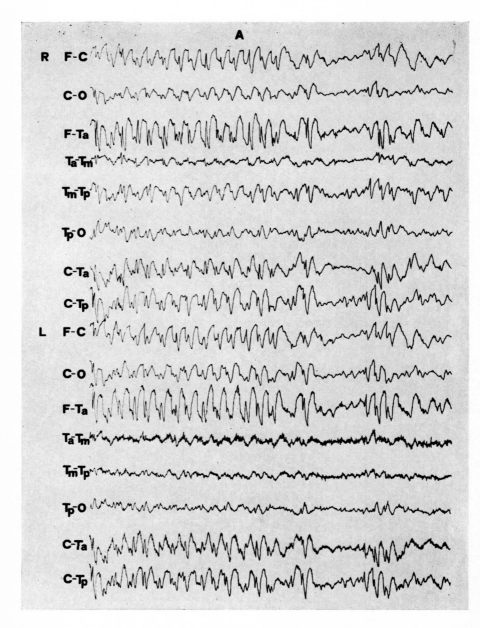

FIG. 4. EEG record of epileptic patient. Disappearance of EEG signs of general-

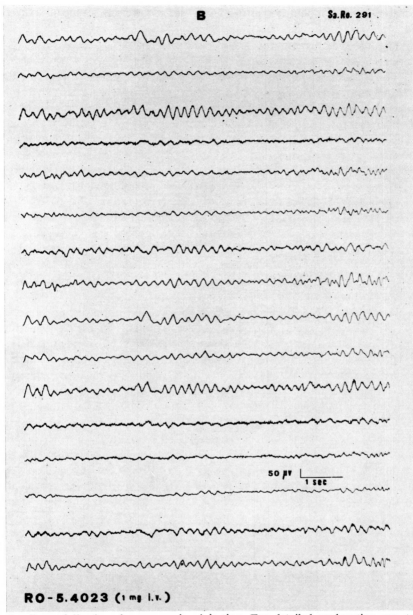

ized epilepsy following clonazepam i.v. injection. For detailed explanation, see text.

chronous epileptic potentials in the EEG—in most cases, atypical spike-and-wave complexes.

The intravenous administration of clonazepam was quite effective in suppressing the EEG generalized epileptic activity. Figure 4 illustrates the EEG record of one patient who showed frequent discharges of generalized atypical 3 cps spike-and-wave complexes. Intravenous injection of 1 mg of clonazepam was followed by the disappearance of the epileptic discharges. This clear-cut antiepileptic effect lasts for about 15 min.

Fifteen patients suffering from daily occurring generalized convulsive or nonconvulsive seizures were treated for at least 4 weeks with clonazepam administered orally. The daily dose varied from 2 to 20 mg. In most subjects, clonazepam was added to the antiepileptic drugs with which the patient had been treated previously for a long time. In 10 of the 15 patients, the effect was good as shown by a significant reduction and even by the disappearance of the seizures, and by the obvious improvement of the EEG record. In the remaining five patients, no signs of improvement could be detected. In addition, the observation was made that, in the patients who benefited from the clonazepam treatment, the antiepileptic action of the drug gradually decreased with the passage of time.

Two of the cases studied represent a particularly interesting illustration both of the very good immediate effect of clonazepam on generalized epilepsy and of its short-lasting action. The subjects were two brothers, 22 and 28 years old, suffering from bilateral massive epileptic myoclonus, classified as Unverricht-Lundborg syndrome. Figure 5A illustrates the generalized EEG epileptic discharges recorded in one of these patients. Figure 5B shows the EEG recorded a few minutes after the intravenous injection of 1 mg of clonazepam: the epileptic discharges have disappeared. The same remarkable effect of the benzodiazepine is also exerted on the convulsive discharges evoked by intermittent photic stimulation, to which both patients were highly sensitive (Fig. 6). The two brothers were given clonazepam orally. The results were excellent in the first week, with 4 mg of the drug; then the antiepileptic action gradually became attenuated. The daily amount of the drug was increased to 6 mg, but at the end of the first month of treatment clonazepam had almost completely lost its capacity for controlling the motor as well as the EEG epileptic manifestations.

B. Effect of clonazepam on partial or focal epileptic activity

1. *Partial epilepsy due to local cortical lesions*

The effect of clonazepam on a simple experimental model of focal

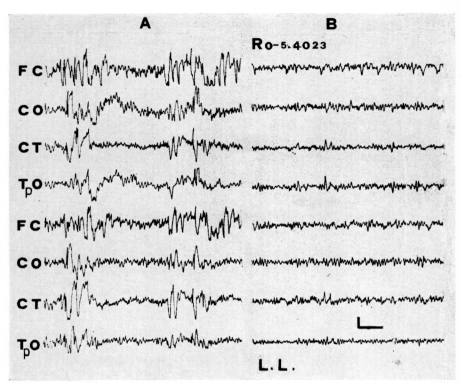

FIG. 5. EEG record of epileptic patient. Disappearance of EEG signs of generalized epilepsy following clonazepam i.v. injection. For detailed explanation, see text.

cortical epilepsy is illustrated in Fig. 7. An epileptic focus had been created in the cat by topical application of penicillin on the left sigmoid gyrus. Spikes were recorded locally (Fig. 7A), and 0.5 mg of clonazepam was then injected intravenously. A sleep-like change of the EEG occurred, thus indicating an obvious effect of the drug on the central nervous system. However, the epileptic spikes persisted almost unmodified (Fig. 7B). The same negative results were obtained in all experiments.

Conversely, the results obtained in epileptic patients did not appear to be uniform. Altogether, the effects of intravenous injection of clonazepam were studied in 21 subjects suffering from partial epilepsy, the epileptogenic zone being located most often in the temporal lobe and, in some cases, in the frontal or parietal lobes. Figure 8 shows the EEG of a girl suffering daily from partial seizures of temporal origin. Independent epileptic potentials were recorded on the temporal and central regions of the scalp on both

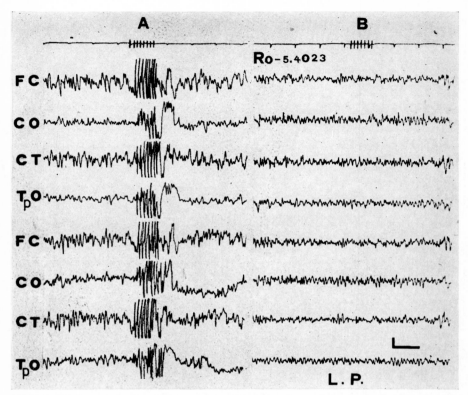

FIG. 6. EEG record of epileptic patient. Disappearance of generalized epileptic discharges evoked by intermittent photic stimulation (above in the figure) following clonazepam i.v. injection. For detailed explanation, see text.

sides (Fig. 8A). Figure 8B shows the EEG recorded some minutes after the intravenous injection of 1 mg of clonazepam: the focal epileptic potentials have disappeared.

In most of our cases, however, the drug was either not capable of suppressing EEG focal intercritical discharges or simply reduced their amplitude and frequency. This is shown in Fig. 9, illustrating the EEG record from a 16-year-old male suffering from partial seizures with elementary motor symptomatology and showing intercritical focal potentials on the left fronto-temporal region (Fig. 9A). As illustrated in Fig. 9B, these discharges are not affected by 1 mg of intravenously injected clonazepam.

The very poor effect of acute administration of the benzodiazepine on intercritical focal epileptic activity was even more apparent when evaluated

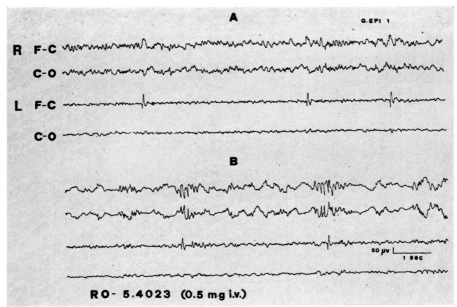

FIG. 7. Cat under urethane anesthesia. Persistence of focal epileptic potentials (L.F-C) following clonazepam i.v. injection. For detailed explanation, see text.

on SEEG recordings. Figure 10 shows a sample of a SEEG from a young woman suffering from frequent partial seizures which were considered to be of temporal origin. Intercritical epileptic potentials are recorded from left temporal and frontal structures; some spikes are present in the left centro-median thalamic nucleus as well (Fig. 10A). One mg of clonazepam was injected intravenously: the intercritical epileptic activity directly recorded from the brain does not seem to be significantly affected (Fig. 10B).

Four patients suffering from *epilepsia partialis continua* enabled us to test the effect of the intravenously injected clonazepam on a much stronger focal epileptic activity producing clear-cut focal motor manifestations. Figure 11A shows the EEG recorded from a patient who had been operated on 1 month before for a small astrocitoma located on the left cerebral hemisphere, at the junction of the temporal, parietal, and occipital lobes. A continuous spike discharge was recorded on the EEG from the left scalp. Corresponding continuous jerks of the right side of the body were present: the jerks of the right arm are recorded in the last EEG channel. Figure 11B illustrates the effects of the intravenous injection of 1 mg of clonazepam: the focal EEG epileptic activity was still present, although amplitude and fre-

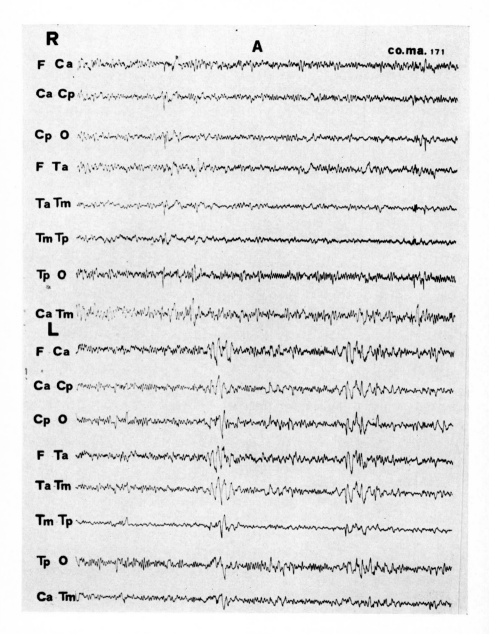

FIG. 8. EEG record of epileptic patient. Disappearance of focal epileptic discharges

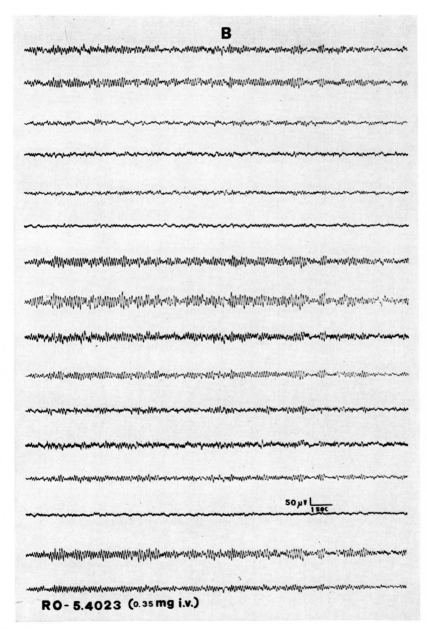

following clonazepam i.v. injection. For detailed explanation, see text.

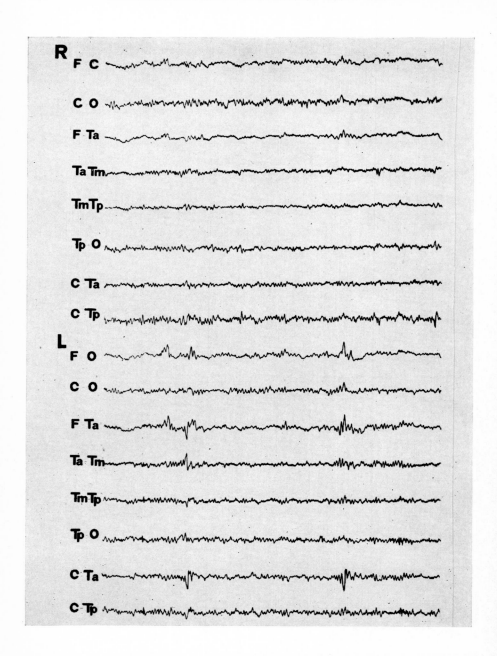

FIG. 9. EEG record of epileptic patient. Persistence of focal epileptic discharges fol-

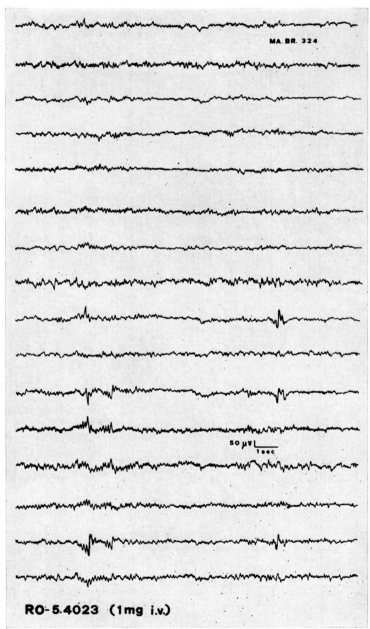

lowing clonazepam i.v. injection. For detailed explanation, see text.

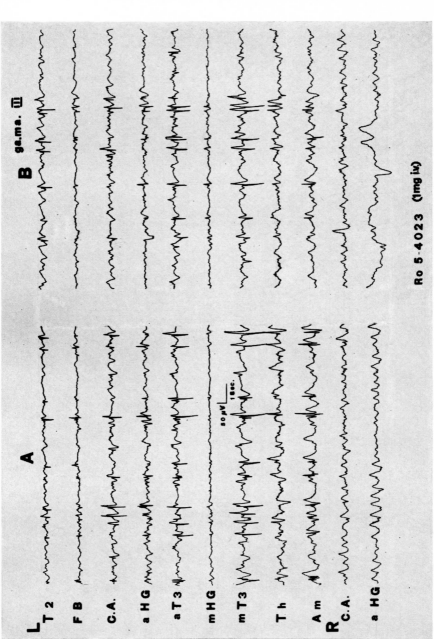

FIG. 10. SEEG record of epileptic patient. Persistence of focal epileptic discharges following clonazepam i.v. injection. For detailed explanation, see text.

L = left; R = right; T2 and T3 = 2nd and 3rd temporal circonvolutions; FB = fronto-basal; C.A. = Ammon's horn; a HG, m HG: anterior and middle hippocampal gyrus; Th = centro-median thalamic nucleus; Am: amygdala.

quency of the epileptic potentials were reduced; the motor jerks had disappeared.

Of the four similar cases examined, only in the one just illustrated could a certain efficacy of clonazepam in controlling *epilepsia partialis continua* be observed. In the other three cases, no effect was produced by the drug either on the motor or on the EEG epileptic manifestations.

Only 12 of our patients suffering from partial epilepsy received protracted oral treatment for a time sufficiently long to evaluate the effectiveness of the drug. Six of them suffered from Jacksonian motor seizures, one from partial sensory seizures, and the remaining five from partial seizures with complex symptomatology of temporal origin. As in the case of patients suffering from generalized epilepsy, in most of the subjects with partial epilepsy, clonazepam also had to be added to the previous antiepileptic therapy. Four to 8 mg of the benzodiazepine was given daily for at least 4 weeks. In seven patients, the initial results appeared satisfactory, as shown by the reduction in the number of seizures as well as by a parallel improvement of the EEG. However, as in the patients suffering from generalized epilepsy, the effect of the drug decreased with the passage of time. The remaining five patients did not appear to benefit at all from the benzodiazepine treatment. Incidentally, it may be worth pointing out that in two of these five cases, one Jacksonian and one temporal epileptic, a paradoxic increase in the number of seizures was observed in the first days of benzodiazepine treatment.

2. *Propagation of focal epileptic activity*

As is well known, the epileptic discharges arising from a given neuronal population show a tendency to propagate to synaptically related cerebral structures. Such a propagation can be favored by several local and general factors. It therefore was considered of interest to study the effect of clonazepam on the propagation of epileptic activity.

The simplest experimental model of epileptic propagation is that of the so-called cortical "mirror focus." Figure 12A shows the EEG recorded from a cat in which epileptic focus was created on the left sigmoid gyrus by means of penicillin; epileptic spikes were recorded locally. Following the injection of 30 mg of bemegride, spikes appeared in the contralateral homologous cortical area (Fig. 12B). Figure 12C shows a recording made a few minutes after the intravenous injection of 0.5 mg of clonazepam: the propagated epileptic discharges were not affected by the drug.

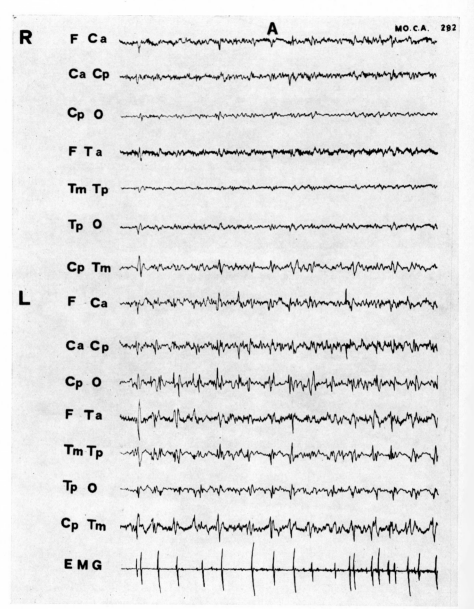

FIG. 11. EEG record and EMG from right arm muscular mass of an epileptic pa-lowing clonazepam i.v. injection. For detailed explanation, see text.

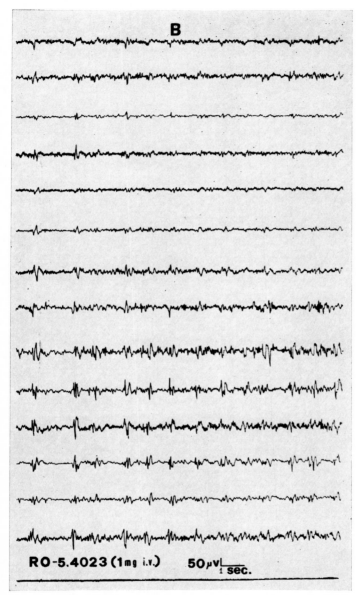

tient. Disappearance of clonic jerks and reduction of EEG epileptic discharges fol-

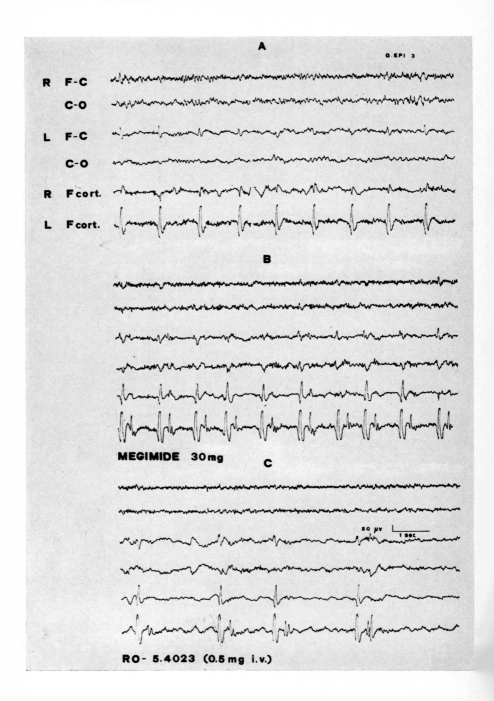

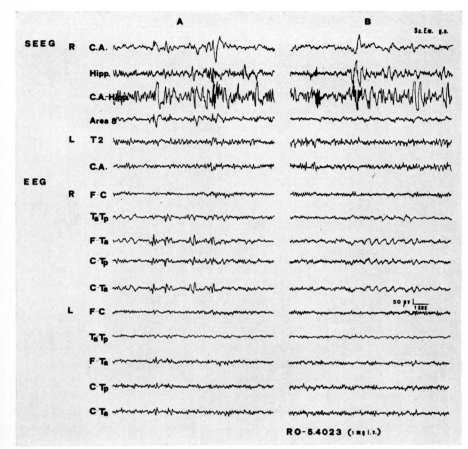

FIG. 13. SEEG and EEG records of epileptic patients. Persistence of epileptic potentials in the deep right temporal structures and their disappearance from frontal structures (area 8) and from the scalp following clonazepam i.v. injection. For detailed explanation, see text. R = right; L = left; C.A. = Ammon's horn; Hipp. = hippocampal gyrus; T2 = 2nd temporal circonvolution.

The results obtained in man appear to be different. An interesting example is illustrated in Fig. 13. In addition to the scalp EEG, depth SEEG recordings from temporal and frontal structures are reproduced. The patient was found to have a right temporal epileptogenic lesion. Figure 13A shows that, coincidentally with the epileptic discharges recorded from the deep

FIG. 12. Cat under urethane anesthesia. Persistence of contralateral propagation of focal cortical epileptic discharges following clonazepam i.v. injection. For detailed explanation, see text.

temporal structures, epileptic potentials appear in the EEG of the frontal lobe and of the right side of the scalp. Figure 13B reproduces the SEEG and EEG, recorded 5 min after the intravenous injection of 1 mg of clonazepam. The epileptic discharges from the deep temporal lobe structures are still present, but no epileptic potentials appear in the frontal lobe or scalp recordings. The most likely interpretation of these findings is that the scalp epileptic potentials have to be regarded as surface propagation of epileptic discharges arising from deep within the brain and that such a propagation has been suppressed by clonazepam.

IV. DISCUSSION

Considered all together, the findings reported above might appear somewhat controversial and make a proper evaluation of the antiepileptic properties of the benzodiazepine derivative tested very difficult. Moreover, we fully realize that both the animal and clinical studies described here are limited. Many other experimental models and clinical trials might and should be considered before drawing any definite conclusions. Nevertheless, we think that in our findings there are pieces of evidence which permit us to gain at least an idea of what the main characteristics of the mode of action of clonazepam on epilepsy are. Since our material on the acute effects of clonazepam is more comprehensive and better documented than that obtained on protracted oral treatment, we will discuss it first.

All the findings obtained show that the intravenous injection of the benzodiazepine derivative RO 5-4023 has a manifest restraining influence on generalized epilepsy. The evidence available from the experiments made both in cats and in epileptic patients, showing various EEG and clinical signs of generalized epilepsy, seems convincing. As is well known, different types of epileptic syndromes are included within the category of "generalized epilepsy" (see the recent classification proposed by an international committee: Gastaut, 1970). Indeed, the patients we had the opportunity to examine suffered from various types of these syndromes, for instance, grand mal seizures, Lennox syndrome, generalized myoclonus, and typical absence. In addition, we observed very good effects of clonazepam in patients presenting generalized epileptic signs which probably had to be considered as a secondary symptom developed from hidden primary epileptogenic zones. The drug proved to be effective in all the cases examined, thereby suggesting a particular sensitivity to the benzodiazepine of the "generalization" phenomenon of the convulsive activity, independently of the mode of origin and development of the phenomenon itself. These findings and conclu sions are in good agreement with those reported in the clinical literature (Swinyard and Castellion, 1966; Gastaut, 1969; Gastaut et al., 1969; Giunta et al., 1969; Poiré and Royer, 1969; Turner et al., 1969*a, b*; Bergamini et al., 1971; Di Rocco et al., 1971), as well as with the results of animal experiments in which an epileptic status was produced (Guerrero-Figueroa, 1969*a, b*; Fariello and Mutani, 1970; Giunta et al., 1970).

It is more difficult to interpret the results concerning the acute effect of intravenous clonazepam on local or partial epileptic activity. What has been found might be summarized as follows.

a. The data obtained from animal experiments and from direct brain SEEG recording in man indicate a very poor effect of the drug in controlling focal epileptic discharges (see also Guerrero-Figueroa et al., 1969*a*, and Giunta et al., 1970, for similar results in experimental epilepsy).

b. The findings obtained in patients suffering from *epilepsia partialis continua* confirm, on the whole, data previously reported. However, it has been shown that in certain patients the drug is capable of suppressing the clinical or motor epileptic manifestations, while leaving the epileptic discharges recorded on the EEG relatively unaffected.

c. The action of the benzodiazepine on the so-called focal EEG epileptic potentials appears to be quite effective in some cases (thus confirming the findings of Gastaut, 1969), while in others it is completely absent (see also Poiré and Royer, 1969). What has been reported by others on the effect of the benzodiazepine derivative clonazepam on the propagation of the epileptic discharges might explain to a certain extent these seemingly contradictory observations. Our findings indicate that epileptic propagation to cerebral structures directly connected with the focus—for instance, propagation from one cortical area to that of the contralateral homologue through callosal connections, or from one hippocampus to the ipsilateral temporal neocortex—is not prevented by the drug. On the other hand, propagation to distant cerebral regions, through long multisynaptic channels, is counteracted by the benzodiazepine. The hypothesis might be advanced that the disappearance of local epileptic EEG abnormalities, following benzodiazepine administration, is due to the fact that such abnormalities do not arise locally, i.e., from the brain site of which a recording was made, but are propagated from other cerebral structures. Conversely, the persistence of the epileptic EEG abnormalities could indicate that they originate from or very close to the recorded brain area. Some evidence favoring the validity of this hypothesis has been obtained. In fact, on the basis of this assumption, we have developed in our Institute a technique of so-called epileptic "activation," in which the use of the convulsant drug bemegride was pre-

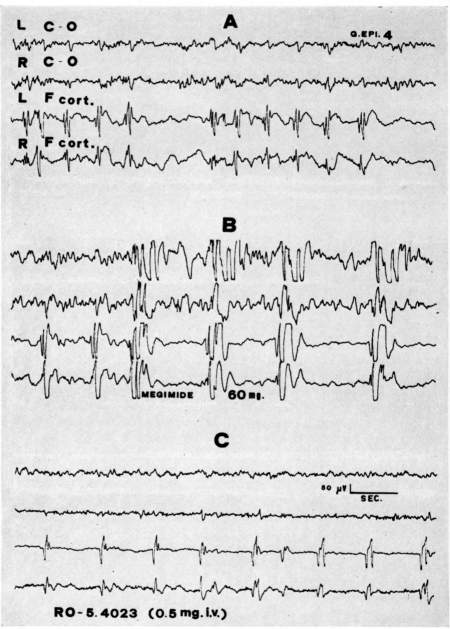

FIG. 14. Cat under urethane anesthesia. Suppression of generalization of epileptic potentials from two frontal foci (L and R F cort.) following clonazepam i.v. injection. For detailed explanation, see text.

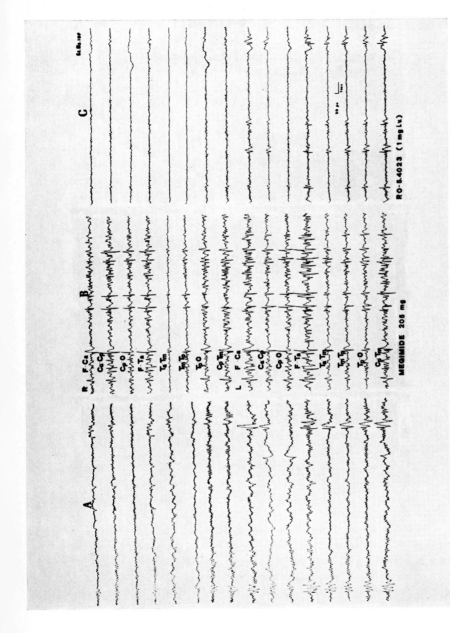

FIG. 15. EEG record of epileptic patient. Suppression of generalization of epileptic potentials from left fronto-temporal focal region following clonazepam i.v. injection. For detailed explanation, see text.

ceded by an intravenous injection of a small amount of clonazepam. This type of "activation" test was applied to several patients who were previously subjected to a classical activation procedure with the convulsant alone. The results of the two tests were then compared. The combined use of clonazepam and bemegride is definitely more apt to bring into evidence focal epileptic EEG activity and to provoke partial epileptic seizures as well than would the use of the convulsant alone (Di Rocco et al., 1971).

We have also found further evidence in favor of the view that the benzodiazepine derivative RO 5-4023 is quite effective in reducing or preventing epileptic generalization while it leaves focal or partial epileptic activity relatively unaffected. Figure 14 illustrates the EEG of a cat in which two symmetrical epileptic foci were created in the right and left frontal cortex. In Fig. 14A, epileptic spikes are confined to the side of the foci. Following the injection of a small amount of bemegride, the epileptic discharges become quickly generalized, as shown in Fig. 14B. The intravenous injection of 0.5 mg of RO 5-4023 immediately suppressed the generalized discharges (Fig. 14C) but did not at all affect the activity of the right and left frontal foci (compare Fig. 14A with Fig. 14C).

Similar examples can be found in our clinical material. Figure 15 reproduces EEG's from a young female suffering from left temporal-lobe epilepsy. Figure 15A illustrates the scalp EEG: spikes are present on the fronto-temporal region of the left side. Figure 15B shows the EEG recorded after the injection of 200 mg of bemegride: the epileptic spikes are very diffuse, being present in almost all recordings from both sides. The EEG reproduced in Fig. 15C was recorded a few minutes after the intravenous injection of 1 mg of RO 5-4023: the epileptic spikes are again limited to the fronto-temporal regions of the left side; the generalization of the epileptic activity is suppressed.

As stated before, the number of patients who received clonazepam for a long period is low and their pathology heterogenous. Furthermore, in most of our cases, the benzodiazepine was combined with other antiepileptic drugs. It is not possible, therefore, to gain any clear-cut idea from our material on the mode of action of the drug when employed for a prolonged time. One can only note that, as far as its effects on generalized epilepsy are concerned, good results appear to have been obtained in many, but not all, of our cases; similar results have been reported in the literature (Gastaut, 1969; Turner et al., 1969; Lison and Fassoni, 1970; Bergamini et al., 1971). Findings and opinions on the effectiveness of prolonged clonazepam treatment in partial epilepsy appear contradictory (see, for instance, Gastaut, 1969). Our results do, however, offer certain evidence that in patients suffering from either generalized or partial epilepsy, the drug shows a tendency to lose its antiepileptic power with the passage of time; this was reported earlier by Bergamini et al. (1971) in some of their patients.

To conclude this chapter, we think it is possible to confirm, on the basis of personal experience, that the benzodiazepine derivative RO 5-4023 has a very powerful antiepileptic effect. We wish to stress that this antiepileptic action is clearly more pronounced—if not limited to—on the processes of generalization and propagation of epileptic activity. The action of the drug on the epileptogenic activity of a relatively limited neuronal population—an epileptic focus or epileptogenic zone—appears to be very poor or, at least, very hard to be convincingly demonstrated.

V. SUMMARY

The anticonvulsant power of the benzodiazepine derivative RO 5-4023 (clonazepam) has been studied in animals and man. The effects of the drug on the motor and EEG signs of different types of epilepsy have been analyzed. Surface as well as depth EEG recordings have been employed. Acute (intravenous) administration of the drug was used in animals; acute as well as protracted (oral) treatments were employed in man.

The results obtained, following acute intravenous injection of clonazepam both in animal and in epileptic patients, indicate: (1) good antiepileptic action on the primarily or secondarily generalized epileptic activity; (2) good effect on propagation of focal epileptic discharges to distant cerebral regions and a poor effect on propagation to cerebral structures directly connected to the epileptogenic zone; and (3) no or very poor effect on focal epileptic activity.

The results obtained with protracted oral treatment in epileptic patients appear controversial. A decrease in the potency of the drug's antiepileptic action, with the passage of time, has been noticed.

REFERENCES

Bergamini, L., Mutani, R., Fariello, R., and Liboni, W. (1970): Elektroenzephalographische und klinische Bewertung des neuen Benzodiazepin RO 5-4023. *EEG EMG*, 1:182–188.

Di Rocco, C., Gentilomo, A., Maira, G., Ottino, C. A., and Rossi, G. F. (1971): Proposal of a new technique of pharmacological "activation" to reveal focal epileptic activity. In: *Abstracts of Papers Presented at the Fourth European Congress of Neurosurgery*, Prague, June 28–July 2, 1971. Avicenum, Czechoslovak Medical Press.

Fariello, R., and Mutani, R. (1970): Valutazione sperimentale dell'efficacia del nuovo farmaco anticomiziale RO 5-4023. *Rivista di Neurologia*, 40:174–183.

Gastaut, H. (1969): The exceptional antiepileptic properties of a new benzodiazepine. *Excerpta Medica International Congress Series No. 193*, p. 5. Excerpta Medica, Amsterdam.

Gastaut, H. (1970): Clinical and electroencephalographic classification of epileptic seizures. *Epilepsia*, 11:102–113.

Gastaut, H., Catier, J., Dravet, C., and Roger, J. (1969): Mise en évidence, par une méthode de "screening," des propriétés anti-epileptiques exceptionnelles d'une benzodiazepine nouvelle. *Revue Neurologique*, 120:402–407.

Giunta, F., Ottino, C. A., Rossi, G. F., and Tercero, E. (1970): Studio sperimentale della azione antiepilettica di un nuovo derivato benzodiazepinico (RO 5-4023). *Rivista di Neurologia*, 40:212–223.

Giunta, F., Rivano, C., Rossi, G. F., and Turella, G. (1969): Effetti di una nuova benzodiazepina sulle manifestazioni cliniche ed elettroencefalografiche della mioclonoepilessia familiare. *Bollettino della Società Italiana di Biologia Sperimentale*, 45: 1473–1475.

Guerrero-Figueroa, R., Rye, M. M., and Heath, R. G. (1969a): Effects of two benzodiazepine derivates on cortical and subcortical epileptogenic tissues in the cat and monkey. I. Limbic system structures. *Current Therapeutic Research*, 11:27–39.

Guerrero-Figueroa, R., Rye, M. M., and Heath, R. G. (1969b): Effects of two benzodiazepine derivates on cortical and subcortical epileptogenic tissues in the cat and monkey. II. Cortical and centrencepalic structures. *Current Therapeutic Research*, 11:40–50.

Lison, M. P., and Fassoni, L. F. (1970): Estudio clinico—electroencefalografico longitudinal en pacientes epilépticos tratados con RO 5-4023. *Arquivos de Neuro-Psiquiatria* (Sâo Paulo), 28:25–36.

Poiré, and Royer, J. (1969): Etude électrographique expérimentale comparée des propriétés antiepileptiques d'un nouveau dérivé des benzodiazépines le RO 5-4023. *Revue Neurologique*, 120:408–410.

Swinyard, E. A., and Castellion, A. W. (1966): Anticonvulsant properties of some benzodiazepines. *Journal of Pharmacology and Experimental Therapeutics*, 151:369–375.

Turner, M., Cantlon, B., Cordero Funes, J. R., Aspinwall, R., Fejerman, N., and Schuguzensky, E. (1969): The clinical EEG evaluation of the antiepileptic action of a new series of benzodiazepine derivatives. *Excerpta Medica International Congress Series*, No. 193, p. 6. Excerpta Medica, Amsterdam.

Turner, M., Shugurensky, M. E. E., and Perea, R. A. (1969): Electroencephalographic evaluation of the anti-epileptic action of MOGADON and other benzodiazepinic derivatives (RO 5-4023). *Electroencephalography and Clinical Neurophysiology*, 27:672.

The Benzodiazepines, Raven Press, New York © 1973

ELECTROPHYSIOLOGICAL ANALYSIS OF THE ACTION OF FOUR BENZODIAZEPINE DERIVATIVES ON THE CENTRAL NERVOUS SYSTEM

R. Guerrero-Figueroa, D. M. Gallant, C. Guerrero-Figueroa and J. Gallant

Department of Psychiatry and Neurology, Tulane University School of Medicine, New Orleans, Louisiana, and Southeast Louisiana Hospital, Mandeville, Louisiana

I. INTRODUCTION

Benzodiazepine derivatives that have anxiety-reducing properties are now used extensively in a wide variety of neuropsychiatric disorders. The benzodiazepine derivatives available for clinical use exhibit antianxiety behavioral effects in man and other mammals and produce specific electrographic changes in spontaneous and evoked electrical activities of the CNS. The electrophysiological properties of chlordiazepoxide (7-chloro-2-methylamino-5-phenyl-3H-1,4-benzodiazepine-4-oxide hydrochloride), diazepam (7-chloro-1-methyl-5-phenyl-3H-1,4-benzodiazepin-2-[1H]-one), RO 5-3350 (7-bromo-1,3-dihydro-5-[2-pyridyl]-2H-1,4-benzodiazepin-2-one), and ORF-8063 (1-methyl-5-phenyl-7-trifluoromethyl-1H-1,5-benzodiazepin-2, 4-[3H, 5H]-dione) are complex and difficult to analyze adequately in a relatively brief presentation (Fig. 1). It would be impossible to cover all of the electrographic data reported from the clinical and experimental evaluations of these compounds here; therefore we will refer only to those studies that seem to be of special significance for the experimental results presented in this chapter.

The first two electrophysiological studies of interest concern the effects of chlordiazepoxide (first benzodiazepine derivative discovered) on the CNS (Schallek and Kuehn, 1960; Randall, 1961). These reports stated that i.p. administration of 5 mg/kg of chlordiazepoxide in the cat produced significant slowing of the EEG spontaneous electrical activity recorded from the septal region, amygdaloid complex, and hippocampal formation; the cortical tracing only showed slow electrical activity when the chlordiazepoxide dose was doubled. By the end of 1964, a rapidly increasing number of clinical and experimental electrographic papers (over several hundred) de-

489

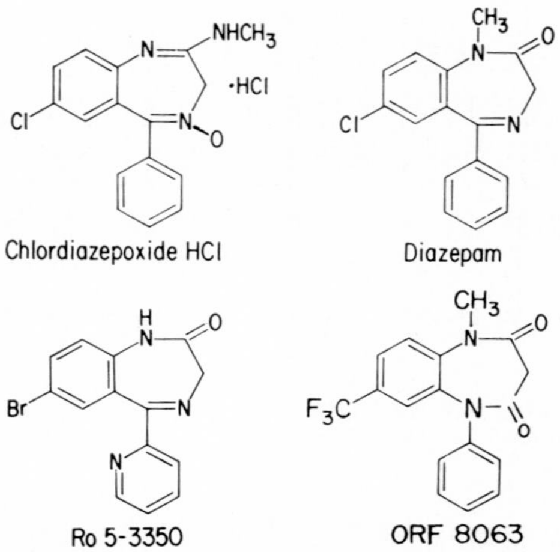

FIG. 1. Chemical structure of four benzodiazepine derivatives.

scribed the suppressant effect of parenteral (intravenous, intraperitoneal, and intracarotid) administration of chlordiazepoxide, diazepam, and other chemically related compounds upon normal and epileptic activities generated from the mesencephalon, diencephalon, and limbic system structures. This inhibitory effect is specifically associated with the intracarotid and intravenous injection of benzodiazepine agents. However, less attention has been devoted to the substantial electrophysiological changes in CNS structures produced by oral administration of the benzodiazepine derivatives. These EEG differences in oral versus parenteral routes of administration are directly related to the discrepancy in the strong antiepileptic effectiveness of parenteral administration of benzodiazepine derivatives, as contrasted with the weak action produced by the oral administration of these agents upon primary and secondary cortical and subcortical epileptogenic tissues.

In the present study, we evaluated the specific electrophysiological changes produced by parenteral and oral administration of chlordiazepoxide, diazepam, RO 5-3350, and ORF-8063 on normal and epileptogenic neuronal populations of cortical and subcortical structures of the CNS of intact conscious cats and monkeys. Of particular interest in this investigation was the evaluation of the effects of benzodiazepine derivatives on the electrical activities generated from the thalamic nuclei of the diencephalon and the midbrain reticular formation of the mesencephalon (MRF) (structures that play a critical role in the mediation of aversive behavioral responses) and preoptic area, lateral hypothalamus, septal region, and hippocampal formation (structures which have a facilitatory action on rewarding behavior).

II. METHODS

Fifty-two aggressive adult cats (2.2 to 3.6 kg body weight) and 12 rhesus monkeys (2.1 to 3.2 kg body weight) were used for chronic implantation of chemical irritants (aluminum oxide, 6-hydroxy-dopamine, or cobalt) and multipolar gross and microelectrodes. Surgical procedures were carried out under pentobarbital anesthesia. Electrode and chemical irritant implantations and stimulating and recording methods of local evoked potentials (LEP) were similar to those described in detail in previous papers (Guerrero-Figueroa et al., 1964; Guerrero-Figueroa et al., 1965). Macroelectrodes were made with six stainless steel wires insulated with enamel or Teflon to within 100 μ of their tips. Tungsten microelectrodes were made according to the Hubel technique (1957) and implanted using the method described by the same author (Hubel, 1959). Electrodes were implanted in the following structures: posterior hypothalamus, different levels of the mesencephalon, intralaminar thalamic system of the thalamus, septal region, preoptic area, lateral hypothalamus, median forebrain bundle, basolateral part of the amygdaloid complex, different pyramidal fields of the Ammon's horn in the hippocampus, and the sensory-motor cortical areas. Twenty-eight of the animals (20 cats and eight monkeys) were also implanted by a technique of Guerrero-Figueroa et al. (1964) for creating discrete subcortical or cortical chemical irritative lesions. The substances used were aluminum oxide (A-583 or A-591 Fisher), 6-hydroxy dopamine, cobalt powder (C-363 Fisher), or penicillin powder and were implanted in the following structures: MRF (four cats and two monkeys), intralaminar thalamic system (two cats and one monkey), amygdaloid complex (four cats and two monkeys), hippocampus (four cats and two monkeys), septal region (two cats), and cortical areas (four cats and one monkey). In addition, two intramuscular hook stainless steel bipolar electrodes were implanted bilaterally into the neck muscles to record tonic electromyographic activity (EMG). All leads were soldered to a 25-prong male plug (Cannon 25 PLI), and the plug was permanently attached with cranioplastic cement to the skull over the frontal sinus. Experimental sessions were initiated 1 month after electrodes and chemical irritants were implanted.

LEP's and extracellular unit activity were recorded on a dual-beam oscilloscope (Tektronix, Type 502A) through a Model P5, Grass preamplifier and (for the extracellular unit activity) a Grass cathode follower (Model P511). A Grass Model S4 stimulator and matching isolation unit were used to deliver electrical stimuli (rectangular pulses at 1 cycle per second were always used during the recording of LEP) to the specific subcortical or cortical structures. In each experimental session, the voltage threshold was determined for the local cortical and subcortical evoked potentials. LEP and extracellular activities were photographed on Kodak film with a Grass kymography camera, Model C4. The oscilloscope tracings of 60 to 100 LEP superimposed sweeps were photographed. The various cortical and subcortical spontaneous activities were recorded on an ink-tracing EEG apparatus (Grass Model 6A). In some animals, spontaneous cortical and/or subcortical activities were recorded at 3-3/4 in/sec on a FR-1300 tape recorder, and the percentage of time spent at each frequency (the low- and high-frequency filters on the Grass EEG machine were set at 0.5 and 70 Hz, respectively) was estimated according to the Saltzberg et al. technique (1968), using the TD-100 special-purpose digital computer. The epochs of data analyzed were 10 min of baseline recording prior to the administration of benzodiazepine agent and two 10-min recordings beginning at 1 hr and at 2 hr after parenteral or oral administration of chlordiazepoxide, diazepam, RO 5-3350, or ORF-8063.

Parenteral doses of chlordiazepoxide hydrochloride, RO 5-3350, and ORF-8063 were given of 1, 2, 3, 4, and 5 mg/kg, and of diazepam 0.2, 0.5, 1, and 2 mg/kg. The composition of the solvent of all the agents was propylene glycol (40%), ethanol absolute (anhydrous) alcohol U.S.P., (10%), benzyl alcohol (2%), and normal saline (48%). Chlordiazepoxide and RO 5-3350 were administered by oral route at doses of 5, 15, 30, and 40 mg/kg; diazepam and ORF-8063 at doses of 2, 4, 5, 8, 10, and 12 mg/kg. Each agent was administered orally in a gelatin capsule.

The relationship between behavioral changes and EEG patterns was studied at baseline, during and after drug administration. To confirm the placement of electrodes, every animal was sacrificed and an intracarotid perfusion of saline followed by 10% formalin was performed on each animal. The brain was removed, fixed in formalin, and subsequently cut in frozen sections, 4 μ in thickness; every fifth section was stained with hematoxylin-eosin or Kluver's stain in order to identify electrode tracts. Only positively identified locations are discussed in this chapter.

III. RESULTS

A. General remarks

LEP were elicited by single-shock stimulation applied 100 to 200 μ away from the recording site. The configuration of the LEP and the number of deflection or wave components of the response depended upon the

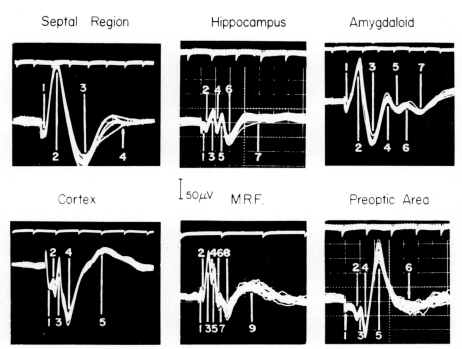

FIG. 2. Electrographic tracings of local evoked potentials (LEP) (superimposition of 60 sweeps) recorded from septal region, hippocampus, amygdaloid complex, cortex, MRF, and preoptic area. Designation of response components are indicated by vertical lines. In this and following figures, positivity is downward. Calibration: 10 msec between peak to peak.

placement of recording and stimulating electrodes (Fig. 2). The short latency deflection or wave 1 of the response appeared to be the primary physiological event and was usually surface positive, and this component was due to impulses approaching the recording electrodes. According to their sequence, designation of the various deflections will be called components 1 through 9 (Fig. 2 follows the convention used for the cortical evoked response as in Schoolman and Evarts, 1959, and Widén and Ajmone Marsan, 1960). Component 1 in the LEP recorded from the septal region and amygdaloid complex and components 1 and 2 in the LEP recorded from the hippocampus, cortical area, MRF, and preoptic area may correspond to neuronal axon responses (presynaptic events) activated by the stimuli. Components 2 through 7 of the septal and amygdaloid responses and components 3 through 9 of the cortical, preoptic area, and MRF may reflect postsynaptic events from the elements of the neuronal pool synchronously triggered by a stimulus. The characteristics of LEP recorded from the intralaminar thalamic system, posterior hypothalamus, and other structures of the CNS have been described previously by Guerrero-Figueroa et al. (1965), and Guerrero-Figueroa et al. (1966). Implanting a chemical irritant (aluminum oxide, 6 hydroxydopamine, penicillin, or cobalt) in subcortical or cortical structures of the CNS in cats and monkeys caused constant irritation in surrounding local neuronal populations (primary epileptogenic tissues), which was manifested by the development of focal epileptiform activity (primary epileptiform discharges). Focal primary epileptiform discharges in subcortical or cortical structures (septal region, amygdaloid complex, hippocampal formation, and sensory-motor cortical areas) of one hemisphere frequently induced electrical activity changes in the neuronal populations of the homologous area of the opposite hemisphere and in interconnecting areas one neuron or more away from the primary focus (secondary epileptogenic tissues). These neuronal changes were manifested by the appearance of electrographic epileptiform activity (secondary epileptiform discharges). It was observed that, in an initial stage, secondary epileptiform activity was related to the epileptiform activity originating from the primary focus (dependent secondary epileptiform discharges) but, after a period of time, secondary epileptogenic tissues acquired the capacity for developing self-sustaining epileptiform activity (independent secondary epileptiform discharges).

B. Effects in the MRF, posterior hypothalamus, and intralaminar thalamic system

Parenteral and/or oral administration of chlordiazepoxide, diazepam, RO 5-3350, or ORF-8063 produced marked electrographic changes in the amplitude of the LEP and in the firing rates of the extracellular activity as well as in the amplitude and frequency of the EEG spontaneous activity recorded from the neuronal populations of the MRF, posterior hypothalamus, and intralaminar thalamic system. These electrographic changes were characterized by: (1) a diminution in the amplitude of all the postsynaptic components (Fig. 3) of the LEP (the earliest or presynaptic components of the responses remained relatively resistant to the effects of the four benzodiazepine derivatives); (2) a decrease in the firing rate of the extracellular electrical activity (Fig. 4); and (3) an increase in the amplitude of voltage and an increase in slowing of the spontaneous EEG activity (Fig. 5) recorded at dosage levels considerably below those causing a decrease in motor activity, ataxia, and drowsiness. Diazepam was found to

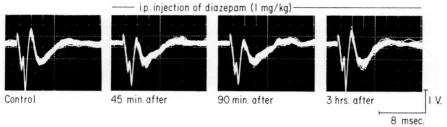

FIG. 3. Local MRF evoked potentials recorded from a normal cat before and following i.p. injection of diazepam (60 superimposed sweeps in each tracing). Note the decreased amplitude of the postsynaptic components of the responses 45 and 90 min after the agent administration.

be a more potent inhibitor of the brainstem, thalamic reticular nuclei, and posterior hypothalamic neurons than chlordiazepoxide, RO 5-3350, or ORF-8063. However, the duration of action of the ORF-8063 depression effect was longer than for diazepam, chlordiazepoxide, or RO 5-3350.

C. Effects in the septal region, preoptic area, hippocampus and amygdala

Parenteral administration of chlordiazepoxide, diazepam, RO 5-3350, or ORF-8063 produced LEP, extracellular unit activity, and EEG spon-

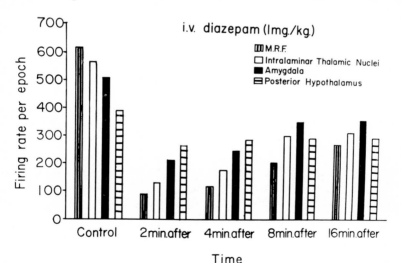

FIG. 4. Effects of i.v. administration of diazepam (1 mg/kg) on the firing rate of the MRF, intralaminar thalamic, amygdala, and posterior hypothalamic neurons. Each epoch = 100 sec. Note the suppressant action of the compound on spontaneous electrical neuronal activity.

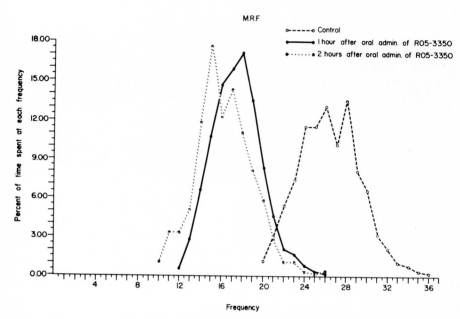

FIG. 5. Percentage of time spent at each MRF (of a normal monkey) frequency (total analysis time 10 min) before and after 1 and 2 hr of oral administration of RO 5-3350 at a dose of 15 mg/kg. The term frequency represents the reciprocal of the time between zero-cross which is usually called *major period* by Saltzberg, et al. (1968).

taneous-activity changes in the septal region (Fig. 6), preoptic area, hippocampal formation and amygdaloid complex similar to those observed in the evoked and spontaneous electrical activities recorded from the MRF, thalamic reticular nuclei, and posterior hypothalamus. In contrast, oral administration of any of the four derivatives always produced augmenta-

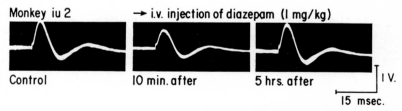

FIG. 6. Electrographic tracings of LEP (superimposition of 60 sweeps) recorded from the septal region of a normal monkey. Note the diminution in the amplitude of the postsynaptic components of the responses 10 min after i.v. administration of diazepam at doses of 1 mg/kg.

oral administration of Diazepam (2 mg./kg.)

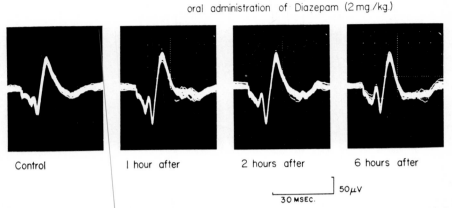

| Control | I hour after | 2 hours after | 6 hours after |

50 μV

30 MSEC.

FIG. 7. Four electrographic tracings of LEP (superimposition of 60 sweeps) recorded from the preoptic area of a normal monkey. Note the enhancement of the amplitude of the postsynaptic components of the responses during the effects of a single dose (2 mg/kg) of diazepam administered by oral route.

tion in the amplitude of the postsynaptic components of the LEP (Figs. 7 and 8) and an increase in the extracellular firing rates of the septal region, preoptic area, and hippocampal formation neuronal populations in association with a slight diminution in the amplitude of the postsynaptic components of the LEP (Fig. 9) and a decrease in the firing rates of the neuronal activity recorded from the amygdaloid complex. The facilitatory action of oral ORF-8063 and diazepam upon the septal region, preoptic area, and hippocampal formation of neuronal populations was usually more intense and more prolonged in time than those changes associated with the administration of chlordiazepoxide or RO 5-3350.

In the spontaneous EEG activity recorded from the septal region, preoptic area, hippocampal formation, and amygdaloid complex, all four benzodiazepine derivatives (by oral or parenteral routes) produced a marked increase in slowing and in the amplitude of voltage (Fig. 10).

D. Effects in cortical areas

RO 5-3350 or ORF-8063 by parenteral or oral routes produced a weak depressing action upon the extracellular cortical electrical activity and a slight diminution in the amplitude of the postsynaptic components of the LEP (Fig. 9) recorded from sensory motor cortical areas. There were no changes in the amplitude of the presynaptic components of the responses.

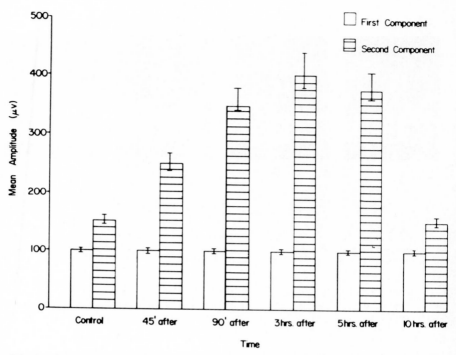

FIG. 8. The mean values of the amplitude of presynaptic (first component) and post-synaptic (second component) potentials of the septal region LEP recorded from normal monkeys. Bar graphs summarize results of 50 experimental sessions before and after oral administration of RO 5-3350 at dose level of 30 mg/kg. Vertical lines indicate the range of variability changes of these amplitude values before and after agent administration. Amplitudes of the potentials were measured from positive wave peak to negative wave peak in each presynaptic and postsynaptic potential of the response.

It appeared that the amplitude of the presynaptic components remained relatively resistant to the effects of both agents at dose levels used in this study. Administration of chlordiazepoxide or diazepam by oral or parenteral routes produced no marked changes in the amplitude of the presynaptic or postsynaptic components of the LEP and in the firing rate of the electrical activity recorded from cortical neuronal populations. Only at high dose levels did chlordiazepoxide, diazepam, RO 5-3350, and ORF-8063 produce changes in the EEG cortical spontaneous activity. These changes were characterized by: (1) an increase in synchronization and an increase in amplitude of voltage and (2) increases in slowing and in superimposed fast activity.

i.p. administration of ORF-8063 (4mg./kg.)

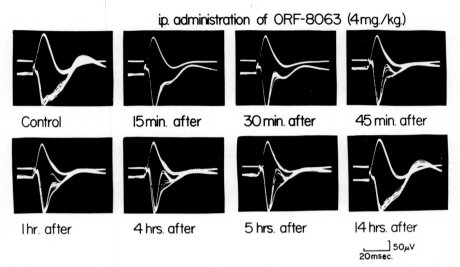

Control   15min. after   30min. after   45 min. after

1hr. after   4hrs. after   5 hrs. after   14hrs. after

]50μV
20msec.

FIG. 9. Eight electrographic tracings (superimposition of 60 sweeps) of LEP recorded from the amygdaloid complex (top beam) and cortical area (bottom beam) of a normal cat. Observe the slight diminution in the amplitude of the postsynaptic components of the responses recorded from the amygdaloid complex and the diminution in the amplitude of the secondary or postsynaptic components of the LEP recorded from cortical area during the effects of i.p. administration of ORF-8063.

E. Effects in primary and secondary epileptogenic tissues

Parenteral administration of each of the four benzodiazepine derivatives produced strong suppression in the EEG epileptiform discharges, the extracellular activity (Fig. 11), and the amplitude of the postsynaptic components of the LEP (Fig. 12) recorded from secondary subcortical and cortical epileptogenic tissues. This suppressant effect was less marked upon the EEG epileptiform discharges, extracellular, and evoked activities recorded from the primary subcortical or cortical epileptogenic neuronal populations. Diazepam and ORF-8063 were found to be more potent inhibitors of secondary and primary subcortical and cortical epileptogenic neurons than RO 5-3350 and chlordiazepoxide. Diazepam, RO 5-3350, and ORF-8063 (by oral route) produced weak suppressant effects upon the EEG epileptiform discharges recorded from secondary subcortical or cortical epileptogenic tissues and produced no effects upon the EEG epileptiform discharges recorded from subcortical or cortical primary epileptogenic tissues (Fig. 13).

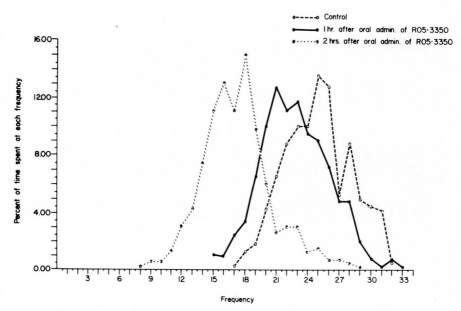

FIG. 10. Percentage of time spent at each amygdaloid complex (of a normal monkey) frequency (total analysis time 10 min) before and after 1 and 2 hr of oral administration of RO 5-3350 at a dose level of 15 mg/kg.

F. Behavioral observations

In aggressive cats and monkeys, all four benzodiazepine derivatives (chronic administration by oral or parenteral routes) produced a marked taming effect and muscle relaxation. This calming effect on aggression and abolishment of fear was always longer in duration after administration of ORF-8063. After administration of diazepam, muscle relaxation, in association with a strong diminution in the amplitude of voltage and frequency of muscular activity (EMG), persisted for a longer period of time as compared with those EMG activity changes observed after administration of chlordiazepoxide, RO 5-3350, or ORF-8063. At higher dose levels, all four benzodiazepine derivatives can produce marked ataxia, drowsiness, and light sleep. Muscle relaxation was never associated with relaxation of the nictitating membrane in the cats.

The behavioral state of muscle relaxation produced by the benzodiazepine derivatives (at dose levels considerably below those causing ataxia, drowsiness, and light sleep) was occasionally preceded by an initial (5 to 15

i.c. Diazepam (200 μg./kg.)

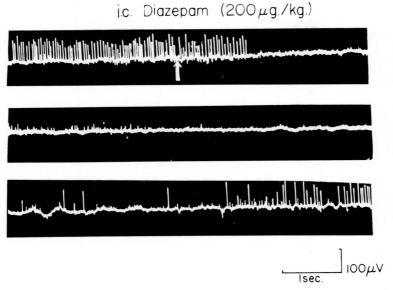

100 μV

1 sec.

FIG. 11. Extracellular unit activity recorded from secondary MRF epileptogenic neurons of an epileptic monkey. Note the strong depression of neuronal activity following intracarotid (i.c.) administration of 200 μg/kg of diazepam.

min) increase in alertness and motor hyperactivity. This brief behavioral state was followed by a relatively long period of: (1) a decrease in alertness and muscle relaxation; (2) a decrease of behavioral responses to a variety of sensory stimulation; (3) an appearance of behavioral state of playfulness; and (4) a marked calming effect on aggression and abolishment of fear behavior.

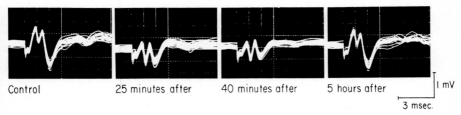

Control   25 minutes after   40 minutes after   5 hours after   1 mV

3 msec.

FIG. 12. Four electrographic tracings (superimposition of 60 sweeps) of LEP recorded from secondary hippocampal formation epileptogenic tissues of a cat with primary epileptogenic focus located in the homologous area of the opposite hemisphere. Note the strong diminution in the amplitude of the postsynaptic components of the responses 25 and 40 min. after i.p. administration of 1 mg/kg of diazepam.

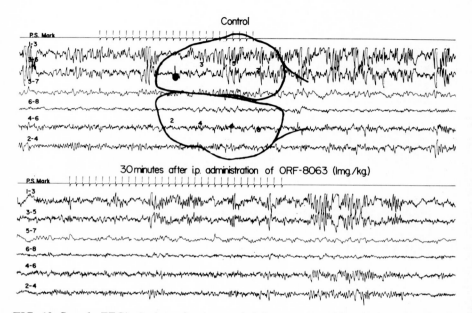

FIG. 13. Sample EEG's (two tracings) recorded from a cat (with primary epileptogenic focus localized in right cortical hemisphere) during relaxed wakefulness, with primary and secondary dependent and independent cortical epileptiform discharges, before (upper tracing) and after (lower tracing) i.p. administration of 1 mg/kg of ORF-8063. Note (lower tracing) the insignificant effects produced by the benzodiazepine derivative upon primary and secondary (dependent and independent) EEG epileptiform discharges recorded from cortical areas.

IV. DISCUSSION

These experimental results provide electrographic evidence for the existence of neuronal excitability changes in specific CNS structures (reticular nuclei of the mesencephalon, intralaminar nuclei of the thalamus, hypothalamus, limbic system structures, and cortical areas) of conscious, freely moving, normal, and epileptic cats and monkeys during the effects of four benzodiazepine derivatives (chlordiazepoxide, diazepam, RO 5-3350, and ORF-8063). In addition to EEG alteration of subcortical and cortical spontaneous activity, analysis of LEP and spontaneous unit activity indicates that chlordiazepoxide, diazepam, RO 5-3350, and/or ORF-8063 administered parenterally may have a different effect upon electrical events of specific areas of the brain than the same substances do by oral route. It should be noted that an alteration of evoked or spontaneous electrical events in specific neuronal area organizations by a psychoactive drug does not necessarily prove that the compound's therapeutic action in man is mediated through this particular brain center.

Experimental data obtained from animals by Delgado et al. (1954) and from human subjects by Heath (1954) show that increases in local neuronal activity (by electrical stimulation) of some subcortical structures of the brain, principally the intralaminar thalamic nuclei of the diencephalon, midbrain reticular nuclei of the mesencephalon, and posterior hypothalamus, play a critical role in the mediation of aversive behavioral responses (including conditioning of anxiety, agitation, tension, and other symptoms observed in psychoneurotic disorders). These four benzodiazepine derivatives by parenteral or oral routes exert a depressant effect upon the neuronal populations of the MRF, posterior hypothalamus, intralaminar thalamic nuclei, and amygdaloid complex. This depressant action was manifested by a decrease in the firing rate of spontaneous unit discharges and in the amplitude of the postsynaptic components of the LEP in association with marked calming effect on aggression, agitation, tension, and abolishment of fear behavior.

Simultaneous electrographic recording of evoked and spontaneous cellular activities from neuronal populations of subcortical and cortical areas has shown a parallelism between shape and amplitude of the evoked potentials and the firing rate of unit discharges (Fox and O'Brian, 1965; Guerrero-Figueroa et al., 1965; Hernández-Peón and Guerrero-Figueroa, 1965; Dreifuss et al., 1968). Therefore, it seems legitimate to admit that changes in the amplitude of the LEP reflect corresponding variations of excitability in the elements of the neuronal pool synchronously triggered by the stimulus. However, although an enhancement of the amplitude of the LEP appears to be consistently correlated with an increase of neuronal excitability, a diminution may result from either occlusion or active inhibition. Active inhibition would result in a decrease in the firing rate of spontaneous neuronal activity as well as in the amplitude of the evoked responses, whereas during occlusion spontaneous neuronal firing rate would be increased (Huttenlocher, 1961). The decrease in the amplitude of the LEP, in the present study, was always associated with a decrease in the firing rate of spontaneous unit activity. This observation would, therefore, favor "active" inhibition as a mechanism for reduction in the amplitude of LEP recorded from the MRF, posterior hypothalamus, intralaminar thalamic nuclei, and amygdaloid complex neuronal populations during the effects of chlordiazepoxide, diazepam, RO 5-3350, and ORF-8063. Decreases in arousal and in responsiveness to sensory stimuli (with a decrease of emotion of responsibility usually described as placidity, which involves a loss of aggression that is particularly striking in wild animals) are the most frequently studied phenomena to be linked with the MRF, posterior hypothalamus, and amygdaloid complex. Moruzzi and Magoun (1949) developed the hypothesis that the EEG and behavioral manifestations of arousal are due to a phasic increase of activity within the ascending reticular system. This hypothesis has been confirmed and further analyzed (in several physiological, pharmacological, and clinical studies) in chronic preparations with lesions placed in this area (Woods, 1964). However, this EEG and behavioral activation can also be elicited by electrical or chemical stimulation of the posterior hypothalamus and amygdaloid complex (Gastaut et al., 1951; Kaada, 1951; Feindel, 1961). It is not surprising, therefore, that the amygdaloid complex and posterior hypothalamus are very important in the regulation of such behavioral changes but are not themselves essential for the initiation of these behavioral associated EEG changes. The inhibitory actions induced by benzodiazepine derivatives, following oral or parenteral administration, upon the electrical activity of the neuronal populations of the MRF, intralaminar thalamic nuclei, posterior hypothalamus, and amygdaloid complex have been presented in a previous paper (Guerrero-Figueroa and Gallant, 1971). These inhibitory actions upon specific neuronal organizations may be one of the fundamental neurophysiological mechanisms by which some benzodiazepine derivatives decrease the level of anxiety, tension, nervousness, aggression, apprehension, and fear, which are frequently observed symptoms in psychoneurotic disorders. The same authors suggested that the periods of light sleep observed during the effects of benzodiazepine derivatives may be due to the inhibitory action of these compounds upon the neuronal populations of these arousal centers.

Physiological and psychological investigations in humans, monkeys, cats, rats, and other mammals by Heath (1954), Olds and Milner (1954), and Delgado et al. (1954) suggested that a neuronal system involving parts of the hippocampus, septal region, preoptic area, lateral hypothalamus, median forebrain bundle, and interpeduncular nuclei of the mesencephalic tegmentum includes the essential anatomic-physiological structures implicated in the development of memory and learning and in the expression of emotion and feeling, as well as components of a rewarding behavior system. Anatomical work by Nauta (1958) has demonstrated that the limbic system structures and midbrain area have complex and largely multisynaptic connections, in which the anterior hypothalamus forms the most important link. Neuropharmacological studies by Randall (1961), Schallek and Kuehn (1960), Hernández-Peón et al. (1964), Zbinden and Randall (1967), and Guerrero-Figueroa (1968, 1969), showed that parenteral administration of chlordiazepoxide, diazepam, RO 5-3350, and other benzodiazepine derivatives produced marked suppressant effects upon evoked and spontaneous electrical events generated in limbic system and the hypothalamic neuronal organization. Our findings with local evoked potentials and spontaneous neuronal firing concur with previous reports and suggest that parenteral administration of antianxiety agents such as chlordiazepoxide, diazepam, RO 5-3350, and ORF-8063 produces an increase or a potentiation of the neurophysiological mechanisms of active inhibition in the limbic-hypothalamic neuronal populations.

Morillo (1962), using the *cerveau isolé* preparation of the cat, observed that amygdala-hippocampal evoked responses were inhibited at a time when an electrographic record from the same place in the ventral hippocampus showed facilitation of the interhippocampal response following administration of chlordiazepoxide and diazepam. Psychological studies of Olds (1966, 1971), using rat preparations, suggested a facilitatory action of chlordiazepoxide and diazepam (at low-dose levels) on limbic-hypothalamic reward behavior. In several papers, Guerrero-Figueroa et al. (1969, 1970) and Guerrero-Figueroa and Gallant (1971) have reported that oral administration of anxiety-relieving benzodiazepine derivatives (at dosage levels considerably below those causing a decrease in motor activity, ataxia, and drowsiness) exerts facilitatory action upon the neuronal populations of the hippocampus, septal region, preoptic area, and lateral hypothalamus of normal and epileptic conscious, freely moving cats and monkeys. In human subjects with implanted subcortical and cortical electrodes, electrographic activity recorded from the septal region and hippocampal formation were facilitated by oral diazepam and inhibited during parenteral administration of the same agent (R. G. Heath, *personal communication,* 1971). Our present findings concur with previous reports and support the suggestion of Guerrero-Figueroa and Gallant (1971) that the action of anxiety-relieving benzodiazepine drugs (oral route) may be a result of their ability to activate the anatomic-physiologic structures involved in the rewarding system, in association with inhibition of the structures responsible for aversive behavior.

The effects of benzodiazepine derivatives on the EEG spontaneous activity recorded from subcortical and cortical structures of the CNS have been studied by many workers; Schallek and Kuehn (1960) and Randall (1961) were probably the first to show that chlordiazepoxide at low-dose levels in cats caused a marked increase in synchronization, increase in the amplitude of voltage, and increase in slowing of the electrical activity recorded from the hippocampus, septal region, preoptic area, MRF, and amygdaloid complex without marked changes in the EEG spontaneous activity recorded from cortical areas. Only at the highest dose levels of the agent do the cortical EEG tracings exhibit changes that resemble those recorded from subcortical structures. We have been able to demonstrate similar effects in our animal preparations and in human subjects (Guerrero-Figueroa et al., 1970), as have Zbinden and Randall (1967), following the administration of other benzodiazepine derivatives. It is not yet possible to define what may be the neurophysiological significance of these EEG changes in the neuronal populations of subcortical mesencephalic, hypothalamic, and limbic system structures. Although a considerable amount of work has been done on the pharmacological influence upon the EEG spontaneous activity of subcortical and cortical structures, relatively little is known about drug effects on the basic rhythmic mechanism itself. This is not surprising because more interest has been focused upon the alterations of the EEG by afferent stimulation than upon the rhythmic mechanism *per se.*

In another set of experiments, electrographical differences between oral and parenteral administration of diazepam and ORF-8063 upon the evoked and spontaneous events were recorded from subcortical CNS structures (Guerrero-Figueroa et al., 1969; Guerrero-Figueroa and Gallant, 1971). This result restresses the differences in the LEP and unit-activity changes which exist between the effects of oral and parenteral administration of benzodiazepine derivatives upon the normal neuronal populations of the rewarding structures and upon subcortical neuronal populations with epileptic disorders. These data make it logical to suggest that electrophysiological differences between oral and parenteral routes may reflect variations in the pharmacological activity of the oral and parenteral compounds, after metabolism by the microsomal enzyme system in the liver. The subsequent metabolic products, after passage through the hepatic system, appear to be entirely new pharmacologic agents.

Histological, physiological, and neurochemical differences between primary and secondary cortical and subcortical epileptogenic tissues have been demonstrated by Guerrero-Figueroa et al. (1964), Wilder and Morrell (1967), and Morrell (1969). Experimental and clinical neuropharmacological studies of parenteral administration of diazepam, ORF-8063, RO 5-4023 (7-nitro-5-(2-chlorophenyl) 3H-1, 4-benzodiazepine-2 (IH) upon centrencephalic and limbic system primary and secondary epileptogenic tissues have revealed strong inhibitory actions in suppressing secondary epileptiform activity and in depressing the spread of primary epileptiform discharges without producing significant changes on epileptiform activity generated by the effect of a chemical irritant (aluminum oxide, 6-hydroxydopamine, penicillin, or cobalt powder) in the primary cortical or subcortical focus (Zbinden and Randall, 1967; Guerrero-Figueroa et al., 1969*b, c,* 1970; and Guerrero-Figueroa and Gallant, 1971). In contrast, in studies of the effects of these oral benzodiazepine derivatives in temporal lobe epileptic patients and in animals (cats and monkeys), only insignificant suppressant or inhibitory action upon primary and secondary epileptiform discharges has been reported (Guerrero-Figueroa et al., 1968, 1969*a, b, c,* 1970). Parenteral administration of these benzodiazepine derivatives effectively suppresses secondary epileptiform activity (generated from tissues with normal vascularization) and depresses the spread of primary epileptiform discharges to interconnected neuronal organizations without producing significant changes on epileptiform discharges originating from primary epileptogenic tissues (with abnormal vascularization). This finding suggested that (1) the importance of studying drug metabolism in CNS by the route that will be used in man; (2) between primary and secondary epileptogenic neuronal populations there exist, at least, two different pathophysiological mechanisms which may involve different neurochemical changes; (3) parenteral administration of benzodiazepine derivatives produces a very high initial concentration of inhibitory action upon CNS secondary epileptogenic tissues that possess a normal vascularization, without affecting or producing significant inhibitory action in CNS primary epileptogenic tissues (that may have vascular damage produced by the implantation of a chemical irritant); and (4) experimental animal preparations of chronic primary and secondary epileptogenic tissues may prove to be an important system for studying alterations produced by antiepileptic drugs in epileptic neuronal metabolism.

V. SUMMARY

The effects of four benzodiazepine derivatives (chlordiazepoxide, diazepam, RO 5-3350, and ORF-8063) upon spontaneous EEG activity, LEP, and extracellular unit activity recorded from subcortical and cortical structures of the CNS were evaluated in chronically implanted, unrestrained, and unanesthetized normal and epileptic preparations (52 cats and 12 monkeys). Of particular interest in this study was the comparison of the differing effects of oral and parenteral administration of the compounds on the electrical activity generated from the mesencephalon (MRF), posterior hypothalamus, intralaminar thalamic nuclei of the diencephalon and amygdaloid complex (structures that play a critical role in the mediation of aversive behavioral responses), and preoptic area, septal region, lateral hypothalamus, and hippocampal formation (structures which have a facilitatory action on rewarding behavior).

Oral administration of the four benzodiazepine derivatives produced a marked augmentation in the amplitude of the postsynaptic components of the LEP and an increase in the firing rate of the unit activity recorded from the rewarding behavior structures in association with a diminution in the amplitude of the postsynaptic components of the LEP and a decrease in the firing rate of the unit activity recorded from the aversive behavior structures of normal and epileptic animals. In contrast, parenteral administration of all four compounds produced a marked diminution in the amplitude of the postsynaptic components of the LEP and a decrease in the firing rate of the unit electrical activity recorded from both the rewarding and aversive behavior structures. By parenteral and/or oral routes, only RO 5-3350 and ORF-8063 produced a slight decrease in the amplitude of the postsynaptic components of the LEP and a slight decrease in the firing rate of unit activity recorded from the cortical neuronal organizations.

Increases in slowing, in the amplitude of voltage, and in synchronization of the EEG spontaneous activity recorded from both aversive and rewarding structures were always obtained during the effects of lowest and highest effective dosages of each of the four compounds, administered by oral and/or parenteral routes. Similar changes were obtained in the cortical EEG tracings but only at highest dosage levels of each compound.

In primary and secondary epileptogenic tissues, all four compounds administered by parenteral route exert a strong inhibitory action upon the epileptiform activity recorded from secondary subcortical and cortical epileptogenic tissues in association with a weak inhibitory action upon the epileptiform discharges generated from primary subcortical epileptogenic foci. No marked inhibitory effects upon primary epileptiform discharges were recorded from primary cortical focus. On the other hand, all four compounds by oral route possess a weak inhibitory effect upon epileptiform activity recorded from secondary subcortical and cortical epileptogenic tissues and no effect upon the electrical epileptiform discharges recorded from primary subcortical or cortical epileptogenic foci.

The data suggests: (1) the possibility of a mode of action of anxiety-relieving benzodiazepine compounds, which implicates a balance of neuronal excitability between the rewarding behavioral system and the aversive behavioral system; (2) the parenteral effectiveness of the inhibitory action of these four benzodiazepine derivatives upon secondary and primary subcortical epileptogenic tissues; (3) the oral ineffectiveness of all four compounds in inhibiting primary and secondary epileptogenic tissues; (4) the important differences in action and metabolism of the benzodiazepine derivatives, depending upon the specific routes of administration; and (5) the importance of using animals as preliminary research subjects for the evaluation of electrographic and metabolic differences produced by different routes of administration of psychopharmacologic agents.

ACKNOWLEDGMENTS

We wish to thank Drs. Phillip Rye and Merrill Rye for their criticisms and comments on this project and Mildred J. Kennedy and Rivers F. Galatas for their technical assistance during the course of this investigation.

This project was supported in part by the Louisiana State Department of Hospitals, and in part by U.S. Public Health Service Grant 2-R10-MH-03701-13 from the National Institute of Mental Health, and in part by grant of the Behavioral Science Research Foundation.

REFERENCES

Delgado, J. M. R., Roberts, W. W., and Miller, N. E. (1954): Learning motivated by electrical stimulation of the brain. *American Journal of Physiology,* 179:587–593.

Dreifuss, J. J., Murphy, J. T., and Gloor, P. (1968): Contrasting effects of two identified amygdaloid efferent pathways on single hypothalamic neurons. *Journal of Neurophysiology,* 31:237–248.

Feindel, W. (1961): Response patterns elicited from the amygdala and deep temporoinsular cortex. In: *Electrical Stimulation of the Brain,* edited by D. E. Sheer. University of Texas Press, Austin, pp. 519–533.

Fox, S. S., and O'Brian, J. H. (1965): Duplication of evoked potential waveform by curve of probability of firing of a single cell. *Science,* 147:888–890.

Gastaut, H., Vigouroux, R., Corriol, J., and Badier, M. (1951): Effets de la stimulation électrique (par électrodes à demeure) du complexe amygdalien chez le chat non narcosé. *Journal de Physiologie et de Pathologie Générale,* 43:740–746.

Guerrero-Figueroa, E., Guerrero-Figueroa, R., and Heath, R. G. (1966): Trigeminal evoked responses during chemical stimulation of cortical and some subcortical structures. *Bulletin of the Tulane University Medical Faculty,* 25:323–335.

Guerrero-Figueroa, R., Barros, A., Heath, R. G., and Gonzalez, G. (1964): Experimental subcortical epileptiform focus. *Epilepsia,* 5:112–139.

Guerrero-Figueroa, R., and Gallant, D. M. (1971): Electrophysiological study of the action of a new benzodiazepine derivative (ORF-8063) on the central nervous system. *Current Therapeutical Research,* 13:747–748.

Guerrero-Figueroa, R., Gallant, D. M., Guerrero-Figueroa, C., and Rye, M. M. (1970): Electroencephalographic study of diazepam on patients with diagnosis of episodic behavioral disorders. *Journal of Clinical Pharmacology,* 10:57–64.

Guerrero-Figueroa, R., Lester, B., and Heath, R. G. (1965): Changes of hippocampal epileptiform activity during wakefulness and sleep. *Acta Neurologica Latinoamericana,* 11:330–349.

Guerrero-Figueroa, R., Rye, M. M., and Gallant, D. M. (1969*a*): Estudio electrofisio-lógico de cuatro derivados de la benzodiazepina sobre las estructuras del sistema nervioso central. *Tribuna Médica (Colombia)*, 35:157–158.

Guerrero-Figueroa, R., Rye, M. M., and Guerrero-Figueroa, C. (1968): Effects of di-azepam on secondary subcortical epileptogenic tissues. *Current Therapeutic Research,* 10:150–166.

Guerrero-Figueroa, R., Rye, M. M., and Heath, R. G. (1969*b*) Effects of two benzo-diazepine derivates on cortical and subcortical epileptogenic tissues in the cat and monkey. I. Limbic system structures. *Current Therapeutic Research,* 11:27–39.

Guerrero-Figueroa, R., Rye, M. M., and Heath, R. G. (1969*c*): Effects of two benzo-diazepine derivates on cortical and subcortical epileptogenic tissues in the cat and monkey. II. Cortical and centrencephalic structures. *Current Therapeutic Research,* 11:40–50.

Heath, R. G. (1954): The theoretical framework for a multidisciplinary approach to human behavior. In: *Studies in Schizophrenia. A multidisciplinary approach to mind-brain relationships,* edited by R. G. Heath. Harvard University Press, Cambridge, Mass., pp. 9–55.

Hernández-Peón, R., and Guerrero-Figueroa, R. (1965): Modifications of local amygda-loid evoked responses during wakefulness and sleep. *Acta Neurológica Latinoamerica-na,* 11:224–233.

Hernández-Peón, R., Rojas-Ramirez, J. A., O'Flaherty, J. J., and Mazzuchelli-O'Flaherty, A. L. (1964): An experimental study of the anticonvulsive and relaxant actions of VALIUM. *International Journal of Neuropharmacology,* 3:405–412.

Hubel, D. H. (1957): Tungsten microelectrode for recording from single units. *Science,* 125:549–550.

Hubel, D. H. (1959): Single unit activity in striate cortex of unrestrained cats. *Jour-nal of Physiology* (London), 147:226–238.

Huttenlocher, P. R. (1961): Evoked and spontaneous activity in single units of medial brainstem during natural sleep and waking. *Journal of Neurophysiology,* 24:451–468.

Kaada, B. R. (1951): Somato-motor autonomic and electrocorticographic responses to electrical stimulation of "rhinencephalic" and other structures in primates, cat, and dog. A study of responses from the limbic, sub-callosal, orbito-insular, piriform, and temporal cortex, hippocampus-fornix, and amygdala. *Acta Physiologica Scandinavica,* 24(*Suppl.* 83):1–285.

Morillo, A. (1962): Effects of benzodiazepines upon amygdala and hippocampus of the cat. *International Journal of Neuropharmacology,* 1:353–359.

Morrell, F. (1969): Physiology and histochemistry of the mirror focus. In: *Basic Mechanisms of the Epilepsies,* edited by H. H. Jasper, A. A. Ward, and A. Pope. Little, Brown and Co., Boston, pp. 357–370.

Moruzzi, G., and Magoun, H. W. (1949): Brainstem reticular formation and activation of the EEG. *Electroencephalography and Clinical Neurophysiology,* 1:455–473.

Nauta, W. J. H. (1958): Hippocampal projections and related neural pathways to the midbrain in the cat. *Brain,* 81:319–340.

Olds, J., and Milner, P. (1954): Positive reinforcement produced by electrical stimula-tion of septal area and other regions of rat brain. *Journal of Comparative and Physio-logical Psychology,* 47:419–427.

Olds, M. E. (1966): Facilitatory action of diazepam and chlordiazepoxide on hypotha-lamic reward behavior. *Journal of Comparative and Physiological Psychology,* 62: 136–140.

Olds, M. E. (1971): Comparative effects of amphetamine, scopolamine, and chlordiaze-poxide on self-stimulation behavior (*submitted for publication*).

Randall, L. (1961): Pharmacology of chlordiazepoxide (LIBRIUM). *Diseases of the Nervous System,* 22:7–15.

Saltzberg, B., Edwards, R. J., Heath, R. G., and Burch, N. R. (1968): Synoptic analysis of EEG signals. *Data Acquisition and Processing in Biology and Medicine,* 5:267–307.

Schallek, W., and Kuehn, A. (1960): Effects of psychotropic drugs on limbic system of cat. *Proceedings of the Society for Experimental Biology and Medicine,* 105:115–117.

Schoolman, A., and Evarts. E. V. (1959): Responses to lateral geniculate radiation stimulation in cats with implanted electrodes. *Journal of Neurophysiology,* 22:112–129.

Widén, L., and Ajmone Marsan, C. (1960): Unitary analysis of the response elicited in the visual cortex of cat. *Archives Italiennes de Biologie,* 98:248–274.

Wilder, B. J., and Morrell, F. (1967): Cellular behavior in secondary epileptogenic lesions. *Neurology* (Minneapolis), 17:1193–1204.

Woods, J. W. (1964): Behavior of chronic decerebrate rats. *Journal of Neurophysiology,* 27:635–644.

Zbinden, G., and Randall, L. O. (1967): Pharmacology of benzodiazepines: Laboratory and clinical correlations. *Advances in Pharmacology,* 5:213–291.

The Benzodiazepines, Raven Press, New York © 1973

A MODULATING EFFECT OF CHLORDIAZEPOXIDE ON DRUG-INDUCED PGO SPIKES IN THE CAT

M.-A. Monachon, M. Jalfre, and W. Haefely

Department of Experimental Medicine, F. Hoffmann-La Roche & Co., Ltd. Basel, Switzerland

I. INTRODUCTION

A number of recent reports—including some in this volume (Taylor and Laverty, 1969; Corrodi et al., 1971; Taylor and Laverty, 1972; Fuxe et al., 1972; Bartholini et al., 1972)—implicate monoamine neurons of the brain with the central actions of benzodiazepines. There is no experimental evidence that the benzodiazepines affect central monoaminergic neurons directly, as do a number of other psychotropic agents. It seems reasonable to assume, however, that benzodiazepines, if they modify the activity of monoaminergic neurons at all, will do so indirectly, by affecting primarily neuronal structures involved (*inter alia*) in the control of monoaminergic systems. The results presented here were obtained in experiments designed to test this hypothesis of an indirect action of chlordiazepoxide on a central monoamine neuron system using electrophysiological methods.

A first difficulty was to find a characteristic and specific as well as quantifiable electrical event in the brain which would serve as an indicator of the activity of a distinct monoamine neuron system. We eventually decided on the so-called ponto-geniculo-occipital (PGO) spikes which proved to be, under our experimental conditions, a suitable criterion for evaluating the activity of a certain serotoninergic and a noradrenergic neuron system. The investigation of this phenomenon was even more rewarding because it is generally recognized to underlie cyclic processes which are of great importance for the normal functioning of the brain. The PGO spikes are characteristic potential changes which can be recorded with macroelectrodes in the pontine reticular formation, in the lateral geniculate bodies, and in the occipital cortex; they seem to initiate the phases of so-called "paradoxical" or

513

"rapid eye movement" (REM) sleep and to occur throughout these REM phases.

Jouvet and collaborators (Delorme et al., 1965, 1966) have found that certain drugs are able to induce in the cat the occurrence of PGO spikes disconnected from the sleep-wakefulness cycle; one such drug is reserpine, which interferes with the storage capacity of central serotonin and catecholamine neurons for their transmitter, and another is *p*-chlorophenylalanine (PCPA), which inhibits the synthesis of serotonin at the step of hydroxylation of tryptophan. In the last two years, we have used these pharmacologically induced PGO spikes in the analysis of a great number of drugs (Jalfre et al., 1970). The method we employ is briefly described below.

II. METHODS

Cats are prepared for the proper experiment under ether anesthesia; bipolar stainless steel electrodes are implanted stereotaxically in both lateral geniculate bodies; the trachea, a femoral vein, and a femoral artery are cannulated for artificial respiration, drug injection, and recording of arterial blood pressure, respectively. After infiltration of wound edges and pressure points with a local anesthetic, the ether is withdrawn and the animal immobilized with gallamine triethiodide throughout the experiment. The rectal temperature is carefully maintained at 37°C. PGO spikes from the lateral geniculate bodies are recorded continuously on a polygraph and are counted simultaneously by an electronic device.

PGO spikes were induced in the following two ways: (1) *Ro 4-1284*, a benzoquinolizine with reserpine-like monoamine depleting action with rapid onset and short duration, was injected intraperitoneally, in a dose of 20 mg/ kg, 1 hr after completion of the surgery. Two hours after Ro 4-1284, the drugs to be investigated were injected intravenously in cumulative doses, the interval between the single injections being 30 min. The number of PGO spikes recorded during the second hour following the application of Ro 4-1284 was taken as control or "pre-drug" value; the number of PGO spikes during the 30 min after a drug injection is the "post-drug" value expressed as percent of the "pre-drug" value. (2) *PCPA* was injected twice intraperitoneally, in doses of 300 mg/kg, 3 and 2 days before the proper experiment. Usually, the effect of a compound on PCPA-induced PGO spikes is expressed as the number of PGO spikes during the first hour after drug injection as compared to that recorded during the hour preceding the injection.

Figure 1 shows typical original polygraphic records before and after

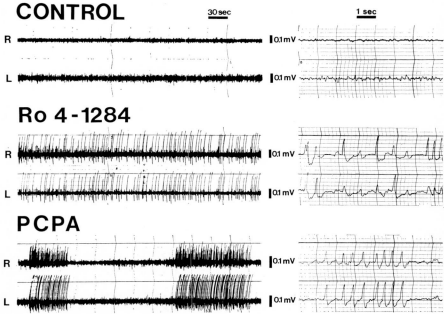

FIG. 1. Original polygraph records (obtained at two different speeds) illustrating typical electrical activity (PGO spikes) in the right and left lateral geniculate bodies of an untreated cat (upper pair of tracings) and of animals given either Ro 4-1284 (middle pair) or PCPA (bottom pair).

Ro 4-1284. The PGO spikes induced by Ro 4-1284 are rather regular monophasic potential changes of 40 to 100 msec duration and amplitudes between 100 and 500 μV which occur synchronously in both lateral geniculate bodies. The PGO spikes induced by PCPA (Fig. 1) differ from Ro 4-1284-induced spikes by their typical occurrence in bursts lasting between several seconds and several minutes, interrupted by the virtual or complete absence of spikes during several minutes.

From Fig. 2 it can be seen that the physiological amino acid precursors of serotonin, L-tryptophan, and DL-hydroxytryptophan (5-HTP) reduce the number of Ro 4-1284-induced PGO spikes in a dose-dependent manner. An intravenous dose of 10 mg/kg of 5-HTP completely abolishes the PGO spikes. In contrast, chlordiazepoxide increases the number of PGO spikes induced by Ro 4-1284. A maximal effect is obtained already with 3 mg/kg. The shape of the individual PGO spikes is unaltered by chlordiazepoxide. The increase of PGO spikes by the benzodiazepine is even more regularly obtained in PCPA-treated animals (Fig. 3 and 4). The most prominent effect

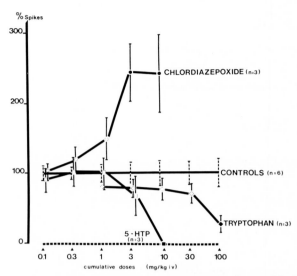

FIG. 2. Effect on Ro 4-1284-induced PGO spikes of chlordiazepoxide HCl, DL-5-HTP, and L-tryptophan administered intravenously in cumulative doses. The effect of each dose is expressed as percent (mean ± SEM) of the number of PGO spikes in control animals injected with saline. *n* represents the number of animals. Statistically significant ($p <$ 0.05) differences from the control level are indicated by solid symbols.

of chlordiazepoxide is to change the discharge pattern of PGO: instead of occurring grouped in irregular bursts, PGO spikes are distributed more regularly after chlordiazepoxide and become virtually identical with those produced by Ro 4-1284. Figure 3 illustrates (1) the large scattering of the absolute number of PGO spikes induced by PCPA in individual cats, (2) the fact that saline injection did not alter the PGO spikes, and (3) the fact that chlordiazepoxide drastically increased PCPA-induced PGO spikes except in a few cases where the absolute number of spikes was extremely high before chlordiazepoxide. If the two animals having PGO spikes above 1000/hr are eliminated, the mean increase of spikes above the pre-drug value by chlordiazepoxide was approximately 600/hr (Fig. 4).

The first idea which comes to mind is that chlordiazepoxide might antagonize the central action of serotonin, especially since we had found that some central serotonin antagonists increase PGO spikes induced by Ro 4-1284 and PCPA, and are even able to induce PGO spikes by themselves (Monachon et al., 1972). This possibility was investigated on PGO spikes induced by Ro 4-1284. Figure 5 shows that chlordiazepoxide did not antagonize the spike-reducing effect of either tryptophan or 5-HTP. An ef-

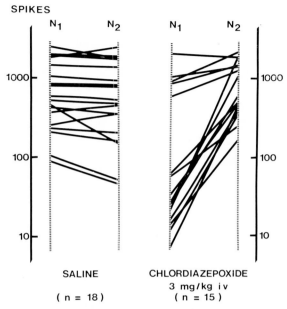

FIG. 3. Effect of saline (left) and of chlordiazepoxide (right) on PCPA-induced PGO spikes. The absolute number of PGO spikes counted over the hour preceding (N_1) and the hour following (N_2) the injection is reported on a logarithmic scale. Injection was performed between 75 and 78 hr after the first administration of PCPA. n indicates the number of observations.

fect of chlordiazepoxide on serotonin receptors involved in the suppression of PGO spikes can therefore be excluded.

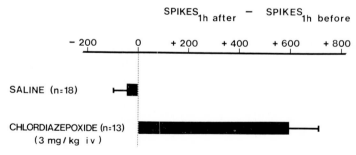

FIG. 4. Effect of saline and chlordiazepoxide on PCPA-induced PGO spikes. Each column represents the mean difference (with SEM) between the number of PGO spikes counted 1 hr after and 1 hr before administration of saline or chlordiazepoxide (3 mg/ kg, i.v.) in the individual experiments reported in Fig. 3.

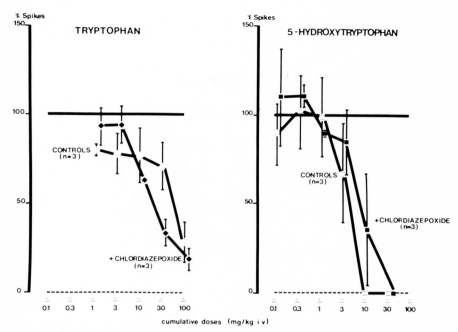

FIG. 5. The effect of DL-5-HTP and L-tryptophan on Ro 4-1284-induced PGO spikes without (open symbols) and after chlordiazepoxide, 10 mg/kg (solid symbols), given intravenously 30 min before. The effect of the amino acids alone is expressed as the percent of the number of PGO spikes counted in animals which received saline injections. The effect of the amino acids after a previous injection of chlordiazepoxide is expressed as the percentage of the number of PGO spikes recorded in cats which obtained only the same single dose of chlordiazepoxide.

We have recently provided evidence for the involvement—although to a quantitatively lesser degree—of catecholamines in the regulation of drug-induced PGO spikes, in the sense that noradrenaline inhibits PGO spike genesis in a way similar to serotonin (Jaffre et al., 1972). Figure 6 shows that three inhibitors of noradrenaline synthesis, one acting at the level of tyrosine hydroxylation (a-methyl-p-tyrosine, a-MPT) and two (Ro 8-1981 and disulfiram) at the step of the dopamine β-hydroxylation, as well as the a-adrenergic receptor blocking agent, phenoxybenzamine, all increase the number of PGO spikes induced by PCPA. The β-adrenergic receptor blocking agent propranolol (0.1 to 3 mg/kg, i.v.) was inactive. The mean increase of PGO spikes produced by chlordiazepoxide was of the same order of magnitude as after the synthesis inhibitors and the a-adrenergic receptor blocking agent. We tentatively assumed that chlordiazepoxide could dimin-

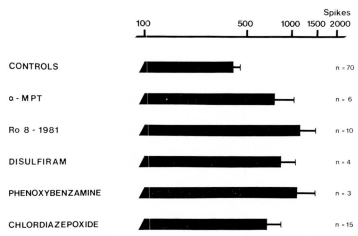

FIG. 6. Effects on PCPA-induced PGO spikes of chlordiazepoxide and different agents acting on the central catecholaminergic mechanisms by blocking either noradrenaline synthesis or noradrenaline receptors. Each column represents the absolute number of PGO spikes ($m \pm$ SEM) counted during the 75th hr after the first administration of PCPA. L-αMPT, 50 mg/kg, was injected intraperitoneally 21 hr, 4 hr, and 1 hr before counting; Ro 8-1981 was given orally in the dose of 300 mg/kg, 20 hr before counting; disulfiram, 100 mg/kg, was administered intraperitoneally 4 hr, 2 hr, and immediately before counting; phenoxybenzamine was injected intravenously 1 hr (1 mg/kg) and again immediately before counting (3 mg/kg); chlordiazpoxide was given in the single dose of 3 mg/kg intravenously immediately before counting. n indicates the number of animals used for each agent. All drugs increased significantly ($p < 0.05$) the absolute number of PCPA-induced PGO spikes as compared to saline-treated animals. However, there is no statistical difference between the effects of the five drugs studied.

ish in some way the activity or effectiveness of central noradrenergic neurons.

It seemed of interest, therefore, to study the effect of the destruction of a noradrenergic neuron system on PCPA-induced PGO spikes. In cats pretreated with PCPA and set up as unanesthetized curarized preparations, bilateral electrolytic lesions were placed with bipolar concentric electrodes in the region of the nucleus loci coerulei (Fig. 7). The histological control revealed a destruction of the posterior part of the locus coeruleus in three cats. The coagulation of the locus coeruleus was carried out 77 hr after the first injection of PCPA. In four control cats, which did not undergo coagulation, the number of PGO spikes had a clear tendency to decrease during the 2 hr corresponding to the time period in which the effect of coagulation was studied in the other three cats. Successful bilateral coagulation of the locus coeruleus produced a marked increase of PGO spikes in two cats and a slight increase in the cat which already had a high number of spikes. We

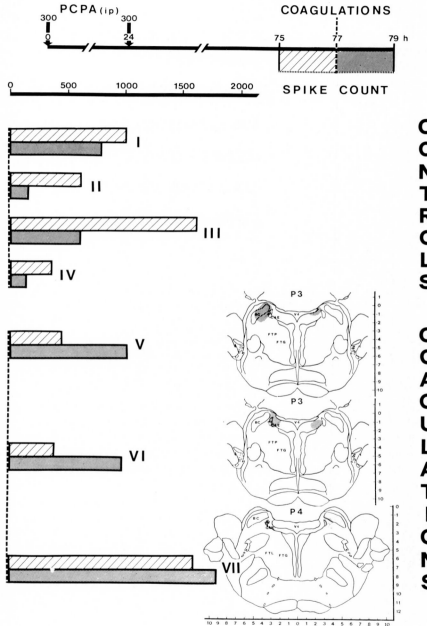
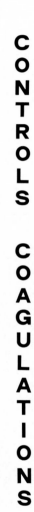

note that the acute destruction of the noradrenergic cell bodies in the locus coeruleus has the same increasing effect on PGO spikes as have the inhibitors of noradrenaline synthesis and the α-adrenolytic agent phenoxybenzamine, and that chlordiazepoxide mimics the effect of all these agents and the destruction of the locus coeruleus.

In view of the marked effect of benzodiazepines on the limbic system— as shown, for example, by the reduction of amygdalo-hippocampal evoked responses (Jalfre et al., 1971)—we suspected that the effect of chlordiazepoxide on PGO spikes could be mediated through forebrain structures. In a few preliminary experiments, where the PGO spikes were recorded in the pontine reticular formation, we found that, after the elimination of the forebrain by transection at the level of the midbrain, the effect of chlordiazepoxide on PGO spikes was abolished. This finding led us to study the effect of chlordiazepoxide in cats with more circumscribed lesions in the forebrain; in this way, the PGO spikes could be recorded in the lateral geniculate bodies as in the aforementioned experiments. Figure 8 summarizes the results obtained in this series of experiments. The difference between the number of PCPA-induced PGO spikes recorded over the hour preceding and the hour following the intravenous injection of saline or chlordiazepoxide is indicated. In intact animals, saline did not significantly change the number of PGO spikes, whereas chlordiazepoxide (3 mg/kg, i.v.) markedly increased the spikes, as already described. When chlordiazepoxide was given to cats which had undergone bilateral electrolytic lesions either of the medial forebrain bundle, the septum, or the amygdala at least 10 days prior to the experiment, the number of PGO spikes remained essentially unaltered. Histological examination showed that the lesions affected the medial forebrain bundles at the mesodiencephalic border, the ventral parts of the septal area, and the anterior parts of the amygdaloid complex (Fig. 9). We have to conclude that, for its PGO-spike-increasing effect, chlordiazepoxide depends on the intactness of the amygdala, of the septum, and of fibers within or in the

---

◄■

FIG. 7. Effect on PCPA-induced PGO spikes of acute bilateral electrolytic lesions in the region of the locus coeruleus. The coagulations were performed with bipolar concentric electrodes using a DC current of 4 mA during 45 sec. The light columns represent the numbers of PGO spikes in individual cats during the 2 hr preceding the coagulation (75th to 77th hour after the first administration of PCPA) ; the dark columns show the numbers of PGO spikes during the following 2 hr. Cats I to IV were used as controls (no coagulation). Cats V to VII were coagulated. The plane of the maximal lesion is presented for each animal.

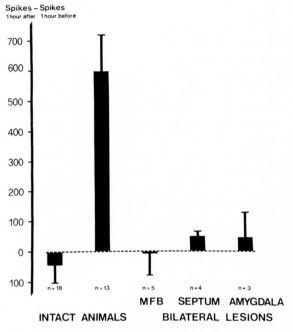

FIG. 8. Effects of chlordiazepoxide on PCPA-induced PGO spikes in intact animals and in animals with chronic bilateral lesions in one of the three basal forebrain structures indicated on the graph (MFB = medial forebrain bundle). Each column represents the difference ($m \pm$ SEM) between the numbers of PGO spikes counted 1 hr after and 1 hr before administration of chlordiazepoxide or saline (3 mg/kg). n indicates the number of animals.

vicinity of the medial forebrain bundle, possibly descending from the limbic structures to the lower brainstem.

We found, furthermore, that the effect of chlordiazepoxide could be prevented not only by circumscribed surgical lesions in the forebrain and in the locus coeruleus, but also by pharmacological means. In fact, chlordiazepoxide completely lost its effect on PGO spikes after atropine sulfate, both in Ro 4-1284- and PCPA-treated animals. This antagonistic effect of atropine in PCPA-treated animals is shown in Fig. 10: whereas chlordiazepoxide (3 mg/kg, i.v.) alone increased the PGO spikes in the mean by 600/hr, no increase at all was seen when the same dose of chlordiazepoxide was preceded by a total of 440 μg/kg of atropine.

Atropine itself affected PGO waves in a peculiar way (Fig. 11).

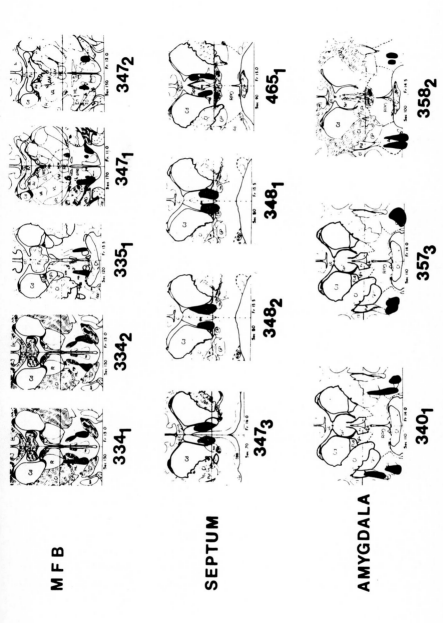

FIG. 9. Planes showing the maximal extension of chronic lesions in the individual animals reported in Fig. 8. The coagulations of the medial forebrain bundle (MFB) were performed with monopolar electrodes; the larger lesions in the ventral septal area and in the amygdaloid complex of each side were obtained by means of two and four electrodes, respectively (3 mA, 45 sec).

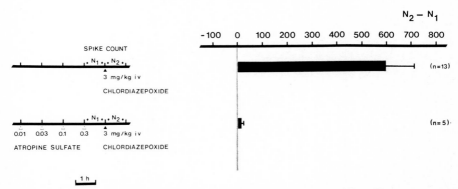

FIG. 10. Effect of chlordiazepoxide, 3 mg/kg, on PCPA-induced PGO spikes in the absence of an additional drug and when preceded by atropine sulfate in cumulative doses. The columns represent the difference (mean ± SEM) of spike numbers between the hour following and the hour preceding the administration of chlordiazepoxide. n is the number of animals used.

Ro 4-1284-induced PGO spikes were not altered at all by atropine in doses up to 440 μg/kg, PGO spikes induced by PCPA, however, were markedly depressed in a dose-dependent manner. Ro 4-1284- and PCPA-treated animals differ mainly in the serotonin and noradrenaline contents in the brain and especially in the pons-medulla. After Ro 4-1284, the noradrenaline content was reduced by 80%, after PCPA by less than 50% (Keller, 1972). In order to obtain comparable amine levels in Ro 4-1284- and PCPA-treated cats, a group of PCPA-treated animals were given in addition the inhibitor of dopamine-β-hydroxylase, Ro 8-1981. As can be seen in Fig. 11, this pretreatment prevented atropine from reducing PGO spikes. On the basis of these results, one could assume that a certain level of noradrenaline is required in order to allow atropine to affect PGO spikes. Since the same requirements must be fulfilled to enable chlordiazepoxide to act on the PGO spikes in a way opposite to atropine, the latter might well be acting somewhere between the primary site of action of chlordiazepoxide in the forebrain and the affected noradrenaline neuron system in the locus coeruleus.

IV. DISCUSSION

We shall now give a summary and attempt an interpretation of the results presented by using a schematic diagram (Fig. 12). This diagram will admittedly appear to be rather hypothetical, because some functional and morphological data supporting parts of it cannot be mentioned here.

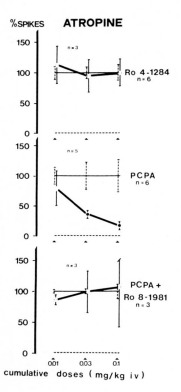

%SPIKES **ATROPINE**

cumulative doses (mg/kg iv)

FIG. 11. Effect of intravenous cumulative doses of atropine on PGO spikes induced by: (1) Ro 4-1284, 20 mg/kg, injected intraperitoneally 2 hr before the first dose of atropine or saline (top); (2) PCPA, 300 mg/kg, given twice intraperitoneally 75 and 51 hr before the first dose of atropine or saline (middle); (3) PCPA, as before, plus the inhibitor of dopamine-β-hydroxylase Ro 8-1981, 300 mg/kg, administered orally 21 hr before the first dose of atropine or saline (bottom). The results are expressed in the percentage (mean $\pm$ SEM) of saline controls. Statistically significant ($p < 0.05$) differences from the control level are indicated by solid symbols.

The PGO spikes originate in the pontine reticular formation (Jeannerod and Kiyono, 1969). We propose the following scheme for the generation of PGO spikes: diffusely located reticular neurons with spontaneous, irregular activity serve as "PGO-pacemaker cells." They are interconnected and are perhaps able to form reverberating circuits. The "PGO-pacemaker cells" impinge on "PGO-generator cells," cells which are not spontaneously active and which require a high excitatory synaptic input to reach the critical level of depolarization for a discharge. If the individual "PGO-pacemaker cells" are discharging at a low frequency, there is little chance that the "PGO-generator cells" receive the number of synchronous synaptic inputs required to reach the critical depolarization threshold. With increasing discharge frequency of the individual "PGO-pacemaker cells," the probability for peaks of simultaneous discharges increases and therewith the chance of driving the "PGO-generator cells." If we assume that the "PGO-generator cells" are again interconnected, it becomes possible to understand how a great number of "PGO-generator cells" can be brought to discharge simultaneously and transmit their synchronous activity through yet unknown pathways to rostral structures, such as the geniculate bodies and the occipital cortex. There is sufficient evidence for the assumption that the excitability of "PGO-pacemaker cells" and/or "PGO-generator cells" is damped by a serotoninergic neuron system originating in the raphé nuclei. Destruction of these raphé nuclei (Jouvet et al., 1966), inhibition of serotonin synthesis by PCPA (Delorme et al., 1966), depletion of serotonin by reserpine (Delorme et al., 1965) and Ro 4-1284, or blockade of serotonin receptors (Monachon et al., 1972) release the "PGO-pacemaker" and "PGO-generator cells" from serotoninergic inhibition and, therefore, induce PGO spikes. In addition to this serotoninergic inhibition, a noradrenergic system with inhibitory function is postulated (Jalfre et al., 1972). Noradrenaline cell bodies have been shown by histofluorescence microscopy to send axons to the midline raphé nuclei and to make synaptic contacts with serotonin cell bodies (Olson and Fuxe, 1971). Whether the noradrenaline released from these endings inhibits or excites the raphé cells is not yet certain, and our experiments provide no information on this point. We postulate, furthermore, that noradrenaline neurons of the locus coeruleus—either the ones projecting also to the raphé or different ones—make inhibitory synaptic contacts with reticular "PGO-pacemaker" and/or "PGO-generator cells." The noradrenaline released from these endings depresses the excitability of "PGO-pacemaker cells," but, perhaps more important, the excitability of "PGO-generator cells." By the postulated inhibitory action on "PGO-generator cells," we could explain the findings that noradrenaline seems to be more important for the pattern of PGO spikes (periodicity) than for their occurrence (note the appearance of PGO spikes in bursts after prevalent depletion of serotonin and the more regular pattern after additional removal of noradrenaline). Chlordiazepoxide increased PGO spikes in Ro 4-1284-treated animals to some extent, and more drastically in cats with PCPA. All the experimental evidence indicates that chlordiazepoxide depresses the noradrenergic neuron system of the locus coeruleus involved in the control of PGO spikes. A direct effect of chlordiazepoxide on these neurons is highly unlikely; the elimination of the chlordiazepoxide effect after lesions of the amygdala, the septum, and the medial forebrain bundle confirmed the expected mode of action, namely, a primary effect on forebrain structures apparently involved with the control of the noradrenergic system of the locus coeruleus. Our results lead us to assume that chlordiazepoxide activates an inhibitory pathway from limbic structures to the locus coeruleus. We furthermore assume that

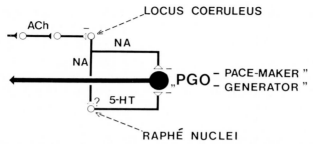

FIG. 12. Hypothetical scheme of the neuronal systems controlling the activity of generator and of pace-maker cells for PGO spikes. For explanation see text.

this inhibitory pathway contains at least one muscarinic excitatory synapse which can be blocked by atropine. This simplest assumption could explain how atropine antagonizes the effect of chlordiazepoxide and, furthermore, why atropine decreases PGO spikes induced by PCPA, but not those induced by Ro 4-1284 and the combination of PCPA and an inhibitor of dopamine-β-hydroxylase. When the noradrenergic system is nearly intact—as after PCPA—the interruption of the inhibitory pathway to the locus coeruleus will increase the activity of the noradrenergic system and thereby reduce the PGO spikes. When the noradrenergic system is impaired—as after Ro 4-1284 and the combined treatment with PCPA and a dopamine-β-hydroxylase inhibitor—the interruption of the descending inhibitory pathway can be expected to have no influence on the amounts of noradrenaline released.

In conclusion, the results of our investigation—although providing largely indirect evidence—seem to indicate that, even at low doses, chlordiazepoxide inhibits a rather distinct noradrenergic neuron system of the brain, which is concerned with the control of an important biological phenomenon, the PGO spikes. It is important to realize that, in the present investigation, the PGO spikes were studied in a rather simple and artificial experimental situation, suitable for the neuropharmacological analysis of a central monoamine neuron system; it must not be inferred from our results that chlordiazepoxide would influence in the same way PGO spikes or REM sleep in the normal animal or man.

V. SUMMARY

Ponto-geniculo-occipital (PGO) spikes induced by the central monoamine depletor Ro 4-1284 (a benzoquinolizine) and the inhibitor of tryptophan hydroxylase, *p*-chlorophenylalanine, were recorded in the lateral geniculate bodies in curarized cats. Chlordiazepoxide was observed to increase the number of these PGO spikes. It is unlikely that the effect of chlordiazepoxide is due to a depression of serotoninergic neurons or blockade of serotonin receptors, since the benzodiazepine did not affect the decrease of PGO spikes observed with 5-hydroxytryptophan and tryptophan. An inhibitory influence of chlordiazepoxide on the noradrenergic system of the locus coeruleus is postulated, since the effect of chlordiazepoxide greatly resembles that of inhibitors of noradrenaline synthesis (α-methyltyrosine and inhibitors of dopamine-β-hydroxylase) and of lesions of the locus coeruleus. Moreover, a cholinergic mechanism seems also to be implicated in the effect of chlordiazepoxide on PGO spikes, since the latter was blocked by atropine sulfate. The modulating effect of chlordiazepoxide on PGO spikes appears to be mediated by structures in the limbic system and the hypothalamus: lesions in either the septum, the amygdala, or the medial forebrain bundle prevented the increase of PGO spikes induced by this benzodiazepine.

ACKNOWLEDGMENT

The authors are grateful to Mrs. I. Kubli and Mrs. S. Peter for technical assistance with the experiments, and to Mr. E. Rund for the histological work.

REFERENCES

Bartholini, G., Keller, H. H., Pieri, L., and Pletscher, A. (1972): The effect of diazepam on the turnover of cerebral dopamine. *This Volume.*

Corrodi, H., Fuxe, K., Lidbrink, P., and Olson, L. (1971): Minor tranquillizers, stress and central catecholamine neurons. *Brain Research,* 29:1–16.

Delorme, F., Froment, J. L., and Jouvet M. (1966): Suppression du sommeil par la p-chlorométamphétamine et la p-chlorophénylalanine. *Comptes Rendus de la Sociéte de Biologie,* 160:2347–2351.

Delorme, F., Jeannerod, M., and Jouvet, M. (1965): Effets remarquables de la réserpine sur l'activité EEG phasique ponto-géniculo-occipitale. *Comptes Rendus de la Société de Biologie,* 159:900–903.

Fuxe, K., Corrodi, H., Lidbrink, P., and Olson, L. (1972): The effects of benzodiazepines and barbiturates on central monoamine neurons. *This Volume.*

Jalfre, M., Monachon, M. A., and Haefely, W. (1970): Pharmacological modifications of benzoquinolizine-induced geniculate spikes. *Experientia,* 26:691.

Jalfre, M., Monachon, M. A., and Haefely, W. (1971): Effects on the amygdalo-hippocampal evoked potential in the cat of four benzodiazepines and some other psychotropic drugs. *Naunyn-Schmeidebergs Archiv für Pharmakologie,* 270:180–191.

Jalfre, M., Monachon, M. A., Keller, H. H., and Haefely, W. (1972): The effects of catecholamines on drug-induced geniculate spikes. *First International Congress, Association for the Psychophysiological Study of Sleep,* Bruges, June 19–23, 1971 (*in press*).

Jeannerod, M., and Kiyono, S. (1969): Décharge unitaire de la formation réticulée pontique et activité phasique ponto-géniculo-occipitale chez le chat sous réserpine. *Brain Research,* 12:112–128.

Jouvet, M., Bobillier, P., Pujol, J. F., and Renault, J. (1966): Effets des lésions du système du raphé sur le sommeil et la sérotonine cérébrale. *Comptes Rendus de la Société de Biologie,* 160:2343–2346.

Keller, H. H. (1972): Depletion of cerebral monoamines by *p*-chlorophenylalanine in the cat. *Experientia,* 28:177.

Monachon, M. A., Burkard, W., Jalfre, M., and Haefely, W. (1972): Blockade of central 5-hydroxytryptamine receptors by methiothepin. Naunyn-Schmiedebergs Archiv fur Pharmakologie, 274:192–197.

Olson, L., and Fuxe, K. (1971): On the projections from the locus coeruleus noradrenaline neurons: The cerebellar innervation. *Brain Research,* 28:165–171.

Taylor, K. M., and Laverty, R. (1969): The effect of chlordiazepoxide, diazepam and nitrazepam on catecholamine metabolism in regions of the rat brain. *European Journal of Pharmacology,* 8:296–301.

Taylor, K. M., and Laverty, R. (1972): The interaction of chlordiazepoxide, diazepam, and nitrazepam with catecholamines and histamine in regions of the rat brain. *This Volume.*

The Benzodiazepines, Raven Press, New York © 1973

EFFECT OF DIAZEPAM ON DEVELOPMENT OF STRETCH REFLEX TENSION

Ute Brausch, H.-D. Henatsch, C. Student, and K. Takano

Department of Physiology II, University of Göttingen,
Göttingen, German Federal Republic

I. INTRODUCTION

A great obstacle in analyzing the effects of psychotropic agents, such as diazepam, is presented by the difficulty of obtaining objective information about psychic processes, as evidenced by the great number of different analytic methods used at present. In neurophysiological studies, there have been numerous trials to uncover the underlying mechanisms responsible for the effects of diazepam found in clinical use. In many cases, electrical activity of special regions of the brain was investigated (Randall et al., 1963; Hernández-Peón et al., 1964; Pletscher, 1964; Guérrero-Figueroa et al., 1969a, b). There are several authors who have reported on the effect of diazepam upon spinal and/or supraspinal activity, using various types of preparations under different conditions (Randall et al., 1961; Ngai et al., 1966; Schmidt et al., 1967; Crankshaw and Raper, 1968; Hudson and Wolpert, 1970; Nakanishi and Norris, 1971).

The present study started with the idea that there are close relations between the psyche and the motor system (e.g., Hoff, Tschabitscher, and Kryspin-Exner, 1964; Tan and Henatsch, 1968). Since muscular reflex tension is a very important component of motor activity, a great part of this investigation deals with the measurement of the tonic reflex tension of the precollicular decerebrate cat. This method has the advantage of being able to integrate information from the spinal cord and that from the higher central nervous system in the mechanical response of a muscle to its extension. As will be shown, this response can be submitted to further analysis (see Discussion). On the other hand, this method might be considered more "natural" than the usual electrical stimulation tests. Furthermore, we did not need to take into account any interaction between the drug under study and anesthesia since our preparations were decerebrated. Although it is true that the state of precollicular decerebration of the cat is somewhat unnatural, it is

531

more related to that of conscious human patients than studies of animals under deep barbiturate or chloralose anesthesia.

II. METHODS

Experiments were performed on 25 adult cats weighing 2.0 to 4.0 kg. Preceding ether anesthesia, 0.1 mg/kg of atropine was given in order to reduce secretion in the respiratory tract during the first period of the operation. The animals were decerebrated by precollicular suction under transient ether anesthesia which was stopped immediately after decerebration. The triceps surae muscles in both hind limbs were separated from the surrounding tissue and the two Achilles tendons were connected by steel links to the two levers of a double-strain gauge myograph. All nerves innervating the hind limb muscles, except those under study, were sectioned. The right and left hind limbs were rigidly fixed at the hips, knees, and feet by steel pins. The muscles were simultaneously stretched linearly 14 mm at a constant rate of 2 mm/sec. The changes in muscle length were obtained by measuring the excursion of the stretcher through a rod-potentiometer. The initial length of the muscle under study was set at a point corresponding to a resting (or static, Takano, 1966) tension of 20 g.

The active tension developed in the left triceps surae muscle was recorded directly by elimination of the passive tension with a compensation method (Takano and Henatsch, 1964). The principle of this method is as follows: the two homolog muscles on both sides, the left with an intact reflex arc, and the right having been totally denervated, were stretched simultaneously. The difference in the tensions developed in both sides, which represents the active tension in the experimental side, was obtained by a special Wheatstone's bridge connected to a carrier amplifier (Tektronix, Type 3C66). For further details see Takano and Henatsch (1971). The tension-extension diagrams were traced directly on transparent A 5 paper, using an X-Y pen recorder (Hewlett-Packard, Moseley 7035AM). The muscles were stretched at intervals of 3 min in order to avoid interactions of subsequent stretches.

In some experiments, laminectomy was performed, and all ventral roots from L_5 to S_1 were cut to eliminate any fusimotor influences on the muscle spindles. Unit discharges from the primary ending of a muscle spindle were recorded. The spindle discharges, their instantaneous frequency, and the change of muscle length were registered on running film, using an oscilloscope (Tektronix, Type 565) and a frequency meter (Tönnies). In experiments with muscle spindles, the muscle was stretched at a rate of 10 mm/sec, with intervals of 15 sec.

In some instances, a nerve-muscle preparation *in situ* was also used to test any direct effect of the drug on this preparation. Supramaximal stimulation (square pulses, duration 0.25 msec) was applied to the distal cut end of the sciatic nerve every 10 sec, and twitch tension of the triceps surae muscle was recorded on running paper.

Diazepam (VALIUM 10, Roche, 10 mg in 2 ml solvent) was injected into the external jugular vein. At doses less than 0.5 mg/kg, the drug preparation was diluted in Ringer's solution (10 mg diazepam in 10 ml Ringer's). As the substance is only poorly soluble in water, the dilution was prepared shortly before injection and was well mixed before injected. The injection was performed very slowly (1 min) to avoid any influence of this drug on blood pressure. Because active tension is very dependent on blood pressure, the latter was observed continuously; when needed, sufficient plasma expander was given to keep the blood pressure almost constant. Animals with blood pressure below 90 mm Hg were not used in this study.

The rectal temperature was measured in earlier experiments of this study. The direct correlation between active tension and body temperature was very small compared to that between active tension and blood pressure. For this reason we have omitted temperature registration in the later part of the study.

III. RESULTS

When the muscle is stretched at high speed and the stretch length is maintained for some time, three components of the active tension can be observed, the phasic, the tonic, and the static (Takano, 1966). The phasic is due to the synchronous firing of many neuromuscular units and shows a twitch-like tension development in the muscle. The tonic component lasts several seconds, probably resulting from polysynaptic tonic reflex activity. These two components are rate dependent. The phasic is very prominent in a pale phasic muscle (e.g., tibialis anterior) and the tonic in a red tonic muscle (e.g., soleus). The static component is not rate dependent; at least part of it is not a true stretch reflex, but is due to the fact that an active muscle (independent of the origin of this activity) produces higher tension on extension than in the passive state (pseudoreflex, Pompeiano, 1960). The active state of the muscle may well be produced, for instance, by a constant supraspinal input.

At the slow rate of 2 mm/sec, which was normally used in this study, there is little chance that the phasic component appears (Takano, *unpublished data*), whereas the tonic component may well develop. The phasic component is mainly due to monosynaptic reflex firing. Probably the tonic component plays the dominant role in the active tension which has been examined in this study, while the static component is also involved. The obtained active tension is, therefore, the sum of reflex and pseudoreflex tension.

A. Effect of diazepam upon maximal active tension

Figure 1 shows the effect of a single dose of diazepam (1 mg/kg)

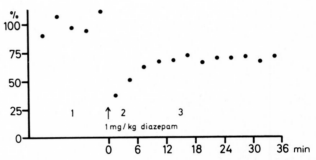

FIG. 1. Effect of diazepam on the maximal active tension of the triceps surae muscle of the decerebrate cat. Ordinate shows the maximal active tension (in percent of controls) of the muscle reached during stretch (rate 2 mm/sec, amplitude 14 mm); 3-min stretch intervals. Average control tension was 1,000 g. At zero time: 1 mg/kg diazepam, i.v. 1, control period; 2, "dip" phase; 3, "plateau" phase.

upon the development of active stretch tension. Each point is the maximum tension obtained at about the end of an individual muscle extension (rate 2 mm/sec, amplitude 14 mm).

According to the graph, the effect of an individual dose of diazepam may be divided into two phases: one is the "dip" directly after the injection, the other is the "plateau." At the end of the injection, the active tension fell promptly, the maximal effect of the drug being obtained at latest 6 min after the end of the injection, and in most instances within 3 min. The depth of this dip was about 70% of the preinjection level in this case, and between 20 and 80% in other experiments, and was relatively independent of the applied dose.

The effects of repeated, increasing doses from 0.01 to 8.0 mg/kg, applied every 18 min, are shown in Fig. 2. Here, the dip after most of the

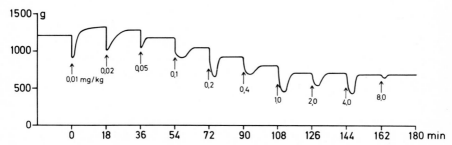

FIG. 2. Effect of repeated doses of diazepam on the maximal active tension. *Ordinate:* Maximal stretch tension (g). *Abscissa:* Compressed time scale (min). Diazepam was injected every 18 min in increasing doses, beginning with the dose of 0.01 mg/kg. Somewhat smaller "dips" and nearly constant "plateau" level at high doses should be noted.

injections is about 25% of the control tension. Sometimes, especially when small doses were administered as in the first cases in Fig. 2, the maximal effect was already observed at the first stretch after drug administration. It was often noted that higher doses induced a smaller dip than lower doses. At very small doses (0.001 mg/kg and lower), no effect was noticed. At small doses (such as 0.01 mg/kg in the case of Fig. 2), the tension recovered after a short time, not only returning to the control level but quite often surpassing it.

It must be stated that the dip is not due to alterations of blood pressure, although it is true that often during and shortly after injection of diazepam, a short drop in blood pressure could be observed. However, the dip could be found even in preparations without any fall in blood pressure. Moreover, the recovery of blood pressure was much faster than that of the dip.

The plateau began 5 to 10 min after the injection and showed a very slow recovery. In contrast to the dip, the plateau level depended on the dose applied: the higher the dose, the deeper the level of the plateau. Even with high doses (cumulative up to 15 mg/kg in 160 min), the active tension could not be depressed completely. It must be emphasized that a total depression was never reached at any dose studied in our preparations. It is of interest that the plateau level in Fig. 2 remains almost constant after injection of 1.0 mg/kg (cumulative doses *ca.* 1.8 mg/kg in 108 min) and of further higher doses. Schmidt et al. (1967) have observed in the spinalized cat anesthetized with sodium pentobarbital that diazepam could not abolish the mono- and polysynaptic reflex with cumulative doses of about 2 mg/kg. On the other hand, Nakanishi and Norris (1971) did not find any effect of diazepam on either mono- or polysynaptic reflex action potentials in

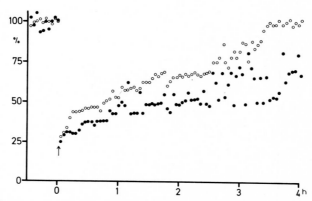

FIG. 3. Prolonged recovery from diazepam effect on the maximal active tension at two different doses. Two different preparations with different doses of diazepam (○, 0.2 mg/kg; ●, 1 mg/kg) which show almost the same initial effect. (See legend of Fig. 1.)

spinalized rats. But they also saw the reduction of these reflexes in decerebrate rats and failed to find a total depression by this drug.

In Fig. 3, diazepam was applied only once and its effect was observed for some time. This figure shows two examples from two different preparations using the doses of 0.2 mg/kg and 1 mg/kg. In both cases the effect of the drug was rather strong, and the active tension was depressed to about 25% of the control level at two quite different doses. The recovery was faster at the lower dose of 0.2 mg/kg: 3.5 hr after the injection, it reached the control level. At the higher dose of 1 mg/kg, its recovery was observed for over 4 hr, and the extrapolation shows that about 8 hr was required for complete recovery. It was our general experience that the higher the dose, the longer the time needed for recovery.

B. Effect of diazepam upon the active tension-extension diagram

The active tension-extension curve (excluding its inital part) is to some extent linear (Granit, 1958; Matthews, 1959*a*; Pompeiano, 1960; Takano and Henatsch, 1971). Changes of this linear course in the X-Y graph can occur in two ways: a slope change or a parallel shift. The slope change in the tension-extension curve can be described by a change of the *gain constant* (Granit, 1958), whereas the parallel shift represents a change of *bias*, as measurable by the difference between the two points at which the two lines cross the X-axis (*cf.* Eldred et al., 1952, for the similar behavior of the frequency-extension diagram of muscle spindle discharges).

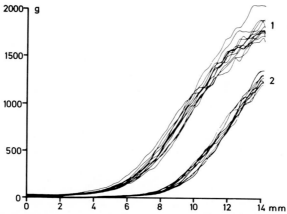

FIG. 4. Effect of diazepam upon the active tension-extension diagram of the triceps surae muscle. *Ordinate:* Active tension (g). Abscissa: Length of muscle extension (mm). Repeated curves were directly recorded and superimposed before (10 curves, 1) and after (10 curves, 2) administration of diazepam (0.5 mg/kg). Note the parallel shift of the curves (2) to the right. The scattering is smaller after diazepam (2) than in the control group (1).

The active tension-extension curves of Fig. 4, resulting from 10 stretch trials in each case before (1) and after (2) administration of 0.5 mg/kg diazepam, were superimposed on the same sheet of paper; 45 min after the injection, the first curve of group 2 was recorded, subsequent recordings following at 3-min intervals. This particular time period after the drug administration (45 to 72 min) had been chosen because there was little change in the maximal active tension. The time represents the second phase of action of a single diazepam dose, which we assume to be the most important one for therapeutic use of the drug. The fluctuation of the curves is an expression of the natural variations of supraspinal activity in our nonanesthetized preparations. When we compare the mean curves in each group, although they are not shown here, it is obvious that a parallel shift of about 3 mm to the right took place following a greater shift which occurred already a few minutes after diazepam injection. There is no noticeable change in the slope; in other words, diazepam only changed the *bias* of the tonic reflex motor system.

Another difference between the two groups of tension-extension curves is represented by the amount of scattering of the curves. After drug administration, it decreased. In this experiment, the scattering of group 2 was about two-thirds of that of the controls (group 1). It appears that diazepam has a stabilizing effect on motor activity.

C. Effect of diazepam on the peripheral motorsensory system

From the changes observed in the active tension-extension curve, we could not get any direct information about the location of action produced by diazepam. Therefore, both peripheral efferent and afferent sides of the reflex arc were tested in 10 preparations.

Nerve-muscle preparation. The question of a peripheral effect of diazepam is a controversial subject in the relevant literature, such as: Crankshaw and Raper (1968), Parkes (1968), Dasgupta and Mukherje (1969), Moudgil and Pleuvry (1970), Feldman and Crawley (1970), and Hudson and Wolpert (1970).

We first tested the efferent skeletomotor side: the twitch tension of the triceps muscle, caused by supramaximal stimulation of the peripheral stump of the cut sciatic nerve, at intervals of 10 sec, was recorded on running film.

However, as has already been stated by Robichaud et al. (1970), it was not easy to come to clear conclusions about the extent of the effect of diazepam on twitch tension. At doses of 2 mg/kg and higher, we found in half of our experiments a slight diminishing (less than 10%) of the twitch tension, which began about 20 min after injection, and sometimes slightly earlier, at very high doses. Recovery to the control level, in some cases, took about 50 min and more. In general, at smaller doses (below 2 mg/kg), we found no constant reduction of twitch tension. On the whole, there were no depressant effects comparable to those observed in the active tension developing during stretch.

Discharges from deefferented muscle spindles. Group Ia spindle discharges under linear stretch (rate 10 mm/sec, amplitude 14 mm, interval 15 sec) were recorded on running paper. There are no distinct differences either in the dynamic peak response of the discharge or in the static discharge before and after the administration of the drug.

Figure 5 shows typical samples of the lack of diazepam effects upon twitch tension and spindle response to a ramp stretch before and after administration of 1 mg/kg of diazepam in each experiment.

IV. DISCUSSION

As already stated above, it had been difficult to decide whether changes in the active tension-extension curves are due to central nervous or peripheral effects of diazepam. Therefore, the influences of the drug on the nerve-muscle preparation and on the muscle spindles were investigated. There was no depressant effect on the deefferented muscle spindle. Some

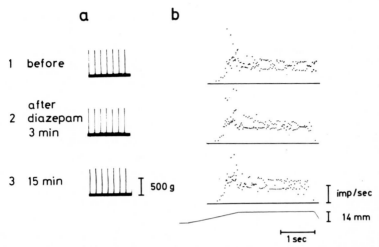

FIG. 5. Lack of peripheral motorsensory effects of diazepam. (a) Twitch tension response to indirect maximal stimulation of triceps surae muscle at intervals of 10 sec. (b) Instantaneous frequency of group I*a* fiber discharges (conduction velocity 105 m/sec) from deefferented muscle spindle (dotted records) in response to muscle ramp stretches (rate 10 mm/sec, amplitude 14 mm). Height of each point represents reciprocal time interval to preceding spindle impulse. Bottom record shows course of length change. In each experiment (two different preparations) 1 mg/kg of diazepam was given: 1, before; 2, 3 min after; 3, 15 min after injection.

effect on the nerve-muscle preparation cannot be entirely ignored, but it is small and inconsistent and, more important, it has quite another time course, so that only in the second phase of diazepam action might there be a slight interference with the registered tension-extension curves. From this, it would appear that the dip and most of the second phase of slow recovery, the plateau of tension, must be due to effects of diazepam on spinal and/or supraspinal structures of the nervous system. The question now remains in which part of the central nervous system the influence of this drug is localized.

It is perhaps useful to consider separately the two temporal phases of the tension effects of diazepam (dip and plateau). Investigations of Takano and Kanda (*unpublished data*) on discharges from muscle spindles with intact gamma-innervation showed that diazepam has a strong depressant effect on the gamma-system, beginning immediately after injection. But the time course of this effect does not show any dip but rather lasts monotonously for a long time. Therefore, it is very doubtful that a transient change in the gamma-system is primarily involved in the dip. It is more probable that the dip depends upon effects on the alpha-system. Since the effect of diazepam on presynaptic inhibition also has another time course (Schmidt et al., 1967), we may exclude it as a factor mainly responsible for the dip. It is also known that the monosynaptic reflex remains nearly uninfluenced after diazepam. For these reasons, we conclude that the main factor causing the dip is not located in the motoneurons themselves. It may depend on depression in interneuronal pathways other than those of presynaptic inhibition, in the spinal cord or in the supraspinal sources of descending inputs. The latter assumption is further supported by the findings of Ngai et al. (1966), who showed that the effect of diazepam on decerebrate cats could be abolished by spinalization. Moreover, Nakanishi and Norris (1971) did not find any influence of diazepam on mono- and polysynaptic reflexes in spinalized rats.

Considering now the second, slow-recovery phase of reduced active tension (plateau) after injection of the drug, we find that the depression of maximal active tension corresponds to a parallel shift of the curves in the tension-extension diagram. Stimulation studies at the cut ventral root with controlled parameters by Brausch, Student, and Takano (*unpublished data*) have confirmed that the tension output from the nerve-muscle preparation has a linear relation to the neural input into this system from the spinal cord, as already previously described in principle by Granit (1958). They further showed that the bias change visible in the parallel shift of the active tension-extension diagram depends on the bias change in the output from the spinal cord. Therefore, the bias change in the active tension-extension diagram depends on that in the spinal and/or supraspinal motor system.

This change of bias could be due to direct as well as indirect input changes to alpha motoneurons. There are several factors that could be responsible for a direct negative alpha-bias change: postsynaptic inhibitory mechanisms seem, according to Schmidt et al. (1967), not to be influenced by diazepam. However, the effect on the presynaptic inhibition described by the same authors could partly cause a change in alpha-bias. We have no quantitative information yet about a possible reduction in postsynaptic facilitation and other potential factors. As a whole, direct changes of alpha-bias do not seem to be strong enough to explain the entire bias change of the active tension-extension curve.

We should mainly consider an alteration of gamma-bias to be an indirect cause for the change of alpha-bias. Investigations of Matthews and Rushworth (1957) and Matthews (1959*b*) have established that a change of gamma-bias does result in a parallel shift of the active tension-extension diagram. According to the literature, there is some indication that diazepam has an influence on the gamma-system (*cf.* Schallek, 1966, cited by Zbinden and Randall, 1967; Hernández-Peón et al., 1964). Indeed, recording the action potentials of I*a*-afferents from the muscle spindle with intact gamma innervation, a striking depressant effect of diazepam on the gamma-system has been observed by Takano and Kanda in our laboratory (*unpublished data*). Although we do not know enough about the proportional factor between I*a*-input to the spinal cord and alpha-output, this finding does support the idea that the gamma-system is probably responsible for most of the effects of diazepam seen in the active tension-extension diagram.

Quite similar depressing effects upon supraspinal gamma (and alpha) drive have been found for a variety of other neuroleptic and tranquilizing substances (Henatsch and Ingvar, 1956; Busch et al., 1960).

Further investigations will be required to find out whether one or both of the two gamma subcomponents (dynamic and static, respectively) are subject to the drug-induced suppression.

V. SUMMARY

(1) The active tension of the triceps surae muscle of the unanesthetized precollicular decerebrate cat, resulting from muscle stretching (rate 2 mm/sec, amplitude 14 mm), was recorded with a compensation method.

(2) The effect of an individual dose of diazepam (0.01 to 8 mg/kg) upon the maximal active tension has two phases: one begins directly after the administration (dip) and lasts not longer than 10 min. Its depth and width are nearly dose independent, and it never leads to a total block. The second phase shows a long-lasting recovery (plateau) which depends on the dose.

(3) After diazepam, the active tension-extension curve, which was directly recorded, shifted parallel to the right without any change in slope of the curve. The scattering of superimposed repeated curves was reduced.

(4) Diazepam caused no consistent change in the twitch tension of the supramaximally excited nerve-muscle preparation and in the afferent responses of muscle spindles, without gamma-innervation, to a ramp stretch.

(5) The study suggests that the main effect of diazepam in the decerebrate cat is a reduction of the gamma-bias. Other possible factors are briefly discussed.

ACKNOWLEDGMENTS

The authors appreciate the technical assistance of Miss K. Junghans and the collaboration of Dr. K. Kanda (guest from Chiba University, Japan) in the last part of this work. Samples of diazepam (VALIUM 10) were kindly supplied by Hoffmann-La Roche A. G., Basel. This study was supported by the Deutsche Forschungsgemeinschaft (SFB 33). Part of the results were presented at the *37. Tagung der Deutschen Physiologischen Gesellschaft, 1970* (Henatsch et al., 1970).

REFERENCES

Busch, G., Henatsch, H.-D., and Schulte, F. J. (1960): Elektrophysiologische Analyse der Wirkungen neuroleptischer und tranquilisierender Substanzen (Phenothiazine, Meprobamat) auf die spinalmotorischen Systeme. *Arzneimittel-Forschung,* 10:217–223.
Crankshaw, D. P., and Raper, C. (1968): Some studies on peripheral actions of mephenesin, methocarbamol and diazepam. *British Journal of Pharmacology,* 34:579–590.
Dasgupta, S. R., Ray, N. M., and Mukherjee, B. P. (1969): Studies on the effect of diazepam (VALIUM) on neuromuscular transmission in skeletal muscles. *Indian Journal of Physiology and Pharmacology,* 13:79–80.
Eldred, E., Granit, R., and Merton, P. A. (1953): Supraspinal control of the muscle spindles and its significance. *Journal of Physiology,* 122:498–523.
Feldman, S. A., and Crawley, B. E. (1970): Interaction of diazepam with the muscle-relaxant drugs. *British Medical Journal,* 2:336–338.
Granit, R. (1958): Neuromuscular interaction in postural tone of the cat's isometric soleus muscle. *Journal of Physiology,* 143:387–402.
Guerrero-Figueroa, R., Rye, M. M., and Heath, R. G. (1969*a*): Effects of two benzodiazepine derivatives on cortical and subcortical epileptogenic tissues in the cat and monkey. I. Limbic system structures. *Current Therapeutic Research,* 11:27–39.
Guerrero-Figueroa, R., Rye, M. M., and Heath, R. G. (1969*b*): Effects of two benzodiazepine derivatives on cortical and subcortical epiloptogenic tissues in the cat and monkey. II. Cortical and centrencephalic structures. *Current Therapeutic Research,* 11:40–50.
Henatsch, H.-D., Brausch, U., and Student, C. (1970): Das Längen-Aktivspannungs-Diagramm bei Extensor-Muskeldehnung als Indicator somatisch-motorischer Wirkungen von "Psychopharmaka" (VALIUM - TOFRANIL - PERVITIN). *Pflügers Archiv,* 316:R 62.
Henatsch, H.-D., and Ingvar, D. H. (1956): Chlorpromazin und Spastizität, eine experimentelle elektrophysiologische Untersuchung. *Archiv für Psychiatrie und Zeitschrift Neurologie,* 195:77–93.
Hernández-Peón, R., Rojas-Ramirez, J. A., O'Flaherty, J. J., and Mazzuchelli-O'Flaherty, A. L. (1964): An experimental study of the anticonvulsive and relaxant actions of VALIUM. *International Journal of Neuropharmacology,* 3:405–412.
Hoff, H., Tschabitscher, H., and Kryspin-Exner, K. (Editors) (1964): *Muskel und Psyche.* S. Karger-Verlag, Basel.
Hudson, R. D., and Wolpert, M. K. (1970): Central muscle-relaxant effects of diazepam. *Neuropharmacology,* 9:481–488.
Matthews, P. B. C. (1959*a*): The dependence of tension upon extension in the stretch reflex of the soleus muscle of the decerebrate cat. *Journal of Physiology,* 147:521–546.
Matthews, P. B. C. (1959*b*): A study of certain factors influencing the stretch reflex of the decerebrate cat. *Journal of Physiology,* 147:547–564.
Matthews, P. B. C., and Rushworth, G. (1957): The relative sensitivity of muscle nerve fibres to procaine. *Journal of Physiology,* 135:263–269.

Moudgil, G., and Pleuvry, B. J. (1970): Diazepam and neuromuscular transmission. *British Medical Journal*, 2:734–735.

Nakanishi, T., and Norris, F. H., Jr. (1971): Effect of diazepam on rat spinal reflexes. *Journal of Neurological Sciences*, 13:189–195.

Ngai, S. H., Tseng, D. T. C., and Wang, S. C. (1966): Effect of diazepam and other central nervous system depressants on spinal reflexes in cats: A study of site of action. *Journal of Pharmacology and Experimental Therapeutics*, 153:344–351.

Parkes, M. W. (1968): The pharmacology of diazepam. In: *Diazepam in Anaesthesia,* edited by P. F. Knight and C. G. Burgess. John Wright & Sons Ltd., Bristol, pp. 1–7.

Pletscher, A. (1964): Zentrale Wirkungsmechanismen von Psychopharmaka. In: *Funktionsabläufe unter emotionellen Belastungen,* edited by K. Fellinger. S. Karger-Verlag, Basel, pp. 25–42.

Pompeiano, O. (1960): Alpha types of "release" studied in tension-extension diagrams from cat's forelimb triceps muscle. *Archives Italiennes de Biologie*, 98:92–117.

Randall, L. O., Heise, G. A., Schallek, W., Bagdon, R. E., Banziger, R., Boris, A., Moe, R. A., and Abrams, W. B. (1961): Pharmacological and clinical studies of VALIUM, a new psychotherapeutic agent of the benzodiazepine class. *Current Therapeutic Research*, 3:405–425.

Randall, L. O., Schallek, W., Scheckel, C., Banziger, R., Boris, A., Moe, R. A., Bagdon, R. E., Schwartz, M. A., and Zbinden, G. (1963): Zur Pharmakologie von Valium, einem neuen Psychopharmakon der Benzodiazepinreihe. *Schweizer Medizinische Wochenschrift*, 93:794–797.

Robichaud, R. C., Gylys, J. A., Sledge, K. L., and Hillyard, I. W. (1970): The pharmacology of prazepam. A new benzodiazepine derivate. *Archives Internationales de Pharmacodynamie et de Thérapie*, 185:213–227.

Schmidt, R. F., Vogel, M. E., and Zimmermann, M. (1967): Die Wirkung von Diazepam auf die präsynaptische Hemmung und andere Rückenmarksreflexe. *Naunyn-Schmiedebergs Archiv für Pharmakologie und Experimentelle Pathologie*, 258:69–82.

Takano, K. (1966): Phasic, tonic and static components of the reflex tension obtained by stretch at different rates. In: *Nobel Symposium I, Muscular Afferents and Motor Control,* edited by R. Granit. Almqvist & Wiksell, Stockholm, pp. 461–463.

Takano, K., and Henatsch, H.-D. (1964): Direkte Aufnahme von Längen-Reflexspannungs-Diagrammen einzelner Muskeln *in situ. Pflügers Archiv,* 281:104–105.

Takano, K., and Henatsch, H.-D. (1971): The effect of the rate of stretch upon the development of active reflex tension in hind limb muscles of the decerebrate cat. *Experimental Brain Research*, 12:422–434.

Tan, Ü., and Henatsch, H.-D. (1969): Wirkungen von Imipramin auf die spinalmotorischen Extensor- und Flexor-Systeme der Katze. *Naunyn-Schmiedebergs Archiv für Pharmakologie und Experimentelle Pathologie*, 262:337–357.

Tschabitscher, H., and Czerwenka-Wenkstetten, H. (1964): Affekt und Muskelspannung. In: *Funktionsabläufe unter emotionellen Belastungen,* edited by K. Fellinger. S. Karger-Verlag, Basel, pp. 188–204.

Zbinden, G., and Randall, L. O. (1967): Pharmacology of benzodiazepines: Laboratory and clinical correlations. *Advances in Pharmacology*, 5:213–291.

The Benzodiazepines, Raven Press, New York © 1973

ACUTE TOLERANCE TO DIAZEPAM IN CATS

Allen Barnett and John W. Fiore

Department of Pharmacology, Biological Research, Schering Corporation, Bloomfield, New Jersey

I. INTRODUCTION

Diazepam has important clinical muscle-relaxant activity, and its primary locus of action appears to be the reticular formation (Ngai et al., 1966). The linguomandibular reflex (LMR) is a polysynaptic reflex which involves interneurons in the reticular formation, and diazepam is a very potent antagonist of this reflex in cats. While studying diazepam-induced depression of the LMR, we found acute tolerance to the LMR-depressant effects of diazepam. Tolerance to the CNS-depressant effects of chlordiazepoxide has been reported after *chronic* administration in rats by Hoogland and co-workers (1966) and in mice by Goldberg and co-workers (1967), presumably by induction of hepatic microsomal enzymes, but there have been no reports of acute tolerance to diazepam or any other benzodiazepines. The present studies were devoted to characterizing the phenomenon of acute tolerance to diazepam, investigating the mechanism, and determining if cross-tolerance exists to methocarbamol (ROBAXIN), another muscle relaxant, or to phenobarbital, a generalized CNS depressant.

II. METHODS

Cats of either sex, weighing 2 to 4 kg were anesthetized with dial-urethane, 0.6 ml/kg, i.p., or by ether inhalation. Cannulae were placed in the femoral vein for i.v. administration and in the femoral artery for measurement of blood pressure. For intraduodenal (i.d.) administration, the duodenum was exposed by laparotomy and punctured with a needle. Each drug response was measured as the peak change (from predrug control) occurring within the first 10 min after i.v. administration and 20 min after i.d. administration.

545

A. LMR in dial-urethane-anesthetized cats

For eliciting the LMR, two needle electrodes were placed on the under-side of the cat's tongue, and electrical stimulation of the tongue then pro-duced a downward force of contraction of the lower jaw which was mea-sured with a Grass FT-10 force-displacement transducer connected to a Grass Model 7 polygraph. Stimulus parameters were of frequency 6/min, duration 500 msec, and sufficient voltage to produce a force of contraction of 400 to 500 mg in each experiment.

B. Flexor reflex in spinal cats

Cats were anesthetized by inhalation of ether and the spinal cord was transsected at the level of the foramen magnum. The ether was removed and cats were maintained on artificial respiration. For eliciting the flexor reflex, the central end of the saphenous nerve was stimulated electrically and con-tractions were recorded from the tendon of the ipsilateral anterior tibial muscle. The parameters of electrical stimulation were the same as previously mentioned for the LMR.

C. EEG in dial-urethane-anesthetized cats

Cats of either sex, 2 to 4.5 kg in body weight, were surgically prepared in the same manner as previously mentioned for the LMR. For measurement of the EEG, a stainless steel screw was placed on the left side of the skull in an area corresponding to the site of the motor cortex. A wire was fastened to each screw and was held in place with dental acrylic. Monopolar EEG's were recorded between the skull screw and an indifferent ear electrode, using a Grass Model 7 Polygraph. EEG amplitude was measured manually and expressed as the mean amplitude per minute for a 10-min period after each drug administration.

D. Ataxia in unanesthetized cats

Male cats, 3 to 4.5 kg in body weight, were placed in individual 3 × 2 × 1½ ft high cages with Plexiglas doors for observation. Ataxia was scored when the cats were taken from the cage and allowed to walk on the floor for approximately 2 min. Scoring was made using a 0 to 8 scale with 0=normal walking, 2=walking with a stagger, 4=staggering and falling but able to get up, 6=inability to rise, and 8=completely prostrate. Drugs were administered intraperitoneally, and ataxia scores were taken before and 30 min after each dose of drug.

III. RESULTS

A. LMR inhibition in dial-urethane-anesthetized cats

Tolerance to diazepam was first observed when complete inhibition of the LMR could not be obtained after i.v. doses of diazepam as large as 1.0 mg/kg (Fig. 1). In these experiments, the time interval between doses

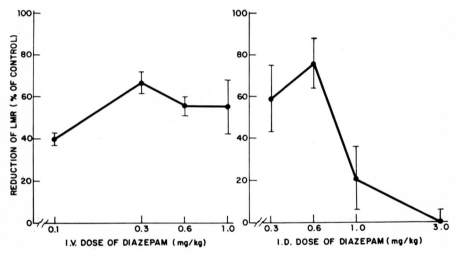

FIG. 1. Dose-response curves for inhibition of the LMR by diazepam. The curve on the left is for diazepam administered i.v. and each point represents the mean ± S.E. of five cats. The one on the right is for diazepam administered i.d. (intraduodenally) and each point represents the mean ± S.E. of three cats. For both curves, doses were given at 60-min intervals.

was 60 min. Tolerance to diazepam was greater with i.d. administration (Fig. 1) in that, after peak LMR inhibition at 0.6 mg/kg, higher doses (1.0 and 3.0 mg/kg) produced progressively less inhibition, whereas with i.v. administration the diazepam dose-response curve plateaued over a dose range of 0.3 to 1.0 mg/kg (Fig. 1).

Since no tolerance was reported in cats by administering graded doses of diazepam at 15-min intervals (Crankshaw and Raper, 1968), diazepam was also administered in the same manner to see if tolerance would result. As indicated in Fig. 2, there was no tolerance, and complete inhibition of the LMR was obtained by administering diazepam at 15-min intervals i.v. and at 30-min intervals i.d. A major difference between administering diazepam at 15-min intervals compared with 1-hr intervals was that, with the shorter interval, the time to complete a dose-response curve was 45 min, whereas in the latter instance, 3 hr was required. To examine the influence of this variable, a second dose-response curve for diazepam at 15-min intervals was obtained 2 hr after the first curve was completed. Under these conditions, tolerance to diazepam was again demonstrated both i.v. and i.d. (Fig. 2).

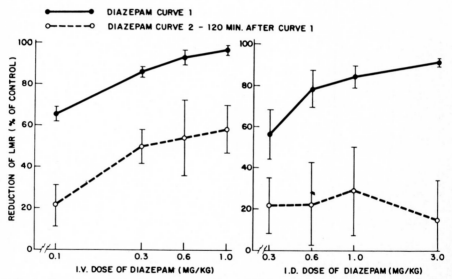

FIG. 2. Cumulative dose-response curves for inhibition of the LMR by diazepam. The curves on the left are for diazepam administered i.v. and the curves on the right are for i.d. administration. For the curves on the left, doses were given at 15-min intervals, and for the curves on the right doses were given at 30-min intervals. In each case, the second dose-response curve (O----O) was obtained 120 min after completion of the first dose-response curve (●——●). Each point represents the mean ± S.E. of four cats for the dose-response curves on the left, and five cats for the curves on the right.

Because of the time lag involved in development of tolerance, the possibility that acute tolerance to diazepam was due to the formation of a less active metabolite was considered. The effects of potential metabolites of diazepam on the LMR were studied by administering doses of these drugs at 15-min intervals. N-desmethyl diazepam was the least active of the potential metabolites tested, being only about one-fifth as potent as diazepam ($p < 0.05$, Table I). It is important to note that when a second dose-re-

TABLE I. *Relative potency of diazepam metabolites for inhibition of the LMR compared to diazepam*

Treatment	No. of cats	Relative potency	95% Confidence limits
Diazepam	4	1	—
N-Desmethyl diazepam	5	0.20	0.05-0.43
3-Hydroxy diazepam	7	0.43	0.16-0.86
Oxazepam	4	0.57	0.21-1.20

All drugs were administered intravenously, with 15 min between doses. Relative potencies and 95% confidence limits were obtained by a parallel line bioassay procedure (Finney, 1964).

sponse was obtained for N-desmethyl diazepam, 2 hr after the first curve, there was no tolerance (Fig. 3).

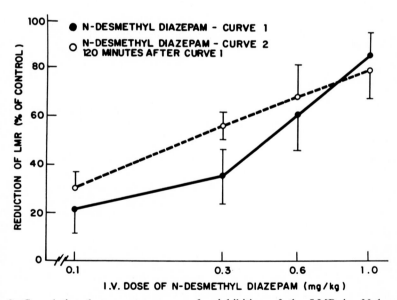

FIG. 3. Cumulative dose-response curves for inhibition of the LMR by N-desmethyl diazepam administered i.v. For each dose-response curve, doses were administered at 15-min intervals. The second dose-response curve (O----O) was obtained 120 min after completion of the first curve (●——●). Each point represents the mean ± S.E. of five cats.

Pretreatment with N-desmethyl diazepam significantly interfered with diazepam-induced inhibition of the LMR (Fig. 4). A comparison of the dose-response curve for diazepam in the presence of its metabolite with diazepam alone (Fig. 4) indicates that the former is only 0.23 times as potent

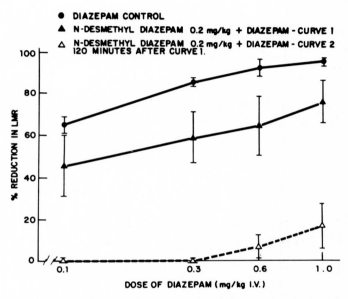

FIG. 4. The effect of N-desmethyl diazepam (0.2 mg/kg, i.v.) on diazepam-induced inhibition of the LMR. The dose-response curves shown are for diazepam alone (●——●), N-desmethyl diazepam plus diazepam curve 1 (▲——▲) and curve 2 (△----△), 120 min after curve 1. Each point represents the mean ± S.E. of four cats for diazepam alone and six cats for the N-desmethyl diazepam plus diazepam curves.

($p < 0.05$, same method as used for data in Table I).

In order to study the tolerance phenomenon further, a simplified method was developed for producing tolerance. A single administration of diazepam, 2 mg/kg, i.v., followed by an interval of 2 hr resulted in a marked reduction of the response to a repeat of the same dose of diazepam (Table II). In experiments with methocarbamol or phenobarbital, each drug was administered 2 hr after the initial dose of diazepam. There was no cross-tolerance to the LMR-depressant effects of either methocarbamol or phenobarbital in diazepam-tolerant cats (Table II). Tolerance to diazepam was verified in each experiment involving phenobarbital or methocarbamol.

TABLE II. *LMR inhibition by methocarbamol and phenobarbital in diazepam-tolerant cats[a]*

Treatment	I.V. dose (mg/kg)	Percent inhibition of the LMR (mean ± S.E.)	
		Normal cats[b]	Diazepam-tolerant cats[c]
Diazepam	2	79.2 ± 6.2	23.9 ± 13.3
Methocarbamol	10	5.2 ± 5.2	21.8 ± 11.8
Methocarbamol	30	75.2 ± 10.7	62.6 ± 7.7
Phenobarbital	3	12.3 ± 11.9	16.2 ± 7.8
Phenobarbital	6	68.1 ± 9.5	69.2 ± 11.0

[a] Tolerance was induced by administering a single i.v. dose of diazepam (2 mg/kg) and waiting for 2 hr.
[b] Seven cats were used in the experiments with diazepam, five cats for methocarbamol, and three cats for phenobarbital.
[c] Seven cats were used in the experiments with diazepam, five cats for methocarbamol, and seven cats for phenobarbital.

B. Inhibition of the flexor reflex in spinal cats

It was of interest to determine if there is acute tolerance to diazepam in another polysynaptic reflex, one that occurs in the spinal cord instead of in the reticular formation. A single administration of diazepam, 2 mg/kg, i.v., followed by an interval of 2 hr resulted in a marked reduction of the response to a repeat of the same dose of diazepam (Table III). There was

TABLE III. *Acute tolerance to diazepam-induced inhibition of the flexor reflex in spinal cats[a]*

Percent reduction of flexor reflex (mean ± S.E.)			
Diazepam control[b]	Diazepam-tolerant[c]	Methocarbamol (30 mg/kg) in diazepam-tolerant cats	Methocarbamol (30 mg/kg) in normal cats[d]
100 ± 0	24.3 ± 19.1	81.1 ± 12.0	96.6 ± 3.4

[a] Five cats were used, each cat receiving two i.v. doses of diazepam (2 mg/kg), 2 hr apart, followed by a single i.v. dose of methocarbamol (30 mg/kg), 1 hr later.
[b] A single i.v. injection of diazepam (2 mg/kg).
[c] Effect of a second dose of diazepam, 2 hr after the first dose (see column 1).
[d] In a separate series of experiments, methocarbamol (30 mg/kg) was administered to six normal cats.

no cross-tolerence to methocarbamol-induced inhibition of the flexor reflex in diazepam-tolerant cats when methocarbamol was administered after the second dose of diazepam. Thus there is also acute tolerance to diazepam-induced inhibition of the flexor reflex as well as the LMR, and no cross-tolerance to methocarbamol with either reflex.

C. Depression of EEG amplitude in dial-urethane-anesthetized cats

Monopolar EEG's were recorded from the left side of the skull in an area corresponding to the position of the motor cortex. It was of interest to investigate whether there was tolerance to a generalized CNS-depressant effect of diazepam, as reflected by depression of EEG amplitude, or whether the tolerance phenomenon was only limited to effects on motor function. Diazepam (2 mg/kg, i.v.) produced marked depression of EEG amplitude (59.9 ± 6.3%). The effect of a second injection of diazepam, 2 mg/kg, 2 hr after the first injection was substantially less (2.1 ± 2.1%) than the first, indicating tolerance to diazepam (Table IV). Since phenobarbital also

TABLE IV. *Depression of EEG amplitude in anesthetized cats*

Treatment	Dose (mg/kg)	% Reduction (% of control) (mean ± S.E.)	
		Normal cats[a]	Diazepam-tolerant cats[b]
Diazepam[c]	2	59.9 ± 6.3	2.1 ± 2.1
Phenobarbital[d]	1	5.0 ± 5.0	22.6 ± 7.3
	3	29.9 ± 12.3	56.4 ± 11.0
	6	50.5 ± 10.2	68.8 ± 11.4
	10	65.9 ± 9.5	76.4 ± 6.3

[a] Before tolerance was produced.
[b] Two hr after diazepam (2 mg/kg), either a repeat dose of diazepam (2 mg/kg) or multiple doses of phenobarbital were administered.
[c] Four cats received diazepam.
[d] Four normal cats and five diazepam-tolerant cats were used.

produces a depression of EEG amplitude, increasing doses of phenobarbital were administered 2 hr after diazepam 2 mg/kg (diazepam-tolerant cats), and to normal cats (no diazepam). As found with the LMR, there was no cross-tolerance to phenobarbital in diazepam-tolerant cats; if anything, there was a slight potentiation (Table IV).

Since there was tolerance to diazepam, the relative potency of N-des-methyl diazepam for EEG-amplitude depression was compared to diazepam. N-desmethyl diazepam was only 0.085 times (95% fiducial limits of 0.036 to 0.155) as potent as diazepam as determined by the parallel line bioassay procedure of Finney (1964).

D. Ataxia in unanesthetized cats

One of the prominent and easily measurable effects of diazepam in unanesthetized cats is ataxia or motor incoordination. The tolerance phenomenon was studied by administering diazepam by several different dose regimens, waiting a total of at least 2 hr and then repeating a single dose of diazepam. Different dose regimens for tolerance had to be tried because in the previous experiments all dosing was intravenous, and this is inconvenient in unanesthetized cats. Neither dose regimen used (Table V) pro-

TABLE V. *Ataxic effects of diazepam in normal versus "tolerant" cats*[a]

	Ataxia score (mean ± S.E.)	
Dose	Normal cats (saline + diazepam)	"Tolerant" cats (diazepam + diazepam)
1 mg/kg i.p.[b]	2 ± 0	3.7 ± 0.3
3 mg/kg i.p.[c]	3.3 ± 1.3	4.0 ± 0.0

[a] N = 3 for each group.
[b] Dose regimen was diazepam 0.25 mg/kg initially, 0.25 mg/kg after 1 hr, and 1.0 mg/kg after 2 hr for "tolerance" versus saline initially, saline after 1 hr, and diazepam 1.0 mg/kg after 2 hr for normal cats.
[c] Dose regimen was diazepam 1 mg/kg initially and diazepam 3 mg/kg after 2 hr for "tolerance" versus saline initially and diazepam 3 mg/kg after 2 hr for normal cats.

duced tolerance to the ataxic effects of diazepam; if anything, the effects of the test dose of diazepam were additive with those of previous doses. At this point, there were at least two possibilities: either the dose regimen of diazepam was not adequate for tolerance or ataxia differs from other effects of diazepam. Since our working hypothesis is that diazepam-induced tolerance is due to the formation of a metabolite(s) that acts as an inhibitor, the ataxic effects of N-desmethyl diazepam were studied alone and in combination with diazepam. N-desmethyl diazepam is 0.336 times as potent as diazepam (Table VI), and this is the smallest potency difference found in these experiments. More importantly, a low dose of N-desmethyl diazepam in combination with diazepam did not antagonize the ataxic effects of diazepam; if anything, there was potentiation (Table VI). Thus for the ataxic effects of diazepam in cats, there is no acute tolerance and N-desmethyl diazepam does not antagonize diazepam.

IV. DISCUSSION

Acute tolerance has been demonstrated using three different pharma-

TABLE VI. *Relative potency of N-desmethyl diazepam to diazepam for producing Ataxia*

Drug	No. of cats with ataxia[a] / No. of cats tested			
Dose (mg/kg i.p.)	0.5	1.0	2.0	4.0
Diazepam	0/6	3/6	4/6	5/6
N-Desmethyl diazepam[b]	1/6	1/6	3/6	3/6
N-Desmethyl diazepam 1 mg/kg + diazepam	2/6	5/6	5/6	5/6

[a] Defined as inability to land on all fours when dropped from a height of 2 ft.

[b] Relative potency (95% fiducial limits) of N-desmethyl diazepam compared to diazepam for producing ataxia is 0.336 (0.146-0.689) as calculated by a variation of the Probit Maximum Likelihood Method (Finney, 1962).

cological effects of diazepam: inhibition of the LMR, inhibition of the flexor reflex, and depression of EEG amplitude. It is likely that this acute tolerance is due to the formation of a less potent metabolite(s), especially N-desmethyl diazepam. The present studies and the work of other investigators (Crankshaw and Raper, 1968; Tseng and Wang, 1971) indicate that there is no tolerance when diazepam is administered i.v. at intervals of 15 min or less, whereas in the present studies there is tolerance at 1- and 2-hr intervals. This is the opposite of what would be expected if tolerance were due to a simple drug-receptor interaction. Moreover, N-desmethyl diazepam, which presumably acts on the same receptor as diazepam, did not produce tolerance to itself. In addition, tolerance was greater after i.d. than i.v. administration, a situation which also favors the hypothesis of metabolite formation. Diazepam given by the former route goes through the hepatic portal shunt after absorption from the duodenum (Coutinho et al., 1970), thus enhancing the chances of hepatic metabolism and tolerance development, if indeed, the latter is due to metabolite formation. Since diazepam accumulates in relatively high concentrations in adipose tissue (Marcucci et al., 1968), a significant percentage of diazepam could be diverted into adipose tissue after i.v. administration when it passes through the general circulation before reaching the liver. This discusson of tolerance due to metabolite formation would be highly speculative were it not for the fact that N-desmethyl diazepam, the major metabolite of diazepam in three different species (Zbinden and Randall, 1967; Van der Kleijn, 1969), is only one-fifth as potent in inhibiting the LMR and one-eleventh as potent in depressing EEG amplitude. These results are summarized in Table VII. In both these mea-

TABLE VII. *Summary of diazepam-induced acute tolerance studies in cats*

Effect of diazepam	Acute tolerance	Inhibition of diazepam by N-desmethyl diazepam	Relative potency[a]	Cross-tolerance (C.T.)
LMR inhibition	Yes	Yes	0.20	No C.T. to methocarbamol or phenobarbital
Flexor reflex inhibition	Yes	—	—[b]	No C.T. to methocarbamol
Depression of EEG amplitude	Yes	—	0.09	No C.T. to phenobarbital
Ataxia	No	No	0.34	—

[a] Diazepam = 1.
[b] There was a very variable dose-response relationship for both diazepam and N-desmethyl diazepam which prevented a relative potency analysis.

sures, there was acute tolerance to diazepam. Ataxia was the only measurement in which there was not acute tolerance to diazepam, and N-desmethyl diazepam did not antagonize the ataxic effect of diazepam. Of all the diazepam effects studied, the differences between diazepam and its N-desmethyl metabolite for producing ataxia was the smallest. The possibility has not been ruled out that tolerance to diazepam may be independent of its metabolites.

Cross-tolerance to either methocarbamol or phenobarbital was not seen in diazepam-tolerant cats. These results suggest that diazepam does not act at the same receptor sites as phenobarbital or methocarbamol. It has been assumed that both methocarbamol and diazepam inhibit polysynaptic reflexes in the spinal cord by depressing interneurons, and yet there was no cross-tolerance to methocarbamol-induced inhibition of the flexor reflex in spinal cats. Similarly, it has been assumed that both diazepam and phenobarbital act at the same sites in the brain to influence behavior and yet there was no cross-tolerance to phenobarbital-induced depression of EEG amplitude. These studies suggest that diazepam has a unique mode of action in affecting motor function in the brain and spinal cord. One of the problems with these cross-tolerance studies was that in each system, at least a three-drug interaction was being studied: diazepam, test drug, and at least one metabolite of diazepam. With all measurements taken, the order of potency is diazepam > N-desmethyl diazepam > phenobarbital or methocarbamol. Since there is presumably enough metabolite formed to inhibit diazepam, it should also be enough to inhibit the less potent methocarbamol or phenobarbital if the latter two drugs were acting at the same receptor.

The acute tolerance phenomenon is a potentially interesting and valuable tool for studying diazepam receptor sites and for differentiating these sites from those of other related CNS drugs.

ADDENDUM

Dr. Morselli and his co-workers have presented data in this volume ("Kinetics of the Distribution of ^{14}C-Diazepam and Its Metabolites in Various Areas of Cat Brain" by Morselli et al.) which indicate that N-desmethyl diazepam is the major metabolite of diazepam in cats. The time course for N-desmethyl diazepam levels in cat brain correlates well with the time course for tolerance development in our own studies. This data strongly supports our hypothesis that N-desmethyl diazepam formation is the major factor in the diazepam acute tolerance phenomenon in cats.

ACKNOWLEDGMENTS

The authors are especially grateful to the North-Holland Publishing Company for allowing the reproduction of Figs. 1 through 4 and Table I from Barnett and Fiore (1971a) and of Table II from Barnett and Fiore (1971b).

The authors gratefully acknowledge the assistance of Mr. Morton Miller in the statistical analyses.

REFERENCES

Barnett, A., and Fiore, J. W. (1971a): Acute tolerance to diazepam in cats and its possible relationship to diazepam metabolism. *European Journal of Pharmacology,* 13: 239-243.

Barnett, A., and Fiore, J. W. (1971b): Acute tolerance to diazepam in cats: Lack of cross-tolerance to methocarbamol or phenobarbital. *European Journal of Pharmacology,* 14:301-302.

Coutinho, C. B., Cheripko, J. A., and Carbone, J. J. (1970): Correlation between the duration of the anticonvulsant activity of diazepam and its physiological disposition in mice, *Biochemical Pharmacology,* 19:363-379.

Crankshaw, D. P., and Raper, C. (1968): Some studies on peripheral actions of mephenesin, methocarbamol, and diazepam. *British Journal of Pharmacology,* 34:579-590.

Finney, D. J. (1964): *Statistical Method in Biological Assay,* 2nd edition, Hafner, New York, pp. 48–64.

Goldberg, M. F., Manian, A. A., and Efron, D. H. (1967): A comparative study of certain pharmacological responses following acute and chronic administrations of chlordiazepoxide. *Life Sciences,* 6:481-491.

Hoogland, D. R., Miya, T. S., and Bousquet, W. F. (1966): Metabolism and tolerance studies with chlordiazepoxide-2-^{14}C in the rat. *Toxicology and Applied Pharmacology,* 9:116-123.

Marcucci, F., Fanelli, R., Frova, G., and Morselli, P. L. (1968): Levels of diazepam in adipose tissue of rats, mice, and man. *European Journal of Pharmacology,* 4:464-466.

Ngai, S. H., Tseng, T.-C., and Wang, S. C. (1966): Effect of diazepam and other central nervous system depressants on spinal reflexes in cats: A study of site of action. *Journal of Pharmacology and Experimental Therapeutics,* 153:344-351.

Tseng, T.-C., and Wang, S. C. (1971): Mechanisms of action of centrally acting muscle relaxants, diazepam and tybamate. *Journal of Pharmacology and Experimental Therapeutics,* 178:350-360.

Van der Kleijn, E. (1969): Kinetics of distribution and metabolism of diazepam and chlordiazepoxide in mice. *Archives Internationales de Pharmacodynamie et de Thérapie,* 178:193-215.

Zbinden, G., and Randall, L. O. (1967): Pharmacology of benzodiazepines: Laboratory and clinical correlations. In: *Advances in Pharmacology,* edited by S. Garattini and P. A. Shore. Academic Press, New York, pp. 213-291.

The Benzodiazepines, Raven Press, New York © 1973

ELECTROMYOGRAPHIC EVALUATION OF DRUG-INDUCED MUSCLE RELAXATION

Paolo Pinelli

Institute of Nervous and Mental Diseases, Catholic University, Rome, Italy

I. INTRODUCTION

Abnormal motor hyperactivity can be observed in psychiatric and neurological patients; in the latter, this condition may be due to three types of disorders: (1) pathological irritation in some points of the nervous system (central or peripheral), (2) disinhibition of lower by superior centers as a result of a dysfunction of the latter, or (3) increase of descending impulses, facilitating activity of lower motor centers.

The problem of muscle relaxation discussed in this chapter will deal primarily with the third group of cases. Since this physiopathological situation cannot be easily identified, we will be concerned not only with the administration of muscle relaxants and their effects, but also with the methodology suitable for detecting the basic functional disturbance to be treated.

II. TYPES OF MUSCULAR HYPERACTIVITY REQUIRING ATTENUATION

Sustained hyperactivity of motor units "at rest" can arise from neurological or psychic disorders. Muscular relaxation has been sought in neurological and allied diseases in conditions such as spasticity, rigidity, and reflex contractures and in states of psychological tension or anxiety occurring in psychoneuroses of different types (Birkmayer, 1970; Bonne et al., 1970).

Therefore, our first aim will be to treat the cause of the disease, since the muscular hyperactivity represents a secondary and, sometimes, collateral effect thereof; however, in some chronic conditions where the lesion is irreversible or the psychoneurosis difficult to relieve, muscle relaxation may constitute the only therapy available.

559

III. EVALUATION OF DIFFERENT ASPECTS OF
MOTOR HYPERACTIVITY

In order to analyze the intensity and, possibly, the nature of tonic hyperactivity "at rest," there are many (1) *elementary* and (2) *complex* phenomena to be explored.

(1) In investigating *elementary* activities, the following tests should be included:

> a. Tendinous reflexes, elicited by a pair of stimuli at different intervals (excitability cycle); more generally, monosynaptic proprioceptive excitability should be evaluated from the degree of diffusion of the responses, the ratio between the amplitude of maximal M response and maximal T response, and the time of the secondary rise of the threshold, 280 to 500 msec after conditioning of the mechanical tendinous stimulus
>
> b. Monosynaptic responses evoked by electrical nerve stimulation (to be differentiated from F responses)
>
> c. Jendrassik effects
>
> d. Silent period, measured according to the methodology described previously (Pinelli, 1965)
>
> e. Tonic reflexes
>
> f. Polysynaptic cutaneous reflexes

(2) With regard to the impairment of automatic and voluntary motor activity in patients with muscular hypertonia, the most important *complex* phenomena to be investigated are represented by the degree of localization of innervation to the muscles involved and by the duration of the after-activity. For evaluating all these latter features, we have adopted the methodology developed by us for studying Exner's prepared reflex and motor learning.

In an investigation carried out on 10 spastic patients, we have observed that the above-mentioned elementary and complex features are correlated with muscular hyperactivity "at rest" at different ratios. The lowest correlation was found for the silent period; the medium for monosynaptic and polysynaptic reflexes and for Jendrassik effects; and the highest for the tonic reflexes.

The test of tonic reflexes is, in fact, a complex investigation, including: determination of body and limb positions which result in muscular electrical silence (or a tendency to it); preliminary trials of normalization to avoid active reactions of the subject; different speeds and range of mobilization for evaluating the threshold of reflex tonic responses; and continuous recording of electrical activity from agonistic and antagonistic muscles in order to investigate phases and features of lengthening reactions.

Such a complete investigation also provides the most important clues for differentiating between neurogenic and psychogenic muscular hyperactivity.

IV. ADMINISTRATION OF MUSCLE RELAXANTS

A. Indications for using muscle relaxants

A correct administration of muscle relaxants in neurological patients presupposes the confirmation of the following two conditions: (1) reflex hyperactivity of motor units constitutes the main difficulty for the patient inasmuch as it prevents normal motor performance, and (2) reduction of this muscular hyperactivity will result in an improvement of motor performance.

It follows that when testing muscular relaxation in clinical practice, it should be determined if the reduction of muscular hyperactivity is a *selected* effect due to which motor performance is also improved.

In cases of psychoneurosis, muscle relaxation can have psychosomatic repercussions; anxiety determines hypertonia which in itself becomes the cause of disagreeable sensations that worsen the psychoneurotic state. On the other hand, we must keep in mind that some drugs can act as muscle relaxants through a primary ansiolitic effect.

B. Doubtful benefits of indiscriminate use of muscle relaxants

The treatment of spasticity due to chronic pyramidal lesions has been the subject of many publications; in our opinion, one reason why the results are rather disappointing may be the fact that muscle relaxation has been used without any critical analysis of what possible benefits can be derived in the different pathological conditions.

It is important to realize that, in the case of spasticity, we are facing a puzzling situation: the main disorder consists of the reduction of pyramidal innervation; the exaggerated stretch reflexes maintain a basic postural mechanism which in itself represents the only remaining type of movement; the few central descending impulses can start this stereotyped diffuse antigravitary activity. In such instances, it would appear evident that muscle relaxation is contraindicated.

In a previous study with diazinotriazine, we obtained a significant reduction of spasticity, corresponding especially to a decreased monosynaptic excitability. The end results, however, were practically useless in most cases of sequelae of acute pyramidal lesions and negative in cases with progressive pathological impairment (Strümpell's paraplegia, Erb's paraplegia; disseminated sclerosis); only in flexor paraplegia of both spinal and cerebral origin was a moderate improvement diagnosed.

In conclusion, it may be said that further research on the conditions requiring reduction of muscular hypertonia is essential in order to develop an appropriate treatment of spasticity with muscle relaxants.

V. ROLE OF GAMMA IMPULSES IN OCCLUSION

A functional situation requiring the reduction of muscular hypertonia can be represented by an abnormal increase in tonic reflex activity, from which some "occlusion" effect of spinal cells arises. In fact, these afferent impulses may block those that descend from the higher motor centers and which are normally capable of eliciting automatic and voluntary activity.

An abnormally high flow of afferent impulses firing through reflex pathways can be best explained by an increase of gamma impulses lowering the threshold of muscle spindle receptors.

In order to detect such a functional situation, two sets of tests can be applied:

(1) *Jendrassik maneuver* which will be ineffective when abnormal gamma hyperactivity preexists. This test is suitable for cases with slight or rather moderate paresis.

(2) *Muscle procainization* which, according to Rushworth (1961), blocks gamma activity at the site of injection, with the result of eliminating or completely settling occlusion effects; voluntary recruitment of motor units will be increased. This test can also be applied in cases of paralysis and severe paresis.

VI. CORRECTION OF OCCLUSION CAUSED BY PROPRIOCEPTIVE IMPULSES

A typical condition where abnormal increase of reflex postural activity contributes to the impairment of motor performance exists in Parkinson's disease. Restoring the amount of dopamine in the corpus striatum by the administration of L-DOPA exerts an inhibitory effect resulting in an attenuation of rigidity; descending impulses, facilitating the abnormal activity of alpha anterior horn cells, directly or through gamma cells, are reduced.

In the hope of elucidating these phenomena, we have studied, in 10 parkinsonian patients with severe akinesia, the changes occurring in reflex, automatic, and voluntary activities after 3 months of treatment with L-DOPA at a dose of 3.5 to 5 g daily.

Monosynaptic proprioceptive excitability appeared reduced only in two out of 10 cases, while in the other eight it was normal or even increased. H response was reduced in three out of 10 cases and it was normal or increased in seven; T/H values were reduced in five cases, without any correlation with the degree of improvement of rigidity, clinically evaluated according to the fivefold scale described in a previous paper (Pinelli, *in press*). The silent period was slightly reduced in four cases. The value of T_j (= amplitude of tendinous reflex during Jendrassik/T = the same without any Jendrassik maneuver) was increased in eight patients.

Tonic stretch reflexes were reduced to the same degree as rigidity (clinically evaluated). Improvement in psychomotor reaction, especially in localization and control of innervation, was a constant finding and, conversely, no significant changes were found in other patients with a slight degree of akinesia or with mental deterioration.

In analyzing these results, we should underline that the improvement induced by L-DOPA has a good correlation with the reduction of tonic stretch reflexes, whereas monosynaptic proprioceptive phasic reflexes are sometimes increased. This paradoxic dissociation may be due to a decrease of tonic stretch reflexes which, before treatment, are so pronounced as to cause occlusion of phasic stretch reflexes; the same degree of occlusion could be thought to occur during central innervation intended for voluntary recruitment of motor units. These conclusions agree rather well with the effects of muscle procainization analyzed by Rushworth (1961) in parkinsonian patients.

VII. EFFECT OF DIAZEPAM IN SPASTIC PATIENTS

By applying the muscle procainization test to 67 patients with spasticity, selected by using the Rushworth test, we obtained relaxation with enhanced phasic reflexes only in six cases. An improvement in the voluntary innervation in these cases was observed only when procainization was performed simultaneously in the antagonistic muscles. The patients in question were suffering from hemiparesis due to chronic cerebrovascular disease. They revealed atypical features of spasticity since more continuous reflex responses were electromyographically recorded during tonic stretch.

These six cases were all treated with diazepam by mouth at the dose of 15 mg daily during at least 5 days. At this low dosage, no adverse collateral effects such as drowsiness occurred. Three cases with the same degree of spasticity of cerebrovasculopathic origin but with a negative Rushworth test underwent the same treatment as controls.

Tonic rest activity appeared slightly reduced in all the nine cases; but clinical improvement with increased motor voluntary strength was observed only in the six Rushworth-positive cases: the EMG during preparatory and executive phases of psychomotor reactions showed reduced motor unit activity in antagonistic muscles, while after-activity and cessation times were shortened. These results correlate well with the clinical improvement. On the other hand, no significant shortening of reaction times was found.

It may be concluded that diazepam, at relatively low doses, appears useful in some cases of spasticity, characterized by continuous motor unit activity "at rest" and during tonic stretch; diazepam may improve the motor performance of these patients mainly through a reduction of the abnormal activity present in antagonistic muscles.

REFERENCES

Birkmayer, W. (1970): Corrélation entre tonus musculaire et psychisme. In: *Détente Neuropsychique: Aspets Thérapeutiques Nouveaux*. Ciba, Basel.

Bonne, A., Heymans, W., and Rosselle, N. (1970): Electromyographic tests for the evaluation of effects of myotonolytic drugs. *Electromyography*, 10:219–225.

Pinelli, P. (1959): Terapia antispastica con un derivato della diazinotriazina. *Rivista di Neurologia*, 29:1–28.

Pinelli, P. (1965): Apporti della elettromiografia allo studio dei disturbi cerebellari. Supplemento: Atti XV Congresso Italiano di Neurologia, Simposi, *Rivista di Patologia Nervosa e Mentale*, (Suppl.): 278–335.

Pinelli, P., and Valle, M. (1966): Studio fisiopatologico dei riflessi muscolari nelle paresi spastiche. *Archivio Scienze Mediche*, 109:1–128.

Rovetta, P. (1970): Controllo elettrofisiologico d'effetti periferici della L-DOPA nella malattia di Parkinson. *Sistema Nervoso*, 22:191–195.

Rushworth, G. (1961): The gamma-system in parkinsonism. *International Journal of Neurology*, 2:34–50.

The Benzodiazepines, Raven Press, New York © 1973

CLINICAL USE OF DIAZEPAM IN INTENSIVE THERAPY: INDICATIONS AND RESULTS

Marialuisa Bozza-Marrubini and Adriana Selenati

Department of Intensive Therapy and Reanimation, Ospedale Maggiore di Milano, Milan, Italy

I. INTRODUCTION

The value of diazepam in anesthetic practice is now well established. Premedication and, in selected cases, induction of anesthesia are its main indications in this field. For both these uses, diazepam has proven to be a safe and effective drug, with powerful sedative and muscle-relaxant effects.

In intensive therapy, diazepam may be a good substitute for narcotics and for peripheral muscle relaxants. Five years of experience has shown that it may be the drug of choice in the following conditions: (1) naso- or oro-tracheal intubation in conscious or semiconscious patients; (2) adaptation of conscious or comatose patients to the mechanical respirator's rhythm; (3) psychomotor agitation in brain injuries and in patients with multiple injuries; (4) sedation in convulsive disorders due to status epilepticus, eclampsia, or poisoning; and (5) relaxation of tonic muscle spasms due to brainstem lesions, tetanus, or other causes.

II. CLINICAL INDICATIONS AND RESULTS

A. Naso- or oro-tracheal intubation

Intubation is often required as a first aid in an emergency to overcome respiratory obstruction or to treat respiratory insufficiency requiring mechanical artificial ventilation.

The patients to be intubated are in most instances uncooperative due to either a state of anxiety or one of delirium and restlessness; unconsciousness with increased muscle tone or convulsive or spastic fits is also frequently associated with airway obstruction or respiratory insufficiency requiring in-

565

tubation. In these circumstances, deep sedation and muscle relaxation are required before tracheal intubation can be performed correctly, quickly, and without trauma.

The usual anesthetic routine for tracheal intubation, i.e., the intravenous barbiturate-curare sequence, is as a rule contraindicated in these emergency cases and may be dangerous since vital functions are apt to be very unstable and general anesthesia may easily precipitate a fatal cardiocirculatory collapse.

Local surface anesthesia is equally unsuitable in these cases because it is too time consuming, and requires the cooperation of the patient.

In our experience, intravenous diazepam (10 to 20 mg in adults) produces deep sedation with relaxation of the jaw muscles; the reflexes of the upper airways are greatly depressed. The peak effect, if the injection is made slowly, appears after 2 to 3 min, lasts about 10 min, and is not accompanied by a significant depression of vital functions; in such a case, blind nasal or direct-vision naso- or oro-tracheal intubation may be performed with ease and minimal trauma. In resistant cases, surface anesthesia of the larynx with 2% lidocaine may readily block residual reflex activity. Intravenous meperidine (20 to 50 mg) may be a good supplement if sedation is insufficient; the association of diazepam with a narcotic may, however, lead to transient respiratory depression.

B. Adaptation to the mechanical respirator's rhythm

In intensive therapy, artificial mechanical respiration with intermittent positive pressure (IPPR) is applied in a number of cases. To achieve a good intrapulmonary distribution of gases and a satisfactory respiratory exchange, the patient must be well "adapted" to the respirator's rhythm; in other words, the patient's respiratory activity must be totally abolished and the thorax must passively follow the inflating and deflating movements produced by the pumping action of the machine. If the patient maintains a spontaneous respiratory rhythm or if he "fights" against the respirator, the treatment may fail completely.

Good adaptation is usually obtained when the acid-base balance of the blood is maintained at the normal level or slightly on the alkaline side, and when, at the same time, a satisfactory level of PO_2 is assured. No attempt to "adapt" the patient to the respirator by the use of drugs should be made until every possible effort has been made to achieve optimal blood-gas levels and acid-base conditions. In many cases, this cannot be accomplished due to pulmonary, circulatory, metabolic, or neurologic pathology. In other instances, adaptation is difficult to obtain or to maintain, even when the above requirements have been met, as a result of disturbing factors such as anxiety, fear, or pain.

For many years, we have observed that diazepam is the drug of choice in all these conditions. A single intravenous dose of 5 to 10 mg is usually quite effective in relaxing the patient and producing a long-lasting adaptation to the respirator. Good results have been obtained in nearly every type of pathology requiring prolonged IPPR; no untoward side effects have been observed even in subjects with very unstable circulatory conditions (for example, after cardiac surgery, in cases of crushed chest, or after a recent severe myocardial infarct).

The effective dose can be repeated every 4 to 6 hr. In our experience, 30 to 40 mg every 24 hr can be given for many days with no apparent harm. Tolerance does not seem to develop. In very severe cases, it may be necessary to supplement diazepam with narcotics or peripheral muscle-blocking drugs. However, even in these extreme conditions, diazepam is very useful in reducing the doses of the latter drugs and in prolonging their effects.

C. Psychomotor agitation, convulsive fits, tonic muscle spasms

In the conditions referred to above, regardless of their cause, experience has led us to use diazepam as the drug of choice. Other sedatives or muscle relaxants have been used in association with diazepam only in those cases where diazepam alone proved to be ineffective or insufficiently active.

The main reasons for our choice lie again in the absence or mildness of undesirable side effects, such as respiratory and circulatory depression, or of toxic effects on the liver and kidney. In most of the instances listed above, the vital functions are often already severely disturbed and, in cases of poisoning, any additive toxic load must obviously be avoided.

For these last three conditions, higher doses than those used for intubation or for adapting a patient to the respirator's rhythm are generally necessary.

III. SUMMARY

Table I summarizes the above-mentioned clinical indications for using

TABLE I. *Diazepam in intensive therapy*

Indications	Dosage range in adults (mg)		Association with other drugs[a]	
	Single dose	Daily dose	Narcotics or barbiturates	Curare
Tracheal intubation	10-20	-	+ surface anesth. or + meperdine (20-50 mg)	-
Adaptation to the ventilator's rhythm	10	40-60	+ meperdine (20-50 mg)	+ in very severe cases
Psychomotor agitation	10-20	50 or more	++ meperidine, methadone, etc., if pain present	-
Status epilepticus	10-20	60	++ phenobarbital, diphenylhydantoin, tiopentone	+ in very severe cases
Eclampsia	10-20	60	+ tiopentone	-
Poisoning by MAO inhibitors, dibenzazepine compounds, organphosphorus insecticides, etc.	10	40	-	-
Decerebrate fits	10-20	60-80	-	-
Tetanus	20	80-120	-	+++ in severe cases tubocurarine
Poisoning by 500 mg strychnine (recovery in 12 hr)	10	55 in 4 hr	thiopentone (2,850 mg in 12 hr)	-

[a] + Useful, seldom necessary; ++ often necessary; +++ necessary in most cases

diazepam, the dosage range employed in adults, and the circumstances in which it may be necessary to use other drugs in association with diazepam.

It must be pointed out that no clear relation between dosage of diazepam and body weight or age of the patients could be found. In all the cases presented here, it is now our usual practice to administer the drug by slow intravenous injection in units of 5 mg in adults and 1 to 2 mg in children until the total dose is reached that gives the required effect. If a satisfactory and/or lasting result is not obtained after a single total dose of 20 to 25 mg in adults and 5 to 10 mg in children, diazepam is supplemented with another suitable drug as indicated in the Table. As a rule, the injection of diazepam is not repeated at less than 3-hr intervals.

The Benzodiazepines, Raven Press, New York © 1973

BENZODIAZEPINES IN THE TREATMENT OF ALCOHOL WITHDRAWAL STATES

S. C. Kaim

Alcohol and Drug Dependence Service, Veterans Administration Central Office, Washington, D.C.

On the occasion of the first symposium on LIBRIUM, which was held in Galveston, Texas, in 1959, Kaim and Rosenstein (1960) reported the first clinical use of chlordiazepoxide in the management of epilepsy. We reported that chlordiazepoxide effectively controlled or reduced the frequency of seizures in eight of our nine cases, and that among the improved patients the drug appeared to be at least as effective as any of the anticonvulsant regimens these patients had previously used. The abnormal EEG's in the series all changed for the better, and the pentylenetetrazol (METRAZOL) threshold was, in most cases, increased over that prevailing while the patients had been taking other anticonvulsants.

At that initial symposium, Ticktin and Schultz (1960) described the first use of the benzodiazepine chlordiazepoxide in acute alcoholism. Among the 28 patients they studied, nine had delirium tremens and one had "alcoholic shakes." Among the symptoms exhibited by these patients were hallucinations, tremulousness, hostility, confusion, restlessness, and convulsions. Of the 28 alcoholic patients, 16 excellent and six good results were obtained.

Kaim and Rosenstein (1961) reported on an expanded series of 30 patients with seizures treated with chlordiazepoxide. Of this series, eight were associated with alcoholism. As in the previous study, most of the patients had better control of their seizures while taking chlordiazepoxide than with prior drug regimens. The EEG again showed marked improvement after chlordiazepoxide therapy had been instituted, and the pentylenetetrazol threshold was increased in 14 of 16 patients who underwent this procedure. Among the alcoholics in this series, chlordiazepoxide appeared to be especially effective in the cases with frank or impending delirium tremens and associated convulsive seizures. Chlordiazepoxide at a dose of 200 to 300 mg daily brought prompt control of both the psychotic and convulsive phenomena.

Rosenfeld and Bizzoco (1961) cited marked or moderate improvement

571

in 22 of 30 chlordiazepoxide-treated cases of alcohol withdrawal, compared to eight of 30 controls given placebos. The difference between experimental subjects and controls was highly significant.

Kurtzke and Kaim (1962) described the first use of chlordiazepoxide given intravenously for multiple seizures associated with acute alcoholism. The slow intravenous administration of 1 mg/kg body weight led to prompt cessation of seizure activity.

Kissen (1962) found in a study of 80 alcoholic patients that 21 of 40 on chlordiazepoxide received optimal ratings in the areas of sleep, appearance, tremor, appetite, and nervousness versus only one of 40 on a variety of other drugs and placebo.

Chambers and Schultz (1965) reported a double-blind study of chlordiazepoxide, promazine, and diazepam in the treatment of 103 males suffering from delirium tremens, impending delirium tremens, acute alcoholic hallucinosis, and acute alcoholic brain syndrome. They found that both promazine and chlordiazepoxide produced good to excellent effects in 70% of their respective groups versus 49% effectiveness in the diazepam group. However, chlordiazepoxide showed slightly better results than promazine in reducing the severity of tremor, delirium, and delusions, and markedly better effects on convulsions. They concluded that chlordiazepoxide was the drug of choice for the management of the severe phase of alcohol withdrawal.

Jaffe (1965) cited the very low degree of cross-dependence with alcohol shown by the phenothiazines, contrasted with the high degree of cross-dependence with alcohol shown by chlordiazepoxide. He cautioned: "since the development of delirium tremens always carries with it a . . . risk of a fatal outcome, it seems appropriate to treat all but the mildest cases of alcoholic withdrawal with agents that show cross-dependence with alcohol."

Greenberg and Pearlman (1967) speculated on an alternate rationale for the use of chlordiazepoxide to prevent delirium tremens. They reported that dreaming was suppressed during periods of increasing blood levels of alcohol. In their study, withdrawal led to an increase in stage 1 rapid eye movement sleep, with 100% stage 1 found just before the development of delirium tremens. Barbiturates and some of the other hypnotic agents also suppress dreaming. Greenberg and Pearlman postulated that with the use of chlordiazepoxide, which does not suppress dreaming during its short-term employment, the dream deficit can be made up during sleep rather than delirium.

Another popular physiological theory has been outlined by, among others, Guerrero-Figueroa et al. (1970). Their studies showed evidence for the view that alcohol has an inhibitory action on cortical and subcortical structures of the brain. They felt that it was probable that chronic alcohol intoxication causes physiological and pathological neural inhibitory processes that decrease, retard, suppress, or arrest the activity of normal and epileptic neuronal elements of cortical and subcortical structures. They go on to speculate that if the integrative action of the CNS is the result of a balance between inhibition and excitation in normal neuronal populations and if chronic administration of alcohol produces a prolonged, sustained, and abnormal potentiation of inhibition, then a sudden abstinence from alcohol will produce a release of CNS inhibition associated with a generalized abnormal increase of excitation. This abnormal increase of excitation may produce an increase of motor activity, an initiation of seizure activity, and the disorganized behavior that is exhibited in man by mental confusion and hallucinations. They propose that this abnormal level of neurophysiological excitation produces an imbalance in the integrative action of the CNS, which is represented by the withdrawal syndrome. In previous work, these authors had reported that diazepam acts by increasing or producing inhibitory actions in the neuronal populations of mesencephalic reticular formation and limbic system structures. The absence of the withdrawal syndrome in animals administered diazepam 6 hr after the last administration of alcohol demonstrated the effectiveness of intravenous diazepam in preventing the development of the expected CNS stage of excitation following alcohol withdrawal.

Victor (1966) has expressed the opinion that no adequate data have yet been reported to prove that any one of the newer psychoactive agents is effective in preventing the development of delirium tremens during the withdrawal state. The Veterans Administration, in 1968, conducted a double-blind comparative evaluation of four commonly employed drug treatments of the alcohol withdrawal syndrome in a cooperative study among 23 of its hospitals. In this study, only 2% of the patients treated with chlordiazepoxide had delirium tremens and/or convulsions, versus an incidence of 16% in the chlorpromazine group, 10% in the hydroxyzine, 11% for thiamine, and 12% for placebo. Kaim et al. (1969), who reported this study, concluded that chlordiazepoxide appeared to be the drug of choice among those tested in the prevention of delirium tremens and convulsions during the acute alcohol withdrawal state.

Victor (1970) later also found chlordiazepoxide preferable to the phenothiazines, "which probably should not be used because of their epileptogenic properties and other serious side effects." Victor states that, although drugs relieve symptoms and provide rest and sleep, there is no evidence that they shorten the duration of withdrawal symptoms once these symptoms have become established. He also finds their usefulness limited in fully developed delirium tremens. He feels that the purpose of medication in this indication is simply to blunt psychomotor overactivity, to lessen exhaustion, and to aid medical and nursing care. He recommends that recurrent withdrawal seizures be treated similarly to any other type of status epilepticus, with intravenous administration of diazepam. Parenthetically, this is very similar to the intravenous use of chlordiazepoxide which the author (1962) recommended for the treatment of this condition.

The Veterans Administration, as reported by Kaim and Klett (1971), has very recently completed a comparative evaluation of chlordiazepoxide, paraldehyde, pentobarbital, and perphenazine in the treatment of frank delirium tremens in 202 cases in a cooperative study conducted by 17 of its hospitals. The hypothesis was that the drugs cross-dependent with alcohol (chlordiazepoxide, paraldehyde, and pentobarbital) would be more effective than perphenazine which is not cross-dependent with alcohol. Although there were no significant differences in the results obtained in the patients treated with these four agents, both paraldehyde and chlordiazepoxide showed strong trends toward a better outcome; therapy had to be terminated in only one case in each of these two groups because of worsening symptoms, as compared to three cases each in the other two groups. In addition, the combined scores of severity of symptoms and duration of the episode tended to favor paraldehyde and chlordiazepoxide over perphenazine and pentobarbital. Since in this study all four groups of patients had a favorable outcome, and since the one quality shared by all four drugs was that of sedation, one might conclude that sedation is, indeed, an important consideration in the management of delirium tremens and that this latter may be a time-limited syndrome, the duration of which may not be amenable to significant reduction.

REFERENCES

Chambers, J. F., and Schultz, J. D. (1965): Double-blind study of three drugs in the treatment of acute alcoholic states. *Quarterly Journal of Studies on Alcohol,* 26:10–18.

Greenberg, R., and Pearlman, C. (1967): Delirium tremens and dreaming. *American Journal of Psychiatry,* 124:133–142.

Guerrero-Figueroa, R., Rye, M. M., Gallant, D. M., and Bishop, M. P. (1970): Electrographic and behavioral effects of diazepam during alcohol withdrawal stage in cats. *Neuropharmacology,* 9:143–150.

Jaffe, J. H. (1965): Drug addiction and drug abuse. In: *The Pharmacological Basis of Therapeutics,* edited by L. S. Goodman and A. Gilman. The Macmillan Co., New York, 3rd edition, p. 304.

Kaim, S. C., and Klett, C. J. (1972): Treatment of delirium tremens. *Quarterly Journal of Studies on Alcohol (in press).*

Kaim, S. C., Klett, C. J., and Rothfeld, B. (1969): Treatment of the acute alcohol withdrawal state: A comparison of four drugs. *American Journal of Psychiatry,* 125: 54–60.

Kaim, S. C., and Rosenstein, I. N. (1960): Anticonvulsant properties of a new psychotherapeutic drug. *Diseases of the Nervous System,* 21 *(Supplement)*:46-48.

Kaim, S. C., and Rosenstein, I. N. (1961): Experience with chlordiazepoxide in the management of epilepsy. *Journal of Neuropsychiatry,* 3:12–17.

Kissen, M. D. (1962): cited in Ban, T. A., Lehmann, H. E., Matthews, V., and Donald, M. (1965): Comparative study of chlorpromazine and chlordiazepoxide in the prevention and treatment of alcohol withdrawal symptoms. *Clinical Medicine,* 72:59–67.

Kurtzke, J. F., and Kaim, S. C. (1962): The acute treatment of convulsive disorders. *Medical Clinics of North America,* 46:373–382.

Rosenfeld, J. E., and Bizzoco, D. H. (1961): cited in Ban, T. A., Lehmann, H. E., Matthews, V., and Donald, M. (1965): Comparative study of chlorpromazine and chlordiazepoxide in the prevention and treatment of alcohol withdrawal symptoms. *Clinical Medicine,* 72:59–67.

Ticktin, H. E., and Schultz, J. D. (1960): Librium, a new quieting drug for hyperactive alcoholic and psychotic patients. *Diseases of the Nervous System,* 21 *(Supplement)*: 49–52.

Victor, M. (1966): Treatment of alcoholic intoxication and the withdrawal syndrome. *Psychosomatic Medicine,* 28:636–650.

Victor, M. (1970): The alcohol withdrawal syndrome: Theory and practice. *Postgraduate Medicine,* 47:68–72.

The Benzodiazepines, Raven Press, New York © 1973

SLEEP LABORATORY AND CLINICAL STUDIES OF THE EFFECTS OF BENZODIAZEPINES ON SLEEP: FLURAZEPAM, DIAZEPAM, CHLORDIAZEPOXIDE, AND RO 5-4200

Anthony Kales and Martin B. Scharf

Sleep Research and Treatment Facility and Departments of Psychiatry and Pharmacology, The Milton S. Hershey Medical Center, The Pennsylvania State University, Hershey, Pennsylvania

I. INTRODUCTION

During the last five years, in our Sleep Research and Treatment Facility, we have conducted a wide variety of sleep laboratory, drug-evaluation studies. In these studies, we have emphasized the following: (1) evaluating the effectiveness of hypnotic and other psychotropic drugs in inducing and maintaining sleep as well as studying the length of this effectiveness, i.e., whether tolerance develops; (2) determining the effects of various drugs on sleep stages and if any drug-induced sleep alterations were accompanied by significant clinical phenomena; and, (3) by utilizing the information from areas (1) and (2), improving the design of both sleep laboratory drug-evaluation studies and traditional drug studies relying solely on subjective reports and/or direct observations.

Investigations of benzodiazepine drugs have been a major part of our sleep laboratory and clinical drug-evaluation program. We became fascinated with this group of drugs for several reasons. In our initial studies of one of these drugs, we found that the drug was extremely effective in inducing and maintaining sleep for a lengthy period of drug administration. Because of the potential significance of this drug as a unique hypnotic agent, we proceeded systematically to evaluate its effectiveness both in the sleep laboratory and in clinical studies. We also found that the benzodiazepine drugs produced little alteration in rapid eye movement (REM) sleep either with drug administration or following drug withdrawal. At the same time, we discovered that some of the benzodiazepines produced marked decrements in stage 4 sleep. Because of this latter finding, we initiated several clinical

577

studies to determine the potential clinical significance of this sleep stage alteration.

To date, we have conducted 14 separate sleep laboratory studies, as well as a variety of clinical studies with four benzodiazepine agents, in varying doses. The general design and primary aims of these studies are briefly summarized in Tables I and II.

TABLE I. *Benzodiazepine sleep laboratory studies with flurazepam*

Dose (mg)	Total nights of study	No. of consec. drug nights	Type of subjects*	Evaluation of:
30	8	3	N	sleep stages
30	8	3	I	short-term efficacy and sleep stages
30	22	14	I	long-term efficacy and sleep stages
30	17	5	I	short-term efficacy with large sample and sleep stages
15	6	3	I	short-term efficacy and sleep stages
60	3	1	N	sleep stages

* N = normal; I = insomniac.

In all of these studies, the general methodology was consistent. Subjects were screened to exclude those with significant medical illnesses, any allergies, or current as well as recent drug use. During the experiments, they were instructed to continue to abstain from drugs, including alcohol, to maintain their usual level of physical activity, and to avoid any naps.

The subjects slept in temperature-controlled, sound-attenuated rooms and were continuously monitored for electroencephalographic (EEG), electromyographic (EMG), and electrooculographic (EOG) recordings. All EEG sleep records were scored according to recently modified criteria (Rechtschaffen and Kales, 1968): the scoring was done independently of any knowledge of the experimental conditions.

II. DRUG STUDIES

A. Flurazepam

1. Flurazepam, 30 mg, 8 consecutive nights, normals and insomniacs

In separate studies, the effects of 30 mg of flurazepam (DALMANE) were evaluated on four normal and four insomniac subjects (Kales et al., 1970a, b). On each night, the subjects were given single gelatin capsules containing either a drug or a placebo. The protocol was as follows, where P

TABLE II. *Benzodiazepine sleep laboratory studies*

Drug	Dose (mg)	Total nights of study	No. of consec. drug nights	Type of subjects*	Evaluation of:
Diazepam	10	8	3	N	sleep stages
	5	12	7	I	short-term efficacy and sleep stages
	5–20	39	27	enuretic	effects on enuretic frequency and sleep stages
	5 and 10 (chronic use)	3	3	I	effects of chronic use on sleep stages
Chlordiazepoxide	50	8	3	N	sleep stages
RO 5-4200	2	10	3	I	short-term efficacy and sleep stages
	1	10	3	I	short-term efficacy and sleep stages
	0.25	14	7	I	short-term efficacy and sleep stages

* N = normal; I = insomniac.

= placebo and **D** = drug: **P P P D D D P P**. The first three nights (placebo) allowed for adaptation to the laboratory (nights 1 and 2) and obtaining baseline values (night 3). The next three nights (drug) allowed for evaluation of the initial and short-term effects of the drug. On the last two nights (placebo), the effects of drug withdrawal, if any, were evaluated. The capsule for each night was administered immediately before "lights out."

In both studies, flurazepam administration produced a slight but not significant REM suppression on the second and third drug nights, with no REM rebound following drug withdrawal. Stage 4 sleep was significantly suppressed. Beginning with the second drug night, the decrease was progressive and marked and continued into the withdrawal period.

In the insomniac subjects, drug administration resulted in a decrease in both sleep latency and wake time after sleep onset. Favorable trends in improving sleep induction and sleep maintenance were also noted in the normal sleepers (Kales et al., 1970b). We did not feel that hypnotic drug effectiveness could be determined in normal sleepers since, by definition, they had little difficulty in falling or staying asleep. Other hypnotic agents, studied in normal sleepers, produced little change in sleep latency and wake time after sleep onset (Kales et al., 1970b). Thus, we were surprised by the flurazepam results in the normal sleepers. These results suggested to us the possibility that many commonly used hypnotic agents might not be very effective. From these deliberations the concept developed of 22-consecutive-night protocols with insomniac subjects and sleep laboratory studies to determine rigorously and objectively a drug's effectiveness and, by extending the period of drug administration, to determine if decreased effectiveness develops.

2. Flurazepam, 60 mg, 3 consecutive nights, normals

In order to obtain a preliminary idea of the effects of 60 mg of flurazepam on sleep stages, a three-night study was employed with normal subjects, following adaptation to the laboratory, where placebo, drug, and placebo were given in that order on the three consecutive nights (Kales et al., 1969b, 1970b). The percentage of REM sleep markedly decreased with drug administration (23.2 to 16.5, $p < 0.01$). Following drug withdrawal, the percentage of REM sleep returned to the baseline and did not exceed this level; i.e., no REM rebound occurred. Changes in REM sleep were accompanied by reciprocal shifts in stage 2 sleep with no significant changes noted in stages 3 and 4 sleep.

3. Flurazepam, 30 mg, 22 consecutive nights, insomniacs

In this study (Kales et al., 1970*a*), we selected four individuals who by history had moderate to severe difficulty falling asleep—that is, they required more than 45 min at least four times per week.

The experimental protocol was as follows:

Nights	Substance	Administration
1–4	Placebo	Baseline in laboratory
5–7	Active drug	Taken in laboratory
8–15	Active drug	Taken in home
16–18	Active drug	Taken in laboratory
19–22	Placebo	Withdrawal in laboratory.

The first night (1 and 16) in each laboratory series was allowed for adaptation and readaptation, respectively, to the laboratory. Each morning, whether in the laboratory or at home, subjects were asked to evaluate their previous night's sleep.

The results of this study showed that 30 mg of flurazepam was effective in insomniac subjects both in inducing and maintaining sleep throughout the 2-week administration period and for several withdrawal nights as well. Following withdrawal of flurazepam, there remained a carry-over effect in terms of sleep induction and sleep maintenance through several withdrawal nights (Kales et al., 1970*a*).

Flurazepam administration produced a marked decrease to absence in stage 4 sleep, similar to that seen in our short-term studies. In contrast to REM sleep rebound following withdrawal of REM suppressant drugs, no rebound in stage 4 sleep was noted following withdrawal of flurazepam. Flurazepam produced only minimal changes in REM sleep in this study, whereas it resulted in slight REM suppression in the short-term studies. In neither the eight- nor the 22-night study was there a REM rebound following withdrawal of flurazepam.

4. Flurazepam, 30 mg, 17 consecutive nights, insomniacs

In this study (Kales et al., 1971*c*), the primary aims were (i) to evaluate further the hypnotic effectiveness of flurazepam with a larger sample; (ii) to evaluate the relationship between the polygraphic data and the subjects' qualitative estimates of sleep; and (iii) to determine the effects of placebo administration per se on sleep stages and sleep induction and maintenance.

The subjects were 12 male insomniacs, ages 21 to 52 years (mean 30.7). They had no medical illness and had not used sleep medication or other drugs for at least three months. The criterion for insomnia was either a history of difficulty in falling asleep (sleep latency more than 45 min at least four times per week) or difficulty in staying asleep (less than 6 hr of sleep at least four times per week). Of the 12 subjects, eight had primary difficulty in falling asleep. Subjects were assigned randomly to one of three groups of four. Drug and placebo administration was carried out on a double-blind basis. Each group was studied for 17 consecutive nights, as shown below: (0 = no treatment, P = placebo, and D = flurazepam 30 mg)

Group	1–4	5–9	10–14	15–17
A	0	D	P	0
B	0	P	D	0
C	0	P	P	0

The first laboratory night was used for adapting the subjects to the laboratory. Baseline measurements were obtained for all subjects on nights 2 to 4. Short-term drug withdrawal effects were studied on nights 10 to 14 (Group A) and 15 to 17 (Group B). Nights 15 to 17 for Group A allowed for evaluation of extended withdrawal.

Upon awakening each morning, subjects completed a questionnaire which was a self-evaluation of the previous night's sleep and included estimates of (1) time to fall asleep, (2) total time spent asleep, (3) degree of drowsiness in the morning, (4) soundness of sleep, and (5) overall quality of sleep with respect to whether sleep was generally better or poorer than sleep at home.

The results of this study (Kales et al., 1971c) corroborated our previous findings that flurazepam, 30 mg, resulted in marked shortening of sleep latency and decreased wake time and number of awakenings. Such findings were noted for Groups A and B during the five nights of drug treatment (Table III). The greatest changes in these parameters were not seen on the first drug night but on the subsequent drug nights. Analysis of the drug-withdrawal period for both groups confirmed previous findings of a drug carry-over effect for the first two or three nights. This was clearly demonstrated in Group A because sleep latency, wake time after sleep onset, and the number of awakenings remained well below baseline levels on the first two placebo-withdrawal nights and approximated baseline on the last placebo-withdrawal night as well as on the final three, no-treatment-withdrawal nights.

The drug carry-over effect was also noted within the drug administra-

TABLE III. *Effects of placebo and flurazapam on sleep induction and maintenance*

Group	Nights	Condition*	Sleep latency (min)	Wake time after sleep onset (min)	Total wake time (min)	No. of wakes
A	2–4	O	29.9	35.2	65.1	12.7
	5–9	D	13.1	11.2	24.3	5.7
	10–14	P	18.7	31.5	50.2	10.8
	15–17	O	26.2	38.1	64.3	11.4
B	2–4	O	63.4	40.2	103.6	8.0
	5–9	P	69.3	38.3	107.6	6.3
	10–14	D	27.0	15.6	42.6	4.5
	15–17	O	32.2	24.9	57.1	5.8
C	2–4	O	46.9	21.3	68.2	7.8
	5–9	P	42.7	31.6	74.3	8.5
	10–14	P	37.7	33.2	70.9	8.9
	15–17	O	41.0	23.8	64.8	6.7

* O = no medication administered; P = placebo administration; D = active drug (flurazepam, 30 mg), administration.
Reprinted from Kales et al. (1971c).

TABLE IV. Effects of placebo and flurazepam on sleep stages and related parameters

Group	Nights	Condition*	Stage 1	Stage 2	Stage 3	Stage 4	REM**	Sleep onset to 1st REM (min)	No. of REM periods
A	2–4	O	9.2	56.1	6.1	4.1	24.5 (107.8)	134.6	3.9
	5–9	D	5.0	63.3	7.0	0.8	23.9 (116.2)	106.1	4.4
	10–14	P	8.6	61.8	4.5	0.1	25.0 (113.8)	105.6	4.0
	15–17	O	10.9	58.4	5.8	0.1	24.8 (108.3)	94.2	3.9
B	2–4	O	7.1	64.5	3.8	1.7	22.9 (94.3)	92.3	3.8
	5–9	P	7.8	62.6	3.9	1.3	24.4 (98.3)	86.0	3.6
	10–14	D	4.4	69.5	3.5	1.0	21.6 (100.4)	101.7	4.2
	15–17	O	8.2	67.4	1.7	0.0	22.7 (101.7)	109.9	3.6
C	2–4	O	7.7	57.7	8.6	3.4	22.6 (101.3)	87.3	4.2
	5–9	P	9.4	58.2	8.6	2.4	21.4 (94.4)	86.4	4.5
	10–14	P	10.1	56.0	9.5	2.7	21.7 (95.0)	91.6	4.4
	15–17	O	9.1	53.5	8.7	5.2	23.5 (106.0)	69.5	4.3

* O = no medication administered; P = placebo administration; D = active drug (flurazepam, 30 mg) administration.
** Numbers in parentheses represent the absolute values in minutes of REM sleep.
Reprinted from Kales et al. (1971c).

tion period. On certain drug nights, the subjects were falling asleep within very short periods of time, i.e., less than 10 min. These abbreviated sleep latencies could not be attributed to the drug given on that night but may have been related to a carry-over effect. This raises the question of how the drug affects daytime levels of performance and alertness, a critical consideration for all hypnotics. In this study, the subjects did not complain of difficulty during the daytime and, by subjective estimates, indicated they were sleeping better than usual (Kales et al., 1971c).

Our previous studies showed that administration of flurazepam, 30 mg, produced either a slight decrease or no change in REM sleep. In the study reported here, the percentage of REM sleep was not changed in Group A and there was only a slight decrease in Group B (Table IV). Following withdrawal of flurazepam in Groups A and B, there was no evidence of an increase in REM sleep above baseline levels (REM rebound). In our previous study with insomniac subjects, flurazepam was administered for a 2-week period with only five placebo nights following withdrawal of the drug. The number of placebo-withdrawal nights that we allowed would not have been sufficient to detect a delayed REM rebound. In the current study, for Group A, eight withdrawal nights followed the five nights on drug and provided a more satisfactory period for detecting delayed REM rebound. Analysis of the eight withdrawal nights did not show any evidence of immediate or delayed REM rebound.

Flurazepam produced a marked and sustained decrease in stage 4 sleep and, to a lesser extent, in stage 3 sleep (Kales et al., 1971c). Not only was there a marked decrease in stage 4 sleep during and immediately following drug administration, but also the decrease was maintained throughout the extended withdrawal period of Group A (Table IV).

Comparisons between the subjects' qualitative estimates of sleep and the EEG findings for Groups A and B were in agreement when the drug period was compared to the baseline and to withdrawal for all parameters measured. During drug administration, the subjects estimated that they fell asleep faster, and slept longer, which was in agreement with the EEG data. The subjective evaluation of improvement in sleep latency was maintained during immediate withdrawal for both Groups A and B and also to a lesser degree in the no-treatment, extended-withdrawal condition for Group A. This is in agreement with the EEG findings of a drug carry-over effect on sleep latency following withdrawal in Groups A and B.

The eight subjects in Groups B and C showed no significant differences in sleep patterns between no-treatment condition and placebo condition over the entire five-night placebo period (Tables III and IV), as well as when each of the placebo nights was analyzed separately. It should be noted that the number of subjects was small, that they all were men, and that all had been informed that medication would either be a drug or a placebo. The last may have minimized placebo effects. Our data indicate that if subjects are so informed, placebo effects are minimal even on the first night or two in the laboratory and do not extend to the third or fourth consecutive laboratory night, which we have previously recommended for obtaining baseline measurements.

There was concurrence between the EEG and subjective data for Group C. Analysis of the no-treatment and placebo conditions did not indicate clear-cut trends in support of a placebo effect in terms of the subjective estimates or EEG results (Kales et al., 1971c).

5. Flurazepam, 15 mg, 7 consecutive nights, insomniacs

As we indicated previously, we had extensively evaluated 30-mg doses of flurazepam in sleep laboratory studies and found that it was quite effective both in inducing and maintaining sleep. As we gained more experience with this drug on a clinical basis, we noted that in many patients 15 mg of flurazepam was quite sufficient to treat their insomnia.

In order to verify this observation objectively, we decided to evaluate the 15-mg dose in a sleep laboratory study. Four insomniac subjects were studied for seven consecutive nights, four placebo nights followed by three nights of active drug (flurazepam, 15 mg). Because of scheduling problems in the laboratory, we were only able to have three drug nights and thus could only really evaluate the initial and short-term effectiveness of the drug. Also, we were not able to schedule any placebo-withdrawal nights.

The mean values for placebo-baseline and the drug nights are listed below:

	Sleep latency	Wake time after sleep onset	Total wake time	No. of wakes	Percentage of sleep time
Baseline	29.3	57.7	87.0	4.8	81.9
Drug (flurazepam 15 mg)	19.1	10.9	30.1	2.7	93.7

These results confirmed our clinical impressions that a 15-mg dose is also quite effective in inducing and maintaining sleep over a short-term period of drug administration.

The results for this study also showed, in regard to sleep stage parameters, that drug administration produced a slight increase in the absolute amount and percent of REM sleep as well as a moderate decrease in stage 4 sleep.

In insomniac patients who have significant difficulty in either falling asleep or staying asleep or both, flurazepam is our drug of choice based on the effectiveness demonstrated in the sleep laboratory studies (Kales, 1970; Kales and Kales, 1970a; Kales and Cary, 1971). For most patients treatment is initiated with a dosage of 15 mg at bedtime. Our studies have shown that the drug is more effective on the second, third, or fourth night of consecutive use than on the first night, and the patient is so informed so as not to become discouraged, as is often the case with insomniac patients regardless of the treatment. In cases where 15 mg of flurazepam at bedtime does not appear to be improving sleep significantly after one week of therapy, the dosage is increased to 30 mg. In those cases where the patient presents a clear-cut history of severe and chronic insomnia, treatment is initiated with a dose of 30 mg of flurazepam at bedtime (Kales, 1970; Kales and Cary, 1971; Kales and Kales, 1970a).

We have evaluated some 200 insomniac patients with psychological testing (primarily MMPI) (Kales et al., 1969a, 1972; Kales, 1970; Kales and Cary, 1971). We found that an extremely high percentage of these patients demonstrated significant psychological disturbances in these tests and psychiatric interviews. These patients had elevated indexes for depression, anxiety, obsessive-compulsive features, and schizophrenic trends. For these reasons, we have indicated that psychotherapy is an important aspect in the treatment of the insomniac patient. Frequently, psychotherapy in insomniac patients is not suggested by the general practitioner because he either is not trained in this area or does not have the time to elicit a thorough history.

In cases of mild insomnia, or insomnia secondary to situational disturbances, pharmacological therapy alone is most often sufficient. In more severe or chronic cases, best results are obtained with a combined approach of psychotherapy and pharmacological treatment (Kales et al., 1969a, 1972; Kales, 1970; Kales and Cary, 1971).

B. Diazepam

1. Diazepam, 10 mg, 8 consecutive nights, normals

As with our other eight-night studies, the design was P P P D D D P P. Administration and withdrawal of diazepam (VALIUM) produced little

TABLE V. Stage REM alterations following drug administration and withdrawal. Eight-night studies

Group* and drugs	% Stage REM on study nights (Ni)						% Change from baseline night	
	Baseline Ni 3	Ni 4	Drug Ni 5	Ni 6	Withdrawal Ni 7	Ni 8	1st drug night	1st withdrawal night
Group A								
Glutethimide (DORIDEN), 500 mg (N = 5)	27.0	14.3	16.2	18.0	31.4	30.7	−47.0	+16.3
Methyprylon (NOLUDAR), 300 mg (N = 7)	25.3	18.4	18.9	25.3	31.5	31.6	−27.3	+24.5
Secobarbital (SECONAL), 100 mg (N = 2)	20.2	15.3	19.8	19.0	21.1	21.2	−24.5	+ 4.5
Methaqualone (QUAALUDE), 300 mg (N = 5)	23.4	18.8	23.4	21.6	27.6	22.4	−19.7	+17.9
Pentobarbital (NEMBUTAL), 100 mg (N = 4)	21.7	18.3	21.1	20.9	26.4	25.6	−15.7	+21.7
Diphenhydramine (BENADRYL), 50 mg (N = 2)	20.8	17.6	15.6	15.2	26.7	20.1	−15.4	+28.4
Promethazine (PHENERGAN), 25 mg (N = 4)	18.3	18.5	23.0	17.3	33.3	26.2	+ 1.1	+82.0
Group B								
Chloral hydrate (NOCTEC), 1000 mg (N = 5)	23.0	21.7	23.7	22.4	23.3	24.8	− 5.7	+ 1.3
Diazepam (VALIUM), 10 mg (N = 3)	20.1	19.2	20.1	18.8	20.1	20.7	− 4.5	0
Flurazepam (DALMANE), 30 mg (N = 8)	22.5	21.5	18.0	19.1	21.9	22.3	− 4.0	− 2.7
Chloral hydrate (NOCTEC), 500 mg (N = 10)	21.4	21.3	23.0	21.4	18.5	23.5	− 0.5	−13.6
Chlordiazepoxide (LIBRIUM), 50 mg (N = 4)	22.6	23.3	18.6	21.9	21.3	22.7	+ 3.1	− 4.9
Methaqualone (QUAALUDE), 150 mg (N = 5)	21.2	21.9	22.9	24.2	24.6	23.1	+ 3.3	+16.0

* Group A = drugs producing marked alterations with drug administration and withdrawal; Group B = drugs producing no change or minimal change with drug administration and withdrawal; N = number of subjects.

change in REM sleep. Table V summarizes the effects of many drugs including diazepam on REM sleep over this eight-night study design. On baseline, the percentage of REM sleep is 20.1; on drug nights 4, 5, and 6, 19.2, 20.2, and 18.8; and on withdrawal nights 7 and 8, 20.1 and 20.7. Diazepam produced a marked and progressive decrease in stage 4 sleep which extended into the withdrawal period (Kales, 1969; Kales et al., 1970b).

2. Diazepam, 5 mg, 12 consecutive nights, insomniacs

In another study of four insomniac patients with varying degrees of anxiety and depression, we evaluated the effects of giving diazepam, 5 mg, either b.i.d. or t.i.d. during the day combined with a 5-mg dose h.s. The study extended for 12 consecutive days with the following schedule: 1 through 3, placebo; 4 through 10, active drug; and 11 and 12, placebo. The patients were monitored in the laboratory on nights 1 through 5 and 8 through 12. They slept at home on nights 6 and 7. Night 8 in the laboratory was allowed for readaptation.

The results show that throughout the 1 week of drug administration both sleep latency and wake time after sleep onset were markedly reduced. The number of awakenings each night showed only a slight decrease on drug nights as compared to baseline and was lowest on the withdrawal nights.

Administration and withdrawal of diazepam produced little change in REM sleep. Stage 4 sleep was markedly decreased by the drug and this decrease was maintained during the withdrawal period.

3. Diazepam, 5 to 15 mg, clinical use

Prior to the approval of flurazepam by the Food and Drug Administration, we frequently used diazepam in a dose of 5 to 15 mg at h.s. in the treatment of insomnia (Kales et al., 1971a). We found the drug to be effective for both sleep induction and maintenance but not to the degree of flurazepam. However, when patients who are already taking diazepam in the daytime for anxiety present also with insomnia, we recommend that diazepam be given in an increased dose at h.s. rather than adding a hypnotic drug to the treatment regimen.

4. Diazepam, 5 to 20 mg, 39 consecutive nights, enuretic children

Since many (but not all) enuretic episodes in children occur out of stages 3 and 4 sleep, we investigated the effects of diazepam on both sleep stages and enuretic frequency (Kales et al., 1971b). Four enuretic children were studied. After four placebo nights, they were given diazepam at h.s. for 27 consecutive nights. The initial dose was 2.5 mg at h.s., and this was increased weekly if a significant reduction in enuretic events had not occurred.

TABLE VI. *Diazepam and sleep stages (enuretic children)*

Administration (days)	% of time at different stages				
	REM	1	2	3	4
Baseline (3-4)	21.6	3.9	46.8	11.7	16.1
Initial drug (5-9)	24.2	4.4	53.0	7.7	10.7
Long-term drug (30-31)	25.6	3.4	62.0	7.9	1.2
Withdrawal (32-36)	25.8	7.3	54.5	8.7	3.7
Withdrawal follow-up (39)	26.0	8.4	42.9	12.3	10.5

The data in Table VI show that the administration (short and long term) and withdrawal of diazepam produced no significant change in REM; a slight increase was noted in both conditions. There was a slight decrease in both stages 3 and 4 with short-term administration and a marked decrease in stage 4 with long-term administration (16.1 to 1.2) (Kales et al., 1971b). Following initial withdrawal, stage 4 increased slightly toward baseline. About 1 week later, stage 4 sleep approximated about two-thirds of the baseline level.

5. Diazepam, 5 and 10 mg, chronic use, insomniacs

We have studied a number of patients who have chronically used hypnotic and tranquilizing medication for sleep over a period of months to years. Two of these patients had taken diazepam regularly at h.s. and sporadically in the day for 6 months in one patient and 1½ years in the other (Kales, 1969; Kales et al., 1969c). Both patients slept in the sleep laboratory, taking the medication in the accustomed manner. After adaptation, their sleep showed normal amounts of REM sleep, but no stage 4 sleep. This finding of a marked decrease in stage 4 sleep has also been noted with the chronic use of pentobarbital and glutethimide (Kales et al., 1969c).

C. Chlordiazepoxide

1. Chlordiazepoxide, 50 mg, 8 consecutive nights, normals

With an eight-consecutive-night protocol consisting of three placebo, three drug, and two placebo nights, we evaluated the effects of chlordi-azepoxide (LIBRIUM), 50 mg, on sleep patterns. The results for this study are included in Table V. REM sleep showed very little change from baseline to drug to withdrawal conditions. On baseline, the REM percentage was 22.6; on drug nights, 23.3, 18.6, and 21.9; and on withdrawal 21.3 and 22.7. Drug administration produced a moderate decrease in stage 4 sleep, with some return toward baseline on withdrawal nights (Kales et al., 1970*b*).

2. *Chlordiazepoxide, 25 to 50 mg, clinical, psychiatric-insomniac patients*

We have had less experience in treating insomnia in psychiatric patients with chlordiazepoxide than with diazepam. In a dosage usually of 25 mg at h.s., chlordiazepoxide appears to be moderately effective in inducing sleep but not very effective in maintaining sleep. Its overall effectiveness is considerably less in comparison to diazepam.

D. RO 5-4200

1. *RO 5-4200, 2 mg, 20 consecutive nights, insomniacs*

Four insomniac subjects were studied with the following schedule: four placebo, three drug (RO 5-4200, 1 mg at h.s.), and three placebo nights. The results in Table VII show a marked decrease in sleep latency and

TABLE VII. *Sleep induction and maintenance with RO 5-4200 (2 mg)*

Condition	Nights	Sleep latency	Wake time after sleep onset	Total wake time
Baseline	2–4	29.8	37.4	67.2
	3–4	32.1	25.6	57.7
Drug	5–7	15.4	20.8	36.2
Withdrawal	8–10	55.4	30.4	85.8

moderate decrease in wake time after sleep onset with drug administration. The number of awakenings each night were also markedly decreased over the three-night period of drug administration.

Both the percentage and the absolute amount of REM sleep were markedly decreased during the drug-administration period. Following drug withdrawal, both the percentage and absolute amount of REM sleep re-turned to baseline levels but there was no REM rebound. Stage 4 sleep was markedly decreased with drug administration (6.8 to 0.9%); following withdrawal there was a partial recovery in stage 4 sleep (to 4.0%).

2. *RO 5-4200, 1 mg, 10 consecutive nights, insomniacs*

The effects of 1 mg of RO 5-4200 were evaluated in four additional insomniac subjects using the same 10-night protocol. Drug administration produced a marked decrease in total wake time (98.5 to 48.7 min); wake time after sleep onset was decreased to a greater degree than sleep latency (Table VIII). The number of nightly awakenings was also reduced from 9.4

TABLE VIII. *Sleep induction and maintenance with RO 5-4200 (1 mg)*

Condition	Nights	Sleep latency	Wake time after sleep onset	Total wake time
Baseline	2–4	77.7	34.7	112.4
	3–4	72.3	26.2	98.5
Drug	5–7	45.0	3.7	48.7
Withdrawal	8–10	72.0	14.3	86.3

on baseline to 2.7 on drug nights.

Both the absolute amount and the percentage of REM sleep were slightly to moderately increased with drug administration, but this appears to be more a function of the marked decrease in wake time after sleep onset than a direct drug effect in increasing REM sleep. The subjects had essentially no stage 4 sleep on baseline so it was not possible to determine the effects of this dose on this sleep stage.

TABLE IX. *Sleep induction and maintenance with RO 5-4200 (0.25 mg)*

Condition	Nights	Sleep latency	Wake time after sleep onset	Total wake time
Baseline	2–4	63.2	12.9	76.1
	3–4	44.4	6.7	51.1
Drug	5–7	24.5	11.7	36.2
	8–11	26.1	8.0	34.1
	5–11	25.3	9.8	35.1
Withdrawal	12–14	41.5	26.6	68.1

3. RO 5-4200, 0.25 mg, 14 consecutive nights, insomniacs

In this study, the design was as follows: four placebo, seven drug (RO 5-4200, 0.25 mg, at h.s.), and three withdrawal nights. On baseline nights, the four insomniac subjects had difficulty primarily in falling asleep. With drug administration, there was a moderate to marked decrease in sleep latency (Table IX).

The absolute amount and percentage of REM sleep were essentially unchanged from baseline to drug to withdrawal conditions. There was a moderate decrease in stages 3 and 4, with drug administration. Following withdrawal, there was a return of stage 4 sleep essentially to baseline levels.

III. SUMMARY OF RESULTS

A. Sleep stages

Table X summarizes the effects of these four benzodiazepine agents on sleep stages. In general, clinical doses of these drugs produce little change in REM sleep either with drug administration or drug withdrawal. An increase in dosage such as to 60 mg of flurazepam or 2 mg of RO 5-4200 does lead to REM suppression. However, even in these situations, drug withdrawal does not result in an increase in REM sleep above baseline (REM rebound)—only a return to the prior baseline levels.

The above comments refer to the total amount of REM sleep. Other investigators, primarily Dement and his associates (1972, *This Volume*), have stressed the importance of phasic events within REM sleep, such as PGO spike activity and eye movements. For this reason, in our 17-night study, we evaluated the effects of 30 mg of flurazepam on eye-movement density (Allen et al., 1972). With drug administration, we found a moderate decrease in eye-movement density. Following drug withdrawal, there was a return of eye-movement density to baseline levels, without a rebound above baseline.

In general, the benzodiazepine drugs also produced moderate to very marked decreases in stage 4 sleep. These decrements are more pronounced with flurazepam and diazepam. Following withdrawal of flurazepam and diazepam, the decreases to absence in stage 4 sleep are maintained.

Not noted in Table X or in our previous descriptions of these studies are the effects of these drugs on EEG frequency. All produced an increase in fast EEG activity which is more noticeable in REM sleep and stages 1

TABLE X. *Benzodiazepines: effects on sleep stages*

Drug	Dose (mg)	REM Sleep		Stage 4 sleep	
		Drug administration	Drug withdrawal	Drug administration	Drug withdrawal
Flurazepam	30	slight decrease, moderate decrease in eye movement, density	return to baseline, no REM rebound, no rebound in eye movement	progressive marked decrease	decrease maintained over several drug withdrawal periods. No rebound
	15	slight increase	not studied	moderate decrease	not studied
	60	clear-cut decrease	return to baseline. No REM rebound	could not be evaluated—only 1 drug night in protocol	
Diazepam	5–10	no change	no change	progressive marked decrease	decrease maintained over several drug withdrawal periods. No rebound
Chlordiazepoxide	50	slight decrease	return to baseline. No REM rebound	moderate decrease	some return toward baseline
RO 5-4200	2	moderate decrease	return to baseline. No REM rebound	marked decrease	moderate return toward baseline
	1	no change	no change	subjects had essentially no stage 4 sleep; thus could not evaluate	no stage 4 sleep; thus
	0.25	no change	no change	slight decrease	returns essentially to baseline levels

and 2. This increase in fast EEG activity is more marked with flurazepam and diazepam.

B. Efficacy

All of the benzodiazepine drugs studied produced improvements in sleep induction and maintenance to varying degrees. The efficacy of chlordiazepoxide appears to be mild. This is inferred from clinical studies, as sleep laboratory evaluations were not carried out. Diazepam and RO 5-4200 were quite effective in both inducing and maintaining sleep over short-term periods of drug administration. The efficacy of these drugs over longer drug-administration periods has not been tested.

Flurazepam has consistently been the most effective drug we have studied in terms of both sleep induction and maintenance as well as the length of effectiveness (Kales et al., 1970a, 1971c). In our 22-night study (Kales et al., 1970a), there was no evidence of tolerance at the end of the 2 weeks of drug administration. With all of the other hypnotic drugs we have studied with this protocol, significant tolerance developed by the end of the drug-administration period.

From all of our sleep laboratory studies, including the benzodiazepine studies described here, a number of basic principles have evolved that are important to the methodology of all drug evaluation studies both clinical and sleep laboratory (Kales and Kales, 1970b). These include:

(1) allowing for adaptation and readaptation to the sleep laboratory or hospital environment (Scharf et al., 1969);

(2) the use of consecutive nights from condition to condition;

(3) within a given protocol, a second drug can be evaluated only after an adequate withdrawal period has been allowed following administration of the first drug;

(4) evaluating long-term as well as short-term efficacy, since our studies showed that within a 2-week period, decreased effectiveness had developed to most of the drugs studied;

(5) evaluating the withdrawal period. We have demonstrated that significant carry-over occurs with some drugs for sleep induction and maintenance, and, as originally pointed out by Oswald and his associates (Oswald and Priest, 1965), REM rebound accompanied by increases in dream frequency and intensity often occurs during this period (Kales et al., 1969b, 1970c; Kales, 1969).

All of these principles are incorporated in our 22-night protocol which is the primary design that we utilize for evaluating short-term and long-term (2 weeks) efficacy as well as sleep stage effects. As mentioned before, this study consists of 22 consecutive nights, including four placebo, 14 drug, and four placebo nights. An important aspect of any experimental design is the reproducibility of the results obtained from such studies. After our 22-night evaluation of 30-mg doses of flurazepam, we further evaluated the drug in this dose with a larger group of subjects in our 17-night studies. We confirmed all of our original findings in terms of both efficacy and sleep stages. Subsequent to our studies with flurazepam, 30 mg, the drug was evaluated in four different university sleep laboratory studies, using the basic design of placebo-baseline to drug to placebo-withdrawal conditions. In all of these studies, all of our original findings regarding efficacy and sleep stages were confirmed (Kales and Cary, 1971).

In its new guidelines for the evaluation of hypnotic drugs, the Food and Drug Administration has included requirements for sleep laboratory studies incorporating the basic design that we had recommended, i.e., placebo-baseline to drug to placebo-withdrawal conditions, with the recommendation that drug administration in some of these studies extend for at least 2 to 4 weeks (Kales and Cary, 1971). Since we found that most hypnotic drugs produce decreased efficacy at the end of a 2-week drug-administration period, we are strongly recommending that all hypnotic drugs be evaluated in sleep laboratory studies, including drug-administration periods of 2 weeks to 2 months. Although hypnotic drugs may be prescribed for only brief periods of time, they are often prescribed and used for chronic periods extending for many years. If they are going to be used in this fashion, it is obvious that their efficacy must be proven in an objective manner over such intervals of time.

Some may argue that sleep laboratory studies do not evaluate areas which are specific to clinical drug studies, i.e., the subject's own estimates as to how well or how poorly he slept. As we have previously indicated, however, our sleep laboratory studies combine the EEG findings with evaluations of the subjective reports obtained each morning. Sleep laboratory studies now play an important role in the evaluation of hypnotic drug efficacy. This is not only in terms of what can be evaluated in the sleep laboratory *per se*, but also in the ways that sleep laboratory findings may be utilized to improve the design of traditional clinical studies.

ACKNOWLEDGMENTS

This research was supported in part by the Foundation for Psychobiological Research, National Aeronautics and Space Administration grant NGR 39-009-204, and U.S. Public Health Service grant MH-19488 from the National Institute of Mental Health.

REFERENCES

Allen, C., Scharf, M. B., and Kales, A. (1972): The effect of flurazepam (DALMANE) administration and withdrawal on REM density. Presented to the First International Congress of the Association for the Psychophysiological Study of Sleep (APSS), Bruges, Belgium. *Psychophysiology,* 9:92–93.

Dement, W. C., Zarcone, V. P., Hoddes, E., Smythe, H., and Carskadon, M. (1972): Sleep laboratory and clinical studies with flurazepam. *This Volume.*

Kales, A. (1969): Drug dependency. In: *Investigations of Stimulants and Depressants. Annals of Internal Medicine,* 70:591–614.

Kales, A. (1970): The clinical implications of sleep. In: *Psychiatry 1970,* edited by E. Robins. Medical World News, New York, pp. 62–63.

Kales, A., Allen, C., Scharf, M. B., and Kales, J. D. (1970): Hypnotic drugs and their effectiveness. All-night EEG studies of insomniac subjects. *Archives of General Psychiatry,* 23:226–232.

Kales, A., and Cary, G. (1971): Insomnia, evaluation, and treatment. In: *Psychiatry 1971,* edited by E. Robins. Medical World News, New York, pp. 55–56.

Kales, A., and Kales, J. D. (1970a): Evaluation, diagnosis, and treatment of clinical conditions related to sleep. *Journal of the American Medical Association,* 213:1119–1134.

Kales, A., and Kales, J. D. (1970b): Sleep laboratory evaluation of psychoactive drugs. *Pharmacology for Physicians,* 4:1–6.

Kales, J. D., Kales, A., Bixler, E. O., and Slye, E. S. (1971c): Effects of placebo and flurazepam on sleep patterns in insomniac subjects. *Clinical Pharmacology Therapeutics,* 12:691–697.

Kales, A., Kales, J. D., Scharf, M. B., Preston, T. A., Tan, T. L., and Allen, C. (1969a): Electrophysiological and psychological studies of insomnia. *Psychophysiology,* 6:255.

Kales, A., Kales, J. D., Scharf, M. B., and Tan, T. L. (1970b): Hypnotics and altered sleep-dream patterns. II. All-night EEG studies of chloral hydrate, flurazepam, and methaqualone. *Archives of General Psychiatry,* 23:219–225.

Kales, A., Malmstrom, E. J., Scharf, M. B., and Rubin, R. T. (1969b): Psychophysiological and biochemical changes following use and withdrawal of hypnotics. In: *Sleep, Physiology and Pathology,* edited by A. Kales. J. B. Lippincott Co., Philadelphia, pp. 331–343.

Kales, A., Preston, T. A., Scharf, M., and Kales, J. D. (1971a): Treatment of disordered sleep: Laboratory and clinical studies. *Psychophysiology,* 7:343.

Kales, A., Preston, T. A., Slye, E. S., and Bixler, E. O. (1972): Sleep patterns and insomniac subjects: Further studies. Presented to the First International Congress of the Association for the Psychophysiological Study of Sleep (APSS), Bruges, Belgium. *Psychophysiology,* 9:137.

Kales, A., Preston, T. A., Tan, T. L., and Allen, C. (1970c): Hynotics altered sleep-dream patterns. I. All-night EEG studies of glutethimide, methyprylon, and pentobarbital. *Archives of General Psychiatry,* 23:211–218.

Kales, A., Preston, T. A., Tan, T. L., Scharf, M. B., and Kales, J. D. (1969c): Effects of chronic hypnotic use. *Psychophysiology,* 6:259.

Kales, A., Scharf, M., Tan, T. L., Zweizig, J. R., and Alexander, P. (1971b): Sleep laboratory and clinical studies of the effects of TOFRANIL, VALIUM, and placebo on sleep stages and enuresis. *Psychophysiology,* 7:348.

Oswald, I., and Priest, R. G. (1965): Five weeks to escape the sleeping pill habit. *British Medical Journal,* 2:1093–1099.

Rechtschaffen, A., and Kales, A. (Editors) (1968): *A Manual of Standardized Terminology, Techniques and Scoring System for Sleep Stages of Human Subjects.* Public

Health Service Publications No. 204, U.S. Government Printing Office, Washington, D.C.

Scharf, M., Kales, J., and Kales, A. (1969): Repeated adaptation and first night effects of the sleep laboratory. *Psychophysiology*, 6:263.

The Benzodiazepines, Raven Press, New York © 1973

SLEEP LABORATORY AND CLINICAL STUDIES WITH FLURAZEPAM

William C. Dement, Vincent P. Zarcone, Eric Hoddes, Harvey Smythe, and Mary Carskadon

Department of Psychiatry, Stanford University School of Medicine, Stanford, California

I. INTRODUCTION

In this volume (1972) and elsewhere (Kales, 1969; Kales et al., 1969b, c; 1970a, b), Kales and his group have reported sleep laboratory studies with benzodiazepines that suggest these compounds may be effective in inducing and maintaining sleep in insomniacs and that this effectiveness may be uniquely prolonged in comparison with other compounds. In addition to their work with benzodiazepines, their contribution of an operational concept of hypnotic effectiveness has permitted our laboratory and others to conduct meaningful and practically useful studies of hypnotic compounds. Our own drug-evaluation work with benzodiazepines consists of two systematic sleep laboratory studies of flurazepam, together with some clinical observations on insomniac patients.

II. SLEEP LABORATORY STUDIES

The laboratory studies were conducted on male insomniacs. The purpose of the studies was to evaluate the effectiveness of flurazepam (DALMANE) in terms of parameters related to sleep induction and sleep maintenance, and to investigate the effects of the drug on sleep stages, with particular attention to rapid eye movement (REM) and slow-wave non-rapid eye movement (NREM, stages 3 and 4) sleep.

A. Method

Ten subjects ranging in age from 19 to 65 years (mean age: 44 years) were chosen on the basis of their subjective complaint of insomnia. In one study, four subjects received flurazepam, 15 mg, and two subjects placebo only, on a double-blind basis, as a check on the habituation process; in the second study, four subjects received flurazepam, 30 mg. The studies followed a 14-night effectiveness protocol. On the first four and last three nights of the studies, the drug subjects were given inactive placebo, whereas on the interim seven nights, flurazepam was administered. Sleep was monitored polygraphically in the laboratory on each of the nights, with scalp and facial leads recording electroencephalogram (EEG), electrooculogram (EOG), and electromyogram (EMG).

Sleep records were coded and scored blind according to the criteria of Rechtschaffen and Kales (1968). In addition to the polygraphic data, each subject completed a questionnaire on each night and each morning of the studies. The evening questionnaire queried the subject as to his sense of well-being during the day, his general energy level throughout the day, and his feeling of readiness for bed. The morning questionnaire asked him to assess the subjective quality of his sleep, how long he felt it took him to get to sleep, and how long he slept.

During the experiments, the subjects refrained from using alcohol or any other drugs or compounds. They also kept their activities at usual levels and did not take naps.

A two $\times$ four $\times$ three, dose $\times$ treatment phase $\times$ night, analysis of variance was performed on selected sleep-recording variables of the experimental subjects. The treatment phase levels were Baseline = nights 2 through 4, Drug 1 = nights 5 through 7, Drug 2 = nights 9 through 11, and Recovery = nights 12 through 14. There were four subjects per cell and three nights nested within each of the treatment phase levels, which were considered as repeated measures.

B. Sleep parameters

The results obtained in our studies (Table I) are comparable to those of Kales and others (Hartmann, 1968) with respect to the effects of flurazepam on sleep induction and sleep maintenance. Sleep latency was significantly decreased during the drug-administration period, with very low values of 2 to 5 min seen on several nights in a few subjects. As Kales and Scharf (1972) have noted, a sleep latency of so short a duration is not attributable to the immediate effects of a single night's dose of flurazepam, but may represent carry-over effectiveness from previous administration. In the 15-mg study, sleep latency decreased by approximately 75% during the drug-administration period; and in the 30-mg study, the decrease was 48.6% below the baseline mean. This was the only parameter on which there

TABLE I. *Flurazepam: experimental results*

Treatment phase level	Baseline (2–4)	Drug I (5–7)	Drug II (9–11)	Recovery (12–14)
Sleep latency (min)				
15 mg* N = 4	54.8	11.7	14.4	22.3
30 mg* N = 4	32.0	16.6	14.8	31.8
Controls N = 2	18.9	18.6	10.6	4.6
Wake time after onset (min)				
15 mg**	64.6	15.2	11.4	24.7
30 mg	13.2	3.2	7.0	10.0
Controls	65.2	74.2	92.1	100.8
Total sleep time (min)				
15 mg**	324.8	407.6	418.1	388.1
30 mg**	413.0	441.9	436.6	417.5
Controls	366.4	372.4	362.3	359.6
Slow-wave sleep time (min)				
15 mg**	42.4	28.0	6.7	8.3
30 mg**	51.8	48.6	37.5	33.6
Controls	33.6	38.4	39.4	30.9
REM time (min)				
15 mg	69.0	78.8	70.9	74.8
30 mg	89.1	81.0	79.8	87.0
Controls	84.6	88.0	84.4	84.9
Percent REM time				
15 mg**	21.3	18.9	17.3	19.2
30 mg**	20.8	17.9	17.8	20.2
Controls	23.1	23.6	23.0	23.5

* $p < 0.05$.
** $p < 0.01$. The group mean values are listed for each treatment phase level in both studies.

seemed to be any habituation to the laboratory situation by the control subjects, and this effect was not evident until the Drug 2 period.

Wake time after onset was reduced significantly during the 15-mg study, but no significant decrease was found in the 30-mg study. The apparent lack of effectiveness of 30 mg resulted, in part, from the initially low baseline wake times of the subjects; there was, however, an overall trend toward the reduction of wake time after onset during the drug period.

With both dose levels of flurazepam, total sleep time increased significantly from baseline during the drug period. The 15-mg group, which had the lower baseline mean sleep time, evidenced a very great rise with the drug, demonstrating a mean increase of 25.5% during Drug 1 and 28.7% during Drug 2. On the other hand, the increase seen in the 30-mg group, while significant, was much smaller: 7.0% in Drug 1 and 5.7% in Drug 2.

Sleep stage results also indicated a measure of confirmation of the earlier results of previous investigators. Slow-wave sleep (NREM stages 3 and 4) was significantly reduced during the drug-administration period in both studies, and remained low, relative to baseline, during the recovery period. The suppression during the recovery nights indicates that there may have been a carry-over effect of flurazepam after administration was terminated. Here again, the effect of flurazepam was seen more dramatically in the 15-mg group than in the 30-mg group. The mean slow-wave sleep times of the 15-mg subjects were 66.0% of baseline in Drug 1, 15.8% in Drug 2, and 19.6% in Recovery. In the 30-mg group, the mean slow-wave sleep times fell to 93.8% of the baseline mean in Drug 1, 72.4% in Drug 2, and 64.9% during the recovery period.

Our data on REM sleep suggest some disagreement with the results reported by others (Hartmann, 1968; Kales and Scharf, 1972) who have found that there is either a very slight suppression or no change of REM sleep when flurazepam is administered in doses of 30 mg or less. Our data for absolute REM time confirm these results. No significant changes were found, but, in the 15-mg study, there was a trend for REM time to be increased; and in the 30-mg study, there appeared to be slight reduction of REM sleep time. However, when REM was considered as a percentage of total sleep time, a significant decrease from the baseline levels was seen in the Drug 2 periods of the two studies. In the 15-mg study, the REM percentage fell to a mean of 88.7% (Drug 1) and 81.2% (Drug 2) of the baseline mean; and, in the 30-mg study, there was a decrease from the baseline mean to 86.1% (Drug 1) and 85.6% (Drug 2). In neither case was there a rebound in REM sleep with the withdrawal of flurazepam. The question now arises as to whether or not the reduction of the *percentage* of REM sleep seen in these studies has any functional significance when the total amount of REM sleep is maintained.

C. Subjective results

All of the valid subjective data showed very good correlation with the polygraphic results. The subjects found that they slept longer and more soundly, had fewer arousals, and had a greater sense of well-being during the drug-administration period than during either the baseline or recovery periods. In addition, as others have reported in insomniacs (Held et al., 1959; Schwartz et al., 1963; Monroe, 1967; Zung, 1970) or persons who notice a sleep disturbance as a result of drug withdrawal (Lewis, 1969), the estimates of sleep latency were greater than and total sleep time less than those which were actually recorded. Paradoxically, the insomniacs consistently underestimated the number of arousals each night.

III. CLINICAL OBSERVATIONS

Through the ages, many words have been written about the tortures of insomnia. Exaggerated or not, there is no question that it is a very common and often serious complaint. In addition, the problem is complicated by the lack of a clear understanding of the causes of insomnia and the absence of exact physiological and biochemical definitions of what does, and what does not, constitute insomnia. At the present time, several groups of investigators are carrying out a variety of observations on patients who complain of insomnia. An important part of such observations is psychological evaluation. This is well demonstrated by the Kales group (1971) who have given the Minnesota Multiphasic Personality Inventory (MMPI) to over 200 insomniac patients and found that over 80% had one or more of the major scales in the pathological range. Their recommendations for treating insomnia (Kales, 1970, 1971; Kales et al., 1970c; Kales and Scharf, 1972) include excellent strategic considerations concerning psychotherapy. Polygraphic all-night sleep recordings of insomniacs (Monroe, 1967; Jones and Oswald, 1968; Passouant et al., 1968; Rechtschaffen, 1968; Kales et al., 1969a, 1970a; Rechtschaffen and Monroe, 1969; Karacan et al., 1970; Kales, 1971) have helped investigators to understand the condition, but have frequently raised more questions than they have answered. Finally, and possibly most important, is pharmacological evaluation. As previously indicated, we owe to the effectiveness studies of Kales's group the ability to respond to the needs or demands of the patients even before, as is often the case in medicine, we completely understand the illness.

A. Approach to persons complaining of insomnia

At the Stanford University Sleep Disorders Clinic, our first experience with insomniacs came during the newspaper recruitment of subjects for drug evaluation studies. Literally hundreds of poor sleepers volunteered for the studies, with females predominating. All of the volunteers were interviewed, but, due to the contingencies of the various protocols, most of them were not acceptable because of current drug use, inadequate symptoms, or situational factors. Initially, when we were dealing with patients, the actual treatment decisions were often agonizing, uncertain, and confused. However, experience with the volunteers and communication with Kales have enabled us to develop a reasonably efficient approach to insomniacs.

As a guideline in evaluating these patients, a general, somewhat hazy characterization of insomnia is used. In order to qualify as an insomniac, the patient must give some evidence, either subjective or objective, of disturbed sleep, whether it is a prolonged sleep latency, a short sleep time, many awakenings, or early morning awakening. In addition, there must be some complaint about the quality of daytime functioning which the patient relates directly to his perception of disturbed nocturnal sleep. The latter is required because of the reports of persons who normally sleep very little at night, but who are otherwise completely healthy (Jones and Oswald, 1968; Hartmann et al., 1971). Finally, as a rule of thumb, the total sleep time as estimated by the patient should be under 6 hr.

After the patient has voiced his complaint, the evidence for his disorder is gathered by obtaining a general assessment of the patient and some quantitative confirmation of his complaint. The following battery of paper and pencil work is completed by each patient: (1) MMPI; (2) Cornell Medical Index; (3) Sleep Inventory; (4) 2 weeks of sleep diaries; and (5) Stanford Sleepiness Scale (Hoddes et al., 1972). Because the patients are referred, it is expected that the medical work-up will be complete, including, if indicated, an evaluation for such conditions as pheochromocytoma and hyperthyroidism, which may be directly related to insomnia.

After this information is completed, it is assimilated and summarized by one person, often a medical student. The physician, with the advantage of all this information, then sees the patient. At this point, one of three choices can generally be made. First, the patient may seem to be primarily neurotic or psychotic, with the sleep disorder either questionable or obviously symptomatic. As noted above, Kales and other investigators have found some degree of psychopathology in most persons who complain of disturbed sleep (Weiss et al., 1962; Monroe, 1967; Kales, 1971). However, the physician must decide whether it is the insomnia or the psychiatric problem which is primary. If the primary diagnosis is of a psychiatric disorder, the patient is referred to the clinic psychiatrist for further evaluation and treatment of the problem.

The second possible diagnosis is based on the work done on the addictive properties of barbiturates and other hypnotics and their effects on sleep (Oswald, 1965, 1968; Kales, 1969b, 1971). If there is a current history of barbiturate use or other dependency-inducing sleep medications, our diagnosis, by definition, is drug-dependency insomnia. These patients usually complain that they cannot sleep without medication, or that they are having trouble sleeping even while taking the medication. In the event of this diagnosis, the patient is immediately begun on a relatively standard drug-withdrawal program, as has been described by the Kales group (Kales et al., 1970c; Kales, 1971). As Kales suggests, the withdrawal should proceed very slowly to avoid the disturbing physiological changes which occur with more acute withdrawal. The patient is reevaluated when withdrawal is complete, at which time he may or may not complain of insomnia.

Finally, for all remaining patients, a presumptive diagnosis of "true," or idiopathic, insomnia can be made. To date, 47 patients with the complaint of insomnia have been evaluated, according to the foregoing schema at the Stanford University Sleep Disorders Clinic. Of these, nine (19.1%) were referred for psychiatric treatment, 20 (42.5%) were diagnosed as drug dependent, and the remaining 18 (38.4%) had the presumptive diagnosis of "true" insomnia. If the latter diagnosis is made, what then? Does the patient actually have insomnia? What treatment, if any, should be instituted? Should the patient by placed on chronic hypnotic medication? How can the treatment be evaluated? The discussion presented in the following sections will hopefully provide some answers to these queries.

B. All-night sleep recordings in "insomniacs"

Six of the 18 patients in the idiopathic-insomnia group and two of the withdrawn drug-dependent patients received diagnostic all-night sleep recordings. We also had the data from the all-night sleep which was recorded as a part of drug-effectiveness studies in an additional 18 volunteers with the complaint of insomnia. Eight were from the flurazepam studies, four from a doxepin study, and six acted as controls for these studies. These subjects had been placed in the "true" insomnia category subsequent to having been extensively screened in a manner similar to that of the clinic patients. Figure 1 shows the mean total sleep times recorded from two to four baseline nights in these 26 patients and subjects. The mean sleep times are also plotted for all occasions when sleep recordings were continued into a treatment period. It is necessary to emphasize that from all the data we had acquired on these patients before they were recorded—the self-reports, the sleep diaries, sleep inventories, interviews, and sleepiness scales—we were absolutely convinced that each was suffering from a serious, chronic sleep loss. We were surprised, if not astounded, to discover when we recorded them that their sleep times should cover such a wide range.

From the data in Fig. 1, we suggest that 6 hr total sleep time might be regarded as a tentative cutoff for regarding a patient as eligible for chemotherapy. We do not feel that this evaluation should be exclusive, as in cases of

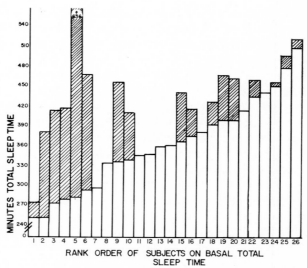

FIG. 1. Total sleep time in persons with presumed insomnia. Subjects numbered 1, 2, 3, 10, 15, 19, 25, and 26 were from the flurazepam studies; subjects 4, 6, 18, and 20 were from the doxepin studies; subjects 8, 12, 13, 17, 21, and 23 were control "insomniacs"; numbers 5, 9, 16, 22, and 24 were patients recorded on baseline and flurazepam; numbers 7, 11, and 14 were patients recorded only on baseline. Unshaded areas represent the mean total sleep times of from 2 to 4 baseline nights. Shaded areas represent the mean total sleep times of from 2 to 14 nights on medication.

people with only excessively long sleep latency or frequent and lengthy nocturnal awakening; however, we believe that some objective criterion is helpful in dealing with "insomniacs." Our suggestion is based on two primary considerations. First, of the eight people who were recorded in both baseline and drug conditions and whose baseline sleep times were under 360 min, the total sleep times while they were receiving medication showed a mean increase of 42.9% over the baseline level. On the other hand, in the nine people whose basal sleep times were greater than 360 min, the increase was 20.5% and less, with a mean increase of only 9.5%. Clearly, the relative effectiveness of sleeping medications declines rapidly as the total sleep time rises above 6 hr. Consequently, the use of such medications in patients with long sleep times is superfluous at best. Secondly, many patients who have long sleep times claim that they have slept a much shorter time than was recorded. Zung (1970) has reported a case of a woman with a 25-year history of insomnia who claimed that she did not sleep at all during four nights of polygraphic recording. The recordings showed that she had a mean total sleep time of 484 min and no trouble falling asleep. In such cases, it might be possible to "cure" the insomnia by simply explaining to the patient that he had obtained adequate sleep, and, if necessary, by going over the polygraphic output page-by-page, showing him exactly when he was asleep.

A case history of one of the patients of the Stanford Sleep Disorders Clinic illustrates a prime example of the usefulness of all-night sleep recordings of insomniacs. A 61-year-old male came to the clinic reporting a 40-year history of severe insomnia. His initial sleep diary record showed an average sleep latency of 119 min, frequent awakenings, an average wake time after onset of 62 min, and a total sleep time of only 241 min. The results of the MMPI were essentially normal with the exception of a slight elevation on the depression scale. The Stanford Sleepiness Scale ratings evidenced much daytime fatigue, with the highest rating a number 4, which states: "a little foggy, not at peak, let down." The patient had been taking diazepam (VALIUM) and secobarbital (SECONAL) occasionally, but had gotten no relief from the insomnia. Four all-night sleep recordings were conducted, two on placebo and two on 30 mg of flurazepam. On every night, the patient had no trouble falling asleep, nor did he awaken excessively during the night. The mean total sleep time on the two placebo nights was 433.3 min. Improvement on flurazepam was minimal, the mean total sleep time rising to only 458.2 min (5.8% increase). These results were shown to the patient and an explanation made that his sleep was essentially normal. The patient now reports that he is sleeping much better, is less worried and less depressed. If no sleep recordings had been conducted on this patient, he would, more likely than not, have been placed on chronic medication for a condition which did not in fact exist.

From these data and general considerations of good management, we have come to believe that no patient should receive chronic chemotherapy for insomnia without an all-night sleep recording. Our findings and those of others (Held et al., 1959; Schwartz et al., 1963; Monroe, 1967; Lewis, 1969; Zung, 1970) have made it apparent that most insomniacs, whether they are unable to or because they hope to receive more immediate attention if they exaggerate their symptoms, do not assess their sleep very accurately. In addition, 12 of the 26 persons we recorded slept longer than 6 hr and, even if one were to be more liberal with the cutoff, using instead 7 hr, nearly 20% would not qualify as insomniacs. Although it is an expensive and cumbersome process, this laboratory procedure is justified because of its extreme usefulness in two areas: (1) if insomnia is indeed present, the sleep recording enables both the patient and the physician to have complete confidence in the validity of the course of action; and (2) when there is adequate sleep, it may save the patient a long history of drug ingestion, clarify the situation, and, in some cases, radically alter a situation which may have existed for years. In the Stanford University Sleep Disorders Clinic, we have reduced the charge to patients of all-night recording so that, at the present time, every insomniac we see can afford to have a sleep recording before any treatment is prescribed.

C. Clinical evaluation of flurazepam

We have accepted the evidence of Kales (Kales et al., 1970*a*; Kales, 1971) and Oswald (1968) that many hypnotics are ineffective, and worse, that they can lead to dependence. Following the lead of Kales (1971), we feel that an effective sleep medication should increase sleep time, lower sleep latency, and decrease wake time in demonstrated insomniacs. In addition, there should be as a result of these changes an increase in the patient's sense of well-being during the day. Finally, there should be no evidence of tolerance. At the present time, our knowledge of sleep laboratory studies suggests that very few hypnotics appear to be effective when judged by these criteria. One of these is the benzodiazepine compound flurazepam.

Of the 47 insomniac patients who have been seen at the Stanford Sleep Disorders Clinic, 22 have been treated with this compound. In all but two of these cases, the treatment was deemed successful, either from sleep-recording data or from subjective reports of improvement by the patients. The subjective reports were obtained in the following manner. For the patients in whom the laboratory determination of insomnia was made, a packet of medications, arranged in a placebo-flurazepam schedule, was prepared. The patient knew that some of the envelopes did not contain active medication, but he was unaware of the exact schedule. While the patient was on the placebo-drug schedule at home, he completed sleep diary forms, sleep evaluation questionnaires, and the Stanford Sleepiness Scale. The resulting data, when compared to the medication the patient was taking, gave a very good picture of how he was responding to the treatment.

Figure 2 presents the total sleep time results for one of the patients who received such a schedule. It is obvious from the histogram that the subjective estimates of total sleep time were considerably higher on the flurazepam nights (shaded area) than on the placebo nights. Note that on the first placebo nights following flurazepam administration, the sleep times remained quite high. This appears to be another example of the carry-over of effectiveness of this compound as Kales (1971) has also noted. However, in this patient, as in the overwhelming majority of patients and subjects in

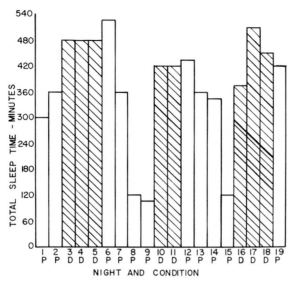

FIG. 2. Total sleep time of patients on home placebo-flurazepam schedule. Unshaded areas represent the patient's estimated total sleep time on placebo (P) nights. Shaded areas are the total sleep time estimates on flurazepam (DALMANE-D) nights.

whom we have used flurazepam, there was no report of any hangover during the day. The other subjective measures recorded in this patient's sleep diary included sleep latency, number of arousals, and daytime feeling of well-being, and followed a similar pattern, thus indicating an excellent response to treatment with flurazepam. For the other patients in whom such a schedule was used, the subjective data also demonstrated that flurazepam was effective in reducing the symptoms of insomnia.

IV. DISCUSSION

Our laboratory and clinical experience with flurazepam has shown the compound to be an effective hypnotic in both inducing and maintaining sleep. Although the drug significantly reduces slow-wave sleep, it does not, as do many hypnotics, have a profoundly suppressive effect on REM sleep. In addition, there appears to be a carry-over of effectiveness from one night to the next and for at least one night after the medication is withdrawn. This carry-over is not reflected in a drug hangover during the day. Furthermore, we found no difference in effectiveness between the doses of 15 and 30 mg.

We also conclude that every effort should be made to obtain all-night sleep recordings of patients who complain of insomnia before chronic sleeping medication is prescribed. Too often in these patients, there is not sufficient sleep loss to justify the use of such medication in their treatment. Finally, we concur with Kales (1971) that, in most cases of demonstrated insomnia, the benzodiazepine compound flurazepam should be an effective treatment.

ACKNOWLEDGMENTS

This work was supported by grant MH 13860 from the National Institute of Mental Health (WCD), grant NGR 05 020 168 from the National Aeronautics and Space Administration (WCD), Research Career Development Award MH 5804 (WCD), grant MH 19071 from the National Institute of Mental Health (VPZ), and Veterans Administration Hospital Research Funds (VPZ).

REFERENCES

Hartmann, E. (1968): Pharmacological sleep studies in man: Pentobarbital (NEMBUTAL), amitriptyline (ELAVIL), chlordiazepoxide (LIBRIUM), and RO 5-6901 (DALMANE). *Psychophysiology*, 4:391.
Hartmann, E., Bakeland, F., Zwilling, G., and Hoy, P. (1971): Sleep need: How much sleep and what kind? *American Journal of Psychiatry*, 127:1001–1008.
Held, R., Schwartz, B. A., and Fischgold, H. (1959): Fausse insomnie. *Presse Médicale*, 67:141–143.
Hoddes, E., Dement, W., and Zarcone, V. (1972): The development and use of the Stanford Sleepiness Scale (SSS). *Psychophysiology*, 9:150.
Jones, H. S., and Oswald, I. (1968): Two cases of healthy insomnia. *Electroencephalography and Clinical Neurophysiology*, 24:378–380.
Kales, A. (1969): Drug dependency. In: *Investigations of Stimulants and Depressants. Annals of Internal Medicine*, 70:591–614.
Kales, A. (1970): The clinical implications of sleep. In: *Psychiatry 1970*, edited by E. Robbins. Medical World News, New York, pp. 62–63.
Kales, A. (1971): Evaluation and treatment of insomnia: A. Introduction to sleep research in modern medicine. B. Sleep laboratory and clinical studies. Reprint from Scientific Exhibit shown at the Annual Convention of the American Medical Association, New Orleans, La., November 28–December 1, 1971.
Kales, A., Allen, W. C., Jr., Scharf, M. B., and Kales, J. D. (1970a): Hypnotic drugs and their effectiveness: All-night EEG studies of insomniac subjects. *Archives of General Psychiatry*, 23:226–232.
Kales, A., and Kales, J. D. (1971): Evaluation, diagnosis, and treatment of clinical conditions related to sleep. *Journal of the American Medical Association*, 213:2229–2235.
Kales, A., Kales, J., Scharf, M., Preston, T., Tan, T. L., and Allen, C. (1969a): Electrophysiological and psychological studies of insomnia. *Psychophysiology*, 6:255.

Kales, A., Kales, J. D., Scharf, M. B., and Tan, T. L. (1970b): Hypnotics and altered sleep and dream patterns: II. All-night EEG studies of chloral hydrate, flurazepam and methaqualone. *Archives of General Psychiatry*, 23:219–225.

Kales, A., Malmstrom, E. J., Kee, H. K., Kales, J. D., and Tan, T. L. (1969b): Effects of hypnotics on sleep patterns, dreaming, and mood state. *Biological Psychiatry*, 1:241.

Kales, A., Malmstrom, E. J., Scharf, M. B., and Rubin, R. T. (1970c): Psychophysiological and biochemical changes following use and withdrawal of hypnotics. In: *Sleep: Physiology and Pathology,* edited by A. Kales. J. B. Lippincott: Philadelphia, pp. 331–343.

Kales, A., and Scharf, M. B. (1972): Sleep laboratory and clinical studies of the effects of benzodiazepines on sleep: Flurazepam, diazepam, chlordiazepam, and RO 5-4200. In: *This Volume.*

Kales, A., Tan, T. L., Scharf, M., Kales, J., and Malmstrom, E. (1969c): Effects of long term and short term administration of flurazepam (DALMANE) in subjects with insomnia. *Psychophysiology*, 6:260.

Kales, J. D., Bixler, E. O., Slye, E. I., and Kales, A. (1971): Effects of placebo and flurazepam on sleep patterns in insomniacs. *Clinical Pharmacology and Therapeutics,* 12:691–697.

Karacan, I., Salis, P. J., Hursch, C. J., and Williams, R. L. (1970): New approaches to the evaluation and treatment of insomnia. Paper presented at the 17th Annual Meeting of Psychosomatic Medicine, Hamilton, Bermuda, November 15–19, 1970.

Lewis, S. A. (1969): Subjective estimates of sleep: An EEG evaluation. *British Journal of Psychology*, 60:203–208.

Monroe, L. J. (1967): Psychological and physiological differences between good and poor sleepers. *Journal of Abnormal Psychology*, 72:255–264.

Oswald, I. (1965): Rebound after sleeping pills. *Electroencephalography and Clinical Neurophysiology*, 19:531.

Oswald, I. (1968): Drugs and sleep. *Pharmacological Review*, 20:273–303.

Passouant, P., Baldy-Moulinier, M., and Gabbai, P. (1968): Insomnie et dualité du sommeil. *Journal de Psychologie Normale et Pathologique,* 3:19–38.

Rechtschaffen, A. (1968): Polygraphic aspects of insomnia. In: *The Abnormalities of Sleep in Man,* edited by H. Gastant, E. Lugaresi, G. Berti Ceroni, and G. Coccagna. Gaggi, Bologna, Italy, pp. 109–125.

Rechtschaffen, A., and Kales, A. (1968): *A Manual of Standardized Terminology, Techniques and Scoring System for Sleep Stages of Human Subjects.* Public Health Publications No. 204, U.S. Government Printing Office, Washington, D.C.

Rechtschaffen, A., and Monroe, L. (1969): Laboratory studies of insomnia. In: *Sleep: Physiology and Pathology,* edited by A. Kales. J. B. Lippincott Company, Philadelphia, pp. 158–169.

Schwartz, B. A., Guilbaud, G., and Fischgold, H. (1963): Etudes électroencéphalographiques sur le sommeil de nuit–"insomnie" chronique. *Presse Médicale*, 71:1474–1476.

Weiss, H. R., Kasinoff, B. H., and Bailey, M. A. (1962): An exploration of reported sleep disturbance. *Journal of Nervous and Mental Disease*, 134:528–534.

Zung, W. W. K. (1970): Insomnia and disordered sleep. *International Psychiatry Clinics*, 7:123–146.

The Benzodiazepines, Raven Press, New York © 1973

BENZODIAZEPINES AND HUMAN SLEEP

I. Oswald, S. A. Lewis, J. Tagney, H. Firth, and I. Haider

*University Department of Psychiatry, Royal Edinburgh Hospital,
Edinburgh, United Kingdom*

I. CLINICAL CONSIDERATIONS

Benzodiazepines have been used both for the relief of daytime anxiety and as hypnotic drugs. Nitrazepam (MOGADON) has been in use for several years in Europe. In addition to being an effective sleep-inducing agent, it has been found to have two outstanding merits. First, it is remarkably safe when taken in overdose (and it should be remembered that self-administered overdose of drugs has become extremely common in Western societies in recent years). Some of our Edinburgh colleagues (Matthew et al., 1969) have attested to this safety, which we are able to confirm. For example, patients who have taken 40 tablets of nitrazepam in an evening are drowsy and ataxic but awake by the morning. Secondly, clinical experience indicates that the drug has a low abuse potential. Instances of people forging prescriptions or similar improper behavior in order to obtain nitrazepam have not been reported; no withdrawal delirium nor self-injection for "kicks" has occurred. When it is taken regularly, patients become dependent upon it for the maintainance of their ease of falling asleep and for satisfactory sleep duration. In addition, as we indicate below, there are withdrawal symptoms and signs, but these features of dependence do not constitute a social problem. The only disadvantage of nitrazepam, although one that has been little explored in clinical practice, would appear to be the slowness of its elimination. Clinical doses of nitrazepam cause drug-induced EEG fast abnormalities to persist rather longer than in the case of sodium amobarbital (Oswald and Priest, 1965). There are also hangover effects, with impairment of psychomotor performance and liability to fall asleep, that may be more persistent after nitrazepam than after sodium amobarbital (Malpas et al., 1970).

II. EFFECTS ON SLEEP

A number of benzodiazepines have been studied in terms of their effects on the electrophysiological features of the sleep of normal volunteers; there have been a few similar studies of patients. The increasing sophistication of methods for research into sleep has been apparent in the publications of recent years, which have examined an increasing number of intrasleep features.

A. Nitrazepam

In an early study of nitrazepam with normal volunteers, the drug was found to reduce the proportion of sleep spent in the paradoxical, or rapid eye movement (REM), phase. Withdrawal after 2 weeks of drug administration caused "rebound" abnormalities of longer paradoxical sleep and its early onset during the night. The rebound took about 4 weeks to disappear (Oswald and Priest, 1965). In further studies of single-night dosage, it has been found that nitrazepam delays the first appearance of paradoxical sleep during the night (Lob et al., 1966; Lehmann and Ban, 1968; Haider and Oswald, 1971), reduces whole-night paradoxical sleep duration (Lehmann and Ban, 1968; Haider and Oswald, 1971), shortens the delay of sleep onset, reduces the time spent in drowsiness during the night (stage 1), reduces time spent in movements, reduces intrasleep restlessness (measured by the number of shifts per hour into stage 1 sleep or wakefulness), increases the total time in stage 2 sleep, and reduces spontaneous electrodermographic (galvanic skin response) activity during sleep (Haider and Oswald, 1971). It was also found that the larger the dose, the larger the effect on the variables measured.

The studies of nitrazepam described above used doses up to 20 mg. In the case of larger doses (self-administered overdoses) of up to 3 mg/kg, it has been found that drug-induced fast activity persists in the EEG for 9 to 10 days. Paradoxical sleep is at first suppressed but then rises to a peak at about 10 days and declines toward normal over a period of a month. Similarly, intrasleep restlessness increases to a peak at around 10 days and declines over the next month (Fig. 1). There is increasing insomnia during this withdrawal period. The insomnia again becomes maximal approximately 10 days after the drug overdose (Fig. 2), with return to normal sleep duration in the subsequent weeks (Haider and Oswald, 1970).

In addition to its effects on paradoxical sleep duration, nitrazepam also has an effect upon its intensity. In common with drugs such as barbiturates (Oswald et al., 1963) and glutethimide (Allen et al., 1968), nitrazepam reduces the frequency of REM per unit time during paradoxical sleep, with a rebound increase above normal after withdrawal (Lewis, 1968). The

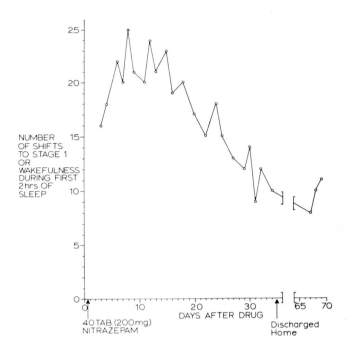

FIG. 1. Intrasleep restlessness rises to high levels from the 8th to the 15th night after a single, very large dose of nitrazepam in a woman of 63 years of age. Restlessness declines gradually in the subsequent 3 weeks. (Reproduced from Haider and Oswald, 1970, with permission of the publisher.)

effects on eye movement profusion have recently been confirmed by one of us (Firth) as Fig. 3 indicates. Paradoxical sleep is the time during which dreaming especially occurs and the profusion of REM relates to the "activity" of the dream events (Berger and Oswald, 1962). In a recent investigation by one of us (Firth), dream reports have been elicited by awakening subjects from paradoxical sleep and tape-recording their descriptions. Awakenings were made during a placebo week, during a first week on nitrazepam, 10 mg nightly, a second week on nitrazepam, 20 mg nightly, and a final week on placebo. There are presently a number of well-recognized techniques for scoring features of dream content. On the scale of Foulkes et al. (1966), the distribution of "bizarre" dreams appeared to be altered by the administration and subsequent withdrawal of nitrazepam (Fig. 4): dreams appear to become less bizarre during drug administration and more bizarre as a consequence of drug withdrawal.

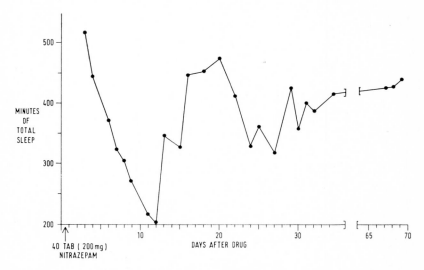

FIG. 2. The same patient as in Fig. 1. The total duration of whole-night sleep is very low 9 to 12 nights post-overdose. Drug-induced EEG fast activity was last observed on the 11th night. (Reproduced from Haider and Oswald, 1970, with permission of the publisher.)

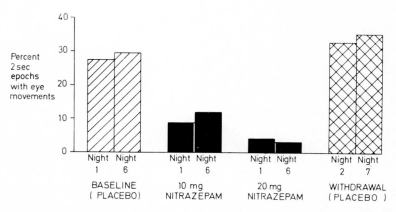

FIG. 3. Nitrazepam reduces the number of REM per unit time during paradoxical sleep ($p < 0.001$), as measured by the proportion of 2-sec epochs during paradoxical sleep that actually contain evidence of eye movement. In this study, nightly placebo for 1 week was followed by nitrazepam, 10 mg nightly, for 1 week, then by nitrazepam, 20 mg nightly, for 1 week, and then by placebo for a final week. The number of tablets per night was constant.

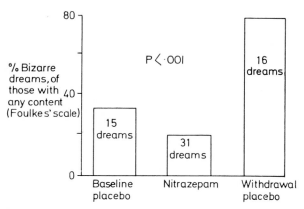

FIG. 4. The administration and withdrawal of nitrazepam influences the quality of dream experiences. The dosage and time schedule were the same as for Fig. 3, and each of the four men was wakened from paradoxical sleep twice each recorded night.

B. Chlordiazepoxide

Chlordiazepoxide has been investigated for its effects on sleep in normal subjects by Hartmann (1967) in a study in which a number of other drugs were also considered. Chlordiazepoxide, 100 mg, was associated with an increased delay of first paradoxical sleep and a decreased whole-night percentage of paradoxical sleep, although neither of these findings reached significance under the conditions of his study, in contrast to 100 mg of pentobarbital. The same comments apply to 30 mg of flurazepam in the same study.

More recently, two of us (Oswald and Tagney) have studied a woman aged 20 who took an overdose (960 mg, or 15 mg/kg) of chlordiazepoxide. She was drowsy and ataxic for 2 days and thereafter appeared clinically normal, but her paradoxical sleep betrayed a prolonged brain recovery process (Fig. 5). As in the recovery after many other drugs, the area under the paradoxical sleep rebound curve very greatly exceeds the area representing suppression. In other words, the paradoxical sleep "compensation" so greatly exceeds the "loss" that it must represent something more than mere compensation for a lost quota of paradoxical sleep time. It is possible that it might provide an index of synthetic processes required for repair after the poisoning. It should be emphasized that slow-wave sleep stages 3 and 4 do not show this pattern of increase after drug overdose; if anything, they are diminished (Haider and Oswald, 1970). Whereas (Fig. 5) the paradoxical sleep percentage is consistently above baseline on nights eight to 34, on those

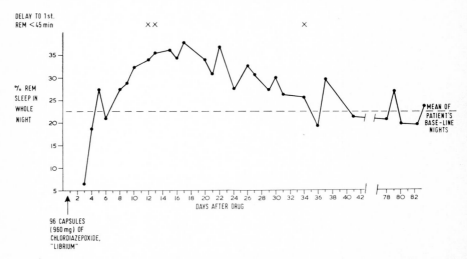

FIG. 5. After an overdose of chlordiazepoxide, a 20-year-old woman appeared clinically normal within 2 days but her paradoxical (REM) sleep revealed a prolonged recovery process lasting several weeks. Her "baseline" mean was assessed by all-night sleep recordings 78, 79, 80, 82, and 83 days after the overdose. On nights 12, 13, and 34, the first paradoxical sleep period of the night began early, as is often seen during paradoxical sleep "rebound." Note that the area below the baseline mean and to the left of the paradoxical sleep curve is very much smaller than the area above the mean and below the paradoxical sleep curve on later nights—the rebound is more than just compensation for lost paradoxical sleep duration and may reflect repair processes after chemical injury.

same nights stages 3 and 4 sleep averaged only 30.8% against 35.7% on the five late baseline nights. Total nightly sleep time varied little and without any marked pattern throughout the entire period of study, ranging from 350 to 494 min.

C. Newer hypnotics

Another benzodiazepine hypnotic has already been mentioned, namely flurazepam. Kales et al. (1970) have reported that 30-mg doses of this drug had little effect on paradoxical sleep but significantly reduced slow-wave sleep stage 4, and that 60 mg of flurazepam caused a significant reduction in paradoxical sleep duration and a delay of its onset. This reduction of stage 4 sleep, generally regarded as the deepest form of orthodox sleep, is of special interest as it has also been found in a study conducted at Edinburgh (Lewis and Firth) of a new benzodiazepine, flunidazepam (RO 5-4200).

In this latter study, normal volunteers are being followed over a period of weeks before, during, and after the administration of the drug and, with an interval of 12 weeks, they are being studied on two different dosages, namely 2 and 4 mg nightly, in a single-blind, cross-over design.

In general, the preliminary results are as would be anticiptated for a hypnotic. There is reduction in the percentage of paradoxical sleep when the drug is first administered. However, after about 3 weeks of drug administration, paradoxical sleep in the second 3 hr of sleep shows an excess over baseline. In other words, there is some internal compensation for paradoxical sleep suppression that took place in the earlier hours of sleep. On withdrawal of the drug, there is a rebound excess of paradoxical sleep, particularly noticeable in the first 3 hr of sleep (Fig. 6). Associated with these changes are corresponding changes in the latency of the first appearance of paradoxical sleep during the night (Fig. 7).

Changes in orthodox sleep are most apparent in the first 3 hr of sleep, as this is the time when the bulk of stages 3 and 4 sleep normally appear. Flunidazepam brings about an initial and continuing decrease in the amount of stages 3 and 4 sleep (Fig. 8). There appears to be no rebound excess.

Intrasleep restlessness is only slightly reduced but there is a marked and immediate increase with drug withdrawal (Fig. 9). The lack of obvious sleep-disturbance decrease with the drug is probably due to the fact that the subjects were healthy young males. The same argument applies to the finding that there was virtually no change in sleep onset latency when the drug was given. During the baseline period, the subjects were taking only 5 to 10 min to fall asleep, so there was little opportunity for change. However, there was an increase in sleep onset latency on withdrawal.

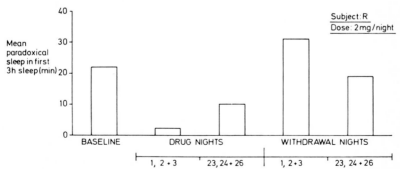

FIG. 6. Flunidazepam (RO 5-4200), 2 mg per night, suppresses paradoxical sleep in the early part of the night when first administered. A rebound excess occurs in the early withdrawal period, as illustrated by this subject. Each block is the mean of several nights, namely 4 baseline (placebo), 3 early, and 3 late drug nights, and the corresponding withdrawal nights.

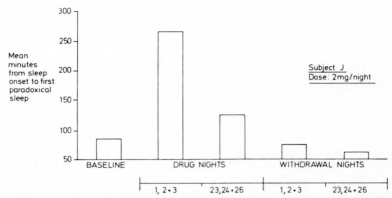

FIG. 7. Flunidazepam (RO 5-4200), 2 mg per night, increases the delay between sleep onset and the first paradoxical sleep period of the night. The blocks are the same as in Fig. 6 in respect to days for which data are represented.

Qualitative changes were also apparent. Drug-induced EEG fast activity similar to that caused by barbiturates was particularly noticeable during paradoxical sleep (Fig. 10), while during stage 2 sleep there was marked enhancement of EEG spindle activity (Fig. 11) of a kind also caused by tricyclic antidepressants (Lewis and Oswald, 1969).

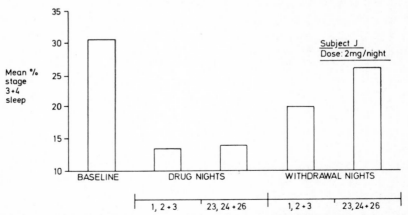

FIG. 8. Flunidazepam (RO 5-4200), 2 mg, causes a persistent loss of stages 3 and 4 orthodox (slow-wave) sleep. The percentages of total sleep are shown (the actual total duration of sleep was little changed in this subject). The blocks are the same as in Fig. 6 in respect to days for which data are represented.

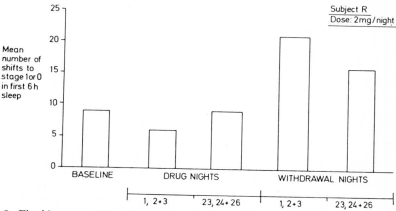

FIG. 9. Flunidazepam (RO 5-4200), 2 mg, reduces intrasleep restlessness in this normal subject, the frequency of shifts to stage 1 sleep or wakefulness (stage 0) from other sleep stages being reduced. Withdrawal of the drug after a month is associated with very restless sleep. The blocks are the same as in Fig. 6 in respect to days for which data are represented.

III. CONCLUSIONS

A general pattern would appear to emerge. Benzodiazepines are safe in clinical use. They diminish restlessness in sleep and reduce paradoxical sleep in duration and intensity, with effects on the concomitant dreaming. Sleep stages 3 and 4 tend to be reduced where there is continuous administration of the drugs over a period of time. Rebound abnormalities occur on withdrawal. One may seek to understand these rebound phenomena in terms of modifications to unknown cerebral components. The modifications must be slowly synthesized during the period of drug administration and are of such a kind as to tend to counteract the drug's effect. These modifications would persist after the drug was withdrawn, so that their effects (opposite in direction to those of the drugs) are then revealed as a rebound that slowly disappears through the inevitable but slow turnover of neuronal protein, the modified cerebral components being presumably replaced again by normal ones. This process apparently requires several weeks. Laboratory studies are most readily conducted on healthy volunteers and, therefore, two of the principal benefits sought from the clinical use of hypnotic drugs—namely, shortened delay and increased duration of sleep—are less clearly apparent than in clinical practice.

The function of sleep is still uncertain but may be related to synthetic processes both for growth and renewal (Oswald, 1969, 1970). Certainly,

Subject T   FLUNIDAZEPAM(Ro5-4200) 4mg

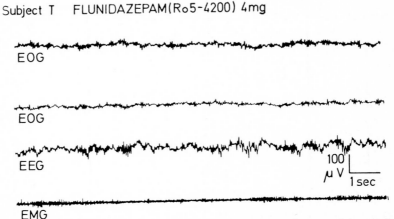

FIG. 10. Flunidazepam (RO 5-4200), 4 mg, provokes EEG fast activity during the paradoxical sleep in this subject who had been taking the drug for 26 nights. Each channel is a bipolar record. There are two EOG channels, the electrode pair being lower outer canthus to contralateral frontal boss (T.C. = 0.1; H.F. = 25 Hz); one EEG channel recording between midline frontal scalp and midline centroparietal scalp (T.C. = 0.3; H.F. = 25 Hz); and an EMG channel with the electrodes placed over the submental region (T.C. = 0.03; H.F. = 0 Hz).

slow-wave sleep stages 3 and 4, which benzodiazepines appear to suppress, can be presumed to be associated with general synthetic activity, since it is now known that the large nocturnal secretion of growth hormone is contingent upon these sleep stages (Honda et al., 1969; Parker et al., 1969; Sassin et al., 1969). Growth hormone increases tissue uptake of amino acids (Riggs and Walker, 1960) and stimulates the incorporation of amino acids into protein (Kostyo, 1964; Korner, 1965), an important tissue-renewing function, since 90% of the organic matter of tissue is protein. Paradoxical sleep has been the subject of a great deal of study in recent years but its precise function is still not known.

The distortion of the normal sleep pattern by benzodiazepines must be presumed to have some adverse consequences but, just as the studies of the effects on sleep tend to indicate some degree of tolerance during chronic use, so one must suppose that the central nervous system and other bodily systems that may be involved manifest tolerance or adaptive processes during the chronic administration of these drugs that prevent any serious or obvious deleterious effects from becoming apparent during their chronic administration. There is clearly a need for much more intensive study of the effects of drugs of this nature on bodily synthetic activities during chronic administration. In any case, such adverse effects as might be presumed to follow from

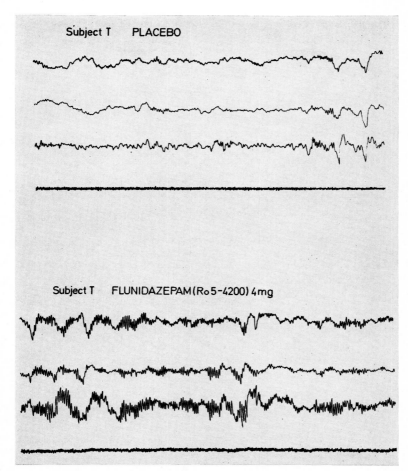

FIG. 11. Enhancement of EEG sleep spindle activity by flunidazepam (RO 5-4200), 4 mg, is exemplified by this subject. Electrode pairs as in Fig. 10. Upper record is taken from the baseline (placebo) period and the lower record from night 26 of the drug period.

these distortions of sleep must be set against the benefits of a more nearly normal sleep duration in the case of patients who would otherwise suffer from some degree of insomnia. There have been many studies of total sleep deprivation of several days, but there is a dearth of precise knowledge about the adverse effects of long-continued restriction of sleep to, say, about 5 hr per night. Such indications of the effects of sleep deprivation that are now available would suggest that there are considerable distortions of sleep processes and impairment of daytime mental performance and attention. The effects on general bodily functioning are unknown and uninvestigated.

ACKNOWLEDGMENTS

Miss J. Tagney is a Medical Research Council Scholar. Grants were received from the Scottish Hospital Endowments Research Trust, the Medical Research Council, and Roche Products Ltd.

REFERENCES

Allen, C., Kales, A., and Berger, R. J. (1968): An analysis of the effect of glutethimide on REM density. *Psychonomic Science,* 12:329–330.

Berger, R. J., and Oswald, I. (1962): Eye movements during active and passive dreams. *Science,* 137:601.

Foulkes, D., Spear, P. S., and Symonds, J. D. (1966): Individual differences in mental activity at sleep onset. *Journal of Abnormal Psychology,* 71:287–299.

Haider, I., and Oswald, I. (1970): Late brain recovery processes after drug overdose. *British Medical Journal,* 2:318–322.

Haider, I., and Oswald, I. (1971): Effects of amylobarbitone and nitrazepam on the electrodermogram and other features of sleep. *British Journal of Psychiatry,* 118:519–522.

Hartmann, E. (1968): The effect of four drugs on sleep patterns in man. *Psychopharmacologia* (Berlin), 12:346–353.

Honda, Y., Takahashi, K., Takahashi, S., Azumi, K., Irie, M., Sakuma, M., Tsushima, T., and Shizume, K. (1969): Growth hormone secretion during nocturnal sleep in normal subjects. *Journal of Clinical Endocrinology and Metabolism,* 29:20–29.

Kales, A., Kales, J. D., Scharf, M. B., and Tan, T. L. (1970): All night EEG studies of chloral hydrate, flurazepam, and methaqualone. *Archives of General Psychiatry,* 23:219–225.

Korner, A. (1965): Growth hormone control of biosynthesis of protein and ribonucleic acid. *Recent Progress in Hormone Research,* 21:205–236.

Kostyo, J. L. (1964): Separation of the effects of growth hormone on muscle amino acid transport and protein synthesis. *Endocrinology,* 75:113–119.

Lehmann, H. E., and Ban, T. A. (1968): The effect of hypnotics on rapid eye movement (REM). *International Journal of Clinical Pharmacology, Therapy and Toxicology,* 5:424–427.

Lewis, S. A. (1968): The quantification of rapid eye movement sleep. In: *Drugs and Sensory Functions,* edited by A. Herxheimer. J. and A. Churchill, Ltd. London, pp. 287–298.

Lewis, S. A., and Oswald, I. (1969): Overdose of tricyclic anti-depressants and deductions concerning their cerebral action. *British Journal of Psychiatry,* 115:1403–1410.

Lob, H., Papy, J. J., and Gastaut, H. (1966): Action du RO 4-5360 (MOGADON) sur le sommeil nocturne. *Revue Neurologique (Paris),* 115:545–546.

Malpas, A., Rowan, A. J., Joyce, C. R. B., and Scott, D. F. (1970): Persistent behavioural and electroencephalographic changes after single doses of nitrazepam and amylobarbitone sodium. *British Medical Journal,* 2:762–764.

Matthew, H., Proudfoot, A. T., Aitken, R. C. B., Raeburn, S. A., and Wright, N. (1969): Nitrazepam—a safe hypnotic. *British Medical Journal,* 3:23–25.

Oswald, I. (1969): Human brain protein, drugs and dreams. *Nature,* 223:893–897.

Oswald, I. (1970): *Sleep.* Penguin Books Ltd., Harmondsworth, Middlesex.

Oswald, I., Berger, R. J., Jaramillo, R. A., Keddie, K. M. G., Olley, P. C., and Plunkett, G. B. (1963): Melancholia and barbiturates: A controlled EEG body, and eye movement study of sleep. *British Journal of Psychiatry,* 109:66–78.

Oswald, I., and Priest, R. G. (1965): Five weeks to escape the sleeping-pill habit. *British Medical Journal,* 2:1093–1099.

Riggs, R. R., and Walker, L. M. (1960): Growth hormone stimulation of amino acid transport into rat tissues *in vivo. Journal of Biological Chemistry,* 235:3603–3607.

Sassin, J. F., Parker, D. C., Johnson, L. C., Rossman, L. G., Mace, J. W., and Gotlin, R. W. (1969): Effects of slow-wave sleep deprivation on human growth hormone release in sleep: Preliminary study. *Life Sciences* 8 *(Part I)*:1299–1307.

Sassin, J. F., Parker, D. C., Mace, J. W., Gotlin, R. W., Johnson, L. C., and Rossman, L. G. (1969): Human growth hormone release: Relation to slow-wave sleep and sleep waking cycles. *Science,* 165:513–515.

DISCUSSION SUMMARY

G. Reggiani

Clinical Research, F. Hoffmann-La Roche & Co. Ltd., Basel, Switzerland

MUSCLE RELAXATION

Takano (Brausch et al., *this Volume*) reported that diazepam reduces the excitability of the muscle spindles, which elicited the question whether diazepam acts directly or indirectly. He had found a reduction of muscle spindle activity only when the anterior roots were intact, indicating that the effect of diazepam is indirect. A corollary to this comment is provided by the following from Takano.

It is known that diazepam has a good therapeutic effect upon tetanus rigidity. How can we give a theoretical background for this treatment? Brooks, Curtis, and Eccles have shown that tetanus toxin diminishes or abolishes several types of inhibition on alpha motoneurons. Dr. Kano and I found at Chiba University, Japan, that this blocking or abolishing of inhibition might occur also on gamma motoneurons. Further, we measured the bias change in the muscle spindle discharges; namely, we measured the change in gamma bias in the tetanus-intoxicated state and concluded that the gamma system plays a rather greater role in the rigidity by tetanus intoxication than the alpha system.

As I have discussed in our report, diazepam may strongly affect the gamma motor system—and the gamma system is activated by tetanus toxin—this could be the explanation for the good effect of diazepam upon the tetanus-convulsive state.

The papers presented by Barnett and Fiore and by Pinelli raised the following comments. Besides an effect on supraspinal structures, diazepam also acts at a spinal level.

Polc reported that in spinal cats he found a reduction of gamma fiber activity and of mono- and polysynaptic reflex potentials when stimulating muscle and cutaneous nerve roots. Dorsal potentials taken as a measure of presynaptic inhibition were increased.

Hürlimann questioned the conclusions drawn by Barnett since, in the described experimental procedure, the second dose of the drug was given in a situation which was not comparable with that of the first dose. Haefely also brought up this point and suggested that the so-called acute tolerance might be due to the disputable experimental procedure.

Since diazepam affects reflex pathways in the spinal cord at different levels and by a number of actions, the effect of the drug greatly depends on the actual

627

functional state of the various pathways interfering with the reflex. Increasing stimulus strength before giving the second dose of drug to obtain the same effect as in the control state in no way compensates for the complex effects of the first dose.

Barnett was aware of the complexity of the action of diazepam and the problem of comparing the effects of the second dose with those of the first. However, since he found a marked difference between diazepam and nordiazepam, he thought these conclusions to be valid.

ALCOHOLISM

Goldberg reported on the interaction between benzodiazepines and alcohol, and Kaim spoke on the use of the benzodiazepines in treating alcohol-withdrawal states. The benzodiazepines (chlordiazepoxide in Goldberg's report) seem to offer to the physician who is looking for a new approach to the treatment of the alcoholic the type of adjunctive support which should accompany psychotherapeutic, counselling, social, and rehabilitative measures.

The ideal medication to treat the anxiety and frustration generated by a patient's being required to give up a cherished escape and to confront problems which had previously been evaded, as well as for the treatment of acute alcohol-withdrawal states, should both calm the patient promptly and effectively and possess anticonvulsant action. Also, it should neither aggravate existing pathology nor create physical or psychological dependence. The benzodiazepines appear to come quite close to fulfilling these requirements.

The fact that alcoholism is only one of many types of addiction and dependence has probably encouraged the research of Bliding, research physician at the Psychiatric Clinic of the University of Lund.

Bliding had questioned whether benzodiazepines with more stable plasma levels would be less prone to abuse than those benzodiazepines with shorter periods of high blood concentrations; this was noted by Goldberg, Kaim, and Hollister. The attention of the panel was focused on the abuse liability of the benzodiazepines.

When drug abuse occurs frequently and when the economic and social consequences affect not only the abuser himself but also his environment, the problem becomes a matter of concern for the the public health authorities, and it must be decided if measures should be taken to restrict the abuse. However, as long as drug abuse occurs only occasionally or sporadically, when there is little or no tendency of its spreading to other individuals, and if the adverse effects are confined to the individual itself, then abuse is not a public health problem. In such cases, adequate information and medical supervision represent the proper means of control.

To evaluate the abuse liability of the benzodiazepines in relation to their therapeutic use, the members of the panel had followed up in detail the world literature ever since the introduction of the first benzodiazepine (chlordiazepoxide) on the market in 1960. They have treated many patients with high psychological vulnerability for any tendency toward excess and/or abuse, those with serious character disorders, neuroses, or psychoses, living in a rather permissive society. They have in addition performed experiments for the purpose of evaluating the qualitative aspects of artificially induced dependence.

Based on the experimental and clinical results available, the panel concluded that there is an extremely low incidence of people who spontaneously increase the doses of these drugs in comparison with the number of legitimate users.

Regarding the qualitative aspect of dependence, Hollister described the appearance of an abstinence syndrome after withdrawal of high doses of benzodiazepines. Abstinence symptoms which appear after abrupt withdrawal are proof of the development of physical dependence. For barbiturates, the development of physical dependence with a characteristic abstinence syndrome, and occasionally fatal consequences, has clearly and often been proven clinically. Although from some points of view the abstinence syndrome, after withdrawal of high doses of benzodiazepines administered for a long period, can be compared to the abstinence syndrome caused by the barbiturates, there are striking differences between them with regard to type, severity, development, duration, and recovery of the syndromes. Also, it remains to be seen to what extent the so-called benzodiazepine withdrawal symptoms can be stimulated by rebound effects of the underlying disease. Finally, it is virtually impossible to commit suicide with a benzodiazepine taken alone, whereas this is not the case for the barbiturates. The phenomenon of physical dependence described by Hollister has no relation to the psychological dependence and must clearly be distinguished from reinforcing psychic factors which are observed in the abuse of barbiturates.

These differences, as well as the small number of benzodiazepine abusers, clearly mediate in favor of a very low risk of abuse, from the quantitative and above all from the qualitative point of view.

SLEEP STUDIES

Following the presentation of the papers in which agreement was reached on the effect observed in the three sleep disorder clinics where benzodiazepines were used for sleep induction and maintenance, the discussion was centered on the interpretation of the modifications these drugs caused on the sleep patterns and which were clearly defined by EEG and EOG manifestations. The total time of sleep is increased and there is no development of tolerance. However, all benzodiazepines, at doses which differ for each of them, reduce the amount of REM sleep and the amount of time spent in slow-wave sleep (stages 3 and 4). The facts that these changes in orthodox sleep in no way disturb the behavior of the normal volunteer nor seem to alter the subject's own favorable report in cases of insomnia and sleep disorders of different origin, gave rise to the question of the significance of the various phases of sleep, and of the relationship of dreams and the phase of sleep in which they occur to that of normal mental activity.

Generally accepted was Dement's conclusion that although this question still remains problematic, the benzodiazepines are the safest drugs presently at our disposal for the induction and maintenance of sleep.

The Benzodiazepines, Raven Press, New York © 1973

BEHAVIORAL AND PSYCHOSOMATIC EFFECTS OF BENZODIAZEPINES: INTERACTIONS WITH OTHER DRUGS

Adriano Marino

Faculty of Medicine and Surgery, University of Bari, Bari, Italy

I. CLINICAL STUDIES: TREATMENT OF PSYCHOSOMATIC DISEASE WITH AMITRIPTYLINE AND CHLORDIAZEPOXIDE

In the rational therapeutic application of psychopharmacology to psychosomatic medicine, the effects of psychotropic drugs can be classified as:

1. *Somatic*, acting either directly on the various organs of the body or indirectly through their nervous regulation; these effects may influence both psychogenic and nonpsychogenic somatic diseases.
2. *Somatopsychic*, producing psychic repercussions of psychogenic and nonpsychogenic somatic diseases.
3. *Psychosomatic*, acting on the psychogenic factor which determines, precipitates, or contributes to the development of a somatic disease.

Psychogenic cases of essential hypertension may show simultaneous somatic, somatopsychic, and psychosomatic components. The somatic component is represented by generalized vasoconstriction, the somatopsychic repercussion by anxiety and psychomotor agitation, and the psychic substrate by chronic hostility with depression. Thus, the ideal psychotropic treatment for this condition should exert vasodilating, tranquilizing, and antidepressant effects.

Adopting this criterion, 29 patients with essential hypertension and depressive-anxiety syndrome were treated with amitriptyline (25 to 60 mg, orally, per day, for between 2 and 6 months) in a double-blind, cross-over experiment. In the cases of severe anxiety, amitriptyline has been combined with chlordiazepoxide (10 to 20 mg, orally, per day). All the patients treated improved significantly after the therapy from both the psychological and cardioangiologic points of view. Before treatment, the mean values of

631

blood pressure were as follows: maximum 209 ± 3.75; minimum 109 ± 2.98; differential 100 ± 2.77. After treatment, we found the following values: maximum 157 ± 2.4; minimum 88 ± 1.9; differential 69 ± 2.19. Therefore, the treatment reduced the control values, respectively, by 25 to 20 to 31% ($p < 0.01$) (Marino et al., 1967).

In another clinical investigation (Marino and Gambardella, 1967), amitriptyline and chlordiazepoxide were administered to 11 patients with *anorexia nervosa,* a psychosomatic syndrome in which anxiety and depression are often present simultaneously. In all these patients in whom previous pharmacological and psychotherapeutic treatments had proven ineffective, a dramatic improvement in their psychological and somatic states (anorexia, body weight) was observed after amitriptyline and chlordiazepoxide administration.

Finally, the antichlolinergic somatic effect of amitriptyline, combined with its antidepressant and tranquilizing action, has been useful in the treatment of *spastic gastrointestinal disease,* especially when the thymoleptic was associated with chlordiazepoxide (Marino et al., 1968).

II. EFFECTS OF BENZODIAZEPINES ON AVOIDANCE CONDITIONING IN THE GUINEA PIG

Before 1960, the guinea pig had not been largely employed in psychological and psychopharmacological experimental research. In 1960, however, we started using this animal, which can be subjected to EKG records without any anesthesia, in order to study the influence of psychological stress on the specific cardiotoxicity of drugs (Marino, 1960, 1961a, 1964, 1969) (see section III). In our experience, the guinea pig has been shown to be susceptible to psychological procedures, such as avoidance conditioning (Marino, 1960, 1961a, 1964, 1969; Sansone, 1967; Sansone and Bovet, 1968) and the conflict situation (Marino, 1960, 1961a, b, 1964), in animals previously conditioned to avoidance.

The deconditioning activity of chlorpromazine, three benzodiazepines, and four thymoleptics has been investigated in guinea pigs. The percentage of conditioned response reduction for each dose of the drugs and the ED_{50} of each drug were evaluated (Sansone and Marino, 1969; Sansone et al., 1969).

As far as we know, no data have been published up to 1969 on the action of these drugs pertaining to the avoidance conditioning of the guinea pig. On the contrary, it is known that in other animal species commonly used in psychopharmacology, chlorpromazine and the other neuroleptics cause a selective depression of avoidance behavior, whereas the deconditioning effects of thymoleptics and minor tranquilizers are still under discussion or require particular devices in doses and techniques to be detected (Hanson, 1961; Randall et al., 1961, 1965; Dews, 1962; Cook and Catania, 1964; Vaillant, 1964; Morpurgo, 1965; Dureman and Henriksson, 1967).

From the results summarized in Table I, diazepam and nitrazepam showed a higher potency of deconditioning activity as compared to chlordiazepoxide. Chlorpromazine was more active than the latter, but clearly less active than the former two compounds. A definite dose-effect relationship resulted for each drug tested.

In comparing the present results with those previously obtained in rats (Randall et al., 1961; Morpurgo, 1965), the guinea pig seems to be more sensitive than the rat to the deconditioning activity of benzodiazepines. Indeed, the results showed chlordiazepoxide to be only slightly less potent than chlorpromazine, whereas diazepam and nitrazepam showed a deconditioning activity definitely greater than that of chlorpromazine.

In regard to its clinical implications, the present experimental results could support the authors (Lambert et al., 1962; Marino et al., 1966) who consider diazepam and, especially, nitrazepam as "middle or major tranquilizers," closer to neuroleptics, such as chlorpromazine, than to minor tranquilizers, such as chlordiazepoxide and meprobamate.

The results, concerning the effects of the four tricyclic thymoleptics on avoidance conditioning of guinea pigs, show a clear difference of activity between the two different series of tricyclic derivates. The two dibenzocycloheptene compounds amitriptyline and nortriptyline cause an evident depression of avoidance behavior, with a definite dose-effect relationship. Amitriptyline appears more active than its demethylated derivative. Conversely, with the two dibenzazepine compounds imipramine and desipramine much higher doses are required to depress the conditioning avoidance responses and the

TABLE I. *Deconditioning action of some psychotropic drugs in the guinea pig*

Drugs	ED_{50}	95% Fiducial limits
Diazepam	0.80	0.35– 1.84
Nitrazepam	0.84	0.59– 1.39
Chlorpromazine	2.30	0.88– 5.98
Chlordiazepoxide	5.20	2.73– 9.88
Amitriptyline	5.70	3.6 – 8.8
Nortriptyline	12.70	7.9 –20.3
Desipramine	38.50	6.7 –92.4
Imipramine	41.80	22.0 –79.4

Doses in mg/kg, i.p.

dose-effect relationship is not so evident. No significant difference was noticed between imipramine and its demethylate derivative.

In the dibenzazepine derivatives, the thymoleptic component is predominant, but in the dibenzocycloheptene derivatives, both components (thymoleptic and tranquilizing) are almost equipotent (Holtz and Westermann, 1965; Marino et al., 1968). Thus, it appears that the depression of avoidance behavior in guinea pigs, exerted by the thymoleptics tested in this study, can probably be ascribed to their tranquilizing activity.

We may conclude that, in the guinea pig, all the drugs tested exert an evident deconditioning effect on avoidance learning. The potency of the effect decreases in the following order: diazepam and nitrazepam-chlorpromazine-chlordiazepoxide and amitriptyline-nortriptyline-desipramine and imipramine. On the whole, the higher activity of chlorpromazine, compared to that of the thymoleptics, parallels the previous data observed in other animal species, whereas diazepam and nitrazepam are the most active tranquilizers of the benzodiazepine group in the guinea pig, which seems to be a peculiarity of that species.

III. EFFECTS OF BENZODIAZEPINES ON EXPERIMENTAL SOMATOPATHIES INDUCED BY PSYCHOSTRESS AND ORGANOTOXIC DRUGS

The experimental investigation of the psychosomatic effects of drugs calls for special technical and psychosomatic experience.

Our method (Marino, 1960, 1961a, 1964, 1969) is based on a combination of psychological and pharmacological techniques aimed at bringing the experimental approach closer to that of clinical observation. An experimental "organ situation," which could be compared to the "organ inferiority" of psychosomatic medicine, is experimentally evoked by means of pharmacological treatment specifically toxic for that organ. On the other hand, psychological stress is induced by a conflict situation in animals previously conditioned to an avoidance situation. In this way, the influence of psychological stress on the target organ is evaluated.

Furthermore, the protective effects of psychotropic and somatotropic drugs, acting either on the psychogenic (psychological stress) or somatic (organ inferiority) factor, can be investigated with respect to the resulting somatopathy.

So far, we have demonstrated that the psychological stress was able to potentiate the specific cardiotoxicity of such drugs as emetine, strophanthin, vasopressin, and hypertensine (Marino et al., 1960 to 1969).

In one of our published studies (Marino et al., 1965), psychostress was shown to potentiate the lethal, toxic, and electrocardiographic effects of epinephrine (single i.p. doses of 1, 2, 5, or 10 mg/kg) in the guinea pig; chlordiazepoxide (10 mg/kg, i.m., daily for 9 days) showed high protection against these effects in the stressed animals, but much less in the normal ones; pentaerythritol tetranitrate (5 to 25 mg/kg orally, daily for 9 days)—a coronary dilator—did not exert any protection in either normal or stressed animals, nor did it interact with the protective effect of chlordiazepoxide. Since chlordiazepoxide, at the dose used, has no peripheral antiadrenergic activity, these results should be attributed to its psychotropic action and may provide an experimental indication for the clinical use of the drug in psychosomatic cardiopathies.

Other research we have in progress seems to indicate a protective effect of benzodiazepines against the toxicity and cardiotoxicity of nicotine.

IV. EFFECTS OF BENZODIAZEPINES ON FIGHTING AND PENTYLENETETRAZOL-INDUCED CONVULSIONS IN MICE

In our previous studies (Marino and Cosimo, 1965), nitrazepam showed a very high activity against *fighting* in mice. Moreover, *fighting* did not show any potentiation of or interaction with the convulsive and lethal effects of pentylenetetrazole (CARDIAZOL) (50 to 500 mg/kg, i.p., sequential doses). On the other hand, nitrazepam (10 mg/kg, i.m., sequential doses) was able to give a high protection against those effects of pentylenetetrazole both in normal and in fighting mice.

V. EFFECTS OF BENZODIAZEPINES IN MAN

In our clinical investigations of benzodiazepines, we found (Marino et al., 1966) that nitrazepam was effective in treating insomnia and anxiety, without having any antipsychotic or antipsychosomatopathic action. However, in all the patients treated, a peculiar "socializing effect" was shown by the drug. This effect was particularly useful in order to render patients more susceptible to psychotherapy and to other, more appropriate psychotropic agents.

VI. EFFECTS OF BENZODIAZEPINES ON THE "HEAD DROP" INDUCED BY TUBOCURARINE IN THE RABBIT

From our studies in rabbits (Sansone et al., 1966) on the effects of psychotropic drugs (1, 10, and 100 STD. i.v. = 1, 10, and 100 times greater than the single therapeutic dose per kilo used in human therapy) on the "head drop" induced by *d*-tubocurarine chloride, i.v. (60 gamma/kg/min), benzodiazepines proved to be highly active in potentiating the effect of tubocurarine. In comparing various psychotropic drugs, the magnitude of the potentiating effect decreases in the following order:

1. Chlorprothixene-chlorpromazine-chlordiazepoxide-diazepam-nitrazepam
2. Amitriptyline-nortriptyline-imipramine
3. Hydroxyzine-carisoprodol-meprobamate
4. Tetrabenazine-reserpine-perphenazine-haloperidol

Conversely, the thymeretic monoamine oxidase inhibitors showed antagonism (nialamide-pargyline-tranylcypromine-phenelzine-iproniazide in increasing order of potency) with the tubocurarine effect, or no interaction (isocarboxazide).

VII. CONCLUSIONS

In our experimental and clinical approach to the evaluation of *somatic, somatopsychic,* and *psychosomatic* effects of psychotropic drugs, we found that benzodiazepines, combined with other psychotropic or somatotropic drugs, offered protection against experimental and clinical psychogenic somatopathies. Our clinical studies showed chlordiazepoxide, combined with amitriptyline, to be useful in the treatment of the somatic and psychic components of psychosomatic disease, such as anorexia nervosa, spastic gastroenteric disorders, and essential hypertension.

Nitrazepam, *per se,* showed efficacy against anxiety and insomnia without any antipsychotic or antipsychosomatopathic effects. However, the drug was able to render patients susceptible to antipsychotic agents because of its own "socializing" effect. Benzodiazepines protected against the specific organotoxicity of various drugs in experimental animals, especially when potentiated by psychostress. Moreover, nitrazepam was able to antagonize the convulsive and lethal effects of pentylenetetrazol, both in normal and in fighting mice. Finally, the benzodiazepines, as compared with other psychotropic drugs, have been shown to be the most active in potentiating the "head drop" induced by *d*-tubocurarine in the rabbit.

Benzodiazepines have been used in combination with other treatments, both physical (electroconvulsive therapy of agitated depressions) and pharmacological. The therapy involving a combination of benzodiazepines with somatropic or psychotropic drugs was made more rational by basing it on an integration of the data obtained in our pharmacological and psychosomatic studies. Benzodiazepines may be useful in combination with: (1) anticonvulsant drugs in the treatment of convulsive disorders, (2) general anesthetics in anesthesia, (3) antihistamines in reducing itching, and (4) stilbestrol in controlling the libidinal urges of homosexuals.

In internal medicine and in psychiatry, we believe that benzodiazepines represent the drugs of choice to be combined with amitriptyline in the treatment of depression and of psychosomatic diseases with depressive substrate and anxious repercussion. This combination, in our opinion, is preferable to that with monoamine oxidase inhibitors.

VIII. SUMMARY

In our experimental and clinical approach to the evaluation of somatic, somatopsychic, and psychosomatic effects of psychotropic drugs, benzodiazepines combined with other psychotropic or somatotropic drugs afforded protection against experimental and clinical psychogenic somatopathies. In particular, protection against the specific organotoxicity of various drugs was shown in experimental animals when potentiated by psychostress. Moreover, the benzodiazepines, as compared with other psychotropic drugs, have been shown to be the most active in the potentiation of the "head drop" induced by *d*-tubocurarine in the rabbit.

Finally, in our clinical research, chlordiazepoxide combined with amitriptyline proved to be useful in the treatment of the somatic and psychic components of psychosomatic disease, such as anorexia nervosa, spastic gastroenteric disorders, and essential hypertension. Results and their clinical implications are discussed with a view to a rational application of benzodiazepines in psychiatry, internal medicine, and anesthesiology.

REFERENCES

Cook, L., and Catania, C. A. (1964): Effect of drugs on avoidance and escape behavior. *Federation Proceedings*, 23:818–835.

Dews, P. B. (1962): A behavioral output entraining effect of imipramine in pigeons. *International Journal of Neuropharmacology*, 1:265–272.

Dureman, E. J., and Henriksson, B. G. (1967): Effects of tricyclic antidepressive compounds on operant behaviour in pigeons. In: *Proceedings of the First International Symposium on Antidepressant Drugs*, edited by S. Garattini and M. N. G. Dukes. Excerpta Medica Foundation, Amsterdam, pp. 141–148.

Hanson, H. M. (1961): The effects of amitriptyline, imipramine, chlorpromazine, and nialamide on avoidance behavior. *Federation Proceedings*, 20:396 (abstr.).

Holtz, P., and Westermann, E. (1965): Psychic energizers and antidepressant drugs. In: *Physiological Pharmacology*, Vol II, edited by W. S. Root and F. G. Hofmann, Academic Press, London, pp. 201–254.

Lambert, A., Charriot, G., Vermorel, M., Versmee, A., and Midenet, M. (1962): Note sûr un tranquillisant majeur: RO 4-5360. *Annals of Medical Psychology*, 2:268–270.

Litchfield, J. T., and Wilcoxon, F. (1949): A simplified method of evaluating dose-effect experiments. *Journal of Pharmacology and Experimental Therapeutics*, 96:99–113.

Marino, A. (1960): Factors increasing emetine cardiotoxicity. *Pharmacologist*, 2:73 (abstr.).

Marino A. (1961a): Psychological stress and emetine cardiotoxicity. *Experientia (Basel)*, 17:117–119.

Marino, A. (1961b): Electrocardiographic and behavioral effects of emetine. *Science*, 133:385–386.

Marino, A. (1964): The influence of psychological stress on the specific cardiotoxicity of drugs. *Nature*, 203:1289–1291.

Marino, A. (1969): Pharmacology and psychosomatic medicine: The experimental and clinical approach to a psychosomatic evaluation of psychotropic drugs. In: *Proceedings of an International Symposium on Psychotropic Drugs in Internal Medicine*, edited by A. Pletscher and A. Marino. Excerpta Medica Foundation, Amsterdam, pp. 47–56.

Marino, A., Colucci D'Amato, F., Miracco, A., and De Marino, V. (1967): Psicodinamica e psicofarmacoterapia della ipertensione essenziale. *Clinica Terapeutica*, 41:219–222.

Marino, A., and Cosimo, V. (1965): Effetti di un nuovo tranquillante benzodiazepinico—1'1,3-diidro-7-nitro-5-fenil-2-h-1,4-benzodiazepin-2-one (MOGADON)—sulla letalità e convulsività da CARDIAZOL nel topo normale o combattente. *Rassegna di Medicina Sperimentale*, 12:121–126.

Marino, A., and De Marino, V. (1968): Ricerche sull'associazione amitriptilina-clordiazepossido (LIMBITRYL) nel trattamento delle colopatie funzionali a impronta psicogena. *Rassegna Internazionale di Clinica e Terapia*, 47:255–272.

Marino, A., De Marino, V., and Aulisio, G. A. (1968): L'associazione di un timolettico (amitriptilina) con un tranquillante (clordiazeposside) nel trattamento delle sindromi depressive e psicosomatiche. *Rassegna Internazionale di Clinica e Terapia*, 47:127–146.

Marino, A., De Marino, V., Mastursi, M., and Salvatore, D. (1965): Interferenze dello stress psicologico, del clordiazepossido e del pentaeritrolo-tetranitrato con gli effetti tossici ed ecgrafici della adrenalina nella cavia. *Rassegna di Medicina Sperimentale*, 12:45–60.

Marino, A., D'Errico, A., Turchiaro, G., and Parrella, F. (1966): Farmacologia sperimentale e clinica del nitrazepam, nuovo derivato benzodiazepinico con attività euipnica e tranquillante. *Quaderni OPIS*, 1:283–323.

Marino, A., and Gambardella, A. (1967): Ricerche preliminari sulla psicofarmacologia delle anoressie psicogene. *Clinica Terapeutica*, 41:51–53.

Marino, A., Tortora, R., Robertaccio, A., and Sallusto, L. (1964): Effetto della streptomicina sul comportamento spontaneo e condizionato. Interferenza fra tossicita streptomicinica e stress psicologico. *Rassegna di Medicina Sperimentale*, 11:173–179.

Morpurgo, C. (1965): Drug-induced modifications of discriminated avoidance behavior in rats. *Psychopharmacologia (Berlin)*, 8:91–99.

Randall, L. O., Heise, G. A., Schallek, W., Bagdon, R. E., Banziger, R. F,. Boris, A., Moe, R. A., and Abrams, W. B. (1961): Pharmacological and clinical studies on Valium, a new psychotherapeutic agent of the benzodiazepine class. *Curr. Ther. Res.* 3:405–425.

Randall, L. O., Schallek, W., Scheckel, C., Bagdon, R. E., and Rieder, J. (1965): Zur Pharmakologie von MOGADON, einem Schlafmittel mit neuratigem Wirkung-mechanismus. *Schweizerische Medizinische Wochenschrift,* 95:334–337.

Sansone, M. (1967): Azione della nicotina sul condizionamento di evitamento nella cavia. *Atti dell'Accademia Nazionale Lincei,* 43:401–404.

Sansone, M., Aulisio, G. A., and Marino, A. (1966a): Tubocurarina e psicofarmaci 1. Tranquillanti minori. *Bolletino della Società Italiana di Biologia Sperimentale,* 43:236–239.

Sansone, M., Aulisio, G. A., and Marino, A. (1966b): Tubocurarina e psicofarmaci 2. Tranquillanti maggiori. *Bolletino della Società Italiana di Biologia Sperimentale,* 43:240–243.

Sansone, M., Aulisio, G. A., and Marino, A. (1966c): Tubocurarina e psicofarmaci 3. Antidepressivi timolettici. *Bolletino della Società Italiana di Biologia Sperimentale,* 43:243–246.

Sansone, M., Aulisio, G. A., and Marino, A. (1966d): Tubocurarina e psicofarmaci 4. Antidepressivi timeretici. *Bolletino della Società Italiana di Biologia Sperimentale,* 43:246–249.

Sansone, M., and Bovet, D. (1968): Onset and offset of the light as conditioned stimulus in guinea pigs. *Communications in Behavioral Biology,* 2:107–111.

Sansone, M., Bovet, D., and Marino, A. (1969): Thymoleptics in avoidance conditioning in guinea pigs. *Pharmacological Research Communications,* 3:311–315.

Sansone, M., and Marino, A. (1969): Benzodiazepines in avoidance conditioning in guinea pigs. *Pharmacological Research Communications,* 1:122–126.

Vaillant, G. E. (1964): A comparison of chlorpromazine and imipramine on behavior of the pigeons. *Journal of Pharmacology and Experimental Therapeutics,* 146:377–384.

The Benzodiazepines, Raven Press, New York © 1973

INTERACTION OF BENZODIAZEPINES WITH WARFARIN IN MAN

Donald S. Robinson and Ellsworth L. Amidon

University of Vermont College of Medicine, Burlington, Vermont

I. INTRODUCTION

The problems of drug interactions in clinical medicine are receiving increasing recognition. Drug interactions with anticoagulants frequently result in bleeding reactions with significant morbidity and mortality.

The propensity for most sedative-hypnotic drugs to interact with the coumarin class of drugs often complicates the regulation of anticoagulant therapy. Barbiturates, glutethimide (MacDonald et al., 1969; Robinson et al., 1970), ethchlorvynol (Cullen and Catalano, 1967; Johansson, 1968), and dichloralphenazone (Breckenridge and Orme, 1971), when administered in usual therapeutic dosages, have been shown clearly to antagonize warfarin effect in man by inducing its metabolism. Methyprylon, another commonly used hypnotic, also appears to decrease the half-life and hypoprothrombinemic effect of warfarin in normal subjects (Robinson, *unpublished data*). Trichloroacetic acid, the major metabolite of chloral hydrate, potentiates warfarin in some patients by competition with and displacement of the anticoagulant drug from shared plasma protein-binding sites (Sellers and Koch-Weser, 1970). The antianxiety agent meprobamate may be capable of accelerating warfarin metabolism in some patients (Hunninghake and Azarnoff, 1968).

The benzodiazepines, a major class of antianxiety, sedative-hypnotic agents, are receiving scrutiny as possible alternatives to those drugs known to interact with warfarin. Lack of significant effect of chlordiazepoxide administration on the anticoagulant action of warfarin in man has been reported by several investigators (Lackner and Hunt, 1968; Robinson et al., 1970; Solomon et al., 1971), even though animal studies employing substantially higher doses of chlordiazepoxide have suggested an inducing effect on several liver microsomal enzymes (Hoogland et al., 1966). Diazepam has also been reported not to influence the anticoagulant response to warfarin (Solomon et al., 1971). The importance of identifying an effective hypnotic agent compatible with warfarin prompted the following studies of flurazepam in normal subjects and in outpatients on long-term anticoagulant therapy.

II. METHODS

A. Studies in normal subjects

Eight healthy male volunteers, ages 21 to 30, were selected for study. The subjects were given a single, 45-mg oral dose of warfarin. Fifteen, 40, and 64 hr after ingestion of warfarin, blood specimens were obtained to measure plasma warfarin concentrations and prothrombin time. Thereafter, the subjects received 30 mg of flurazepam at bedtime for 28 days. On days 14 and 28, the single 45-mg oral dose of warfarin was repeated and the same measurements made. The subjects were cautioned to take no other medications during the study. Prothrombin times were measured by the Quick one-stage method, utilizing a single lot of reagent and control plasma throughout the experiment. Plasma warfarin concentration was determined by the method of Corn and Berberich (1967).

B. Studies in outpatients

Twelve outpatients, 11 male and one female, ages 48 to 69 years, who had received chronic warfarin therapy for periods of 3 months to 10 years, were selected as subjects for this study. Patients were considered suitable for admission if they had previously demonstrated stable anticoagulant control and were taking no other drugs capable of interaction with warfarin. Six patients received 30 mg of flurazepam and six patients received identical-appearing placebo at bedtime for 28 days, in addition to their usual medications. One patient experienced excessive daytime sedation on this dosage of flurazepam; therefore, on the 15th day, his dose was reduced to 15 mg nightly. Blood specimens were obtained at 1-week intervals during the 4 weeks of flurazepam or placebo administration and for 2 additional weeks after discontinuance of the test drug. Prothrombin times and plasma warfarin concentrations were determined on all specimens.

III. RESULTS

Flurazepam administration had no significant effect on the plasma warfarin concentrations or plasma half-lives in the eight normal subjects (Fig. 1). The mean plasma warfarin half-lives were 53.5 ± 2.7 hr for the

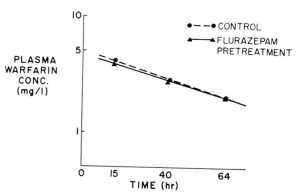

FIG. 1. Mean plasma warfarin concentrations (mg/liter) of eight subjects are shown for 15, 40, and 60 hr after ingestion of a single 45-mg dose of warfarin. The average plasma warfarin concentrations and the plasma warfarin half-lives before and after 28 days of flurazepam administration were not significantly different.

control period, 60.4 ± 5.2 hr after 14 days, and 56.4 ± 3.6 hr after 28 days of flurazepam administration. However, a progressive decrease in the mean hypoprothrombinemic response to the single dose of warfarin was observed after 14 and 28 days of flurazepam administration (Table I). This diminished response to warfarin was not marked, but the differences in prothrombin times between the control values and after 28 days of flurazepam administration did reach statistical significance ($p < 0.05$). This apparent disparity between unchanged plasma warfarin concentrations and half-lives and the diminished pharmacologic response to warfarin led to a second experiment, to test the effect of flurazepam administration to outpatients receiving chronic warfarin therapy. As shown in Fig. 2, there were no significant changes in mean prothrombin times or plasma warfarin concentra-

TABLE I. *Effect of flurazepam administration on the hypoprothrombinemic response to single dose of warfarin*

Group	Mean prothrombin time (sec)	
	40 hr	64 hr
Control	18.6 ± 0.5	17.3 ± 0.6
Flurazepam pretreatment for 14 days	17.4 ± 0.8	15.8 ± 0.8
Flurazepam pretreatment for 28 days	16.8 ± 0.5*	15.1 ± 0.5*

Mean prothrombin times measured at 40 and 64 hr after a single, 45-mg oral dose of warfarin in eight normal subjects are shown before and after 14 and 28 days of pretreatment with flurazepam, 30 mg, daily.

* Difference significant at $p < 0.05$ by Student's t test for paired data comparing the control and post-drug treatment values for each subject.

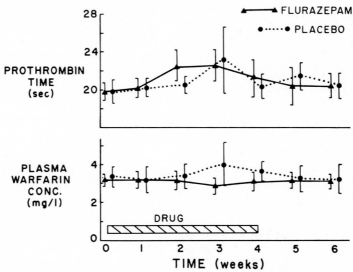

FIG. 2. Mean prothrombin times and steady-state plasma warfarin levels are shown during 4 weeks of flurazepam (six patients) or placebo (six patients) administration, and for 2 weeks following discontinuance of the test drugs. Warfarin dosage remained constant throughout the experiment. No significant changes in either prothrombin times or plasma warfarin levels were observed during flurazepam administration, and the flurazepam and placebo groups did not differ significantly from each other.

tions during administration of flurazepam, nor did the values for the placebo and flurazepam groups differ significantly from each other. Mean warfarin dosage remained constant throughout the period of the experiment.

IV. DISCUSSION

At least 35 classes of compounds, involving well over 100 clinically useful drugs, have been implicated as potential interactors with coumarin anticoagulants in man (Koch-Weser and Sellers, 1971). The nearly routine use of sedative-hypnotic drugs, particularly in the hospital setting, poses a significant problem for the patient receiving warfarin. An impressive number of the available sedative-hypnotics have been demonstrated to interact with warfarin and other coumarin drugs. The implicated drugs include practically all barbiturates, glutethimide, ethchlorvynol, methyprylon, and chloral hydrate and its various complexes such as dichloralphenazone and chloral betaine. Therefore, identifying an alternative hypnotic agent to these interacting drugs is of clinical importance.

Several previous studies have demonstrated that the sedative antianx-

chlordiazepoxide 40 mg/kg/day for 4 days. On the fifth day, the following measurements were made in control and in treated animals: (1) pentobarbital sleeping time; (2) zoxazolamine paralysis time; (3) plasma half-life of ^{14}C-pentobarbital; and (4) maximal velocity (V_{max}) and apparent affinity constant (K_m) of N-demethylation of ethylmorphine by rat liver microsomal enzymes. Details of these methods have been given elsewhere (Breckenridge et al., 1969).

III. RESULTS

A. Human studies

1. *Long-term studies*

Nitrazepam. Administration of nitrazepam, 10 mg nightly, to three patients caused no change in steady-state plasma warfarin concentration in any patient or in anticoagulant control. Table I shows the mean plasma warfarin concentration and thrombotest for the 14 days prior to nitrazepam administration and for the last 14 days in each patient (Fig. 1). In none of these patients was there any significant change in plasma warfarin concentration or in anticoagulant control when nitrazepam was administered.

Chlordiazepoxide. Administration of chlordiazepoxide, 5 mg three times daily, to one patient caused a slight fall in steady-state plasma warfarin concentration from 2.26 ± 0.2 to 1.86 ± 0.10 μg/ml and a corresponding slight change in anticoagulant control (Fig. 2). In none of the other four patients, even on the larger dose of chlordiazepoxide, was there any significant change in either of these parameters. Plasma warfarin concentration and anticoagulant control before and during chlordiazepoxide treatment is shown in Table II.

Diazepam. As shown in Table III, the administration of diazepam, 5 mg three times daily, caused no change in plasma warfarin concentration or in anticoagulant control in the three patients studied.

TABLE I. *Change in plasma warfarin concentration and anticoagulant control produced by administration of nitrazepam, 10 mg nightly, for 30 days*

Patient	Plasma warfarin (mg/liter)		Thrombotest (%)	
	Control	Drug	Control	Drug
(1)	0.65 ± 0.15	0.86 ± 0.26	7	8
(2)	1.91 ± 0.16	2.08 ± 0.20	8	7
(3)	1.45 ± 0.13	1.18 ± 0.12	16	20

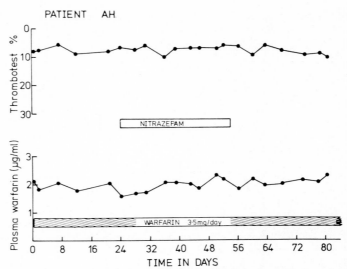

FIG. 1. Change in plasma warfarin concentration and anticoagulant control in a patient given nitrazepam, 10 mg nightly, for 30 days.

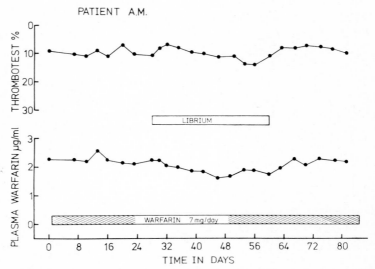

FIG. 2. Change in plasma warfarin concentration and anticoagulant control in a patient given chlordiazepoxide, 15 mg daily, for 30 days.

TABLE II. *Change in plasma warfarin concentration and anticoagulant control produced by administration of chlordiazepoxide, 15 or 30 mg daily, for 30 days*

Patient	Plasma warfarin (mg/liter)		Thrombotest (%)	
	Control	Drug	Control	Drug
(8)*	0.87 ± 0.15	0.82 ± 0.08	7	7
(9)*	0.76 ± 0.10	0.79 ± 0.09	9	10
(10)*	2.26 ± 0.2	1.86 ± 0.10	8	11
(11)	0.76 ± 0.21	1.03 ± 0.47	7	7
(12)	3.4 ± 0.2	3.7 ± 0.02	6	7

* 15 mg chlordiazepoxide/day.

2. Plasma warfarin half-life

The plasma warfarin half-life in patient 4 in the control period was 70.6 ± 17 hr and when taking diazepam, 15 mg daily, 76.9 ± 8.4 hr. In patient 11, the control plasma warfarin half-life was 29.6 ± 6.4 hr, and, when taking chlordiazepoxide, 30 mg daily, it was 29.9 ± 5.3 hr. In neither of these patients did the administration of the benzodiazepines alter the plasma warfarin half-life.

B. Animal studies

Tables IV-VI show the changes in pentobarbital sleeping time, zoxazolamine paralysis time, half-life of ^{14}C-pentobarbital and V_{max} for ethylmorphine N-demethylation produced by nitrazepam, chordiazepoxide, and diazepam. There was no significant change in any of these parameters from controls in animals given nitrazepam or diazepam, but pretreatment with chlordiazepoxide 40 mg/kg/day for 4 days caused a significant ($p < 0.05$) shortening in each parameter measured. Thus, chlordiazepoxide, in the dose used in rats, acts as an inducer of hepatic drug-metabolizing enzymes.

IV. DISCUSSION

These studies form part of an investigation designed to find hypnotic and sedative drugs which can be given to patients on long-term oral anticoagulants, without changing the degree of anticoagulant control. We have previously shown that dichloralphenazone is an inducer of hepatic drug-metabolizing enzymes, whereas chloral hydrate, through its metabolite trichloracetic acid, displaces warfarin from binding on plasma albumin and thus may potentiate its anticoagulant effect (Breckenridge et al., 1971). Amo-

TABLE III. *Change in plasma warfarin concentration and anticoagulant control produced by administration of diazepam, 15 mg daily, for 30 days*

Patient	Plasma warfarin (mg/liter)		Thrombotest (%)	
	Control	Drug	Control	Drug
(4)	2.35 ± 0.81	2.03 ± 0.20	9	10
(5)	2.10 ± 0.09	1.99 ± 0.36	7	7
(6)	0.88 ± 0.12	0.84 ± 0.3	6	7
(7)	1.68 ± 0.26	1.54 ± 0.19	6	5

barbital also acted as an enzyme inducer, while a dose-dependent effect was observed with secobarbital (Breckenridge and Orme, 1971). The studies reported here suggest that the clinical effect of the benzodiazepines on warfarin metabolism is minimal in man. No effect on steady-state plasma warfarin levels was observed with nitrazepam given in the usual therapeutic dose of 10 mg nightly, and we regard this drug as one of the safest hypnotics for patients on long-term warfarin. In a dose of 40 mg/kg/day given for 4 days, nitrazepam had no effect on hepatic drug metabolism in the rat. Similarly, diazepam had no effect on steady-state plasma warfarin concentrations in man. Like nitrazepam, diazepam had no effect on drug metabolism in the rat.

The results with chlordiazepoxide were variable. In one patient, there was a fall in plasma warfarin concentrations and a change in the thrombotest, although this never fell outside the therapeutic range. In four other patients, however, neither the steady-state plasma warfarin concentrations nor the plasma warfarin half-life showed any significant change when chlordiazepoxide was given. We also found that chlordiazepoxide stimulated hepatic drug-metabolizing enzymes in the rat, as reported previously (Hoogland et al., 1966). In man, Lackner and Hunt (1968) showed that the administration of chlordiazepoxide, 10 mg twice daily, to patients for 14 days did not alter their anticoagulant response or the dose of warfarin required.

TABLE IV. *Changes in drug metabolism in rats produced by nitrazepam, 40 mg/kg daily, for 4 days*

Measures	Control	Test
Sleeping time (min)	94.7 ± 22.3	125.5 ± 41.9
Paralysis time (min)	301.2 ± 147.8	232.0 ± 137.2
^{14}C-Pentobarbital plasma half-life (min)	191.5 ± 33.3	189.7 ± 87.5
Ethylmorphine N-demethylation V_{max} (nmoles/mg protein/10 min)	36.0 ± 2.5	41.7 ± 2.9

TABLE V. *Changes in drug metabolism in rats produced by chlordiazepoxide, 40 mg/kg daily, for 4 days*

Measures	Control	Test
Sleeping time (min)	118.7 ± 22.9	15.29 ± 6.5
Paralysis time (min)	262.8 ± 62.3	118.5 ± 13.4
^{14}C-Pentobarbital plasma half-life (min)	83.1 ± 19.1	27.5 ± 7.2
Ethylmorphine N-demethylation V_{max} (nmoles/mg protein/10 min)	75.18 ± 4.4	109.33 ± 4.6

Robinson and Sylwester (1970) found no effect on plasma warfarin half-life in eight volunteers given chlordiazepoxide in a mean dose of 50 mg/day for 21 days. The rate of metabolism of another coumarin anticoagulant, ethyl biscoumacetate, was not increased by chlordiazepoxide administration (Van Dam and Overkamp, 1967). The reason for the individual variations obtained in our studies is not clear. We have previously shown that the extent of hepatic enzyme induction by dichloralphenazone is related to the plasma half-life of its inducing component phenazone, and that the same is true for secobarbital. Since chlordiazepoxide is itself oxidized by hepatic enzymes, this differential effect may be related to its rate of degradation.

V. SUMMARY

Three benzodiazepines, nitrazepam (10 mg/day), diazepam (15 mg/day), and chlordiazepoxide (15 and 30 mg/day), have been investigated in man for their ability to stimulate the activity of hepatic drug-metabolizing enzymes. None of these drugs produced a consistent fall in steady-state plasma warfarin concentration, a change in anticoagulant control, or a change in plasma warfarin half-life.

In the rat, chlordiazepoxide, in a dose of 40 mg/kg for 4 days, significantly shortened the pentobarbital sleeping time, the plasma half-life of

TABLE VI. *Changes in drug metabolism in rats produced by diazepam, 40 mg/kg daily, for 4 days*

Measures	Control	Test
Sleeping time (min)	109.8 ± 21.8	112.2 ± 18.05
Paralysis time (min)	379.4 ± 95.4	376.7 ± 97.8
^{14}C-Pentobarbital plasma half-life (min)	122.3 ± 50.5	155.2 ± 40.7
Ethylmorphine N-demethylation V_{max} (nmoles/mg protein/10 min)	69.7 ± 7.7	75.04 ± 1.8

^{14}C-pentobarbital, and the duration of paralysis produced by zoxazolamine. It also increased the V_{max} of ethylmorphine N-demethylation by isolated rat liver microsomes. Neither nitrazepam nor diazepam given in a similar dose produced these changes.

ACKNOWLEDGMENTS

We are grateful to Mrs. W. Watts for technical assistance and to Roche Pharmaceuticals for financial assistance.

REFERENCES

Breckenridge, A., Davies, D. S., Orme, M., and Thorgeirsson, S. (1969): Screening methods for measuring liver microsomal enzyme induction by drugs. *Journal of Physiology*, 202:15–16.

Breckenridge, A., and Orme, M. (1971): Clinical implications of enzyme induction. *Annals of the New York Academy of Sciences*, 179:421–431.

Breckenridge, A., Orme, M., Thorgeirsson, S., Davies, D. S., and Brooks, R. V. (1971): Drug interactions with warfarin. *Clinical Science*, 40:351–364.

Hoogland, D. R., Miya, T. S., and Bovsquet, W. F. (1966): Metabolism and tolerance studies with chlordiazepoxide-2-^{14}C in the rat. *Toxicology and Applied Pharmacology*, 9:116–123.

Lackner, H., and Hunt, V. E. (1968): The effect of librium on haemostasis. *American Journal of Medical Science*, 256:368–372.

Lewis, R. J., Ilnicki, L. P., and Carlstrom, M. (1970): The assay of warfarin in plasma or stool. *Biochemical Medicine*, 4:376–379.

Robinson, D. S., and Sylwester, D. (1970): Interaction of commonly prescribed drugs and warfarin. *Annals of Internal Medicine*, 72:853–856.

Van Dam, F. E., and Overkamp, M. J. H. (1967): The effect of some sedatives on the rate of disappearance of ethyl biscoumacetate from the plasma. *Folia Medica Neerlandica* (Haarlem), 10:141–145.

The Benzodiazepines, Raven Press, New York © 1973

COMPARATIVE STUDY OF THE EFFICACY OF DIAZEPAM, PENTOBARBITAL, AND FENTANYL-DROPERIDOL (THALAMONAL) AGAINST TOXICITY INDUCED BY LOCAL ANESTHETICS IN MICE

H. Wesseling

Institute for Clinical Pharmacology, State University of Groningen, Groningen, The Netherlands

I. INTRODUCTION

Various drugs are used as part of the premedication for ear-nose-and-throat surgery under local anesthesia. Some of these (barbiturates) are well-known anticonvulsants, and their activity is usually considered to be an advantage because convulsions are a common sign of overdosage with a local anesthetic (Kilian, 1959; Goodman and Gilman, 1970). Other drugs are not supposed to have this effect but share with the barbiturates their sedative properties (e.g., neuroleptic agents), a number of them being analgesic compounds (meperidine, fentanyl); sometimes fixed combinations are used (e.g., droperidol-and-fentanyl).

Occasionally death occurs during a minor operation under local anesthesia (Elder, 1963; De Deacock, 1964; Kruizinga, 1969). It is possible that some of these deaths are caused either by one of the local anesthetics or by one of the many drugs used for premedication. Safer premedication might be achieved by using a single drug with antagonistic effects on toxicity (i.e., convulsions, induced by local anesthetics). Diazepam might be such a drug, and, therefore, we have examined its interactions with two widely used local anesthetics, lidocaine and tetracaine, and compared them with those of pentobarbital and THALAMONAL, a combination of fentanyl and droperidol (Jansen, *undated*).

A part of this study, the comparison between diazepam and pentobarbital, has been published elsewhere (Wesseling et al., 1970).

655

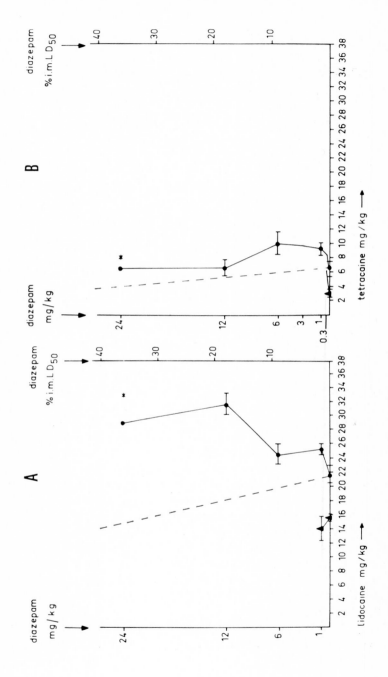

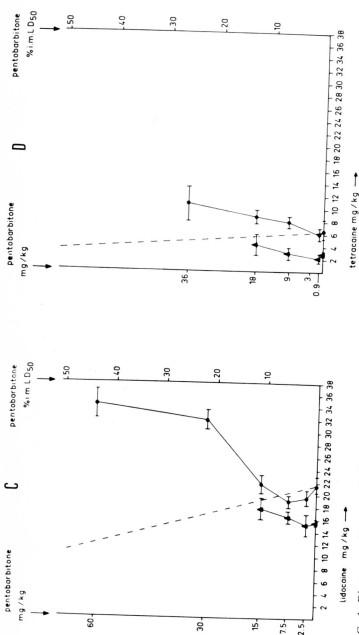

FIG. 1. Diazepam and pentobarbital compared as antagonists of local anesthetics. Isoboles of the LD_{50} ($\bullet$—$\bullet$) and ED_{50} ($\blacktriangle$—$\blacktriangle$) values in mice for combinations of (A) diazepam and lidocaine, (B) diazepam and tetracaine, (C) pentobarbital (pentobarbitone) and lidocaine, and (D) pentobarbital (pentobarbitone) and tetracaine.

The doses of the protecting drug are plotted on the left side of the ordinate and in % of their own i.m. LD_{50} on the right side. The local anesthetics are plotted on the abscissa. Confidence limits are presented as horizontal bars. If the lethal effects of the two drugs were simply additive, the LD_{50} isobole would be a straight line, joining the LD_{50} values of the individual drugs, as indicated by the dotted lines (the "addition lines").

* No confidence limits given.

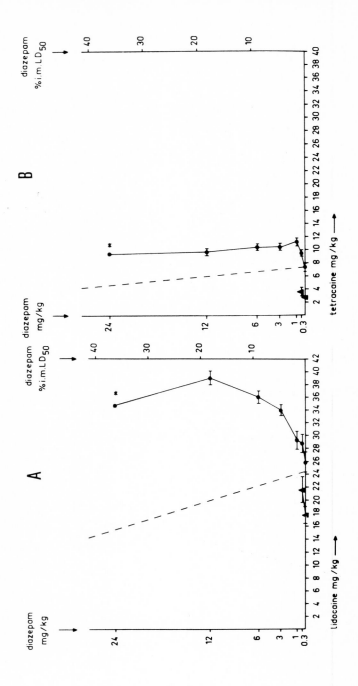

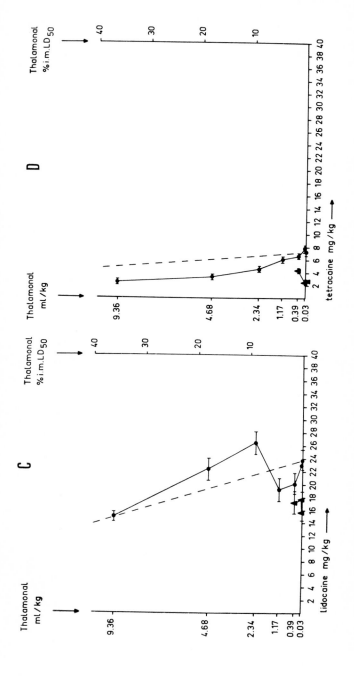

FIG. 2. Diazepam and THALAMONAL compared as antagonists of local anesthetics. Isoboles of the LD$_{50}$ (●—●) and ED$_{50}$ (▲—▲) values in mice for combinations of (A) diazepam and lidocaine (B) diazepam and tetracaine, (C) THALAMONAL and lidocaine, (D) THALAMONAL and tetracaine. For the rest of the legends, see Fig. 1.

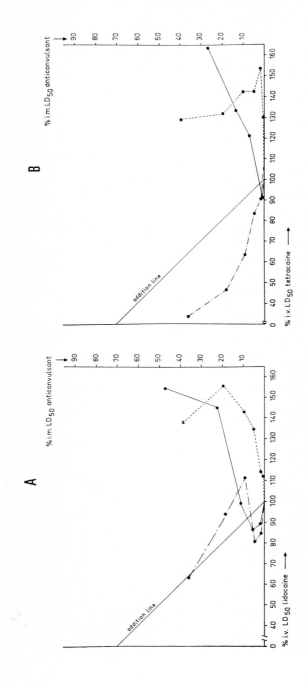

FIG. 3. Influence of three sedative drugs on lethality due to lidocaine and tetracaine. Isoboles of the LD$_{50}$ values in mice for combinations of lidocaine (A) and tetracaine (B) with: diazepam (·---·), pentobarbital (·—·), and THALAMONAL (-·-·-) The doses of the protecting drugs and lidocaine are plotted in % of their i.m. LD$_{50}$ and i.v. LD$_{50}$, respectively, on the ordinate and abscissa.

The figure is, for geometric reasons, "open" on the left side. For reasons of surveyability, only the more recent diazepam values (winter 1970-1971) are presented while confidence limits and ED$_{50}$ values are omitted.

II. MATERIALS AND METHODS

ED_{50} and LD_{50} values for various local anesthetic/protective drug-combinations were estimated in mice (the ED_{50} is that which causes convulsions).

The results are plotted as isoboles (Loewe, 1953). The doses of the sedatives are plotted as a percentage of their i.m. LD_{50}; the ED_{50} and LD_{50} values of the local anesthetics, in the various combinations, are given in absolute values (Figs. 1 and 2) and the LD_{50} values also as a percentage of their "pure" LD_{50} (Fig. 3).

The detailed experimental design, including the solutions of local anesthetics, diazepam, and pentobarbital, has been described earlier (Wesseling et al., 1971). THALAMONAL was obtained from the University Hospital Pharmacy, Groningen, in ampoules containing 2.5 mg droperidol and 0.05 mg fentanyl/ml. The volume of the i.m. THALAMONAL injections was that of the 60% value of the i.v. LD_{50}; different concentrations were prepared by dilution with saline. Saline up to 40 ml/kg injected i.m. and i.v. was given to judge volume influences.

III. RESULTS

Table I gives the ED_{50} and LD_{50} values for all the drugs used. The lethality of tetracaine was about three times that of lidocaine; as a convulsant, tetracaine was about five times more potent than lidocaine. Pentobarbital, as expected, and THALAMONAL did not cause convulsions. We were surprised to find that i.v. diazepam caused convulsions (although it did not do so when given i.m.). This effect appeared to be due to the solvent, which turned out to be a convulsant with an intravenous ED_{50} of 3.4 ml/kg. As the i.v. ED_{50} of diazepam was 15.8 mg/kg, corresponding to a 3.16 ml/kg standard diazepam solution, the convulsive action of i.v. diazepam might be largely due to the solvent. (In low doses, however, the solvent antagonized—in a few experiments we performed—convulsions due to lidocaine.) The intravenous LD_{50} of the solvent was 5.6 ml/kg and that of diazepam itself 19.6 mg/kg; this value corresponds to 3.92 ml/kg standard diazepam solution. Thus the solvent also seemed to contribute considerably to the lethality of diazepam.

The solvent for diazepam was analyzed at the Pharmacy of the State University Hospital, Groningen (Prof. Dr. T. Huizinga) and was shown to contain water, propylene glycol, sodium benzoate, and about 6% ethyl alcohol.

TABLE I. ED_{50} and LD_{50} values for diazepam, pentobarbital, fentanyl-droperidol, lidocaine, and tetracaine in mice

Substance	Route of admin.	LD_{50} or ED_{50}	Results from winter 1969/1970 (mg/kg)	Results from winter 1970/1971 (mg/kg)
Diazepam	i.v.	LD_{50}	19.6 (18.8–20.3)	19.8 (18.3–21.4)
	i.m.	LD_{50}	65 ($\pm$5.0)	62.0 (51.0–67.4)
	i.v.	ED_{50}	16.0 ($\pm$1.4)	
	i.v.	ED_{50}*	15.8 ($\pm$1.3)	
	i.m.	ED_{50}	NC	
Solvent for diazepam	i.v.	LD_{50}	5.6 ml/kg	
	i.v.	ED_{50}	3.4 ml/kg	
Pentobarbital	i.v.	LD_{50}	81 ($\pm$4)	
	i.m.	LD_{50}	135	
	i.v.	ED_{50}	NC	
	i.m.	ED_{50}	NC	
Fentanyl-droperidol	i.v.	LD_{50}		7.8 ml/kg
	i.m.	LD_{50}		26.0 ml/kg
	i.v.	ED_{50}		
	i.m.	ED_{50}		
Lidocaine	i.v.	LD_{50}	21.5 (21.4–23.1)	21.8 (20.3–23.4)
	i.v.	LD_{50}*	21.4 (20.3–22.3)	
	i.v.	ED_{50}	15.6 ($\pm$0.5)	15.7 ($\pm$1.3)
Tetracaine	i.v.	LD_{50}	6.6 (5.2–8.3)	7.3 ($\pm$0.6)
	i.v.	ED_{50}	3.0 (2.5–3.6)	2.8 ($\pm$0.3)

NC, no convulsions occurred.
The 95% confidence limits are given in parentheses.
* Reversed lighting

Seasonal and other variations may influence ED_{50} and LD_{50} values considerably. Because the experiments with THALAMONAL were performed a year after those with pentobarbital, the reference estimations with diazepam and the local anesthetics had to be repeated. Our data show good agreement between the results of both periods.

Physical reactions of animals are known to be modified by circadian rhythm (Halberg et al., 1958; Haus and Halberg, 1959; Lutsch, 1967). In our experiments, no interference caused by reversed lighting could be demonstrated.

The volume of saline used had no effect on its own.

Figure 1 compares diazepam with pentobarbital, Fig. 2 diazepam with THALAMONAL. Figure 3 shows the influence of the three sedative drugs on the LD_{50} values of lidocaine and tetracaine, respectively, plotted as percentage of their LD_{50}. From these figures it can be seen that diazepam, in low doses, considerably increased the LD_{50} of lidocaine and tetracaine; this increase was of the same order of magnitude (Fig. 3).

Figures 1A and B and 2A and B further show that doses of diazepam of about 1.5 mg/kg (i.e., less than 2.5% of the i.m. LD_{50}) protected more than 50% of the animals against lidocaine and tetracaine convulsions. The results of the experiments of the 1969-1970 winter period are again in good agreement with those of 1970-1971. In contrast with diazepam, pentobarbital, in low doses, increased mortality due to lidocaine; only doses larger than 20 mg/kg (i.e., approximately 15% of the i.m. LD_{50} of pentobarbital) reduced it, and this results in a Z-shaped isobole (Figs. 1C and 3A). Pentobarbital lowered mortality due to tetracaine. Although Fig. 3B also gives an indication of a Z-shaped isobole, the only point that has been determined on the left of the addition line falls within the confidence limits of the LD_{50} of tetracaine (Fig. 1D).

Doses of pentobarbital up to 15 mg/kg had no effect on lidocaine and tetracaine convulsions. This is shown by the vertical course of the isobole for ED_{50} (in Fig. 1C and D). Above this dose, convulsions were progressively antagonized.

THALAMONAL had a dual effect on lidocaine-induced mortality; the resulting Z-shaped isobole is even more pronounced than that of pentobarbital/lidocaine (Figs. 2C and 3A). There is no such effect on the mortality due to tetracaine; THALAMONAL decreases the LD_{50} values (Fig. 3B) (the only value that was determined on the right of the addition line falls within the confidence limits of the LD_{50} of tetracaine; Fig. 2D). THALAMONAL antagonized lidocaine and tetracaine-induced convulsions as effectively as did diazepam (Fig. 2C and D).

IV. DISCUSSION

Our results indicate that convulsions in mice caused by local anesthetics are much more effectively antagonized by diazepam (in both experiments) and THALAMONAL than by pentobarbital. It is not clear how much the phamacological action of the diazepam solvent interferes with the effectiveness of diazepam in this respect. There is other evidence that solvents of benzodiazepines may have important pharmacological action, especially on the central nervous system (Vieth et al., 1968).

Diazepam and pentobarbital decreased lethality caused by local anesthetics to about the same extent. However, higher doses of pentobarbital were needed, especially in the case of lidocaine, where this resulted in a Z-shaped isobole. Conversely, THALAMONAL does not protect against lethality at all, except—and only in high doses—against lidocaine-induced death, where a Z-shaped isobole is again observed.

The explanation of the Z-shaped isoboles is not clear, but a suggestion will be given here. There are indications that in some species benzodiazepines and local anesthetics in low doses have a rather selective action on the same central structures, e.g., the region of the amygdala and hippocampus (Arrigo et al., 1965; Wagman et al., 1967, 1968; Parkes, 1968; Stevens, 1968; Olds and Olds, 1969; Tuttle and Elliott, 1969). Barbiturates and opiates have a more generalized action on many parts of the central nervous system; their depressant effect on the respiratory center is pronounced, especially that of opiates (Goodman and Gilman, 1970).

Convulsions caused by local anesthetics are clonic seizures, preceded by a tonic phase, although sometimes this tonic phase is not followed by convulsions. Respiration is impaired during the tonic phase. We observed that only those mice that had an immediate respiratory arrest died from lidocaine; mice that showed even shallow respiratory movements survived.

It is suggested that barbiturates and opiates in low doses add their depressant effect on the respiratory center to the respiratory impairment caused by the tonic phase of the convulsions that are not yet sufficiently suppressed. Higher doses of the anticonvulsant drug, however, suppress the tonic convulsions and protect the animal.

This hypothesis offers an explanation for the Z-shaped isoboles, especially in the case of THALAMONAL, which contains the strong opiate fentanyl. It also explains why diazepam exerts its action even in the lowest doses. Z-shaped isoboles were far more pronounced in lidocaine antagonism than with tetracaine. This supports the clinical impression (Adriani, 1966) that death due to tetracaine is not caused by failure of the central nervous system.

The results of this study encouraged us to reconsider the routine premedication in ear-nose-and-throat surgery. A subsequent clinical trial showed that diazepam might be a good alternative for this kind of premedication (Wesseling et al., *in press.*)

V. SUMMARY

The effects of diazepam on the toxicity of lidocaine and tetracaine in mice were compared with those of pentobarbital and fentanyl-droperidol (THALAMONAL). Isoboles were constructed for each antagonistic pair of drugs. Diazepam and THALAMONAL prevented seizures more effectively than pentobarbital. However, the solvent of diazepam, in high doses, was itself a convulsant. Diazepam markedly decreased the lethality of lidocaine. So, too, did higher doses of THALAMONAL and pentobarbital, but low doses increased the lethality of lidocaine. Diazepam and higher doses of pentobarbital markedly decreased the lethality of tetracaine whereas THALAMONAL increased it.

REFERENCES

Adriani, J. (1966): Reactions to local anesthetics. *Journal of the American Medical Association*, 196:405–408.

Adriani, J., and Campbell, D. (1956): Fatalities following topical application of local anesthetics to mucous membranes. *Journal of the American Medical Association*, 162:1527–1530.

Arrigo, A., Jann, G., and Tonali, P. (1965): Some aspects of the action of VALIUM and LIBRIUM on the electrical activity of the rabbit brain. *Archives Internationaux de Pharmacodynamie*, 154:364–373.

De Deacock, A. R., and Simpson, W. T. (1964): Fatal reactions due to lidocaine. *Anaesthesia*, 19:217–221.

Elder, J. L., and Smith, W. G. (1963): *British Medical Journal*, 1:1416–1417.

Goodman, L. S., and Gilman, A. (1970): *The Pharmacological Basis of Therapeutics*. Macmillan, New York.

Halberg, F., Barnum, C. P., Silver, R. H., and Bittner, J. J. (1958): 24-Hour rhythms at several levels of integration in mice on different lighting regimens. *Proceedings of the Society for Experimental Biology and Medicine*, 97:897–900.

Haus, E., and Halberg, F. (1959): 24-Hour rhythms in susceptibility of C mice to atoxic dose of ethanol. *Journal of Applied Physiology*, 14:878–880.

Jansen Pharmaceutica (*undated*): *Droperidol and fentanyl. Clinical aspects and pharmacology*. Beerse, Belgium.

Kruizinga, P. J. W. (1969): *Personal Communication. Groot Weezenland*.

Loewe, S. (1953): The problem of synergism and antagonism of combined drugs. *Arzneimittel Forschung*, 3:285–290.

Lutsch, E. F., and Morris, R. W. (1967): Circadian periodicity in susceptibility to lidocaine hydrochloride. *Science*, 156:100–102.

Olds, M. E., and Olds, J. (1969): Effects of anxiety relieving drugs on unit-discharges in hippocampus, reticular midbrain and preoptic area in the freely moving rat. *International Journal of Neuropharmacology*, 8:87–103.

Parkes, M. W. (1968): The pharmacology of diazepam. In: *Diazepam in Anesthesia*, Wright, Bristol, pp. 1–7.

Stevens, J. R. (1968): Convulsive seizures and lidocaine treatment. *Journal of the American Medical Association*, 205:116–117.

Tuttle, W. W., and Elliott, H. W. (1969): Electrographic and behavioural study of convulsants in the cat. *Anesthesiology*, 30:48–64.

Vieth, J. B., Holm, E., and Knopp, P. R. (1968): Electrophysiological studies on the action of MOGADON on central nervous structures of the cat. A comparison with pentobarbital. *Archives Internationaux de Pharmacodynamie*, 171:323–339.

Wagman, H., De Jong, R. H., and Prince, A. (1967): Effects of lidocaine on the central nervous system. *Anesthesiology*, 28:155–172.

Wagman, H., De Jong, R. H., and Prince, A. (1968): Effects of lidocaine on spontaneous cortical and subcortical electrical activity. *Archives of Neurology*, 18:277–290.

Wesseling, H., Bovenhorst, G. H., and Wiers, J. W. (1971): Effects of diazepam on convulsions induced by local anaesthetics in mice. *European Journal of Pharmacology*, 13:150–154.

Wesseling, H., Kruizinga, P. J. W., and Leezenberg, J. A. (1972): Premedication in ENT-interventions under local anaesthesia. (*in press*).

DISCUSSION SUMMARY

Eppo van der Kleijn

*Sint Radboud Hospital Pharmacy, Department of Clinical Pharmacy,
University of Nijmegen, Nijmegen, The Netherlands*

It has been implied in several papers presented during this conference that the benzodiazepines show little or no side effects during acute and chronic treatment, even after high doses. Although their gross pharmacological characteristics show many similarities, several specific properties can be attributed to each compound. It is very likely that the varying pharmacokinetics are responsible for these different shades of activity.

From Marino's work, it is questionable whether the synergistic effect which the benzodiazepines produce with the antidepressants centrally, as well as with *d*-tubocurarine peripherally can be explained completely by their influence on the central nervous system. There is no reason, however, to suggest that somatic changes caused by the benzodiazepines are responsible for this effect.

One of the more serious threats of multiple drug therapy is the possibility of adverse interactions. Adverse drug effects can be caused by strict pharmacological interactions or may be the result of the biochemical and metabolic changes causing differences in the pharmacokinetics of these drugs. Interactions may also occur on the level of absorption caused directly by interference with the diffusing drug or indirectly through an influence on gastrointestinal tract motility or bile production. No such interactions have been reported for the benzodiazepines. There are two areas where combinations of psychotropic drugs have caused practical problems. The more serious one therapeutically is the interaction of benzodiazepines with anticoagulants. The work of Robinson and Breckenridge and Orme has made it clear that the benzodiazepines, unlike the barbiturates, do not produce severe alterations of the prothrombin time.

A case report by Breckenridge on a decreased anticoagulant level in a patient during chlordiazepoxide treatment was correlated with an increased rate of microsomal metabolism and a shortening of the time course of the pharmacological effects of pentobarbital and of zoxazolamine in rats. This effect reported for chlordiazepoxide after rather high doses did not occur with diazepam. This observation was contradicted by Garattini and Jori, who reported induction of microsomal metabolism in rat liver after a dose of 50 mg diazepam per kg body weight.

The application of the isobole technique allows the detection of synergism, potentiation, and antagonism of drugs that may have different modes of action but manifest a similar measurable effect. Wesseling has studied the antagonism of diazepam with a number of local anesthetics. His method permits the detection of interactions of combinations of drugs that may be potentially harmful to man. Niemegeers warned that the testing of combined drug preparations, for example, THALAMONAL® (droperidol and fentanyl), may give misleading results.

AUTHOR INDEX

669

SUBJECT INDEX